Studies in Polymer Science 7

Polypropylene and other Polyolefins

Polymerization and Characterization

Studies in Polymer Science

Studies in Polymer Science 7

Polypropylene and other Polyolefins

Polymerization and Characterization

by

Ser van der Ven

Billiton Research, P.O. Box 40, 6800 AA Arnhem, The Netherlands

ELSEVIER
Amsterdam — Oxford — New York — Tokyo   1990

ELSEVIER SCIENCE PUBLISHERS B.V.
Sara Burgerhartstraat 25
P.O. Box 211, 1000 AE Amsterdam, The Netherlands

Distributors for the United States and Canada:

ELSEVIER SCIENCE PUBLISHING COMPANY INC.
655, Avenue of the Americas
New York, NY 10010, U.S.A.

Library of Congress Cataloging-in-Publication Data

Ven, Ser van der.
    Polypropylene and other polyolefins : polymerization and
  characterization / by Ser van der Ven.
      p.   cm. -- (Studies in polymer science ; 7)
    Includes bibliographical references.
    ISBN 0-444-88690-7 (U.S.)
    1. Polypropylene.   I. Title.   II. Series.
  TP1180.P68V46 1990
    668.4'234--dc20   90-33336
                                                           CIP

ISBN 0-444-88690-7

This book is printed on acid-free paper.

Printed in The Netherlands

C O N T E N T S

PREFACE

Fascination with the versatility and complexity in the field of polyolefins was the main driving force in writing this book. It is a feeling that many scientists will share when working with performance products of this type. In the field of polyolefins, even in the limited area described in the present book, not many true books are around. Mainly proceedings of conferences are issued - helpful, but they are mostly not reviewing the area. Exceptions are indeed rare, and one has to mention John Boor's book of course, which is still used as a standard on early catalyst work. The present book might thus be a useful addition even when covering only part of the total polyolefins science and technology field - namely only catalysis, polymerization behaviour and (mainly analytical) characterization. It is also hoped that this book will contribute to the knowledge base on these topics, and will help people on the way to reach full understanding and even predictability. And, apart from all the above sweeping statements, one needs perseverance to finish an undertaking like writing the present book. I'll think twice before starting a new one.

I am glad to have received the expert help of many colleagues, of whom above all Brian Goodall has to be mentioned, who wrote the complete chapter 1 on propylene polymerization catalysis. In addition he gave worthwhile advice on the other synthesis chapters, both on the contents and the language. Johan Helms helped greatly in the writing of the section on infrared analysis of polymers and was a fund of ideas in thinking with me on the structure of the book. This of course led to extra work and rearrangement but improved the end product considerably! Apart from this expert help the assistance of Arie Knoester and Arjen Miedema in retrieving data from the past proved invaluable.

It will be clear to anyone that a fair part of the work used in this book is the result of investigations by many workers in the Shell laboratory of Amsterdam. The main extra information came from studies by Jaap Bakkum, Wim Bergwerf, Wout Boog, Rob Bouwman, Ian

Carson, Keith Fulcher, Piet Groeneweg, Nico Groesbeek, Freek ten Haaft, Wouter Hagens, Wilma Hoogervorst, Gerard van Kessel, Bandi Kortbeek, Winnie van der Linden, Martin van Loon, Neil Mayne, Garmt Mulder, Wilma Whincup-Oscamp, Dan Pauk, Piet Rietveld, Cyp van Rijn, Nico Rol, Jaap van der Sar, Mark Swallow, Ruud Segers, Wim Sjardijn, Louis Smeets, Marinus de Vente, Henk van der Vliet, Netty de Waal, Frans Waals, Antonio Wilson, Wim de Wit and Henk Wolters, to all of whom my thanks are due. Special mention is made of Frans Waals' help in the chapter on toughened polypropylene characterization.

I gratefully acknowledge the permission of Shell Research BV to publish this book.

Ser van der Ven
(Billiton Research)
Arnhem, jan 1990

INTRODUCTION

This book deals with polyolefins prepared via Ziegler-Natta catalysis, from a polymer chemists viewpoint, i.e. with emphasis on their preparation and on the basic composition and properties. In the synthesis of polymers, control of composition is the polymer chemist's main aim, since the properties can then be controlled and tailored to specific needs. The usual approach is followed in this book, i.e. studying the interrelation of preparation and the resulting composition and basic properties of the polymer product. To arrive at basic understanding one needs to characterize both the catalyst - very difficult since they are mostly heterogeneous and thus very complex in structure - and the polymers made on these catalysts. These also turn out to be very heterogeneous and of complicated structure. Because, in these studies, the hypotheses are mainly based on evidence from characterization studies which are mostly circumstantial in nature - since, for instance, isolating and experimenting with one specific surface site of a heterogeneous catalyst is impossible - the quality of the characterization methods should be very high and the outcome should be challenged time and again. Hence this book contains, in addition to chapters on catalysts, polymerization behaviour, polymer properties such as tacticity, crystallinity morphology, etc., also a chapter on characterization methods, mainly summarizing known techniques, but describing a few own-developed methods in more detail.

Initially it was our intention to discuss only the commercially important catalysts, however we quickly decided to include the recent developments in homogeneous catalysis as well. The reasons seem obvious: not only can extreme high activity be achieved but also stereospecificity towards various types, such as isotactic and syndiotactic polypropylene. The versatility of these homogeneous organometallic catalysts is proven time and again. Although these catalysts are not yet commercial they might become so fairly quickly, maybe sooner for special (co)polymers. There is only one drawback in including this topic in the book: the rate of development is so fast that the material given could be soon outdated!

A word of caution is necessary otherwise the reader might not find in this book what he expects from reading the above, and some comments are needed to put the above in its proper perspective. For instance the main part of the book is devoted to polypropylene in all its forms (i.e. homopolymer, random copolymer and toughened ("block") copolymer), for which extensive own-experience was present. The other polyolefins are also dealt with, but only by a thorough literature review, using our polypropylene know-how as a basis for judging the importance of the literature. In addition, the choice of characterization studies mentioned and discussed is rather arbitrary, mainly based on personal experience, preference and idea of their relevance and importance. To illustrate this, aspects such as tacticity, crystallinity and dispersion get fair attention but on the other hand mechanical properties, melt flow or crystalline morphology are almost completely left out. Moreover no polymerization process technology is discussed, neither is the processing of polymers dealt with. Admittedly this creates a possible imbalance; additionally there is a huge commercial inte- rest in the general area of polyolefins which ensures that no total freedom yet exists to discuss all topics in the same depth. But as a book of this kind has not appeared before it is thought that worthwhile information has been compiled, which will hopefully stimulate the scientific world to attack the still abundant number of unresolved problems in the area of synthesis and characteriza- tion of polyolefins. A case in point is the understanding of the behaviour of toughened polypropylene in relation to its composi- tion. This book attempts a start by describing a relatively simple system - this certainly needs extension.

It is thought that the near future might see a new era of polyolefin synthesis through the use of tailored homogeneous catalysts, which could very well simplify the investigation of the important relationships of composition and properties.

The book is in the first place intended for scientists active in the field of polyolefins, including catalyst development. On the other hand the book could, either as a whole or in part, be used in academia to illustrate for instance the growth of our understanding in catalysis, kinetics and characterization of a commercially very important class of polymers.

Chapter 1

POLYPROPYLENE; CATALYSTS AND POLYMERIZATION ASPECTS
(by Brian L.Goodall)

1.1 INTRODUCTION

Since its discovery in the mid-1950's polypropylene has grown into one of today's most important commodity polymers. In 1988 the total world capacity for polypropylene manufacture was 11 million tonnes, of which about 30 % was located in the U.S.A., 30 % in Europe and 14 % in Japan. It is timely to review polypropylene manufacture since now, in the latter half of the 1980's, the industry is undergoing the most sweeping changes since its conception in the late 1950's. The propylene polymerization process is being revolutionized in terms of both manufacturing hardware and, most importantly, catalyst technology. This review attempts to survey all types of catalyst used commercially for polypropylene manufacture starting with the landmark discoveries of Karl Ziegler and Giulio Natta, via the highly sophisticated supported catalysts currently being introduced into todays plants, also the remarkable developments in homogeneous catalysis will be mentioned. In chapter 14 our ideas about the future are given, including sections on catalysis.

In addition to the above, co-catalyst chemistry is discussed (including related topics such as isotacticity control agents, chain transfer and molecular weight control) as are selected aspects of the kinetics of propylene polymerization. The enormous number of publications and patent citations makes it essentially impossible to comprehensively review the area, but we have attempted to present a balanced and thorough survey of this fascinating field.

1.2 HISTORICAL BACKGROUND

The origins of the sophisticated catalysts used in today's polypropylene plants are to be found in the early 1950's when Karl Ziegler, investigating his new triethylaluminium-catalyzed synthesis of higher olefins (Fig. 1.1) known as the "Aufbau Reaction", serendipitously discovered the "nickel effect", which was caused by a colloidal nickel contaminant. Instead of the usual oligomeric wax the product comprised almost exclusively 1-butene; nickel having

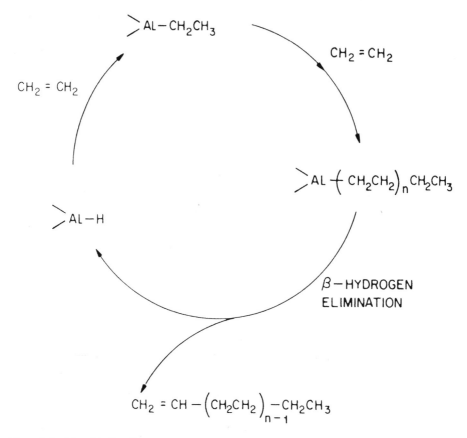

Fig. 1.1: The "Aufbau" reaction (ethylene oligomerization)

catalyzed the displacement reaction (β-hydrogen elimination).

In search of other displacement catalysts Ziegler and co-workers then tested AlEt$_3$ in combination with a wide range of transition metal compounds. A new breakthrough was made when zirconium acetylacetonate was found to catalyze the formation of high-molecular-weight polyethylene at low pressure, i.e. it catalyzed the growth reaction. Other early (group IV to VI) transition metals salts were also found to polymerize ethylene, by far the most active being TiCl$_4$; this catalyst (TiCl$_4$ + AlEt$_3$) was subsequently developed for the manufacture of high density poly ethylene(HDPE). An extensive review of Karl Ziegler's life and scientific contributions was recently published[1].

Although Ziegler's invention formed the cornerstone for the subsequent development and commercialization of present-day polyolefin manufacturing processes, the full scope of the discovery was realized by Giulio Natta's group in Italy. In early 1954 Natta first succeeded in isolating crystalline polypropylene using Ziegler's $TiCl_4$ + $AlEt_3$ catalyst; a mixture of amorphous and crystalline polypropylene was produced, the crystalline material being isolated by solvent extraction. Natta determined that crystalline polypropylene (which he designated "isotactic") comprises extended sequences of monomer units with the same configuration; the amorphous (or random) analogs were dubbed "atactic". At about the same time as Natta's discovery Banks and Hogan (Phillips Petroleum Company) isolated crystalline polypropylene from a mixture obtained using a chromium-on-silica catalyst. In the following we will distinguish different tacticities in polymers made from α-olefins determined by the stereochemical placement of the sidegroup. If they are all identically placed the polymer is called isotactic, in an alternating sequence syndiotactic, and in a random order atactic.

Natta also first recognized the importance of the catalyst crystal structure and particularly the crystal surface in determining the polymer isotacticity, and found that when certain solid titanium chlorides were used instead of $TiCl_4$ highly isotactic polypropylene was produced. These are aspects which will be developed in later sections.

It would be a gross injustice to one of the most colourful and exciting eras in recent chemical history not to expand a little on this "in a nutshell" description of the events which took place in the 1950's. A detailed and very readable account of all the developments in this period and their industrial impact is to be found in the book of McMillan[2]. The exact date (26 October 1953) of the first successful ethylene polymerization and the nature of the follow-up experiments are both described as are the methods by which Ziegler's group, without today's sophisticated techniques, were able to recognize that they had produced, for the first time, a straight chain, high molecular weight polyethylene.

Regarding polypropylene it is noteworthy that only one day after the first successful polyethylene run Ziegler's co-worker, Hans Breil, carried out a polymerization experiment with propylene but surprisingly came to the incorrect conclusion that propylene could not be polymerized in this way. This error probably contributed to Ziegler's decision to concentrate on discovering new, improved catalysts rather than exploring the scope of the new polymerization reaction.

Natta's group elected to study the Ziegler catalysts' performance with other olefins and on 11 March 1954 Paolo Chini discovered that at elevated temperature and pressure it was possible to generate polymer which could be extracted with ether to give a white powder. X-ray diffraction studies carried out within the group revealed this white powder to be highly crystalline, explaining its unexpected high rigidity and high melting point.

Related developments such as synthetic rubbers and the first commercialization of the various new polymeric products are all accurately documented by McMillan.

Recently Pino (a member of the original Natta group) and Moretti[3] have published an account of the discovery of the polymerization of α-olefins in which attention is given not only to the contributions of the Natta group but also to the various other groups active with non-Ziegler catalysts in the 1940's and 1950's. A commentary on the ensuing patent litigation procedures is also given.

1.3 THE ARCHITECTURE OF A CATALYST PARTICLE; CATALYST MORPHOLOGY

All the Ziegler-Natta catalysts used commercially (both $TiCl_3$ catalysts and $MgCl_2$-supported catalysts) consist of the so-called primary crystallites, the average size of which varies from about 5 nm to 20 nm depending on the catalyst type, which are agglomerated into larger particles (typically 1-50 μm). The number of primary crystallites per catalyst particle is enormous; for example in one commercial catalyst possessing a primary crystallite size of 15 nm and particle size 3 μm this number is about 10^7. An impression of the internal structure of a catalyst particle is given in Fig. 1.2. The primary crystallites are held together in a mesh network leaving voids which form the internal pore structure of the catalyst particle. This pore network is essential to high catalyst efficiency since it is through these pores that monomer and co-catalyst can reach the innermost crystallites.

In subsequent sections dealing with catalysts and related chemistry, the catalysts will be discussed in terms of the primary crystallites since it is on the surface of these crystallites that polymerization occurs. Indeed, most of the chemistry involved in the polymerization process can best be understood in terms of these primary crystallites and their interactions with the co-catalyst and olefin monomer. However the morphology (i.e. size, shape

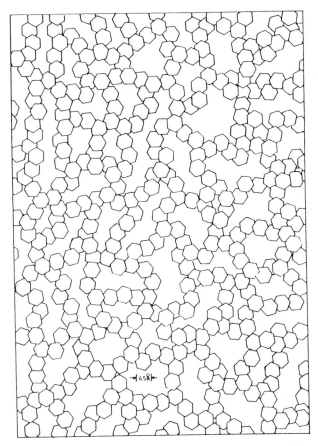

Fig. 1.2: Internal structure of a catalyst [from reference 19. Copied with permission of Harwood Academic Publishers]

and porosity) of the catalyst particles or "agglomerates" is a critical factor in determining process economics. This economic importance stems from the fact that heterogeneous Ziegler-Natta catalysts are uniquely capable of replicating their morphology into the morphology of the resulting polymer particles.

There are a number of reasons why polymer morphology is a major factor in determining process viability. Fine particles pack poorly, limiting reactor capacity and retaining substantial amounts of polymerization solvent on centrifuging the final slurry; they also present a dust explosion hazard on isolation. Zenz and Othmer[4] have indicated that powders with a diameter of less than 100 μm will neither flow well nor pack closely. On the other hand much coarser particles are difficult to keep in suspension and can give problems both in drying and in product homogeneity. Typically polymer powder size in the range >100 μm and <800 μm is aimed at although in recent years some manufacturers (e.g. Himonts "spheripol" process) produce larger particles (1-5 mm) which enables them (for certain end-use applications) to avoid the energy-intensive extrusion step in their process.

Obviously the shape of the polymer particles is also of primary importance, with spherical particles being preferred on account of their high packing efficiency. Particle porosity must be optimized since a certain degree of porosity is necessary in order to allow propylene monomer to reach the innermost catalyst crystallites (active centers), while high porosity results in a low density polymer particle which limits reactor capacity and hence process economics.

The usual method for quantifying the combined effects of powder size, shape and porosity is simply the powder bulk density; this is a very important value since it effectively determines plant capacity; the higher the bulk density, the higher the plant throughput. In the 1960's and early 1970's bulk densities of 0.30 to 0.37 g/ml were considered normal. The state of the art in morphology control is so far advanced that today a bulk density of 0.45 g/ml is commonplace and values as high as about 0.50 g/ml have been reported (the density of polypropylene itself is 0.90 g/ml, which allows a maximum bulk density of about 0.54 with uniform spheres). It is not to be expected that bulk density will be further increased unless extreme (and probably impractical) measures are taken such as developing spherical particles with a bimodal particle size distribution such that the smaller spheres fit into the holes left upon stacking the larger ones. The relatively high bulk densities measured in some polymers of broad particle size distribution is due, at least in part, to the packing of smaller particles into the voids between neighbouring particles.

The morphology of the polymer produced is dependent not only on the catalyst morphology but also on process conditions. For example in those cases where the resulting polymer is soluble in the reaction medium (e.g. in the bulk polymerization of 1-butene) special techniques such as flash drying are used to control the shape of the product and the resulting polymer morphology is therefore independent of the catalyst used. However when the polymerization occurs with the introduction of a suspension of heterogeneous catalyst particles and the formation of insoluble

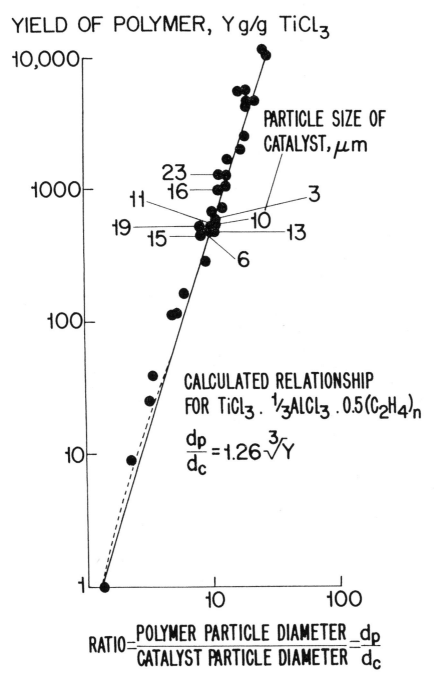

Fig. 1.3: Relationship between polymer particle size and polymer yield

polymer particles (as is the case for virtually all polypropylene manufacturing processes) the phenomenon of replication is usually observed such that the polymer particles formed are essentially enlarged replicas of the catalyst particle used. In the case of a spherical catalyst particle growing into a spherical polymer particle it is possible to predict the final polymer particle size if the polymer yield (per unit mass catalyst) and catalyst and polymer densities are known. An example is given in Fig. 1.3 where the calculated relationship between polymer particle diameter and catalyst diameter ($TiCl_3$ catalyst) is compared[5] with experimental data over a wide range of polymer yields and catalyst particle sizes (all catalysts were spherical and possessed a narrow particle size distribution). Thus, under the condition that a heterogeneous catalyst is used to produce insoluble polymer, the size and shape of the final polymer particle is determined by catalyst morphology and by the yield of polymer on catalyst. The only exceptions occur when polymer agglomeration or break-down occur as a result of wrongly selected process conditions. Poor agitation can lead to agglomeration while the exposure of the catalyst to uncontrolled or abnormally high initial polymerization rates can result in fragmentation (also referred to as "catalyst explosion").

Replication is well-documented both for $TiCl_3$ catalysts as well as for the newer $MgCl_2$-supported catalysts. In the case of $TiCl_3$ catalysts the phenomenon was elegantly demonstrated by Mackie et al[6] in 1967 using a mixture of (near) spherical and needle shaped catalyst particles which after polymerization have grown to a mixture of polymer particles containing again spheres and needles. Much more recently Galli[7] presented a detailed survey of advances in this area using supported catalysts; the ultimate objective (just as with $TiCl_3$ catalysts) being high bulk density spherical catalyst particles, Galli listed five "primary properties" that a catalyst must possess if the polymer granule produced is to be a faithful replica of the catalyst particle:

i)   High surface area of the catalyst particle.

ii)   High porosity of the catalyst particle, with a correspondingly high number of cracks even in the innermost zones.

iii) Free access of the monomer throughout the catalyst particle, including the innermost regions.

iv)   Homogeneous distribution of the active centres throughout the catalyst particle.

v)   Mechanical strength high enough to withstand manipulations without fragmentation.

The first four properties ensure that the whole catalyst particle takes part in the polymerization process while the last aspect refers to eliminating (or at least minimizing) the risk of "catalyst explosion", and hence loss of morphological control, in the initial stages of the reaction.

The phenomenon of replication can best be understood by considering how the polymer particle grows during the polymerization process. As will be discussed in later sections the active centres (where polymerization occurs) are located on the edges of the primary crystallites which comprise the catalyst particle; thus polymer is formed throughout the catalyst particle on each of the almost countless primary crystallites. Indeed, Buls and Higgins[8] have shown that as the polymerization proceeds the catalyst particle is split up into progressively smaller units which become uniformly distributed throughout the growing polymer particle - this concept is often referred to as the "expanding universe model". The catalyst particle is split up by the growing polymer mass and once a sufficient polymer yield has been achieved the catalyst will be homogeneously dispersed throughout the polymer particle in the form of the primary crystallites. This is clearly demonstrated in Fig. 1.4 which is a TEM photograph of a non-deashed homopolymer particle. The yield of polypropylene on catalyst was 100 g/g $TiCl_3$. The black dots were shown to be catalyst crystallites (Ti/Al/Cl analysis) and to possess an average size of 15 nm (X-ray determination) which corresponds exactly with the primary crystallite size of the catalyst used.

In SEM-micrographs of polymer powders very often globules or strands are observed with a diameter of about 1μm, see e.g. Fig. 5.2. From studies by Kakugo et al[9] this appears to be a secondary structure build from a number of smaller particles of about 0.2 to 0.35 μm each containing a primary crystallite from the catalyst.

Boor[10] reviewed the various models which have been proposed to explain exactly how a polymer chain might grow and crystallize about the catalytic centres. As far as replication is concerned the relevant issue is that the polymer acts as "cement" holding the

Fig. 1.4: Transmission electron micrograph of non-deashed homopolymer powder particle with a yield of 100 g/gTiCl$_3$ (particles embedded in resin, microtomed and covered with carbon)

millions of primary crystallites within the expanding matrix of the polymer particle. The phenomenon of catalyst fragmentation or explosion occurs when a catalyst is too fragile to withstand the forces exerted on it during the initial stages of the polymerization and is split up before sufficient of this polymer cement has been formed to hold it together (i.e. the catalyst particle fails to meet Galli's mechanical strength criterion described above). In such cases the polymer "cement" can be synthesized under mild pre-polymerization conditions before subjection to main reactor conditions.

1.4 COMPOSITION AND STRUCTURE OF THE CATALYSTS

In the thirty years since the first commercialization of propylene polymerization three distinct "generations" of catalysts can be identified:

i) The earliest commercial catalysts were essentially titanium trichloride simply prepared by reducing $TiCl_4$ with alkylaluminiums to yield brown (β) $TiCl_3$ which was subsequently heated to convert it to the stereospecific purple (γ) form.

ii) In the 1970's, improved or "second generation" catalysts were developed. The essence of the improvement was that dialkyl ethers (especially di-n-butyl ether and diisoamyl ether) were used, either in the $TiCl_4$ reduction step or as a subsequent wash treatment, to remove the catalyst poisons $AlCl_3$ or $AlEtCl_2$ which are co-crystallized with or absorbed onto the $TiCl_3$ catalyst. Furthermore, excess $TiCl_4$ was used to "catalyse" the phase transformation of β-$TiCl_3$ to the stereospecific γ or δ-form at lower temperatures, leading to smaller and hence more active catalyst crystallites.

iii) The 1980's have heralded the widespread commercial implementation of supported catalysts. These catalysts comprise a $TiCl_4$ "catalyst" on a specially-prepared $MgCl_2$ support. Electron donors of various types are incorporated into the $MgCl_2$ support and are also used in conjunction with the co-catalyst as a "selectivity control agent".

In this section the composition of each of the catalyst generations will be reviewed in turn, the structure of each catalyst-type being described in terms of a single primary crystallite (which in each case represents the elementary unit of the catalyst agglomerates described in section 1.3).

1.4.1 First generation $TiCl_3$ catalysts

$TiCl_3$ can be prepared from $TiCl_4$ by many different routes including reduction with hydrogen, irradiation and the commercially-preferred reduction with alkylaluminiums:

$$e.g. \quad TiCl_4 + 1/3AlEt_3 \longrightarrow TiCl_3 \cdot 1/3AlCl_3 + 1/2C_2H_6 + 1/2C_2H_4$$
$$TiCl_4 + 1/2AlEt_2Cl \longrightarrow TiCl_3 \cdot 1/2AlCl_3 + 1/2C_2H_6 + 1/2C_2H_4$$

12

Natta found these catalysts to exist in four different structural modifications, depending on the reduction process used; he designated these TiCl$_3$ variants the α, β, γ and δ modifications.

The α, γ and δ forms are typified by a deep purple colour and have a layer lattice structure, whereas β-TiCl$_3$ is brown and has a chain-like structure as shown in Fig. 1.5 . The layer lattice (α,

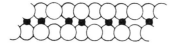

Fig. 1.5: β-TiCl$_3$ chain lattice structure [from reference 19, copied with permission of Harwood Academic Publishers]

γ, δ) structure is best illustrated as in Fig. 1.6. In each individual TiCl$_3$ layer, titanium ions fill 2/3 of the octahedral holes between the two chloride layers (giving the observed TiCl$_3$ stoichiometry). In α-TiCl$_3$ the individual layers are stacked so as to give a hexagonal close packing of the chloride ions, γ-TiCl$_3$ has a cubic close packing arrangement, while the δ-form (a sort of intermediate between α and γ) is more disordered.

Natta and co-workers demonstrated that the TiCl$_3$ catalyst lattice structure determined the stereoselectivity of the catalyst and hence polymer isotacticity; brown (β)-TiCl$_3$ giving low isotac-

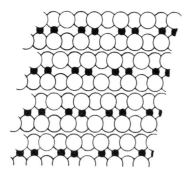

INDIVIDUAL TiCL$_3$ LAYER

Fig. 1.6: α-TiCl$_3$ layer lattice structure [from reference 19, copied with permission from Harwood Academic Publishers]

ticity, while purple (α, γ, δ)-TiCl$_3$ gave high isotacticity. The elegant work of Cossee and Arlman[11] gave a rational and still widely accepted explanation of these facts. An essential feature of the Cossee mechanism is that the catalytically active centres are those Ti ions which possess chloride vacancies, such vacancies on the (α, γ, δ)-TiCl$_3$ surface being necessary to ensure the electro-neutrality of the crystal (Fig. 1.7). These and related aspects will be dealt with in section 1.7 together with various other mechanistic considerations.

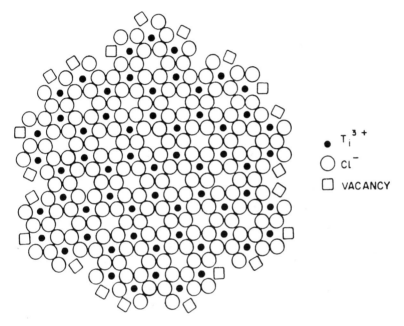

Fig. 1.7: α, γ and δ-TiCl$_3$ layer with edge vacancies emphasized, top view [from reference 19, copied with permission from Harwood Academic Publishers]

The first commercial catalysts were natural extensions of Natta's pioneering work. Typically TiCl$_4$ was reduced at low (sub-zero) temperatures with triethylaluminium in a hydrocarbon diluent, resulting in the precipitation of TiCl$_3$.1/3AlCl$_3$; however this material, being β-TiCl$_3$, is not suitable for isotactic polypropy-lene manufacture. Hence the slurry was slowly heated to 160-200°C for several hours to effect the phase transformation to the purple γ-TiCl$_3$ form necessary for the final catalyst. These first cata-lysts were thus prepared following a two step procedure - low temperature reduction followed by a heat treatment - although the whole process could be conveniently executed in the same reactor since no filtration or solvent switching was required. The reduction step is particularly important in determining catalyst morphology and activity while the heating step (i.e. the phase transformation ---> γ-TiCl$_3$) plays a critical role in controlling the stereoselectivity of the final catalyst.

During the reduction step the TiCl$_3$ which is formed, being insoluble in the reaction medium, precipitates almost instantaneously. Its formation is believed[12] to occur via the decomposition of the soluble intermediate (C$_2$H$_5$)TiCl$_3$ which may occur spontaneously or under the influence of preformed TiCl$_3$ or other reactive species present in the solution. The formation and decomposition of (C$_2$H$_5$)TiCl$_3$ proceeds very much faster in the case of reduction with triethylaluminium than with diethyl aluminium chloride, so much so that the reaction with triethylaluminium is virtually instantaneous at 0°C, whereas that between diethyl aluminium chloride and titanium tetrachloride takes several hours to reach completion at this temperature. This controlled formation of TiCl$_3$ encountered in the latter case leads to a much more regular growth of the precipitated TiCl$_3$, and by variations in reaction conditions spherical catalyst particles of between 10 and 30 μm can be produced. The reaction with triethylaluminium at this temperature yields a fine (1-3 μm) irregularly shaped particle which is in other aspects a very satisfactory catalyst for propylene polymerization.

In the 1970's it was found that it is possible to slow down the reaction between triethylaluminium and titanium tetrachloride sufficiently to allow a controlled precipitation of TiCl$_3$ by carrying out the reaction at much reduced temperatures[13]. In this way catalysts can be prepared which give polymers having an adequate particle size such that the resultant polymer powders flow freely. Fig. 1.8 illustrates how the size and distribution of polymer particles (constant yield on catalyst) vary using catalysts made by reducing TiCl$_4$ under identical conditions but at different temperatures[5].

Further variation of the catalyst preparation recipe allows fine tuning of the catalyst particle size and form. One important variable is the solvent in which the reaction is carried out. This can affect both the solubility of the TiCl$_3$ formed and thus its tendency to precipitate, the degree to which TiCl$_3$ particles agglomerate when they are first formed, and shear forces on particle surfaces as the reaction mixtures are stirred. Fig. 1.9 demon-

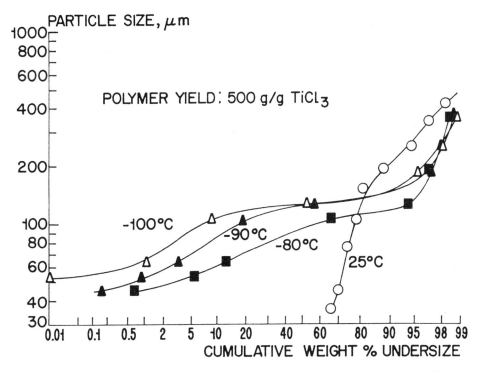

Fig. 1.8: Effect of catalyst preparation temperature on polymer particle size and distribution

strates how catalyst particle size can be varied by choice of solvent used in the catalyst manufacture. Substantial control can be achieved in this way although the range is somewhat limited due to problems encountered in working with very low boiling hydrocarbons and the fact that the solvent used must not freeze at the low temperatures employed. In each case reaction conditions were identical except for the choice of solvent which was always the same for both components ($TiCl_4$ and $AlEt_3$).

Other important variables include the mode of addition of the ingredients ($TiCl_4$ addition to $AlEt_3$ or vice versa), rate of addition and the power (of stirring) input. High stirrer power inputs cause greater shear forces on the surface of the growing particles resulting in the generation of smoother catalyst particles. It is of course essential to supply enough power to maintain the particles in suspension; failure to do so results in very irregular and generally fine catalyst particles.

In the course of this text two types of commercial "first

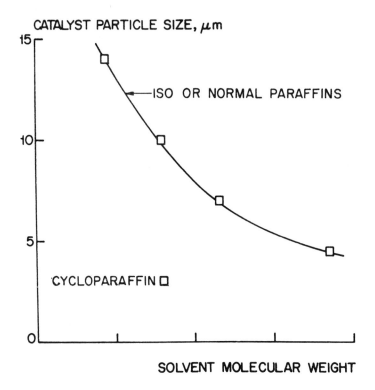

Fig. 1.9: Catalyst (TiCl₃) particle size as a function of hydrocarbon used in the catalyst preparation

generation" catalysts will be discussed, both of which are manufactured employing the above-described precipitation technique. Both catalysts are prepared via addition of $TiCl_4$ to the aluminium alkyl using aliphatic reaction solvents. The first type is made by employing triethylaluminium at low temperatures to afford $TiCl_3 \cdot 1/3AlCl_3$, while the second is is prepared by reducing $TiCl_4$ with diethyl aluminium chloride (DEAC) at about -30°C (this being the lowest temperature at which appreciable reaction takes place between DEAC and $TiCl_4$) to generate $TiCl_3 \cdot 1/2AlCl_3$. The activity and stereoselectivity of such DEAC-reduced catalysts are generally inferior to that observed in catalysts manufactured using triethylaluminium. This is probably attributable to the higher levels of $AlCl_3$ co-crystallized in the catalyst as will be discussed later.

The slurry resulting from the reduction step typically comprises a colourless supernatant and a brown crystalline precipitate of β-TiCl$_3$ containing co-crystallized AlCl$_3$. Simply heating the slurry to a temperature typically in the range 160 to 200°C results in the desired phase transformation of brown β-TiCl$_3$ to the stereoselective purple γ-TiCl$_3$. In practice variables such as the rate of warming-up, hold times and the exact temperature chosen are important in determining the final catalyst performance.

Generally speaking the efficiency and selectivity of first generation catalysts were so poor that there was a clear economic incentive for the development of improved catalysts. It should however be noted that certain catalysts (including proprietary Shell catalysts) enjoy excellent stereoselectivity obviating the necessity for atactic removal procedures. First thrusts in the direction of catalyst performance improvement included :

i) the use of electron donors ("third components") to improve stereoselectivity (see section 1.5) and

ii) ball-milling of the TiCl$_3$.xAlCl$_3$ catalyst. The Stauffer AA catalyst is a commercial example of such a ball-milled TiCl$_3$ catalyst.

The ball-milling technique and its effects on catalyst structure and performance are extensively described by Tornqvist[14], although it was Natta[15] who first studied the grinding of α- and γ-TiCl$_3$ and concluded that a sliding was taking place at the interfaces between pairs of chlorine layers. This sliding of the double ("Cl-Ti-Cl") layers results in stacking disorders causing both α- and γ-TiCl$_3$ forms to approach the disordered δ-TiCl$_3$ form on prolonged grinding. Tornqvist concluded that the activity increase observed on ball-milling is not simply due to an increase in surface area (i.e. particle size reduction) but rather that it can be attributed to reduction in size of the primary crystallites. A similar effect is observed in the ball-milling of MgCl$_2$ which is frequently applied in the preparation of supported catalysts (section 1.4.3).

Some of Tornqvist's data are represented in Fig. 1.10 where catalyst activity (propylene polymerization using AlEt$_3$ as co-catalyst) is plotted against the duration of the ball-milling applied. Some important conclusions drawn were:

18

i)   dry-ball-milling is more effective than wet

ii)   long ball-milling times are required

iii) catalysts containing controlled levels of co-crystallized $AlCl_3$ give higher activity catalysts on prolonged milling (the optimum level of co-crystallized $AlCl_3$ being 0.33 mole per mole $TiCl_3$).

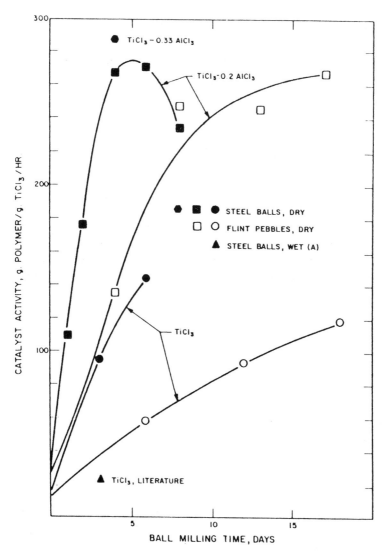

Fig. 1.10: The effect of ball-milling on catalyst activity, TiCl₃ catalysts [from reference 14, copied with permission of the New York Academy of Sciences]

1.4.2 <u>Second generation TiCl$_3$ catalysts</u>

As understanding of the nature of TiCl$_3$ catalysts increased, it became clear that their performance was limited by two constraints:

i) the presence of co-precipitated AlCl$_3$ which leads to the formation of the catalyst poison AlEtCl$_2$ (see below) and

ii) the large size of the catalyst (TiCl$_3$) crystallites resulting in a low proportion of active (surface) Ti sites.

It has long been known that AlEtCl$_2$ acts as a catalyst poison in the TiCl$_3$-catalyzed polymerization of olefins. Its rate-lowering effect can best be described in terms of a dynamic adsorption/desorption equilibrium between active sites and AlEtCl$_2$:

$$Ti \; + \; AlEtCl_2 \quad \longleftrightarrow \quad [TiAlEtCl_2]$$

   active   inactive

Thus removal or complexation of AlEtCl$_2$ shifts this equilibrium to the left resulting in a larger number of active sites and a consequent increase in polymerization activity. The complexation of AlEtCl$_2$ will be dealt with in section 1.5 and the present discussion will be limited to its extraction or removal from the catalyst system.

The AlEtCl$_2$ originates from two sources; it is formed via reaction of the cocatalyst (Et$_2$AlCl) with coprecipitated AlCl$_3$ present in the catalyst,

$$Et_2AlCl \; + \; AlCl_3 \quad \longleftrightarrow \quad 2AlEtCl_2$$

or it is formed as a by-product of the alkylation of the active sites on the TiCl$_3$ catalyst by the cocatalyst:

$$Ti\text{-}Cl \; + \; Et_2AlCl \quad \longrightarrow \quad Ti\text{-}Et + AlEtCl_2$$

The latter source of AlEtCl$_2$ is the minor component and is conveniently "neutralized" via complexation (i.e. use of third components) as described in section 1.5.

It is apparent from the above that most of the AlEtCl$_2$ formation could be eliminated by washing coprecipitated AlCl$_3$ out of the catalyst prior to use. Arlman[16] determined that $AlCl_3$ and $TiCl_3$ are completely miscible in the range $0 < Al/Ti < 0.2$; in this range Al-ions are incorporated in the $TiCl_3$ lattice while any additional $AlCl_3$ (usually in commercial catalysts the Al/Ti ratio is 0.33 to 0.5/1, see section 1.4.1) exists as a separate $AlCl_3$ phase ("coprecipitated $AlCl_3$"). The solubility of $AlCl_3$ in hydrocarbons such as heptane and toluene (see Fig. 1.11) allows the removal of the coprecipitated $AlCl_3$ by washing with these hydrocarbons at elevated temperatures. Treatment of a typical "$TiCl_3 \cdot 1/3AlCl_3$" catalyst in

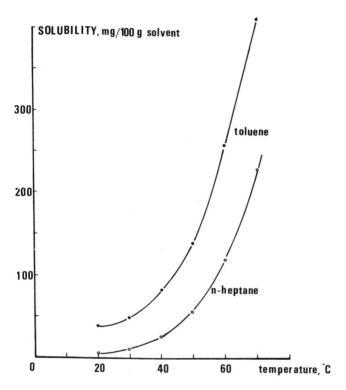

Fig. 1.11: *The solubility of AlCl₃ in heptane and toluene as function of the temperature*

this way gives an activity improvement of about 30%. The investigations of Caunt[17] are also very relevant in this respect.

Alternatively $AlCl_3$ can be extracted using ethers which, being Lewis basic, complex, solubilize and remove the $AlCl_3$. Table 1.1 gives some typical data from our laboratories in which the catalyst was washed with an isooctane solution of dibutylether (DBE) using various DBE/Ti molar ratios. These results show that, up to a DBE/TiCl$_3$ molar ratio of 0.15-0.2 the stereospecificity of the catalyst is barely affected while at higher DBE/TiCl$_3$ ratios the stereospecificity is diminished, as indicated by increasing xylene solubles and the reduced yield stress values of the resulting polymers. We attribute this dependence of final catalyst performance on the DBE/TiCl$_3$ molar ratio applied to the existence of the two types of AlCl$_3$ (co-crystallized and coprecipitated) described above. Hence at low DBE/TiCl$_3$ ratios only the separate (coprecipitated) AlCl$_3$ is solubilized while at the higher DBE levels co-crystallized AlCl$_3$ can also be extracted. When we refer to

TABLE 1.1

The effect of DBE/TiCl3 ratio used in the washing procedure (First generation TEA-reduced TiCl3 catalyst)

DBE/TiCl$_3$ ratio	blank	0.04	0.09	0.14	0.20	0.23	0.40
Al/Ti, molar	0.29	0.29	0.27	0.26	0.24	0.18	0.18
Activity, gPP/ gTiCl$_3$.h.bar	46	54	59	59	65	71	56
Reactor solubles, %m/m	2.2	1.9	3.7	3.0	4.4	5.5	6.1
Yield stress, MPa	37.0	38.0	37.0	35.0	36.0	35.0	32.0
Melt Index, g/10min	4.9	4.6	2.1	3.1	3.2	2.2	2.5

co-crystallized AlCl$_3$ we of course mean that Al-ions are incorporated into the TiCl$_3$ lattice and consequently Al-ions will also be present at crystallographic Ti-sites on the catalytically active faces of the γ- (δ-) TiCl$_3$. Extraction of such Al-ions from the surface of TiCl$_3$ produces active centres of reduced stereospecificity ("atactic centres").

This situation is schematically represented in Fig. 1.12a which shows an idealized γ-TiCl$_3$ surface wherein all sites are stereospecific (i.e. "isotactic centres", see section 1.7). If 'A' is an aluminium ion then the extraction of AlCl$_3$ from the surface will result in the situation shown in Fig. 1.12b with the formation of a new isotactic centre D but the originally isotactic centre B is transformed into an atactic centre containing 2 loosely bound chloride ions and two vacancies. Since it is more difficult to

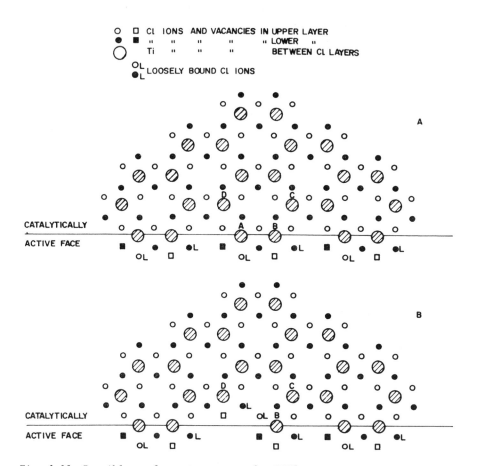

Fig. 1.12: Possible surface structures of γ-TiCl$_3$

remove co-crystallized AlCl$_3$ than AlCl$_3$ which exists as a separate phase mild washing of γ-TiCl$_3$ catalysts will only lead to removal of the non-co-crystallized AlCl$_3$ whereas extensive washing will result in extraction of AlCl$_3$ from the surface of the γ-TiCl$_3$ lattice yielding a catalyst of diminished stereospecificity. Interestingly a similar effect is observed when applying ethers as "third components" in polymerization experiments (see section 1.5).

It was pointed out at the beginning of this section that the performance of first generation TiCl$_3$ catalysts was limited by two factors, the presence of coprecipitated AlCl$_3$ and the large primary crystallite size of the catalysts. The above discussion deals with various approaches which have been used to address the former,

$AlCl_3$ (or rather $AlEtCl_2$), issue. The latter problem, i.e. catalyst primary crystallite size, is in fact one of lowering the temperature required to effect the conversion of brown β-$TiCl_3$ into the stereoselective γ/δ form (purple or violet in colour). Indeed this phase transformation typically required temperatures of 160-200°C; at such high temperatures the primary crystallites can grow substantially in size. The breakthrough came in the early 1970's with the discovery that $TiCl_4$ catalyzes the phase transformation, and that the presence of excess $TiCl_4$ allows the phase transformation to occur at temperatures below 100°C. This finding led to the development and subsequent commercialization of "second generation catalysts". Examples of such catalysts are to be found in a patent filed by Solvay in 1971[18].

The "Solvay catalyst" comprises a highly porous and active δ-$TiCl_3$ catalyst with a high surface area (about 160 m^2/g). The preparative route involves 3 distinct chemical steps:

i)   reduction of $TiCl_4$ with an alkylaluminium to afford a brown precipitate of β-$TiCl_3$ (essentially the same procedure as used for first generation catalysts),

ii)   treating the brown solid with an ether to extract $AlCl_3$ from the solid, and

iii) heating the treated solid (<100°C) with excess $TiCl_4$. This step serves two important functions - to effect the phase transformation (β --> δ-$TiCl_3$) and to extract the ether adsorbed to the solid after step (ii) ($TiCl_4$ is a stronger Lewis acid than $TiCl_3$).

In addition to the above chemical steps the catalyst is also extensively washed with a hydrocarbon solvent to remove any loosely adsorbed ether or $TiCl_4$ which would adversely affect polymerization performance. The washing of these more sophisticated catalysts is also a characteristic of the third generation (supported) catalysts. Indeed these washing procedures can contribute substantially to catalyst manufacturing costs since the large volumes of the hydrocarbons involved require distillation and recycling.

A typical Solvay catalyst can be prepared using the following procedure: $TiCl_4$ and DEAC are reacted at 1°C and heated to 65°C to give a precipitate of β-$TiCl_3$. The brown β-$TiCl_3$ is suspended in a 13 % mixture of diisoamylether in hexane (ether/Ti ratio 0.95/1) at

35°C for one hour followed by washing to give the "treated solid" which contains considerable levels of adsorbed ether but little aluminium. The treated solid is heated to 65°C in the presence of a large excess of $TiCl_4$ (40 % in hexane) to give the final catalyst as a violet coloured (δ-$TiCl_3$) solid.

In a detailed analysis of the Solvay catalyst Nielsen[19] concluded that three factors contribute to its good performance:

i)   smaller crystallite size (due to the lower temperature involved in manufacture)

ii)   removal of $AlCl_3$ by ether extraction. This has the dual effect of lowering $AlEtCl_2$ (catalyst poison) levels in the polymerization and of inducing microporosity in the final catalyst, and

iii) epitactic adsorption of $TiCl_4$ on the solid $TiCl_3$ resulting in active centre formation.

This rather controversial proposal implies that the Solvay catalyst (and related catalysts) is in fact a $TiCl_3$-supported $TiCl_4$ catalyst. There is little evidence in the literature for this last proposal and furthermore it should not be taken to imply a difference in active centre type between first and second generation catalysts because in most first generation catalysts (e.g. the various Shell catalysts described earlier) the final heating step was executed in the presence of slight excesses of (unreacted) $TiCl_4$, again offering the potential for epitactic adsorption of $TiCl_4$ onto the $TiCl_3$ surface. The enormous similarities in polymerization behaviour and kinetics between the first and second generation catalysts argues strongly in favour of the same type of active centre being present in both catalyst types. The improved performance of a Solvay catalyst (typically 4-5 times more active, at a given stereoselectivity, than first generation catalysts) can be attributed to its higher concentration of active centres (smaller primary crystallites, lower $AlEtCl_2$ concentration).

Although endowed with an attractive polymerization performance the Solvay catalyst suffers the disadvantage of involving a complicated multi-step, and hence costly, manufacturing procedure. Subsequent patents from a large number of companies describe simplified methods of preparing catalysts of the same general type. Frequently these catalysts are referred to as "low-aluminium catalysts", although it is implicit that for optimum performance the catalysts must also comprise small primary $TiCl_3$ crystallites.

A typical "low aluminium catalyst" is prepared[20] by reacting excess $TiCl_4$ with triethylaluminium in the presence of a suitable dialkylether (e.g. diisoamylether or di-n-butylether). The reaction ($TiCl_4$ reduction) can be carried out at temperatures between 60°C and 100°C in which case δ-$TiCl_3$ is formed directly, or the reduction is executed at lower temperatures (0-30°C), generating β-$TiCl_3$, followed by heating the resulting slurry (without an intermediate washing procedure) to 60-100°C. The major advantages of such a preparative route are its great simplicity, and that only one final washing step is involved. It should be noted that in these routes a more than stoichiometric amount of $TiCl_4$ is used in order to ensure the presence of $TiCl_4$ in the subsequent heating step; the $TiCl_4$ is of course essential to catalyze the β- to γ/δ-$TiCl_3$ transformation at low temperatures.

The role of ethers in this chemistry is twofold:

i)   complexation and extraction of $AlCl_3$ from the final catalyst,

ii) complexation of triethylaluminium, resulting in a diminished "reducing power" (allowing the $TiCl_4$ ---> $TiCl_3$ reduction to be carried out at ambient temperature or above rather than the -30 to -100 °C typical of first generation systems).

The second point is of critical importance, as reaction of $TiCl_4$ and triethylaluminium leads to overreduction (to $TiCl_2$) and poorly-defined catalysts unless the reaction is carried out at very low temperatures (well below 0°C). By complexing the aluminium alkyl (and in some cases the $TiCl_4$ as well) dialklylethers allow controlled reaction without measurable overreduction even at ambient temperature or above.

An elegant route to these "low-aluminium" catalysts was disclosed by Mitsubishi Chemical[21]. This chemistry involves dissolving a pre-formed solid $TiCl_3$ in a hydrocarbon solvent by complexing it with a suitable (long alkyl chain) dialkyl ether, resulting in a solution of the ether-$TiCl_3$ complex. The solid $TiCl_3$ used can either be in the form of pure (hydrogen-reduced) $TiCl_3$ or in the usual $TiCl_3$.$nAlCl_3$ form in which case the $AlCl_3$ is also solubilized. Excess $TiCl_4$ is then added to the solution; since $TiCl_4$ is a stronger Lewis acid than $TiCl_3$, the $TiCl_3$ is decomplexed and precipitated.

$$ether.TiCl_3 + TiCl_4 \longrightarrow TiCl_3 + TiCl_4.ether$$

By applying appropriate temperatures (60 to 100°C) in the presence of the excess $TiCl_4$ the precipitated $TiCl_3$ is generated in the desired stereoselective δ-$TiCl_3$ form.

1.4.3 Third generation (supported) catalysts

A supported catalyst can perhaps be most easily visualized as a $TiCl_4$ catalyst on a specially-prepared $MgCl_2$ support. Almost invariably an electron donor is used in the catalyst preparation, most frequently aromatic esters. In the last few years there has been a great deal of interest in such catalyst systems, illustrated by the fact that in the last decade hundreds of patent applications (describing propylene polymerization with such catalysts) have been filed by very many different companies.

However, it would be wrong to assume that such supported catalysts are a recent discovery because, in fact, their development has parallelled that of the most conventional $TiCl_3$ catalysts over a period of more than 20 years. The first example of such a supported $TiCl_4$ catalyst[22] dates back to 1955, although this patent only gives examples on ethylene polymerization. In coming to this invention, Petrochemicals Limited reasoned that since it is the "surface layer of each catalyst particle which is effective in promoting the polymerization reaction", it follows that greater or more accessible catalyst surface area will result in higher catalyst activity. They claimed that this could be achieved by supporting $TiCl_4$ on various carriers, including $MgCl_2$ and $MgCO_3$.

Early Shell patent applications in this area include a filing in 1960 describing the first $MgCl_2$-supported $TiCl_4$ catalyst for propylene polymerization[23]. This patent describes the preparation of catalysts by supporting $TiCl_4$ on carriers such as $MgCl_2$ and $CoCl_2$ which could be ground or milled prior to use. The preferred metal halides are those possessing the same layer lattice structure as the stereoselective purple γ-$TiCl_3$ catalysts; furthermore the dimensions of $MgCl_2$ and $CoCl_2$ (ionic radii of Mg^{2+} and Co^{2+} are 0.066nm and 0.072nm respectively) making them optimum supports for $TiCl_4$ (ionic radius Ti^{4+} 0.068nm). These catalysts were demonstrated for propylene polymerization using $AlEt_3$ as co-catalyst; furthermore the use of electron donors, notably carboxylic acid esters such as ethyl acetate, was shown to increase stereoselectivity.

After these initial patents there followed many publications describing $MgCl_2$-supported $TiCl_4$ catalysts for olefin polymerization. Considering only α-olefins other than ethylene (where stereoselectivity plays no role) the following publications are particularly noteworthy:

(a) Mitsui Petrochemical claimed, in 1968, improved catalysts prepared by treating $MgCl_2$ with an electron donor (e.g. ethers, esters) prior to $TiCl_4$ treatment[24],

(b) Montedison issued a series of patents in which $MgCl_2$ "in an active form" was used as catalyst support[25]. The activation of $MgCl_2$ (although not its use as a catalyst support) was first reported in 1967 by Kamienski[26]. This author discovered and reported many routes to "active $MgCl_2$" such as

i) treating $MgCl_2$ with "activating agents" which were electron donors (ethers, alcohols),

ii) ball-milling commercial $MgCl_2$,

iii) reacting Grignard reagents with chlorinating agents such as benzylchloride or chlorine.

Kamienski concluded that "active $MgCl_2$" may have a particular crystal structure, and proposed that X-ray analysis would supply useful information in this regard. It was also observed that some of the activating agents (electron donors) were tightly held by the "active $MgCl_2$", while others were readily removed. Activated magnesium chlorides derived by chlorination of Grignard reagents had previously been described by Bryce-Smith[27]. The resulting "activated catalysts" were found to be active not only in olefin polymerization but also in isomerization, Friedel-Crafts reactions, etc.

Later patents, for example Montedison (1971)[28], contain examples of supported catalysts which combine a high yield (>50 kg polypropylene/g titanium) with a good stereoselectivity (isotactic index > 90%). These catalysts were prepared by milling $MgCl_2$ with $TiCl_4$-electron donor complexes such as $TiCl_4$.ethylbenzoate. In order to combine a high activity with good stereoselectivity, the co-catalyst used was not simply trialkylaluminium but a modified or complexed co-catalyst which is, in fact, a mixture of trialkyl-aluminium and an electron donor. Preferred electron donors were esters of aromatic carboxylic acids such as ethyl anisate and ethyl toluate. One of the major features in these patents is the incorporation of an electron donor, frequently ethyl benzoate, in the solid catalyst components. In the examples, $MgCl_2$ is usually ground in the presence of the electron donor to afford "$MgCl_2$ in an active form".

There is a wide variety of routes to prepare high-activity supported catalysts as is reflected by the abundant patent literature. In addition to $MgCl_2$ and its adducts with various electron donors (in particular aromatic esters and aliphatic alcohols) almost every conceivable magnesium derivative has been used as a starting material in catalyst manufacture, involving many different types of chemical conversion. Various physical techniques such as ball-milling, vibration-milling, spray-drying and precipitation have been claimed as efficient techniques for catalyst synthesis.

One of the first routes to supported catalysts, and the most commonly described preparative method, involves a ball-milling procedure which usually takes the form of a two-step manufacturing process. In the first step anhydrous $MgCl_2$ (frequently exposed to an exhaustive drying procedure) is ball-milled with ethyl benzoate (or another aromatic ester) to afford "active $MgCl_2$", the catalyst support. Crystalline $MgCl_2$ exists in the form of agglomerates of small primary crystallites (similar to crystalline $TiCl_3$); in the case of ball-milled $MgCl_2$ the dimensions of these primary crystallites are much smaller than those of commercially available, anhydrous $MgCl_2$. In the ball-milling procedure ethyl benzoate is thought to stabilize small $MgCl_2$ crystallites by adsorption on to the crystallite surface: grinding experiments have shown that the forces of adhesion between small particles can only be reduced by materials which are chemisorbed on the particle surface[29]. We envisage that the $MgCl_2$ crystallites are broken down during the milling process and that the freshly cleaved surfaces are complexed by ethyl benzoate, thus inhibiting re-aggregation of the resulting small primary crystallites. Owing to the nature of the $MgCl_2$ lattice structure (layer lattice), cleavage occurs predominantly between neighbouring layers of chlorine atoms. Youchang et al[30] also concluded on the basis of their X-ray diffraction studies of ball-milled $MgCl_2$ that shear forces during milling cause the Cl-Mg-Cl double layer to slide over each other producing hexagonal

$MgCl_2$ primary crystallites of only a few layers in thickness.

The dimensions of the primary crystallites can be determined using X-ray diffraction techniques. Fig. 1.13 shows representations of both commercially available anhydrous $MgCl_2$ and ball-milled (with ethyl benzoate) $MgCl_2$ crystallites. As can be seen an efficiently ball-milled sample is reduced in crystallite size in the c-direction (i.e. thickness) to less than 3 nm (i.e. only about 2-4 layers thick).

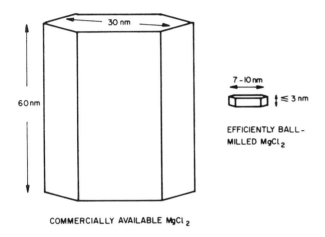

Fig. 1.13: *MgCl$_2$ primary crystallite dimensions [from reference 158, copied with permission from Harwood Academic Publishers]*

The second step of the catalyst preparation involves bringing $TiCl_4$ on to the ball-milled support; this is most frequently achieved either by suspending the catalyst support in hot, undiluted $TiCl_4$ or by further ball-milling the catalyst support in the presence of $TiCl_4$. The resulting solid is then thoroughly washed with hot hydrocarbon to remove soluble titanium complexes. Typically, the resulting catalyst, a pale yellow solid, contains 1-2 %m/m Ti and 5-20 %m/m ethyl benzoate. In this last step of the catalyst preparation some ethyl benzoate (Lewis base) is extracted by $TiCl_4$ (Lewis acid) and $TiCl_4$ becomes adsorbed on to free $MgCl_2$ surface vacancies.

The performance of a ball-milled catalyst is illustrated in Fig. 1.14 as a function of polymer yield and isotactic index. For comparison the performance of an early first generation and a

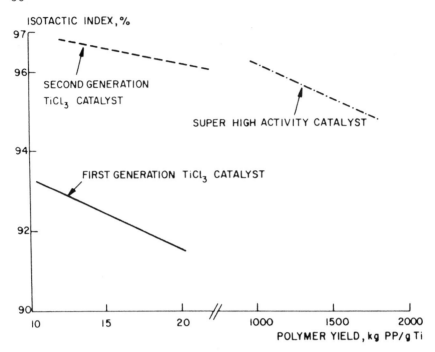

ISOTACTIC INDEX, %

SECOND GENERATION
TiCl₃ CATALYST

SUPER HIGH ACTIVITY CATALYST

FIRST GENERATION TiCl₃ CATALYST

POLYMER YIELD, kg PP/g Ti

Fig. 1.14: Relationship between isotactic index and polymer yield [from reference 158, copied with permission from Harwood Academic Publishers]

second generation (Solvay type) TiCl₃ catalyst are plotted, demonstrating that the polymer yield of the supported catalyst is more than an order of magnitude higher than the TiCl₃ systems.

The supported catalyst polymerization performance illustrated in Fig. 1.14 represents results from 60-minute polymerization runs. Thus, in one hour $1.7*10^6$ g of polypropylene (with an isotactic index of 95%) are produced per g titanium. Even under the assumption that every titanium atom is part of an active centre, it follows that the "turn-over rate" (rate of monomer insertion) averaged over the whole polymerization period is >500 moles propylene per gram atom of Ti per second; in other words, the "change-over time" (interval between two insertion steps) is less than 2 milliseconds. Kaminsky[31] described an even higher rate for ethylene polymerization with a homogeneous zirconium catalyst (change-over time less than 0.3 milliseconds), but it should be pointed out that such rates could only be achieved with enormous excess of alkylaluminium (Al:Zr molar ratio >100,000 : 1) whereas the supported catalyst can be activated with modest levels of alkylaluminium

(Al:Ti molar ratio 50:1 or even less).

While many of these supported catalysts possess exceptionally high initial polymerization rates, many also exhibit a significant rate of decay (activity decline) as discussed in section 1.7. For certain process types (e.g. liquid propylene reactors with short residence times for example) the decay presents no problem but for other processes (e.g. "block" co-polymerization in multi-stage reactors) it is necessary to modify either the process or the catalyst in order to overcome the obstacle of diminishing polymerization activity.

Recently it has been reported that catalyst decay can be greatly reduced by simply modifying the electron donor package used in the catalyst (both solid pro-catalyst and co-catalyst). These alternative electron donors comprise dialkylphthalates (e.g. diisobutylphthalate) in the solid component (replacing ethyl benzoate etc.) and certain silane derivatives (e.g. phenyl triethoxy silane, $PhSi(OEt)_3$) in combination with the triethylaluminium co-catalyst (replacing para ethyl anisate, etc.). These alternative electron donors first appeared in the patent literature[32] and were recently reviewed by Soga and Shiono[33]. These authors pointed out that phthalate-based catalysts (similarly to most ethyl benzoate catalysts) exhibit poor stereoselectivity in the absence of a suitable selectivity control agent (electron donor); in one example using a diisobutyl phthalate-based catalyst with $AlEt_3$ as co-catalyst an isotactic index of 68% was reported, this being improved to 97-98% in the presence of a suitable silane derivative. Soga and Shiono reported the following order of efficiency in selectivity control for a series of these silanes:

$$Ph_2Si(OMe)_2 \geq PhSi(OEt)_3 > PhSi(OMe)_3 \geq Si(OMe)_4 \geq Si(OEt)_4 >$$
$$Si(OPr)_4 > Si(OnBu)_4$$

It is interesting to note that the efficiencies of these silanes is much lower when applied to ethyl benzoate-modified catalysts: in this case the stereoselectivity improvement is less (typically 85-90%) and this is only achieved at the expense of a large drop in catalyst activity (compared to a minimal activity loss in the case of a phthalate catalyst). This difference in behaviour points to differing modes of action for the various electron donors; these aspects and the implications for catalyst kinetics and deactivation are discussed in section 1.7.

Turning now to <u>catalyst</u> <u>characterization</u>: A supported catalyst of the type under discussion exists as a crystalline powder, frequently pale yellow in colour. Just as in the case of $TiCl_3$ catalysts each catalyst particle comprises a large number of primary crystallites - which for a supported catalyst consist of $MgCl_2$ crystallites on the surface of which are adsorbed $TiCl_4$ ("the catalyst") and the electron donor used in the catalyst preparation (in this discussion ethyl benzoate is assumed).

$MgCl_2$ has a layer lattice structure consisting of a hexagonal stacking of sheets of closely packed chloride ions making it very similar to purple (γ) $TiCl_3$. In contrast to $TiCl_3$ (in which 2/3 of the octahedral holes are filled with Ti^{3+} ions) in the case of $MgCl_2$ <u>all</u> the octahedral holes are filled with Mg^{2+}. Another important difference between $MgCl_2$ and $TiCl_3$ is that $MgCl_2$ knows no chain lattice (β-$TiCl_3$ type) configuration; this fact greatly simplifies the art of generating small crystallites (i.e. no heating/phase transformation steps involved). The structural arrangement of Mg ions in an $MgCl_2$ layer is represented in Fig. 1.15. The nature of the surface sites is more clearly illustrated by a two-dimensional representation of a whole lattice layer (1 Mg

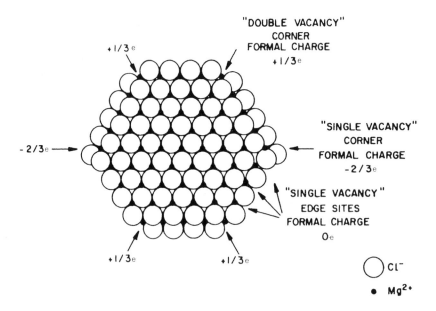

Fig. 1.15: *Model crystallite of $MgCl_2$, one layer approx. 3nm in size, top view [from reference 158, copied with permission from Harwood Academic Publishers]*

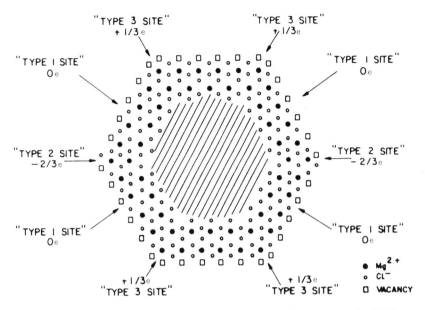

Fig. 1.16: Surface structure and sites of hexagonal $MgCl_2$ crystallites (two-dimensional representation) [from reference 158, copied with permission from Harwood Academic Publishers]

and 2 Cl layers) as shown in Fig. 1.16. In this way the creation of surface Mg ions - which form the adsorption sites for ethyl benzoate and $TiCl_4$ - can be envisaged by building up $MgCl_2$ crystallites step by step and calculating for each crystallite size (assuming electroneutrality for each crystallite) the number of surface Mg ions and the number of Cl vacancies.

From this procedure it follows that three different types of surface Mg ions can be distinguished on the surface of a $MgCl_2$ crystal:

- type 1   on the lateral faces : single vacancy Mg ions having an effective charge of 0e;

- type 2   on the corners: single vacancy Mg ions having an effective charge of -2/3e;

- type 3   on the corners: double vacancy Mg ions having an effective charge of +1/3e.

34

Interestingly, constructing hexagonal MgCl$_2$ crystals of different sizes shows that the number of double vacancy (type 3) sites is always 4, irrespective of the size of the crystal, and that these sites are located on 4 of the 6 corners of the hexagonal crystal. Moreover, the two remaining corner sites are single vacancy sites and have an effective charge of -2/3e (type 2 sites).

Assuming a hexagonal crystal shape as shown in Fig. 1.16 the fraction of double and single vacancy surface Mg ions can be plotted as a function of MgCl$_2$ crystallite size. This is done in Fig. 1.17 and it can be seen that the fraction of double vacancy surface Mg ions (type 3 site, effective charge +1/3e) increases much more rapidly with decreasing crystallite size than the fraction of single vacancy sites (type 1 site, effective charge zero).

Investigation of a series of MgCl$_2$/ethyl benzoate ball-milled samples using infa-red techniques indicated that ethyl benzoate is coordinated to the MgCl$_2$ crystallite surface via the carbonyl function. Very similar IR spectra were obtained over a wide range of ethyl benzoate levels (5-43%), the characteristic carbonyl

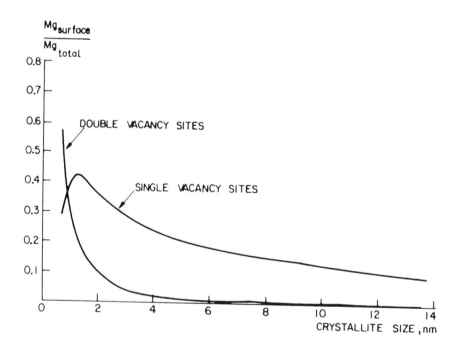

Fig. 1.17: Fraction of surface Mg-ions as a function of MgCl$_2$-crystal size [from reference 158, copied with permission from Harwood Academic Publishers]

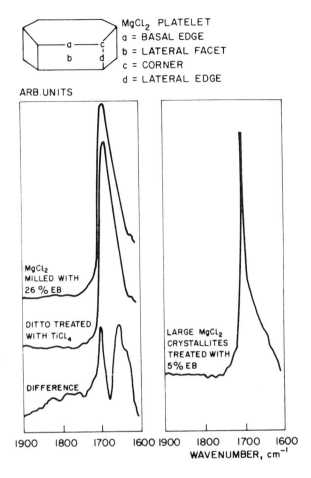

Fig. 1.18: Carbonyl regions in the infrared absorption spectra of catalyst treated with ethyl benzoate [from reference 158, copied with permission from Harwood Academic Publishers]

stretching frequency always falling in the range 1678-1688 cm^{-1} (free ethyl benzoate absorbs at 1721 cm^{-1}). We have assigned the absorption at approximately 1680 cm^{-1} to ester molecules bound to the basal edges and corners of the crystallite (see Fig. 1.18). The broadness of the absorption demonstrates the inhomogeneity of the surface sites.

Adsorption-desorption experiments with ethyl benzoate and various $MgCl_2$ samples allowed us to assign the intensity at about 1700 cm^{-1} to ethyl benzoate coordinated to the lateral faces: Indeed adsorption of ethyl benzoate onto large $MgCl_2$ crystallites exhibits an intense, narrow signal at 1700 cm^{-1}. As expected, the relative intensity at 1700 cm^{-1} was found to decrease with decreasing size of the lateral faces (i.e. with decreasing crystallite thickness).

Treatment of these MgCl$_2$/ethyl benzoate samples with neat TiCl$_4$ at elevated temperatures, followed by thorough hydrocarbon washing, affords the final catalysts. The infrared spectra of the catalysts are very similar to those of the unconverted ball-milled MgCl$_2$ and in each case the carbonyl absorption was to be found at 1680 ± 2 cm^{-1}. IR difference spectra revealed that the TiCl$_4$ treatment results in subtle changes in the carbonyl region, namely preferential losses in intensity at 1700 cm^{-1} and about 1650 cm^{-1} (Fig. 1.18).

On the basis of this data and other IR studies we were able to conclude that:

i)   in ball-milled MgCl$_2$/ethyl benzoate samples ethyl benzoate remains bound to almost all surface sites; and

ii)   during the TiCl$_4$ treatment ethyl benzoate is lost from weakly bonding sites on the lateral faces (1700 cm^{-1} absorption); furthermore, ethyl benzoate is also lost from the most strongly coordinating corner sites (type 3 sites, 1650 cm^{-1} absorption). The fact that desorption also takes place from relatively strong bonding sites suggests that the TiCl$_4$ coordinates to these sites by replacement of the ethyl benzoate molecules.

It is reasonable to assume that the structure of the dialkyl phthalate-modified catalysts is very similar to the above, due to the obvious chemical and steric similarities to the alkylbenzoate system. The potential bidentate or chelate binding ability of the phthalate might be expected to lead to stronger coordination of this electron donor to the catalyst surface. Indeed Soga and Shiono[33] compared two ball-milled MgCl$_2$ catalysts in this regard; one contained di-n-butyl phthalate while the other was prepared using n-butyl benzoate. Both catalysts were treated with a heptane solution of triethylaluminium and then filtered; the benzoate catalyst was found to have lost roughly half of the original butyl benzoate whereas in the second catalyst the butyl phthalate level remained essentially unchanged. On the basis of these observations Soga concluded that the phthalate was more strongly bound to the catalyst and hence more difficult to extract with weak Lewis acids such as triethylaluminium.

Although the logic of these arguments cannot be denied it would appear to be an over-simplification since it is found that the phthalate readily exchanges with, and is replaced on the catalyst surface by, the silane electron donors used as selectivity control agents during the polymerization. This is perhaps most clearly demonstrated in a recent patent application[34] describing the pre-treatment of such phthalate-containing catalysts with mixtures of trialkylaluminiums and selected silane derivatives. In one example a catalyst containing 9.9%m/m di-isobutyl phthalate and 2.3%m/m titanium was treated with a hydrocarbon solution of trie-thyl aluminium and phenyl triethoxy silane (20°C, 1 hour) followed by filtration and washing to afford a catalyst containing essenti-ally the same level of titanium (2.1%m/m) but a greatly diminished level of the phthalate (1.8%m/m). The treated catalyst was also found to contain 1.3%m/m silicon, indicating that the lost phtha-late had been replaced (or displaced) by a roughly equimolar quantity of the silicon species. In other examples the phthalate is almost completely removed by similar treatments. Thus Soga's conclusion must be slightly modified; phthalates, although being strongly bound to the catalyst surface, are readily displaced in the presence of suitable electron donors such as $PhSi(OEt)_3$.

1.4.4 Homogeneous catalysts

In the introduction to this chapter it was stated that the discussion would be limited to commercial catalyst systems. Al-though no homogeneous catalyst has yet found commercial application in the isotactic polymerization of polypropylene, the authors have decided to include this area because of the many new and important developments in this exciting field.

Probably the first claim of a homogeneous catalyst for the stereospecific polymerization of propylene is to be found in the early 1970's when Ballard[35] described how catalysts such as tetra-benzyl zirconium and tribenzyl titanium chloride could be used to generate isotactic polypropylene. The catalysts were first synthe-sized, and tested for polymerization activity by Giannini and co-workers[36]. Doubts have been cast on the true homogeneous charac-ter of these systems by, among others, Soga et al[37] who proposed that polymerization in fact occurs on the surface of colloidal particles arising via decomposition of the tetrabenzyl zirconium "precursor". Although these catalysts attracted considerable attention in the 1970's as ethylene polymerization catalysts no extensive studies on their application in the production of isotactic polypropylene were carried out: this is presumably to be attributed to their mediocre activity, the fact that commercially-acceptable stereoselectivity proved unattainable and the above-described doubts concerning the homogeneity of the systems.

Recent spectacular break-throughs in this field have resulted in the production of highly isotactic polypropylene using very active homogeneous zirconium catalysts. This work can be largely attributed to Kaminsky and co-workers at the University of Hamburg, the origins of the discoveries being identified in the mid 1970's when the group first reported[38] that the activity of soluble catalysts of the type dimethyl titanocene + trimethyl aluminium was dramatically improved by premixing the trimethyl aluminium with controlled amounts of water. As described in chapter 9 these catalysts, particularly the zirconium analogues thereof exhibit ethylene polymerization activity an order of magnitude or more higher than the best heterogeneous catalyst systems. Recently Kaminsky described[39] a productivity of $40*10^6$ g polyethylene/g Zr per hour at 8 bars ethylene pressure.

The reaction of $AlMe_3$ with water is best carried out under carefully controlled conditions such that the two reagents react in the molar ratio 1:1, resulting in the formation of oligomeric methylaluminoxanes containing the repeating unit $-Al(CH_3)O-$. The degree of oligomerization of these aluminoxanes is critical in determining catalyst activity, with high molecular weights being preferred[31]. Although methylaluminoxanes can be prepared via direct reaction with water, the preferred methods involve reacting trimethyl aluminium with the water of crystallization in metal salts, particularly $CuSO_4 \cdot 5H_2O$ and $Al_2(SO_4)_3 \cdot 18H_2O$ (ref 40).

Kaminsky[41,42] and others[43] also reported the performance of the dimethyl zirconocene + methylaluminoxane catalyst system in propylene polymerization. High polymer yields were described (550,000 gPP/g Zr) but the product was found to be completely atactic, demonstrating that this catalyst exhibits no stereocontrol whatsoever. Surprisingly Ewen[44] discovered that in contrast to the dimethyltitanocene + methylaluminoxane catalysts the diphenyltitanium analogue affords isotactic polypropylene with a novel stereoblock microstructure when applied at low temperature (-45°C). Using

this catalyst the polymerization rate, the polymer molecular weight
and isotacticity all peak at -45°C due to the catalyst instability
at higher temperatures.

The remarkable stereoselectivity of the Cp_2TiPh_2/methylalumi-
noxane system at low temperatures is clear from Fig. 1.19 where the
^{13}C-NMR spectra of the methyl pentad region for the polypropylene
produced at various temperatures is compared. At 25°C the polypropylene produced is completely atactic showing 3 regions of peaks
corresponding to the (from left to right) mm- ("isotactic), mr-

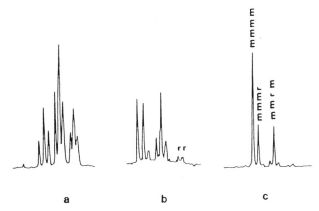

a   b   c

Fig. 1.19: Comparison of ^{13}C-NMR spectra of the methyl pentad region for
polypropylene obtained with $Cp_2Ti(Ph)_2$/methylaluminoxane at three different
temperatures: 25 °C (a); 0 °C (b); -45 °C (c) [from reference 44. Reprinted with
permission form J.Am.Chem.Soc.. Copyright 1984 American Chemical Society]

("atactic") and rr- ("syndiotactic") centred pentads. At 0°C it is
apparent that the stereoselectivity is improved and the virtual
disappearance of the rr-centred pentads is observed. At -45°C an
isotactic polypropylene results although compared to commercial
grades of polypropylene it is clear that the material contains
rather short "stereoblocks" indicated by the significant mmmr and
mmrm pentads (see also chapter 2). In explaining these remarkable
findings Ewen argues that the stereoblock nature of the product is
consistent with a chain-end stereochemical control mechanism, and
draws parallels with the VCl_4/DEAC catalyst system[45] which exhibits
a simlar degree of stereospecificity (in this case to syndiotactic
polypropylene) also at low temperatures. The selectivity of this
vanadium catalyst is also attributed to chain-end control[46]. The
difference between the isotactic-titanium system and the syndiotac-

40

tic-vanadium catalyst is generally attributed to the mode of
insertion of propylene into the growing polymer chain; isotactic
polymerization arising from the a "1-2" insertion with retention of
chain configuration while syndiotactic propagation is explained by
a "2-1" insertion mechanism[47]. This difference in mode of insertion
is most probably due to a difference in polarity between the Ti-C
and V-C bonds. Ewen speculates that the lack of isotactic control
of the corresponding zirconocene catalysts, even at low temperatures, could be accounted for by the large ionic radius of zirconium
(0.79 Angstrom) compared to titanium and vanadium (0.68 and 0.63
Angstrom respectively) resulting in the absence of the van der
Waals' contacts required for the postulated chain-end control.

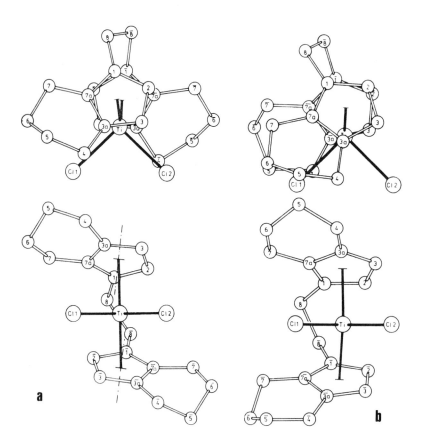

Fig. 1.20: Molecular structure of racemic-(a) and meso-(b) ethylene-bis(4,5,6,7-
-tetrahydro-1-indenyl)titanium dichloride [from reference 48, copied with
permission from Elsevier]

In the same publication[44] Ewen reported the synthesis of isotactic polypropylene using, instead of a simple titanocene derivative, a stereorigid, chiral titanium species racemic-ethylenebis(indenyl) titanium dichloride (again the methylaluminoxane co-catalyst was utilized). This type of novel titanium species was first synthesized by Brintzinger and co-workers[48]. However the polypropylene produced by Ewen contained 37% atactic material as a result of the presence of the achiral meso-ethylene-bis(indenyl) titanium derivative in the catalyst; Ewen reported his catalyst to comprise a mixture of 56% of the chiral racemic- and 44% of the achiral meso-species, the structures of which are shown in Fig. 1.20.

Recently Kaminsky and coworkers[49] have shown that isotactic polypropylene can be obtained in high purity and in high yield using the chiral zirconium analogue of the above-described titanium compounds in combination with the methylaluminoxane co-catalyst. In the case of the zirconium compounds, and in contrast to the titanium complexes, the achiral meso-form is unstable so that pure racemic mixtures of the chiral zirconocenes can be isolated and utilized to produce highly isotactic polypropylene even at relatively high temperatures (room temperature and above). Furthermore the zirconium catalysts are three orders of magnitude more active than their titanium counterparts. This discovery forms the basis of a patent[50].

The molecular weight of the polypropylene produced with these chiral catalysts is highly temperature dependent[51]; increasing the polymerization temperature from -20°C to 60°C causes the molecular weight to decrease by two orders of magnitude as shown in Fig. 1.21. The molecular weight distributions were narrow (M_w/M_n 2 to 2.5), being characteristic of a homogeneous catalyst with only one type of active centre. The temperature dependence of molecular weight parallels that previously reported in the case of polyethylene[41] (Fig. 1.22) although in the case of polypropylene the molecular weight observed is much lower at any given temperature, presumably reflecting the relative ease of β-hydrogen elimination from a tertiary carbon atom (polypropylene) compared to a secondary carbon (polyethylene).

Resolution of the racemic mixture of the zirconium complex affording an optically active pure enantiomer led to the first ever preparation of optically active polypropylene[51]. The chain helix remains stable unless dissolved or subjected to high temperatures,

42

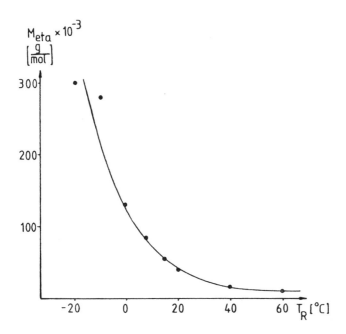

Fig. 1.21: Molecular weight of isotactic polypropylene, prepared with a homogeneous catalyst, in dependence of the polymerization temperature [from reference 51, copied with permission from Hüthig & Wepf Verlag]

when the optical activity is lost due to racemization into helices of both screw senses.

The very high selectivity of these zirconium catalysts and their high activity, which approaches that of the $MgCl_2$-supported catalysts decribed in the previous section, makes them a serious contender for future processes. Indeed it can be expected that the relative ease of "tailoring" of homogeneous catalysts compared to the complicated heterogeneous systems will enable these catalysts to be further improved and exploited in terms of activity and selectivity (isotacticity, molecular weight and distribution). Three major problems remain to be solved before such zirconium catalysts can be considered commercially viable:

i) Catalyst cost. While the polymerization activity of these catalysts is unprecedented in terms of polmer yield per gram of zirconium derivative (making the cost of the often-difficult-to-synthesize zirconocene derivatives almost insignificant) the high

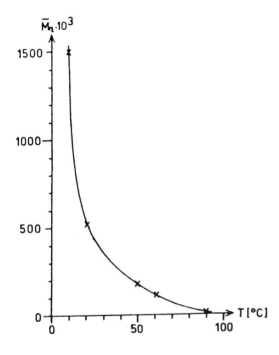

Fig. 1.22: Molecular weight by regulation of temperature in the polymerization of ethylene with a homogeneous catalyst [from reference 41, copied with permission from Harwood Academic Publishers]

levels of the expensive aluminoxane co-catalyst must remain an area of concern.

ii)   The rapid decline in molecular weight at high temperatures must be controlled since polymerization temperatures of >60°C are essential to allow efficient, economic removal of the heat of polymerization, and

iii) Morphology control. Although not disclosed in the literature it is assumed that these homogeneous catalysts afford a very finely divided powdery product; this type of morphology is commercially unacceptable for reasons given in section 1.3. One route to improve morphology would be to support the catalyst on a suitable carrier such as alumina or silica but in this case there remains a risk of losing many of the advantages of these homogeneous and well-defined catalyst systems and their future potential via such a process of heterogenization.

In addition the level of regioselectivity has to improve, as it appears that even at low polymerization temperatures both 2,1 and 1,3 insertion occurs, which lowers the melting point of the polymer unacceptably (for details see chapter 2.6).

In the recent literature it can be seen that advances are being made on some of the above fronts. Regarding item 1, the cost of methaluminoxane, Chien and Wang[52] recently disclosed that 90% or more of the aluminoxane co-catalyst could be replaced with trimethylaluminum (TMA) with no adverse effect on catalyst performance; although the catalyst performance was only demonstrated in ethylene polymerization it can be assumed that similar effects would be seen in the case of propylene. Unfortunately the cost of TMA is still much higher than that of the triethyl or triisobutyl analogues (which are manufactured directly from the appropriate olefins) making the commercial value of this invention somewhat limited.

A recent contribution from Ewen et al.[53] demonstrates that the use of hafnocene derivatives rather than the usual zirconocene analogues affords polymers of substantially higher molecular weight. The compounds studied by Ewen were rac-ethylenebis(indenyl)hafnium dichloride (rac-Et[Ind]$_2$HfCl$_2$) and rac-ethylenebis(4,5,6,7-tetra-hydro-1-indenyl)hafnium dichloride (rac-Et[IndH$_4$]$_2$HfCl$_2$). Some typical data is given in Table 1.2.

TABLE 1.2

Propylene Polymerization Results with Chiral Hf and Zr Metallocenes[*]

metallocene	polym temp. °C	Rate 10^{-6}.Rp,	Mol. wt. 10^{-3}.Mw	MWD Mw/Mn
rac-Et[Ind]$_2$HfCl$_2$	50	26.8	>724	2.2
rac-Et[Ind]$_2$ZrCl$_2$	50	21.2	28	2.1
rac-Et[IndH$_4$]$_2$HfCl$_2$	80	34.8	42	2.4
rac-Et[IndH$_4$]$_2$ZrCl$_2$	80	25.5	9	2.1

[*] All polymerizations were run in propylene/toluene mixtures except the last which was run in bulk propylene. R_p = g/mol.[mon].h

Some recent patent applications[54,55] deal with supporting these metallocene/methaluminoxane catalyst systems on silica to afford spherical catalyst particles which yield polymer (polyethylene in each case) of high bulk density. The resulting catalysts are exposed to a light prepolymerization before being used in olefin polymerization. Yet another patent application to Mitsui Petrochemical[56] deals with the spray-drying and precipitation of methaluminoxane to afford controlled particle size distribution.

In addition to homopolymerization the zirconium/aluminoxane catalysts have been applied in EPDM[57,58] and LLDPE (ethylene/1-hexene and ethylene/1-butene copolymers[51,58]). The living character of the catalysts at low temperatures offers intriguing possibilities in terms of producing true block-copolymers such as PP-EPR-PP. This topic is further discussed in chapters 5 and 9. The growing interest in these homogeneous catalyst systems is illustrated by recent patents and literature articles from Exxon[59], Mitsubishi Petrochemical[60], Mitsui Petrochemical[54,56,61,62] and many others.

Earlier in this section we pointed out that being homogeneous and therefore relatively easy to characterize and modify or "tailor" it must be expected that significant advances will be made using these catalysts both in terms of new polymers as well as in the understanding of how these catalysts function. Indeed it is already clear from the literature that progress is being made in both of these areas.

Using homogeneous catalysts it has already proven possible to control molecular weight distribution and composition to a degree unattainable with heterogeneous catalysts; the resulting polymers could reasonably be referred to as "new" polymers. However in addition to these materials there are an increasing number of references in the literature to truly novel polymeric materials made using these homogeneous catalysts as the following examples illustrate:

i) Kaminsky and co-workers[63] have reported that the zirconocene dichloride/methaluminoxane catalyst system can be used to homopolymerize cyclopentene by 1,2 addition (not ring-opening) polymerization to afford an isotactic polymer with an exceedingly high melting point (420 °C). Norbornene was found to behave similarly to afford a homopolymer which decomposes at about 600 °C. Copolymers with ethylene were also reported.

46

ii)   Idemitsu[64] recently published a series of patents describing
highly syndiotactic polystyrene which can be prepared using titanium compounds (e.g. $CpTiCl_3$ or $TiCl_4$) in combination with methaluminoxane.

iii)   Idemitsu also described[65] modified Kaminsky catalysts for use
in the production of low molecular weight propylene oligomers.

iv)   Ewen[53,66] described modified Kaminsky catalysts which polymerized propylene to afford highly syndiotactic polymer (this material has already been prepared using homogeneous vanadium catalysts[67]
but only at very low temperatures, <-40 °C). The unique catalyst
used by Ewen to invoke this high syndiospecificity was isopropyl(cy-
clopentadienyl-1-fluorenyl) hafnium (or zirconium) dichloride shown
in Fig. 1.23. Even though this result represents a remarkable
discovery, it must be emphasized that the choice of metallocene

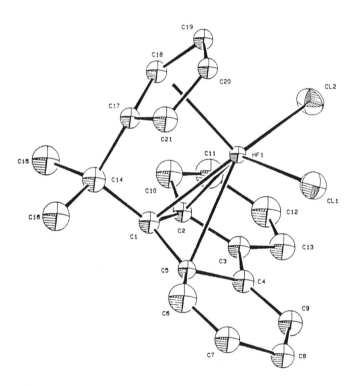

Fig.   1.23:   Structure   of   isopropyl(cyclopentadienyl-1-fluorenyl)hafnium
dichloride   [from   reference   66.   Reprinted   with   permission   from
J.Am.Chem.Soc..Copyright 1988 American Chemical Society]

ligand remains an art rather than a science at the time of preparing this manuscript. This fact was underscored in a recent paper
by a group from Hoechst[68] who described an extensive study aimed at
the synthesis of new ligands of this type, all of which failed to
afford the desired result of high isotacticity in combination with
acceptably high molecular weight.

Although evidence is to be found in much earlier literature it
is particularly in the last three years that it has become recognized that homogeneous Ziegler catalysts (and therefore probably the
heterogeneous catalysts also) involve <u>cationic</u> transition metal
complexes. In this regard it is primarily the work of Jordan et
al.[69,70] which should be mentioned. Jordan discovered that dimethylzirconocene $[Cp_2Zr(CH_3)_2]$ and related compounds react with
silver tetraphenylborate $[Ag(BPh_4)]$, or the analogous ferrocene
tetraphenylborate $[Cp_2Fe(BPh_4)]$, to afford novel cationic zirconocene derivatives $[Cp_2Zr(CH_3)^+]$ according to the following equation:

$$Cp_2Zr(CH_3)_2 + Ag(BPh_4) \xrightarrow{\text{CH}_3\text{CN}} [Cp_2Zr(CH_3)(CH_3CN)]^+[BPh_4]^- + Ag^0 + 1/2\ C_2H_6$$

The reaction probably proceeds via a one-electron oxidation of the
dimethylzirconocene which occurs with loss of a methyl radical
(which is ultimately observed as ethane) to form the new cationic
complex. The bulky and relatively unreactive tetraphenylborate
anion was found to be the counterion of choice; classic counterions
such as PF_6^-, BF_4^-, $AlCl_4^-$ and $CF_3SO_3^-$ were all found to complex
the zirconium species tightly or to react with the resulting
cation. The X-ray structure of the benzyl analogue of the above
methylzirconocene cation is shown in Fig. 1.24.

The most novel chemical behaviour of these cationic compounds
is that they polymerize ethylene (albeit with low activity) without
the presence of an alkylaluminium co-catalyst. The relevance of
this observation to Ziegler catalysis in general, and to Kaminsky
catalysts in particular, is immediately apparent. Indeed in respect
to the Kaminsky catalysts it would appear that an important function of the methaluminoxane co-catalyst is to abstract a methyl
group (or chloride) from the zirconocene catalyst to afford a
highly active and relatively stable cationic zirconocene deriva-

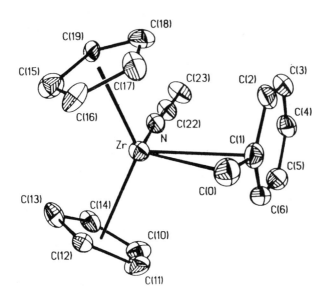

Fig. 1.24: Structure of the $Cp_2Zr(\eta^2-CH_2C_6H_5)(CH_3CN)^+$ cation (hydrogen atoms not shown) [from reference 70. Reprinted with permission from J.Chem.Ed.. Copyright 1988 American Chemical Society]

tive.

Earlier evidence for the possible presence of, such ion pairs in Ziegler catalysts was presented by Eisch and co-workers[71] who isolated a cationic titanium alkenyl species while studying the "classic" homogeneous Cp_2TiCl_2 + $AlMeCl_2$ catalyst system. In these investigations Eisch used a bulky acetylene as a model for an olefin, the benefit of this model being that the reaction halts after one insertion step permitting the facile isolation of the intermediate:

$$Cp_2TiCl_2 + AlMeCl_2 + PhCCSiMe_3 \quad ---->$$
$$Cp_2Ti[C(SiMe_3)=C(Me)Ph]^+[AlCl_4]^-$$

Direct evidence for the presence of cationic titanium species in this same family of titanocene catalysts during actual ethylene polymerizations was reported already in the 1960's by Dyachkovskii and co-workers[72,73].

The above-described Jordan catalysts, or cationic zirconocene systems, despite presenting very elegant models of the highly-active Kaminsky systems, only exhibit a very low polymerization activity. The poor performance of the catalysts can be attributed to the fact that the cation is coordinated by a Lewis base (typically acetonitrile as shown in the above equation, or THF). Being a 14 electron system the cation is readily solvated by such Lewis bases, and it is apparent that the Lewis base occupies a reactive site which could otherwise bind olefin in the polymerization reaction. Since the silver and ferrocene tetraphenylborates are only soluble in polar solvents such as acetonitrile, it is impossible to isolate an electron-donor free cation using this reagents. Turner et al. at Exxon[74,75] recognized this fact and switched to using tributylammonium tetraphenylborate which is soluble in toluene; this finding allowed them to isolate the naked zirconocene cation for the first time. However these species still suffer one more drawback (as reported earlier by Jordan[70]), namely that the cation (even in the solvated state) reacts with the tetraphenylborate anion causing eventual decomposition of the catalyst. Turner et al. found that use of the permethylated zirconocenes prevents anion degradation and, presumably due to their steric bulk, increases the lability of the anion towards olefin resulting in higher catalyst activity:

$$Cp^*_2ZrMe_2 \xrightarrow[-2CH_4]{[Bu_3NH][B(C_6H_4R)_4]} Cp^*_2Zr^{(+)}\text{—}\underset{R \quad H}{\overset{H \quad B^{(-)}(C_6H_4R)_3}{\bigcirc}}\text{—}H \quad + \quad R_3N$$

It should be noted that metallation of one of the rings (meta position) of the anion has occurred making the active Zr-C bond in this system a zirconium-phenyl bond; the phenyl group being part of the borate anion. Interestingly Turner reports[75] that there is no evidence to support the presence of borate end groups in the resulting polymer.

Another suitable group of anions reported by the same authors are the polyhedral carboranes which can be directly reacted with the permethylated dimethyl zirconocene in simple hydrocarbons such as pentane to afford a non-solvated cation, also highly active in olefin polymerization:

$$Cp^*_2ZrMe_2 + C_2B_9H_{13} \dashrightarrow [Cp^*_2ZrMe]^+[C_2B_9H_{12}]^-$$

The use of carborane anions in homogeneous Ziegler systems was also reported in a recent anonymous research disclosure[76] although the activity of these catalysts (up to 17,500 g polyethylene/g metal) is about an order of magnitude lower than that claimed by Turner[74]. In one example a catalyst, $Ti(C_2B_{10}H_{11})_4$, was used in combination with a methaluminoxane co-catalyst; in all other examples diethyla-luminium chloride was applied.

1.5 CO-CATALYST CHEMISTRY

The previous sections describe the macro- and microstructure of the various Ziegler-Natta catalysts used commercially for the production of isotactic polypropylene. Although the term "catalyst" is the generally accepted nomenclature (and is used for convenience in this text) it is in fact a misnomer since the "catalyst" alone is not usually at all effective in catalyzing the polymerization; the "catalyst" must be activated by a co-catalyst comprising a metal alkyl from the groups I to III of the Periodic Table. Perhaps a more correct name for a $TiCl_3$ catalyst or supported catalyst would be the "pro-catalyst". This section describes the various types of co-catalysts used commercially and the interactions occurring between the co-catalyst and the pro-catalyst.

Although most alkyls of the group I to III metals are capable of generating active polymerization systems in combination with a suitable $TiCl_3$ catalyst, only aluminium alkyls are used in commercial operation. The uniqueness of aluminium alkyls can be attributed to both economic and technical factors. Aluminium alkyls such as triethyl aluminium have been commercially available on a large industrial scale since the 1950's, being manufactured via the relatively inexpensive, direct reaction between aluminium, hydrogen and the olefin. Most important however is the fact that by the appropriate choice of aluminium alkyl co-catalyst it is possible to generate highly stereoregular polyolefins, most other metal alkyls give significantly lower selectivity. Natta[77] was the first to report a correlation between polymer isotacticity and the choice of co-catalyst. Indeed Natta disclosed that polymer isotacticity, or in other words the stereoregulating ability of the resulting catalyst, is a function of the ionic radius of the metal in the metal alkyl chosen - the smallest ionic radius giving rise to the highest isotactic index. Natta's data is summarized in Table 1.3.

TABLE 1.3

Relationship between polypropylene isotacticity and
the ionic radius of M in MEtn

metal alkyl	ionic radius nm	isotactic index[*] %
BeEt$_2$	3.5	94 - 97
AlEt$_3$	5.1	80 - 90
MgEt$_2$	6.6	78 - 85
ZnEt$_2$	7.4	30 - 40

[*] : polymerization of propylene at 75°C using α-TiCl$_3$ catalyst

The acute toxicity of beryllium compounds makes the industrial application of BeEt$_2$ unacceptable. It should be noted that while there is obviously a distinct trend between ionic radius and isotactic index, care must be exercised when establishing such correlations since other factors such as the polymer molecular weight can greatly influence the polymer solubility in heptane, and hence its "apparent" isotactic index. Indeed both Boor[78] and Firsov and co-workers[79] reported that the crystallinity of zinc alkyl-derived polypropylene is significantly higher than suggested by Natta's findings but its low molecular weight results in a high solubility in heptane and a consequent low apparent isotactic index. The low molecular weight observed can be attributed to the efficient chain transfer capability of ZnEt$_2$ (see section 1.6).

A much more recent finding in the area of co-catalyst chemistry is that dimethylmetallocenes (e.g. Cp$_2$TiMe$_2$ and Cp$_2$ZrMe$_2$) act as excellent co-catalysts[80-82] with both TiCl$_3$ catalysts as well as supported systems[81]. In the case of a supported TiCl$_4$ on MgCl$_2$ catalyst (no electron donor) Soga[81] reported that using the series Cp$_2$MtMe$_2$ (Mt = V, Ti, Zr, Hf) as co-catalyst the vanadium and titanium systems gave high stereoselectivity (92 and 90% I.I. respectively) while the zirconium and hafnium analogues gave poor selectivity (61 and 69% I.I. respectively). Like Natta, Soga attributes the selectivity relationships to the ionic radii of the cocatalyst metals in the Cp$_2$MtMe$_2$ complexes (V, Ti, Zr, Hf being 7.2, 7.5, 8.5 and 8.6 nm respectively).

The following discussion will concentrate on the commercially significant aluminium alkyl co-catalysts. The differing nature of TiCl$_3$ catalysts and supported catalysts results in different types of co-catalysts being used; generally speaking dialkylaluminium halides, frequently diethyl aluminium chloride (DEAC), are applied with $TiCl_3$ catalysts whereas supported catalysts require the application of trialkyl aluminiums, frequently triethylaluminium (TEA), in order to achieve optimum catalyst activity. In both cases "third components" (electron donors) are frequently employed in order to improve the performance of a given catalyst system.

1.5.1 $TiCl_3$ catalysts

Within the family of aluminium alkyls the substitution of one of the alkyl groups in triethyl aluminiums by a halide ion has a large effect on the stereoselectivity of propylene polymerization. Natta and co-workers[83] and Danusso[84] investigated this phenomenon concluding that when used with any of the purple $TiCl_3$ modifications it was the nature of the alkyl aluminium which determined the product isotacticity. Typical data is given in Table 1.4, where it can be seen that among the ethyl derivatives Et_2AlI (DEAI) affords the highest isotacticity, although the selectivity improvement is only achieved at the cost of a dramatic loss in polymerization

TABLE 1.4

Stereoregularities and molecular weight of the fractions insoluble in boiling heptane of polypropylenes obtained with δ-$TiCl_3$ in conjunction with different aluminium compounds as co-catalysts

Alkylaluminium compound	Triad distribution			Molecular weights	
	mm %	mr %	rr %	$M_n *10^{-5}$	$M_w *10^{-5}$
$AlEt_3$	94	4	2	0.89	6.9
$AlEt_2Cl$	94	4	2	1.3	13.3
$AlEt_2Br$	$\approx$98	>1	<1	1.5	16.9
$AlEt_2I$	$\approx$98	>1	<1	-	-

activity. Indeed it is now generally accepted that there is a reciprocal relationship between isotacticity and activity. Commercial systems invariably comprise $TiCl_3$/DEAC since DEAC offers an acceptable level of both activity and selectivity; furthermore the application of iodide- or bromide- based systems would introduce additional process complications related to corrosion and product (colour) problems. It should be noted however that DEAI is used commercially as a co-catalyst component in the manufacture of

poly-1-butene (see chapter 11). The application of DEAI in polybutene is unavoidable due to the extreme difficulty of achieving
acceptable isotacticity when using 1-butene as monomer.

In an elegant experimental study Keii[85] showed that a $TiCl_3$
catalyst could be endowed with the stereoselectivity of DEAI and
the kinetics/rate of a DEAC system by initiating the polymerization
with DEAI, washing and then continuing the polymerization using
DEAC as co-catalyst.

In contrast to DEAC which is the co-catalyst of choice for
$TiCl_3$ systems, ethyl aluminium dichloride (EADC) is a catalyst
poison. Indeed even small amounts of EADC considerably reduce the
rate of propylene polymerization using γ-$TiCl_3$/DEAC catalyst
systems[17,86-88]. This rate lowering effect can best be explained in
terms of a dynamic adsorption/desorption equilibrium between the
surface $TiCl_3$ sites and EADC :

$$Ti + AlEtCl_2 \quad <---> \quad Ti.AlEtCl_2$$
$$\text{active} \qquad\qquad\qquad \text{inactive}$$

It is apparent that removal or complexation of EADC shifts this
equilibrium to the left, effectively generating a larger number of
active centres and therefore results in a higher polymerization
rate.

The presence of EADC during the polymerization also affects
product molecular weight substantially. This phenomenon is attributed to the same adsorption/desorption equilibria which can be
used to explain deactivation. As illustrated in Fig. 1.25 EADC can
become adsorbed to the active centre by one of two modes of complexation, and once adsorbed two modes of desorption can be envisaged.
One of these desorption modes is effectively a chain transfer
process (equilibria K_{12} and K_{22}; see also section 1.6). To a first
approximation the molecular weight can be described as the ratio
between the ratio of chain propagation (R_p) and chain termination
(R_t) rates:

$$R_p/R_t = k_p[\text{propene}][Ti]/\{k_t[H_2][Ti] + (k_{12} + k_{22})K[Ti][EADC]\}$$

where $K = (k_1 + k_2)/(k_{11} + k_{12} + k_{21} + k_{22})$.

This relationship shows that product molecular weight is inversely
proportional to the EADC concentration - therefore lowering the

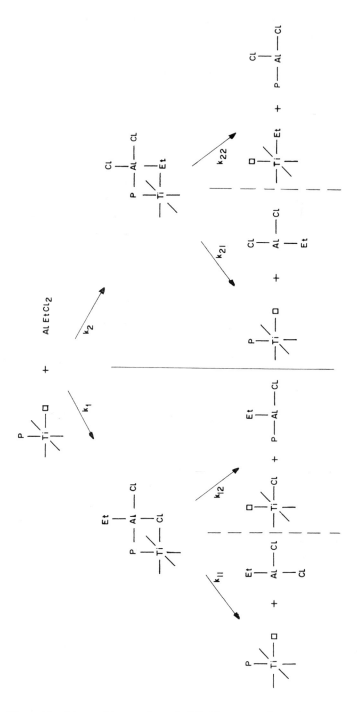

Fig. 1.25: Adsorption/desorption modes of AlEtCl₂ at active sites

EADC concentration by extraction, complexation or simply reducing catalyst concentration results in higher product molecular weights.

In section 1.4 it was pointed out that the activity of $TiCl_3$ catalysts can be significantly increased by treatments (with ethers for example) which extract $AlCl_3$ from the solid and hence reduce the concentration of the catalyst poison EADC during the polymerization. Such catalyst activation procedures involve extra process steps in catalyst manufacture; furthermore the washing of fine catalysts would be exceedingly time-consuming, and hence commercially unattractive, because of the long settling times required after each washing step. For reasons of economy therefore, much research effort has been expended over the years aimed at finding Lewis base/co-catalyst combinations which can improve the activity and/or selectivity of a given $TiCl_3$ catalyst without involving extra catalyst manufacturing steps; the function of the Lewis base ("third component") in these systems is almost invariably to complex EADC, the most Lewis acid component of the co-catalyst. Boor (ref. 10, pages 213-243) argued that the role of Lewis bases is considerably more complex, but it is now generally accepted that for commercially important $TiCl_3$/DEAC systems it is complexation of the catalyst poison EADC which is the single most significant factor.

The effects of some simple Lewis bases on a typical (triethyl-aluminum-reduced) first generation $TiCl_3$ catalyst are shown in Table 1.5. The concentration of the Lewis base applied is of critical importance as is illustrated by the data given for dibutylether (DBE). Using different levels of DBE it was found that the polymerization activity could be improved by as much as 25% with no adverse effect on stereoselectivity (expressed in terms of yield stress) providing that the DBE/$TiCl_3$ ratio was no greater than 0.15; higher ratios resulted in a substantial loss of stereoselectivity. These results parallel the findings from studies on DBE washing of $TiCl_3$ catalysts (section 1.4.2) where a similar relationship was found between the DBE/$TiCl_3$ molar ratio employed and the final catalyst performance. Again it must be assumed that the fraction 0.15 corresponds to that portion of the $AlCl_3$ in the catalyst which exists as a separate phase, being readily soluble and hence reactive under polymerization conditions:

$$AlCl_3 + Et_2AlCl \longleftrightarrow 2\ EtAlCl_2$$

In the presence of a Lewis base such as DBE this equilibrium is
driven back to the left as a result of DBE complexing the strongest
Lewis acid in the system ($AlCl_3$).

TABLE 1.5

The effect of di-butyl ether and other Lewis bases on catalyst activity and
yield stress (First generation TiCl3 catalyst).

Lewis base	base/$TiCl_3$, molar ratio	activity, gPP/g $TiCl_3$.h.bar	yield stress, MN/m^2	melt index, g/10 min
-	-	55	36.0	4.0
dibutyl ether	0.13	64	36.5	3.3
,,	0.15	71	37.0	2.5
,,	0.18	66	35.0	3.4
,,	0.33	68	32.5	1.8
,,	1.0	56	20.0	7.6
tetrahydrofuran	0.15	64	36.5	5.1
diisopropyl ether	0.15	62	36.5	4.0
water	0.15	58	37.0	4.3
triethyl amine	0.15	58	38.0	3.6
tetramethyl-methylenediamine	0.1	54	38.0	3.2
hexamethyl-phosphortriamide	0.05	55	39.0	3.2

Other ethers such as tetrahydrofuran and diisopropylether gave
similar results. Nitrogen donors on the other hand were found to
increase stereoselectivity without affecting catalyst activity;
hexamethylphosphortriamide gave the best results in this respect.

A variation on the theme of "third components" involves the
application of active hydrogen-containing species, such as alco-

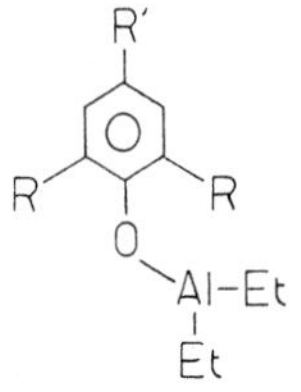

Fig. 1.26: Sterically-hindered phenoxy-substituted aluminium alkyl

hols, which can react with aluminium alkyls to generate new, well-defined co-catalyst species. Particularly good results were obtained by applying sterically-hindered phenols[89]. Danusso[84] reported diethyl aluminium phenoxide (Fig. 1.26, R=R′=H) to be inactive as co-catalyst, the well-known[90] association of this species into dimers and trimers must be considered responsible for this lack of activity. Introducing severe steric hindrance into both ortho-positions (e.g. by reacting TEA with IONOL to afford the species shown in Fig. 1.26 (R=tert.-butyl, R′=methyl) suppresses self association leaving the active monomer species in solution. These sterically-hindered phenoxy co-catalysts were tested in combination with a second generation $TiCl_3$ catalyst prepared by reducing $TiCl_4$ with TEA in the presence of ether followed by heating in excess $TiCl_4$ at 110°C, to afford a dark purple $TiCl_3$ catalyst containing about 5% $AlCl_3$ (compared to first generation catalysts which contain typically 33% $AlCl_3$ - see section 1.4). Typical polymerization data are listed in Table 1.6 where in a standard run (DEAC co-catalyst) the catalyst showed good activity (141 gPP/g $TiCl_3$.bar.h) and reasonable, but not commercially acceptable, stereoselectivity (7.8 %m/m xylene solubles).

TABLE 1.6

Polymerization experiments with Ionol-modified co-catalysts

Ionol, mmol.	AliBu$_3$, mmol.	AlEt$_3$, mmol.	AlEt$_2$Cl, mmol.	AlEtCl$_2$, mmol.	activity gPP/g TiCl$_3$.bar.h	xylene solubles, %m/m
-	-	-	9	-	141	7.8
4	-	4	-	-	158	>20
4	-	4	-	4	178	2.3
4	-	-	8	-	144	1.8
4	4	-	-	-	300	>20
4	4	-	-	4	206	1.3

The new co-catalyst (the reaction product of equimolar amounts TEA and IONOL) was found to improve the polymerization activity somewhat but at the expense of stereoselectivity which decreased drastically (entry 2). Remarkably, mixing equimolar proportions of the species with EADC (catalyst poison!) led to a significant improvement (entry 3) in both activity (by about 30%) and stereoselectivity (2.3%m/m xylene solubles). Reacting IONOL with DEAC (molar ratio 1:2) also gave excellent catalyst performance.

We ascribe the poor stereoselectivity of the TEA/IONOL co-catalyst to the inherent instability of the monomeric species[91] which can undergo a redistribution reaction liberating triethyl aluminium, which is known to result in poorly isotactic polymers :

$$2 (RO)AlEt_2 \longleftrightarrow (RO)_2AlEt + AlEt_3$$

On the other hand we attribute the excellent performance of the EADC modified and DEAC analogues to the presence of chloride bridges which "stabilize" the sterically-hindered species preventing the above-described redistribution reaction (Fig. 1.27).

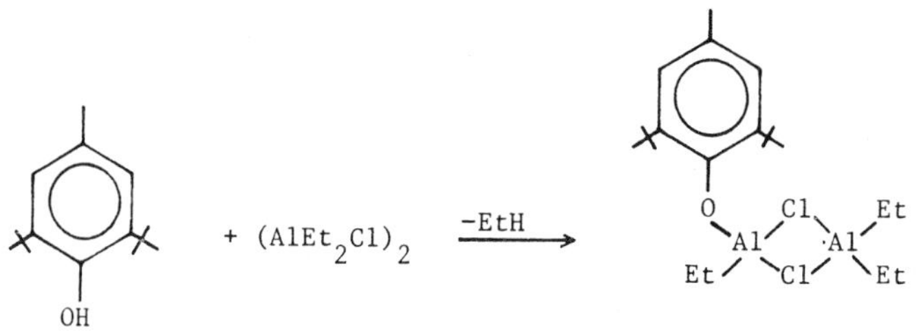

Fig. 1.27: Reaction of hindered phenol with DEAC

An even better result was found when the degree of steric-hindrance was further increased by using triisobutylaluminium instead of TEA. As shown in Table 1.6 (entry 6) application of the IONOL/-triisobutylaluminium/EADC co-catalyst increased catalyst activity (compared to the standard DEAC run) by virtually 50% while at the same time demonstrating unprecedented stereoselectivity.

The outstanding activity achieved with these co-catalysts can be explained by considering that they selectively complex the

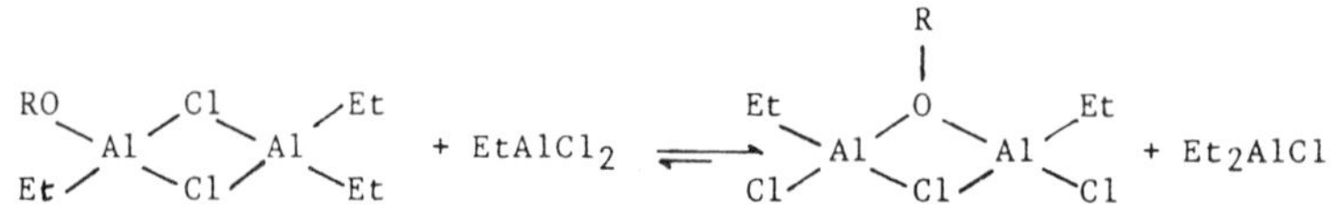

Fig. 1.28: Reaction of phenoxy-aluminium alkyl with EADC

catalyst poison EADC as illustrated in Fig. 1.28. That these species are exceptionally efficient in this respect is due to the Lewis basicity of the phenoxy group combined with its severe steric hindrance which allow it to form strong complexes only with the very least sterically demanding Lewis acid in the system: EADC.

Interestingly these same sterically hindered phenoxy aluminium co-catalysts were found to double the activity of a β-TiCl$_3$ catalyst applied in "high-cis" isoprene rubber preparation[92]. Again the performance improvement must be attributed to efficient complexation of the EADC catalyst poison.

1.5.2 Supported catalysts

The co-catalyst of choice for supported catalysts are trialkyl aluminiums, the most commonly applied co-catalysts being triethyl- and triisobutyl aluminium. Other organometallic compounds such as DEAC or diethyl zinc result in very low polymerization rates.

Although TEA as co-catalyst leads to exceedingly high activities the stereoselectivity achieved is typically very poor (isotactic index 50-70%). Indeed the production of highly isotactic polypropylene requires the application of an electron-donor-modified co-catalyst. In the case of a typical MgCl$_2$/EB/TiCl$_4$ supported catalyst (section 1.4.3) the electron donors or "selectivity control agents" of choice are aromatic esters such as ethyl benzoate or para-ethyl anisate, higher concentrations of these esters, i.e. lower TEA/ester ratios, favouring higher stereoselectivity. Unlike simple Lewis bases such as the ethers or tertiary amines used with TiCl$_3$ catalysts, aromatic esters can undergo irreversible chemical reactions with trialkyl aluminiums under polymerization conditions. It can be learned from the literature[93,94] that the reactions are complex and that consecutive steps involving alkylation, reduction and elimination can lead to a multiplicity of products. The reactive nature of these mixtures makes the co-catalyst chemistry of supported catalysts substantially more complex than that of TiCl$_3$ systems.

After a study of the reactivity towards trialkyl aluminiums of a range of esters and ketones it was found that in the case of aromatic esters a TEA/ester molar ratio of >2:1 is required to give a reaction at typical polymerization temperatures (60-70°C). In the case of a commonly described selectivity control agent, para-ethyl anisate, the reaction was shown to be a clean, two-step alkylation reaction affording a new alkoxy aluminium dialkyl species as shown

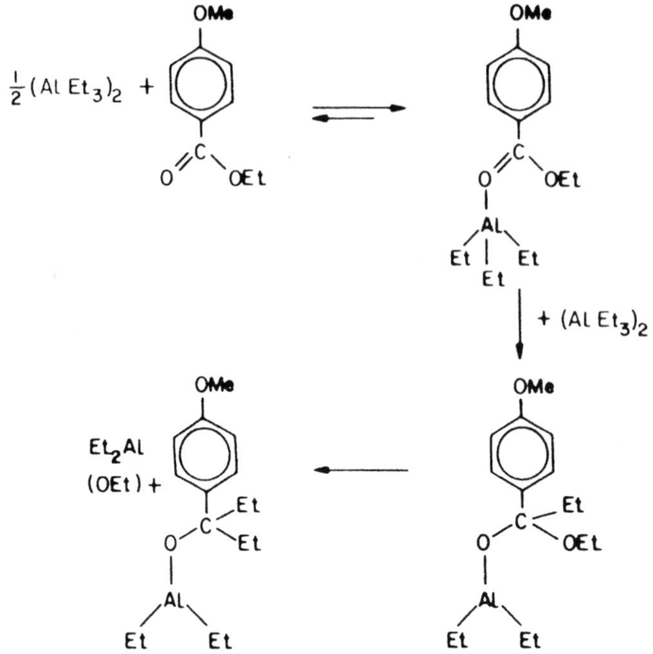

Fig. 1.29: The alkylation of ethyl anisate [from reference 158, copied with permission from Harwood Academic Publishers]

in Fig. 1.29. The reaction was studied under fairly concentrated conditions in perdeuterotoluene allowing the reaction to be monitored by ^{13}C-NMR methods, as illustrated in Fig. 1.30. At ambient temperature a simple adduct of TEA and the ester was formed, as witnessed by a yellow colouration of the solution and small shifts in the ^{13}C-NMR spectrum. If the molar ratio is held at 1:1 the reaction proceeds no further, even after extended heating at 60°C. However at a TEA/ester molar ratio of 3:1 (in polymerization runs with para ethyl anisate a molar ratio of 2.5 to 3.5 :1 is typically used) further reaction occurs, slowly at ambient temperatures, and rapidly at 60°C, to afford a colourless solution of the new alkoxy aluminium species (Fig. 1.29) resulting from double alkylation of the carbonyl function. In the course of these studies it was observed that the reactivity of aromatic esters towards TEA was strongly dependent not only on the molar ratio of the components and temperature, but also on the concentration and solvent used.

Langer and co-workers[95] studied the reaction between TEA and ethyl benzoate using proton NMR techniques, coming to similar

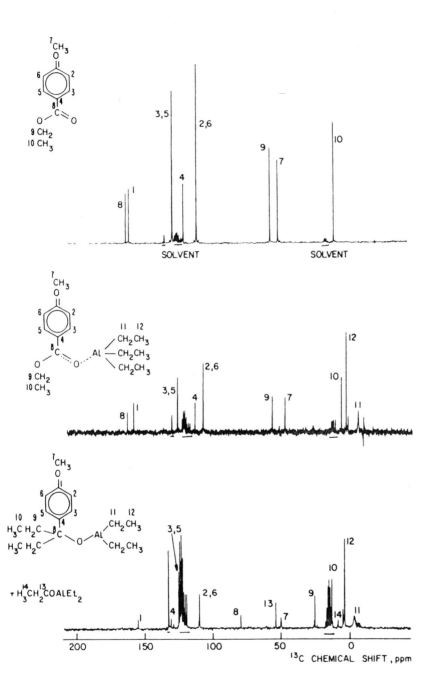

Fig. 1.30: ^{13}C-NMR spectra of ethyl anisate and its reaction products with TEA
[from reference 158, copied with permission from Harwood Academic Publishers]

conclusions regarding the reactions occurring. They reported no alkylation to occur at 25°C and 1:1 stoichiometry, at molar ratios >1 and <3 the reaction is initially rapid but than slows dramatically, while at a TEA/ester ratio of >3 the reaction proceeded to complete conversion (Fig. 1.31). The same authors furthermore discovered that the rate of alkylation could be greatly reduced by using sterically-hindered trialkyl aluminium compounds instead of

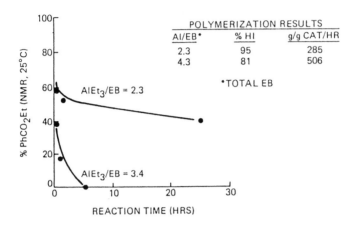

Fig. 1.31: Effect of AlEt$_3$/ethyl benzoate(EB) ratio on consumption of EB as a function of time (proton-NMR). Also shown are the polymerization results carried out at two AlEt$_3$/EB ratio's (%HI=%boiling heptane insolubles) [from reference 95, copied with permission from Plenum Press]

TEA, their results being illustrated in Fig. 1.32. Remarkably it was observed that during polymerization studies at low ethyl benzoate concentration (AlR$_3$: EB molar ratio 4.3) which gives poor stereoselectivity using TEA as co-catalyst (82% isotactic index), much higher selectivity was observed when the sterically hindered trialkyl aluminiums were applied (Table 1.7). The fact that the increasing stereoselectivity parallels the decreasing ester alkylation rates led Langer et al to conclude that it is the ethyl benzoate itself which accounts for the high isotacticity attainable with these catalysts and not the alkoxy aluminium species which are formed during the alkylation process.

Another approach[95] to circumventing the ester alkylation reaction involved the use of sterically hindered esters, the rates of alkylation of which are known[96] to be substantially lower. In

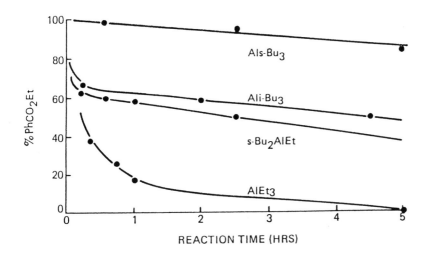

Fig. 1.32: Ethyl benzoate consumption as a function of time for several hindered AlR₃ compounds (proton-NMR) [from reference 95, copied with permission from Plenum Press]

the series ethyl-, n-butyl-, iso-butyl- and tert.butyl benzoate the effectiveness of improving the isotactic index was found to decrease with increasing steric hindrance as shown in Table 1.8. On the basis of these findings the authors concluded that some sort of catalyst-ester interaction is required in order to achieve high

TABLE 1.7

Effect of hindered AlR3's on catalyst performance (from ref. 95)

AlR_3	rate g/g cat.h	II,%m/m (heptane insolubles)	$M_w * 10^{-3}$
$AlEt_3$	506	82	238
$AliBu_3$	520	88	315
sBu_2AlEt	501	89	307
tBu_2AlEt	489	90	313
sBu_3Al	461	92	357

(1.0 mmol AlR_3, 0.2 g catalyst)

isotacticity, and that this interaction is also disfavoured by increasing steric hindrance. A similar argument was used to explain the mediocre isotacticity achieved using 2,6-dimethyl substituted ethyl benzoate (Table 1.8); under NMR concentration conditions and at 25°C this ester was found to be alkylated very slowly indeed (22% in 22 hours).

TABLE 1.8

Hindered esters as third components (from ref. 95)

catalyst type	$AlEt_3$, mmol	ester, mmol	rate, g/g cat.h	II, %m/m	M_w*10
B1	0.4	0.12 $PhCO_2Et$	310	95	378
,,	,,	0.12 $PhCO_2n$-Bu	336	91	319
,,	,,	0.12 $PhCO_2i$-Bu	406	87	302
,,	,,	0.12 $PhCO_2t$-Bu	401	73	233
A1	1.0	0.2 $PhCO_2Et$	285	95	421
,,	,,	0.2 a)	378	89	303

a) 2,6-dimethyl-1-phenyl as a para-substituent on ethylbenzoate

On the basis of the foregoing we assume that the interaction between the catalyst and the ester, which accounts for stereoselectivity improvement, is again the ester alkylation reaction, but involving Ti-C bonds rather than Al-C bonds:

$$Ti-R + O=C \quad ---> \quad Ti-O-C-R$$
$$\text{deactivated site}$$

R is a growing polymer chain or alkyl group.

Indeed it is known that group IV (Ti, Zr) metal alkyls are even more reactive towards esters and ketones than are aluminium alkyls[97,98]. We suggest that the atactic sites are less hindered, or more exposed (Cossee model, see section 1.7) and hence more reactive towards the ester than are the isotactic sites: however this alkylation process also occurs, albeit slower, at the isotactic centres contributing significantly to catalyst deactivation. It would therefore appear that in choosing an aromatic ester as third component a careful balance in properties must be found:

- if the ester is <u>too</u> <u>unreactive</u> towards alkylation (e.g. sterically hindered esters such as tertiary butyl benzoate) little or no
improvement in isotactic index is observed (although less catalyst
deactivation and a higher average polymerization rate is also
seen).

- if the ester is <u>too</u> <u>reactive</u> towards alkylation (e.g. an aliphatic ester, or even ethyl benzoate) then it is either completely
consumed by the TEA co-catalyst during premixing, resulting in no
selectivity improvement, or it rapidly deactivates or poisons the
catalyst by inserting into the Ti-C bonds.

It has been shown[93,94] for aromatic esters that the reactivity
can be steered by suitable substitution of the aromatic nucleus;
electron-donating substituents such as methyl and methoxy were
shown to reduce reactivity, explaining why ethyl anisate and
toluate are preferred to ethyl benzoate. On the other hand the
introduction of too many electron-donating substituents, as in
2,4,6-trimethyl substituted EB for example, so greatly reduces the
reactivity of the ester function that little or no effect on
stereoselectivity is observed. Reviewing the more than three
hundred patents describing stereospecific supported catalysts
(those containing ethylbenzoate or analogues as "internal donor")
for propylene polymerization we found that almost without exception
the third component utilized could be defined as a para-substituted
ester of benzoic acid, with H, CH_3, OCH_3 and OC_2H_5 as possible para
substituents, whilst the ester is either methyl or ethyl.
The fact that aromatic esters undergo irreversible reactions
with the trialkyl aluminium co-catalyst dictates that it is frequently not possible (or at least not advisable!) to mix these
components more than a few minutes before the polymerization
reaction is commenced.

Having discussed why most supported catalysts for propylene
polymerization are used in combination with an aromatic ester third
component, and the chemistry involved as a result, it is perhaps
interesting to consider those systems which form exceptions in that
esters are not applied to improve isotacticity. One series of
exceptions takes the form of early patents[99,100] from Mitsui
Petrochemical which described solid catalyst components containing
ethyl benzoate as internal donor which were tested in polymeriza-

tions using TEA as co-catalyst but no external donor, and yet the selectivity achieved was in excess of 90% in many cases (for example 90.7% isotactic index in one case[100] with a polymer yield of 9.8 kg/g catalyst - 70°C, 3 h at 7 bar propylene pressure). This is probably explained by the fact that no molecular weight control by hydrogen was applied in these cases. Another system where no ester is applied as external donor is to be found in a Mitsubishi Petrochemical patent[101] where remarkably high isotacticity (97.6%) was reported in the absence of an external donor. In this case an apparently standard ball-milled $MgCl_2$/EB/$TiCl_4$ catalyst was treated with ICl_3 or iodine and washed prior to polymerization. It is possible in this case that an iodine (or iodide) ligand is incorporated into the active titanium centres resulting in the observed stereoselectivity improvement; in this case this work is obviously related to the work of Doi and Keii on iodine-modified $TiCl_3$ catalysts (see section 1.5.1).

Other remarkable exceptions include a series of patents of Denki[102,103] describing the use of sodium tetraethyl aluminate as co-catalyst, again in the absence of an external donor. A comparative example[102] reports TEA to give poor results (51.4% isotactic index) while the sodium salt ($NaAlEt_4$) yields high stereoselectivity (94.3% isotactic index) with the same solid catalyst component. This surprising result perhaps implies that the active centre is a cationic species:

$$-Ti-Cl + NaAlEt_4 \longrightarrow -Ti^+[AlEt_4^-] + NaCl$$

A precedent for this concept is to be found in early publications by Russian workers[72,73,104] who demonstrated the presence of cationic titanium species in the homogeneous Cp_2TiCl_2/$AlRCl_2$ system:

$$Cp_2TiRCl \cdot AlRCl_2 \longleftrightarrow [Cp_2TiR]^+ + [AlRCl_3]^-$$

For further discussion on the possible role played by cationic species in Ziegler catalysis the reader is referred back to section 1.4.4.

Using $Cp_2Ti(CH_3)_2$ as sole co-catalyst with internal-donor free $TiCl_4$/$MgCl_2$ (ball-milled) pro-catalyst relatively high isotacticities are found, as reported by Soga et al[81,105]. The stereospecificity can be improved from 90 to 97 % when the titanium content of the pro-catalyst is diminished fivefold, at the expense of activity[82]. Adding an external donor (EB) to such a catalytic system lowers the activity drastically.

It would be impossible to review this field of co-catalyst chemistry without referring to the elegant studies of Langer and co-workers at Exxon's laboratories involving the application of various types of sterically-hindered co-catalysts to typical ball-milled $MgCl_2$/EB/$TiCl_4$ catalysts. Some typical examples[106] are listed in Table 1.9. This remarkable effect on stereoselectivity is, in our opinion, best attributed to the monomeric nature of these co-catalyst species, formation of dimers (the predominant form in the case of TEA) being prevented by the sterically-deman-

TABLE 1.9

Sterically-hindered trialkylaluminium co-catalysts: effects on polymerization activity and selectivity* (from reference 106)

AlR_3	activity, g PP/g cat	selectivity, heptane insolubles, %m/m
$AlEt_3$	244	83.1
$(Me_3SiCH_2)_2AlEt$	182	90.8
$Me_3SiCH_2AlEt_2$	140	92.9
$(neopentyl)_2AlEt$	182	93.1
$neopentylAlEt_2$	227	91.9
$(2-norbornyl)_2AliBu$	247	91.8
$(s-Bu)_2AlEt$	357	93

*:*supported catalyst prepared by ball-milling $MgCl_2$ with ethyl benzoate and treating with $TiCl_4$. Polymerization conditions: 1 bar, 65°C, 1 hour.*

ding alkyl groups. Although not fully understood it would appear that the active centres generated between the surface titanium sites and the monomeric aluminium species possess a high stereospecificity (see also section 1.7). Unfortunately the real value of these systems must remain in some doubt since no data is presented under commercially representative conditions (high pressures, hydrogen for molecular weight control).

Although diethyl aluminium chloride itself is a very poor co-catalyst when used with supported catalysts there are reports in the patent literature[107,108] describing its beneficial effect when used in combination with a typical TEA and aromatic ester co-catalyst system.

As pointed out in section 1.7 we believe that the surface titanium centres in a supported catalyst are four-coordinate. In the case of a conventional TiCl$_3$ catalyst such a titanium centre would have the octahedral, "double vacancy" geometry illustrated in Fig. 1.33a since the titanium ion could be forced into an octahedral environment by the surrounding crystal lattice. However, it would be a misconception to assume the same site geometry for

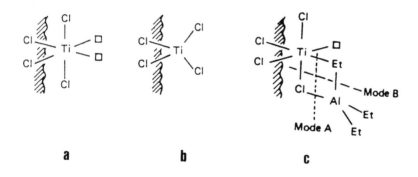

a   b   c

Fig. 1.33: Structure of surface site

supported catalysts, since in this case the TiCl$_4$ is simply adsorbed on to the support surface; as a result the four-coordinate titanium centres will assume the lower energy, tetrahedral geometry (Fig. 1.33b). This site will only assume octahedral geometry when complexed with TEA to give the active polymerization centre (Fig. 1.33c). The resulting active site can cleave in one of two ways (mode A and B in Fig. 1.33c) eliminating either TEA (mode A) or DEAC (mode B).

The whole process of titanium centre complexation and cleavage is represented in Fig. 1.34. The surface site is complexed by TEA (equilibrium i) to form the active site. Decomplexation of TEA

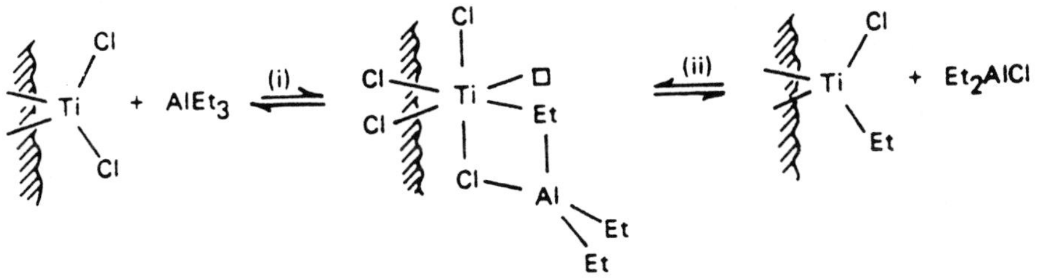

Fig. 1.34: Active centre formation

(i.e. cleavage via mode A) causes the titanium ion to revert to its
original structure (n.b. during the polymerization process Et_2AlR,
where R is the polymer chain, will be eliminated. In other words,
equation i represents the chain transfer process). On the other
hand cleavage via mode B (equilibrium ii) results in elimination of
DEAC and causes the titanium ion to assume an alkylated, tetrahedral structure. While the titanium ions resulting from decomplexation are not deactivated, they are both "non-participating active
centres" due to their tetrahedral, coordinatively-saturated geometry. These sites can only be converted into "participating active
centres" by complexation with aluminium alkyls according to equilibria i and ii in Fig. 1.34.

A homogeneous model of the above concept is to be found in the
well-known homogeneous ethylene polymerization catalyst system
Cp_2TiRCl plus $RAlCl_2$[109]. In this case the pro-catalyst (Fig. 1.35a)

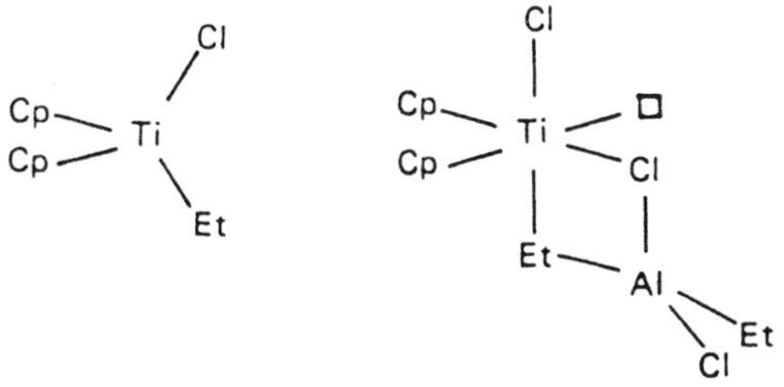

a   b

Fig. 1.35: Active centre formation in Cp_2TiCl_2

is totally inert towards olefins, until it is complexed by the co-catalyst which forces the titanium into an octahedral, coordinatively- unsaturated form (Fig. 1.35b). The careful reader will note the danger of drawing analogies however. In section 1.4.4 we reported the now convincing evidence that these homogeneous metallocene systems are in fact active by virtue of the formation of cationic species, in which the function of the alkylaluminium co-catalyst is to generate the stabilizing (and non-coordinating) anion. It should therefore be borne in mind that it is highly likely that cationic species are also the active centres in heterogeneous catalysts and that rather than representing titanium as a five-coordinate centre (e.g. Fig. 1.34) it would be more accurate to represent it as a three coordinate cation in the presence of an aluminium-based counterion.

Further examination of Fig. 1.34 reveals that the role of DEAC is to push equilibrium ii to the left converting non-participating sites into active centres, thus increasing the number of sites participating in the polymerization process. Of course the TEA concentration (or TEA/Ti molar ratio) has a similar effect of pushing equilibrium i to the right. Hence a co-catalyst comprising a TEA/DEAC mixture assures optimum catalyst performance by converting the maximum proportion of surface sites into active centres; the use of TEA alone results in a significant fraction of the sites being non-participating at any given time due to DEAC elimination.

Summing up, the co-catalyst of choice for application with most $MgCl_2$/aromatic ester/$TiCl_4$ supported catalysts comprises a trialkyl aluminium in combination with an aromatic ester such as para-ethyl anisate (PEA). In its most simple terms the method by which the external donor (PEA) improves stereoselectivity is by inserting preferentially into the Ti-C sigma bonds of atactic polymerization sites giving an inactive titanium alkoxide species. The fact that such alkoxide species can be realkylated by trialkyl aluminiums dictates that a controlled concentration of the external donor must be present throughout the polymerization.

1.6 THE EFFECT OF CATALYST AND PROCESS VARIABLES ON THE MOLECULAR
WEIGHT AND ITS DISTRIBUTION ("CHAIN TRANSFER")

Polypropylene is almost invariably processed in the molten state and hence the quality of the finished product (impact, warping, surface finish, tenacity) as well as the stability of the melt during processing (die swell, melt fracture) are largely controlled by its rheological properties. The rheological behaviour of the melt is a function of the zero shear viscosity (related to the weight average molecular weight) and the molecular weight distribution. Most commercial polypropylenes possess a broad MWD, although for certain applications such as melt spinning a more narrow distribution is required.

Numerous studies have addressed the effect of process conditions on MWD[110-114], although the results were often inconclusive or contradictory. In this section some results of our own studies on this complex topic are presented, molecular weights and MWD being determined using capillary rheometry or gel permeation chromatography techniques. Three categories of molecular weight control can be identified:

i)   catalyst modification

ii)   process conditions, and

iii) modification of preformed polymer

Methods (i) and (ii) will be discussed in this section. Since polymer modification (iii) is essentially independent of the catalyst used it will suffice to simply mention two examples of what is meant by polymer modification:
-   Degradation of high molecular weight polypropylene (e.g. by thermal cracking or by treatment with peroxides) results in lower molecular weights without appreciably increasing the width of the MWD or sometimes slightly decreasing it. This method therefore represents a route to lower molecular weight materials with a less broad MWD. An example is given in reference 115.
-   Extraction of atactic material using, for example, hot xylene causes a substantial improvement in both yield stress and crystallinity but has only a small effect on molecular weight or MWD.

1.6.1 The effect of catalyst modification (TiCl$_3$ catalyst)

Testing a wide variety of both first- and second generation catalysts led to the conclusion that the level of aluminium in a TiCl$_3$ catalyst is critical in determining the MWD. Some examples are given in Table 1.10.

On the interpretation of the rheological data the following : Mieras and van Rijn[116] have shown that the dependence of the melt viscosity (η) of polypropylene on the shear stress σ_t can be adequately described by the semi-empirical relationship

$$(\eta) = (\eta)_0/[1+(\sigma_t/\sigma_c)^n]^{1.7}$$

$(\eta)_0$ is the zero shear viscosity which is independent of the MWD, n is a non-linearity parameter. σ_c is the apparent zero shear elastic modulus characterizing the width of the MWD, σ_c being smaller for polymers with a wider distribution.

The narrowest MWD was found with the second generation "Solvay type" catalyst which is of course characterized by a low aluminium

TABLE 1.10

Rheological properties and GPC-data of polypropylene samples produced with various catalysts under standard polymerization conditions (70°C, 4 bar propylene, DEAC/Ti=2)

Catalyst	GPC-data			Melt flow data		
	$M_w \cdot 10^{-5}$	Q	R	σ_c, N/m^2 10^{-4}	η_o, Ns/m^2 10^{-3}	MI g/10min
TEA reduced TiCl$_3$	2.2	6.9	2.46	1.5	1.5	8.6
DEAC reduced TiCl$_3$	2.7	8.9	2.38	1.8	3.9	3.7
TEA reduced TiCl$_3$	2.6	6.5	2.42	2.3	2.5	3.9
Solvay	2.7	5.7	2.27	3.2	1.8	3.6
Stauffer	3.0	7.7	2.49	2.1	3.9	3.0

level (<0.1 Al/Ti). The MWD increases in broadness with increasing aluminium levels, the broadest MWD being found in the case of the Stauffer AA and the other catalyst which was made by DEAC reduction of TiCl$_4$ giving an Al/Ti ratio of about 0.5. The intermediate TiCl$_3$

catalysts (Al/Ti 0.33, $TiCl_3$ made via reduction with $AlEt_3$ at various sub-zero temperatures) gave an intermediate MWD.

The same trend was observed when a series of catalysts was prepared using different $AlEt_3/TiCl_4$ ratios. The resulting catalysts possessed Al/Ti molar ratios of 0.33 to 0.44/1, and the MWD was found to increase in broadness with increasing Al/Ti ratio; there is also a significant change in the molecular weight, as indicated by variation in the melt index - see Table 1.11. As the Al/Ti ratio in the catalyst increases there is a steady decrease in product molecular weight: Since these polymers were prepared with the same hydrogen concentration and under the same process conditions the decrease in molecular weight must be attributed either to a change in reactivity (propagation/termination rate) of the active sites or to the presence of increasing levels of $AlEtCl_2$ which is known to be a powerful chain transfer agent.

Typically, in laboratory catalyst preparations, the β to γ-$TiCl_3$ conversion was executed at 160°C for 1 hour. Substantially extending this heating time (up to a maximum of 20 hours) results in a considerable reduction in product molecular weight and a narrowing of the MWD.

The above results are indicative of there being a wide variety of different active centers of the surface of a $TiCl_3$ catalyst - although it should be added that we believe these differences to

TABLE 1.11

Effect of Al/Ti ratio in catalyst preparation on molecular weight characteristics of polypropylene.

Al/Ti	MI	LVN	Activity
	g/10min	dl/g	N*
0.33	2.5	2.67	28
0.35	3.5	2.47	46
0.37	7.0	2.26	51
0.39	6.1	2.26	54
0.44	18.2	1.8	47

* N = g polymer/ g $TiCl_3$.bar.h

be of a subtle electronic and/or geometric nature. The different sites, having different propagation rate constants give rise to the observed broad MWD. Soluble Ziegler catalysts, on the other hand afford narrow MWD polyolefins by virtue of the existence of one distinct type of active centre. In the case of heterogeneous $TiCl_3$ catalysts the propagation rates of the various surface sites (edge, corner, crevice,etc.) are almost certainly different. Also the presence of co-crystallized $AlCl_3$ in the $TiCl_3$ lattice will have an electronic effect on the reactivity of the surface Ti-ions.

All these factors lead to a catalyst containing active sites with a distribution in reactivity (and hence a broad polymer MWD). Supporting evidence is given by the narrow MWD polymers produced using low aluminium containing "Solvay type" catalysts, and the increasing MWD observed with the increasing aluminium levels in the catalyst series described above. Furthermore the observation that extended β to γ heat treatment times yields polymers of narrower MWD is attributed to the fact that these conditions promote the formation of larger and more nearly perfect γ-$TiCl_3$ crystallites containing active sites with a narrower range of reactivity distribution. Note that this narrowing of MWD is accompanied by a loss in catalyst activity since larger catalyst crystallites result in less active centres per unit volume of catalyst.

1.6.2 The effect of process conditions

The main process variables which affect the molecular weight and MWD of the polymer produced are the polymerization temperature, the hydrogen concentration and the type of co-catalyst used. Polymerization time can also have an effect on the molecular weight although the literature contains conflicting reports in this respect. For example it has been suggested[114] that the number average degree of polymerization (P_n) increases with polymerization time, becoming virtually constant on reaching steady-state conditions. In the absence of hydrogen and at 70°C the stationary state would be reached after about 3 hours. The large body of the scientific literature on the subject now seems to agree that there is little or no dependence of molecular weight or MWD on polymerization time[37,117] in the presence of hydrogen. A particularly elegant study[118] recently demonstrated that, in the case of a $MgCl_2$-supported catalyst, not only the molecular weight and its distribution of the total polypropylene but also those of the boiling heptane soluble and insoluble fractions were essentially independent of the polymerization time. Even a polymerization time as short as 5 seconds gave the same MWD as observed in the final polymer; this gives a good impression of just how rapid the rates of propagation

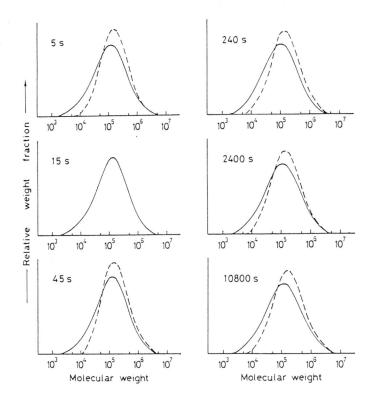

Fig. 1.36: Molecular weight distributions (MWD) of polymers obtained after various polymerization times with the TiCl$_4$/MgCl$_2$/EB//TEA catalyst system. The solid lines and the dashed lines represent the MWD of whole polymers and of fractions insoluble in boiling heptane respectively [from reference 118, copied with permission from Hüthig & Wepf Verlag]

and termination are in propylene polymerization. Fig. 1.36 shows the results of the above study in terms of MWD's as a function of time.

Temperature has a significant effect on molecular weight; for example increasing temperatures of reaction decrease the average life-time of the growing chains such that higher temperatures favour lower molecular weights. The most dramatic examples of this effect are to be found in the work of Kaminsky using homogeneous zirconium catalysts for both polyethylene[38] and polypropylene[51]. Fig. 1.21 and Fig. 1.22 illustrate this effect for both polymers; the lower values of molecular weight at any given temperature in the case of polypropylene is probably a reflection of the relative ease of β-hydrogen elimination (see later in this section).

MWD usually narrows with increasing polymerization temperature. Indeed Chien[110] reported this phenomenon for both polyethylene using a homogeneous catalyst (Cp_2TiCl_2 + Me_2AlCl) as well as polypropylene using a heterogeneous α-$TiCl_3$ catalyst system.

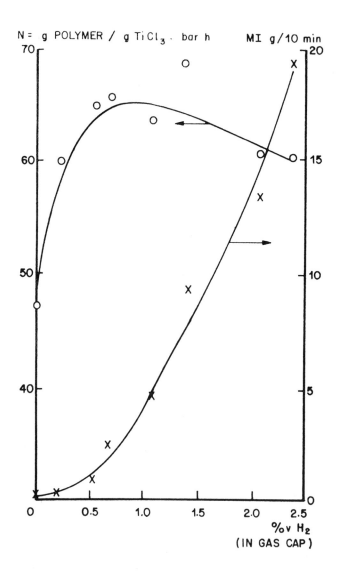

Fig. 1.37: *Effect of hydrogen on performance of an activated-$TiCl_3$-catalyst (which was prepared at low temperature)*

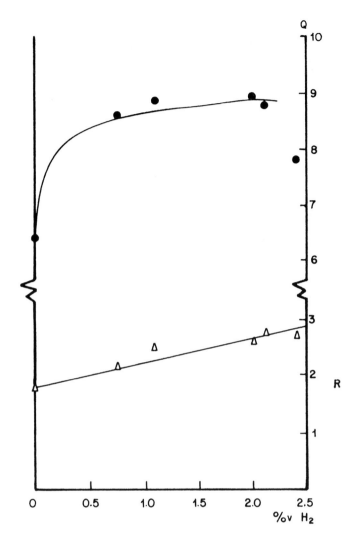

Fig. 1.38: Effect of hydrogen on the molecular weight distribution of polymers generated with an activated "low-temperature"-catalyst

Commercially the most important method of controlling molecular weight is by the use of hydrogen as "chain transfer agent". Table 1.12 lists a series of polymerization runs with a typical $TiCl_3$ catalyst at varying hydrogen concentrations. The data is also plotted in Fig. 1.37 and Fig. 1.38. Fig. 1.37 shows that hydrogen has a large effect not only on the product molecular weight (melt index) but also on the catalyst activity which is doubled (compared

TABLE 1.12

Effect of hydrogen concentration on molecular weight characteristics of polypropylene produced with a typical first generation $TiCl_3$ catalyst (polymerization @ 70°C, 4 bar propylene, DEAC/Ti=2).

% V/V H_2 in gas cap	$M_w \cdot 10^{-5}$	Q	R
0	6.1	6.4	1.82
0.75	3.7	8.6	2.15
1.1	2.6	8.8	2.48
2.0	2.5	8.9	2.54
2.1	2.0	8.7	2.76
2.4	2.0	7.7	2.63

with polymerization in the absence of hydrogen) at high levels of hydrogen. In Fig. 1.38 the Q and R values of the polymers are plotted as a function of the hydrogen concentration used. From this figure it is clear that polymer produced in the absence of hydrogen has the narrowest MWD and that increasing levels of hydrogen results in broader MWD. Hydrogen is also the commercial chain transfer agent of choice in the case of supported catalysts where similar effects on molecular weight and MWD are observed.

Finally the co-catalyst, together with any "third component" used, can have significant effects on the rate of chain transfer and hence molecular weight control. These effects are also mentioned in section 1.5. A commonly used class of third component for both $TiCl_3$ and supported catalysts are aromatic esters such as para-ethylanisate (PEA). The effect of the ester is at least twofold, having a large influence on both kinetics (yield) and molecular weight. Fig. 1.39 shows the effect of PEA on the yields of atactic and isotactic polypropylene with a typically ball-milled ($MgCl_2$ + EB + $TiCl_4$) catalyst containing about 2 %m/m titanium. A sharp drop in atactic activity is already noticeable at low PEA concentrations and this selective deactivation of atactic propagation centres is responsible for the drastic reduction in the percentage of xylene solubles despite the (more gradual) deactivation of isotactic propagation centers. Fig. 1.40 shows the effect of PEA on the MWD and yield of isotactic polymer. For these experiments the isotactic fraction of the polymer was recrystallized from xylene. From this figure it is clear that an increasing amount of PEA shifts the molecular weight gradually to higher values. Three

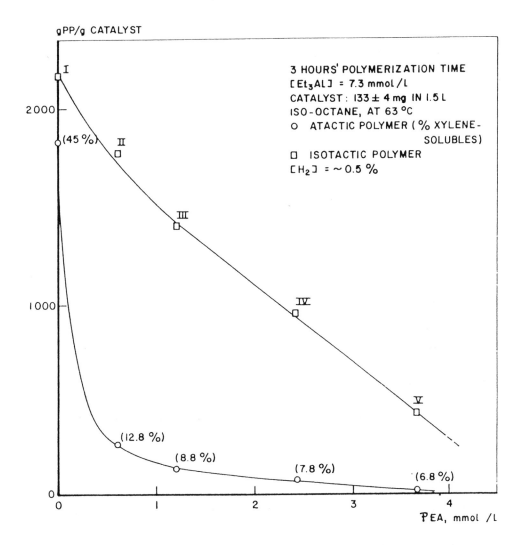

Fig. 1.39: *Effect of ethyl anisate (PEA) on the yield of atactic and isotactic polypropylene*

possible explanations for this phenomenon can be proposed:

i)   preferential reaction of PEA so as to deactivate those propagation centres which tend to produce polypropylene with a relatively low molecular weight (i.e. those centres having a high k_t/k_p ratio, where k_p = propagation rate constant, and k_t = termination rate constant). This could also explain the selective deactivation

DIFF. MWD x RELATIVE POLYMER YIELD (ARBITRARY UNITS)

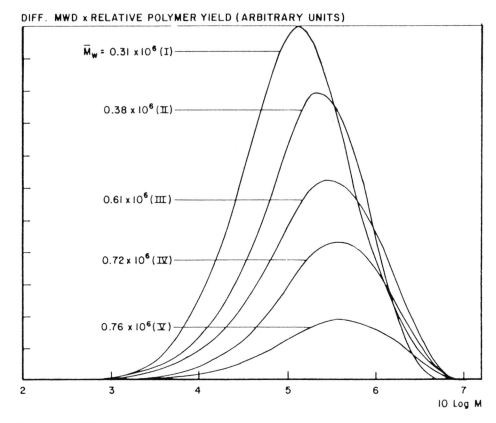

$\overline{M}_w$ = 0.31 x 10^6 (I)

0.38 x 10^6 (II)

0.61 x 10^6 (III)

0.72 x 10^6 (IV)

0.76 x 10^6 (V)

10 Log M

Fig. 1.40: Effect of ethyl anisate on molecular weight distribution and yield of isotactic polymer (xylene residue)

of atactic centres, since atactic polymer invariably has a lower molecular weight (and consequently a higher k_t/k_p ratio).

ii) complexation of AlEt$_3$ by the ester reduces the effective AlEt$_3$ concentration. Since triethylaluminium is known to be a powerful chain transfer agent this will result in an increase in molecular weight.

iii) the presence of a suitable electron donor such as PEA could result in the generation of a new type of active center which yields higher molecular weight polymer. A possible mechanism via which such "new" sites could be formed is presented in section 1.7.

In this context it is useful to consider the effect of PEA (in combination with a triethylaluminum co-catalyst) on polymerization using a $TiCl_3$ catalyst. For this purpose a pure $\alpha-TiCl_3$ (from photochemically prepared $\beta-TiCl_3$) was chosen in order to eliminate any effects of co-crystallized $AlCl_3$ etc.. Upon addition of PEA to the co-catalyst ($AlEt_3$: PEA 3/1) the level of xylene solubles decreased from 40 %m/m (observed using pure $AlEt_3$ as co-catalyst) to 5 %m/m with a concomittant threefold reduction in polymer yield. As shown in Fig. 1.41 a similar substantial increase in molecular weight was observed as was seen when using the supported catalyst.

DIFF. MWD x RELATIVE POLYMER YIELD (ARBITRARY UNITS)

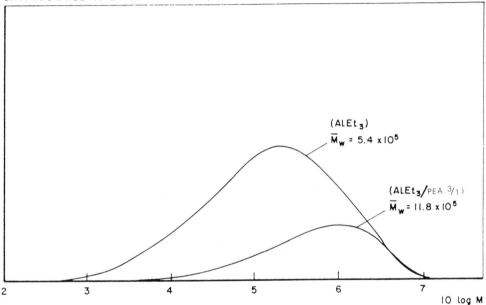

Fig. 1.41: Effect of PEA on the molecular weight distribution and yield of isotactic polymer obtained with $\alpha-TiCl_3-AlEt_3$

In general third components can interact with a catalyst in many ways to affect activity, selectivity and the molecular weight. These aspects are reviewed in detail by Boor[10]; the ways in which these materials influence molecular weight can be listed as follows:

- deactivation of (certain) active sites
- modification of (certain) active sites

- complexation of catalyst poisons, and
- complexation or reaction with the co-catalyst.

In the case of supported polypropylene catalysts, commercially the co-catalyst invariably comprises a mixture of a trialkylaluminium and an electron donor, an example of which is discussed above. Thus for supported catalysts it is not possible to separate molecular weight effects of the co-catalyst from those of the third component in a practical example. Using a supported catalyst in ethylene polymerization Zucchini[119] was not able to find a correlation between MWD and Lewis acid strength in co-catalysts of the type R_2AlX (X=R, H, Cl, OR). In the case of δ-$TiCl_3$ polypropylene catalyst Doi[120] reported that MWD width is a function of the co-catalyst choice and decreases in the order:

$$Et_2AlBr > Et_2AlCl > AlEt_3.$$

1.6.3 On the mechanism of chain transfer

On the basis of the foregoing it is apparent that a wide variety of catalyst and process variables can have a significant effect on the MWD of the polymer produced, and hence on the relative rates of chain propagation (k_p) and termination (k_t) (i.e. on the rate of "chain transfer"). The effect of the catalyst can be simply explained by the fact that a heterogeneous catalyst will inevitably possess a profusity of (slightly) different active centers resulting in a broad spectrum of k_p/k_t and therefore a broad polymer MWD. Indeed, using a homogeneous zirconium catalyst (i.e. only one distinct, well-defined pro-catalyst centre), Kaminsky[49] recently reported the preparation of narrow MWD polypropylene having Q-values of about 2 compared to most commercial grades which possess a Q of about 5 or larger. Thus when attempting to describe the chain transfer processes occurring in the case of a heterogeneous catalyst it is necessary not only to assume the existence of only one, unique type of surface site but also to realize that this is in fact represents a continuum of similar sites (varying slightly both electronically and sterically).

It is now generally accepted that there are essentially four modes of chain transfer in propylene polymerization processes, all of which are given in Fig. 1.42. The above four reactions can be used to explain all observations made concerning chain transfer and polymer MWD. For example, reaction (c) demonstrates that monomer

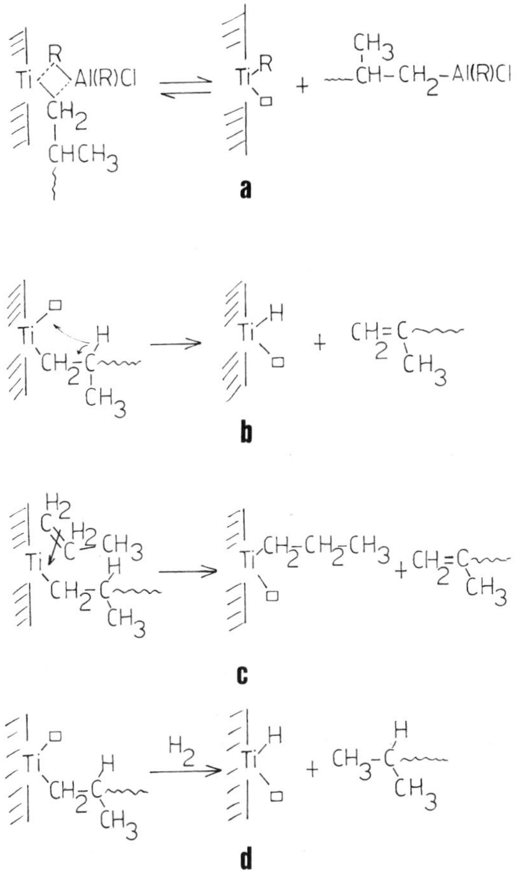

Fig. 1.42: Four modes of chain transfer: chain transfer via co-catalyst interactions (a); β-hydrogen transfer (b); monomer-assisted chain tranfer (c); hydrogenation, when hydrogen is applied as chain transfer agent (d)

concentration can have a significant impact on product MWD, since not only does it effect the propagation rate (section 1.7) but it also can significantly influence the rate of chain termination. In this regard it is interesting to note that α-olefins are the chain transfer agents of choice in the cobalt-catalyzed polymerization of butadiene (to butadiene rubber).

The co-catalyst effect (reaction (a)) is an equilibrium involving adsorption/desorption of the alklylaluminium species, and hence it is only to be expected that the diluent chosen for the polymerization can influence this process. It should be noted that the β-hydrogen transfer process, (b) and (c), results in olefinic endgroups whereas hydrogen (d) and alklyaluminium ((a); after deashing with an alcohol, or on exposure to moist air) result in saturated endgroups.

1.7 MECHANISM AND KINETICS OF PROPYLENE POLYMERIZATION

It would be beyond the scope of this chapter to give a comprehensive review of all the kinetic and mechanistic studies published over the last three decades. Furthermore despite the enormous amount of effort expended both in industry and academia the exact structure of "the" active centre and the mechanism involved in stereospecific polymerization still remain uncertain. Factors which contribute to the mystery of the Ziegler-Natta catalyst systems are

i) the sheer diversity of catalyst systems under study which severely hamper meaningful comparison of results from different research groups,

ii) the large number of process variables which can affect the kinetics and the selectivity of the systems, and

iii) the fact that only a very small percentage of the surface titanium centres are thought to be active in polymerization.

Our approach will be to review the more important aspects affecting reaction kinetics, to summarize various "milestone publications" in terms of mechanism and kinetics and to discuss selected own results and hypotheses which have influenced our research programmes. For a more comprehensive review of the literature the reader is referred to the book of Keii[121], the classic work of Natta[122] and also the work of Tait (see e.g. ref 123). This latter reference also reviews the sizeable literature on active centre determination using such techniques as CO or allene adsorption and radio-tagging with tritiated alcohol. This is an area which we will not cover other than to point out that for $TiCl_3$ catalysts it appears that only of the order of 1% of the contained titanium is active in polymerization while for $MgCl_2$ supported catalysts this level is substantially higher.

1.7.1 On the mechanism of the Ziegler-Natta catalyzed
polymerization of propylene

In the simplest form the propylene polymerization reaction

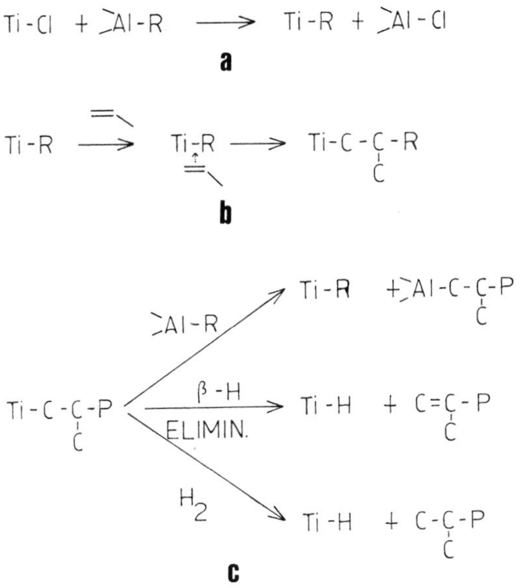

$$Ti-Cl + {\small\rangle}Al-R \longrightarrow Ti-R + {\small\rangle}Al-Cl$$

a

$$Ti-R \xrightarrow{\;=\!\!\backslash\;} Ti{-}R \longrightarrow Ti-C-C-R$$
$$\underset{C}{|}$$

b

$$Ti-C-C-P \begin{cases} \xrightarrow{\;{\small\rangle}Al-R\;} & Ti-R \quad +{\small\rangle}Al-C-C-P \\ \xrightarrow[ELIMIN.]{\beta-H} & Ti-H \quad + \; C{=}C-P \\ \xrightarrow{\;H_2\;} & Ti-H \quad + \; C-C-P \end{cases}$$

c

Fig. 1.43: Simplified scheme for propylene polymerization: active centre formation (a); the growth reaction (b); chain transfer (c)

can be represented as shown in Fig. 1.43. The process can be seen to comprise three reaction steps:

i) Active centre formation. The surface titanium centre (usually a chloride) is alkylated by the co-catalyst generating a titanium-carbon σ-bond, the active centre. In our opinion this is a gross oversimplification, this reaction "step" being better represented by a series of equilibria as illustrated below in which the various aluminium species (R_2AlR, R_2AlCl) are themselves involved in a complex series of monomer-dimer equilibria which is frequently further perturbated by the presence of an electron donor.

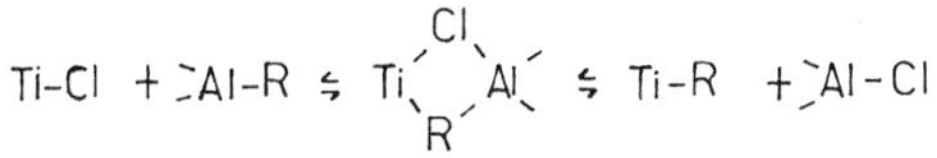

$$Ti-Cl + {\small\rangle}Al-R \; \leftrightarrows \; Ti\underset{R}{\overset{Cl}{\diagdown\!\!\diagup}}Al \; \leftrightarrows \; Ti-R \; +{\small\rangle}Al-Cl$$

ii) The growth reaction. This step represents the essence of the polymerization process; the propylene molecule is assumed to complex the transition metal rapidly followed by insertion into the titanium-alkyl bond. Rapid repetition of this complexation-insertion cycle constitutes the propagation of the polymer chain. The rate of this propagation reaction (i.e. each insertion step) is unprecedented with the possible exception of certain enzyme processes. Indeed in the case of homogeneous zirconium catalysts for ethylene polymerization it has been estimated that the "change over time" (the interval between two insertion steps) is less than 0.3 milliseconds[31]. It is this reaction step and the stereoselectivity thereof which will receive most attention in this section.

iii) Chain transfer. The polymer chain is eliminated or displaced from the active centre via β-hydrogen elimination, hydrogenation or co-catalyst mediated chain transfer. This aspect is dealt with in section 1.6.

(i) TiCl$_3$ catalysts and general aspects

In the 30 years since their discovery many theories have been developed to explain the mechanism involved in polymerizations using Ziegler-Natta catalysts, although it was Natta who first

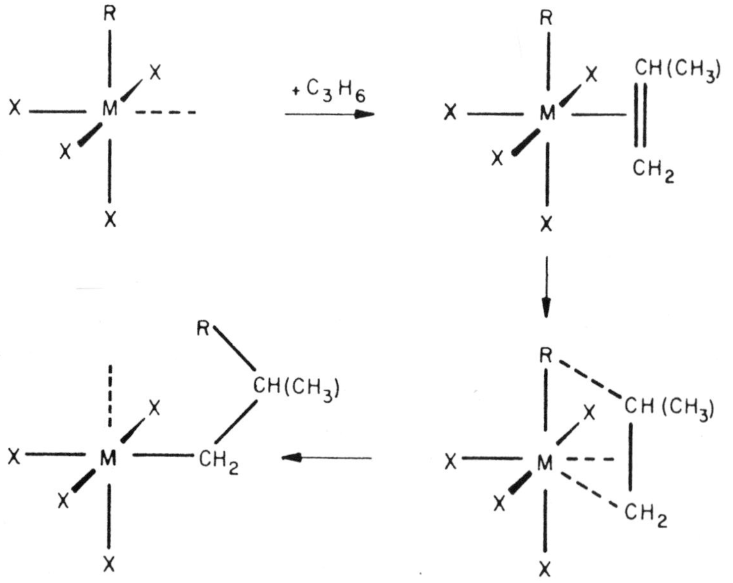

Fig. 1.44: The Cossee mechanism [from reference 158, copied with permission from Harwood Academic Publishers]

recognized the importance of the crystal surface in this respect. The comprehensive and elegant work of Cossee and Arlman[11] in the 1960's led to the wide acceptance of the so-called monometallic mechanism, illustrated in Fig. 1.44. The mechanism developed by Cossee is in part an extension of the work of Boor and Arlman who proposed the existence of chloride vacancies on the $TiCl_3$ catalyst surface, such vacancies being necessary to ensure the electroneutrality of the crystal (see section 1.4). An essential feature of the mechanism is that the active centres are those Ti atoms possessing chlorine vacancies, which are located in the lateral faces of the $TiCl_3$ crystal; and further, that isotactic propagation is intimately related to the geometry of these "exposed" titanium atoms and the surrounding chlorine atoms. Each isotactic centre possesses one chlorine vacancy (a in Fig. 1.45) and is bound to five chlorine atoms, of which four are fixed within the crystal

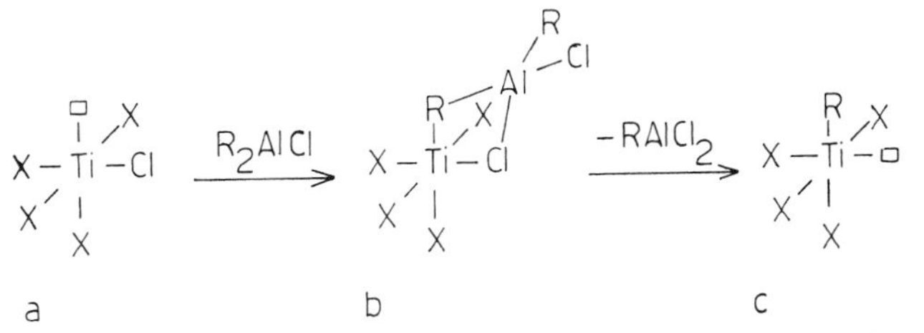

Fig. 1.45: *Active site generation via catalyst/co-catalyst interaction*

lattice (x) and one protrudes from the crystal surface. It is this protruding or "pendant" chlorine which reacts with the co-catalyst and yields the active centre, see Fig. 1.45. One frequently raised objection to the Cossee mechanism is that after monomer insertion the Ti-R bond will have changed places with the vacant site, requiring that the growing polymer chain (Ti-R bond) migrates back to its original position prior to the following monomer insertion step in order to retain the original chirality and hence stereoselectivity. Allegra[124] proposed a similar monometallic model which avoids the Cossee "jump-back" issue by assuming a different mode of olefin coordination to the active centre. The difference is clear from Fig. 1.46; Fig. 1.46a corresponds to the Cossee-Arlman representation in which two possible orientations of the incoming propylene are allowed and which is considered unlikely by Allegra

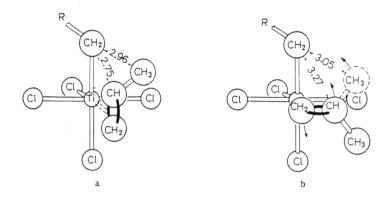

Fig. 1.46: *Different modes of olefin π-complexation on an octahedral Ti-atom with the double bond parallel to a Ti-X bond; distances are given in Angstrom units [from reference 124, copied with permission from Hüthig & Wepf Verlag]*

on the grounds of steric repulsions due to the short interatomic distances indicated. In Fig. 1.46b four possible orientations of the incoming propylene is favoured by Allegra who argued that only the outward <u>trans</u>-arrangement is likely on steric grounds. Again after insertion the growing polymer chain and vacant site will have changed places but in this case the necessity for a "jump-back" is obviated since the next propylene molecule will coordinate the "new" vacant site with the same chiral sense as the last inserted olefin.

In the Cossee mechanism atactic polymer arises from two different types of surface titanium site: in contrast to the isotactic site illustrated above, the atactic sites possess either two vacancies with one pendant chlorine or one vacancy with two

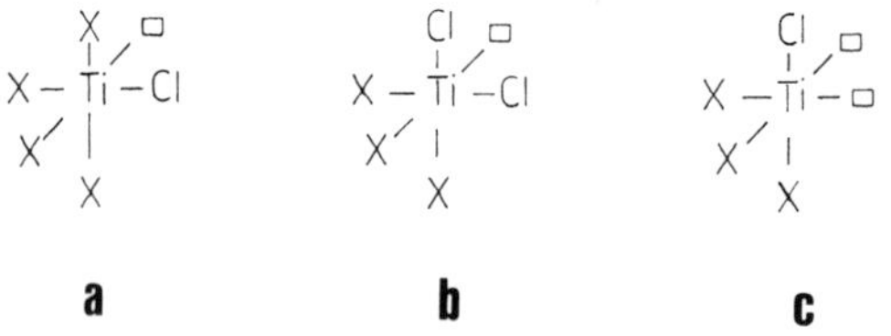

$$
\begin{array}{ccc}
\overset{X}{\underset{X'}{\overset{\mid}{X-Ti-Cl}}}\diagup{}^{\square} & \overset{Cl}{\underset{X'}{\overset{\mid}{X-Ti-Cl}}}\diagup{}^{\square} & \overset{Cl}{\underset{X'}{\overset{\mid}{X-Ti-\square}}}\diagup{}^{\square} \\
\mathbf{a} & \mathbf{b} & \mathbf{c}
\end{array}
$$

Fig. 1.47: *"Cossee sites": α-site (a); β_I-site (b); β_{II}-site (c) (□=chloride vacancy, X=chloride fixed in lattice, Cl=pendant ("alkylatable") chloride)*

pendant (and hence "alkylatable") chlorines. The isotactic site is predominant in α (and γ, δ) $TiCl_3$ and is therefore sometimes referred to as an α-site whereas the chain lattice, unstereoselective β-$TiCl_3$ comprises largely the atactic sites which are referred to as β_I and β_{II} sites (see Fig. 1.47).

The Cossee mechanism has truly stood the test of time and, with the exception of the discussions concerning the issue of whether or not the polymer chain migrates between two coordination positions on the active centre (above), is still widely accepted as the best representation of the mechanism. The most important alternative to the Cossee mechanism is the bimetallic mechanism which was also first proposed in the late 1950's and early 1960's.

The bimetallic mechanism has appeared in several variations. Initially in the early years of Ziegler catalyst it was believed, by analogy with the Aufbau reaction (section 1.2), that the growth reaction occured at the aluminium-carbon bond with the transition metal salt functioning to activate the olefin towards insertion. Based on this background Patat and Sinn[125] and Natta[126] proposed bimetallic mechanisms. The Natta mechanism illustrated in Fig. 1.48 represents the growing polymer chain as a bridging group between titanium and aluminium. This concept was later further developed by Rodriguez and van Looy[127] who proposed the mechanism illustrated in Fig. 1.49; this mechanism bears similarities to the Cossee mechanism except that the growing polymer chain bridges between titanium

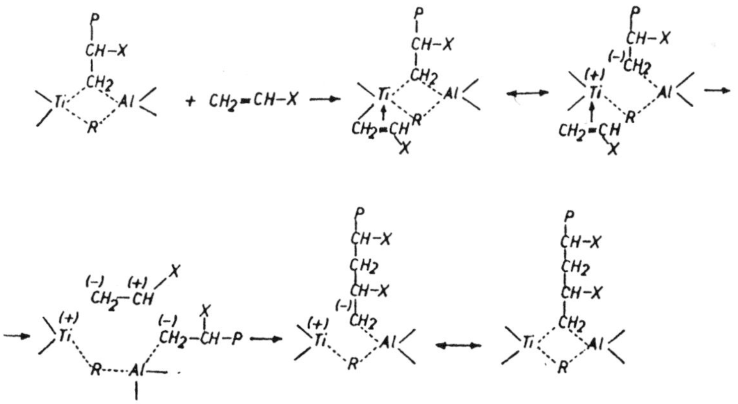

Fig. 1.48: The Natta bimetallic mechanism [from reference 126, copyright 1960, Pergamon Press Plc]

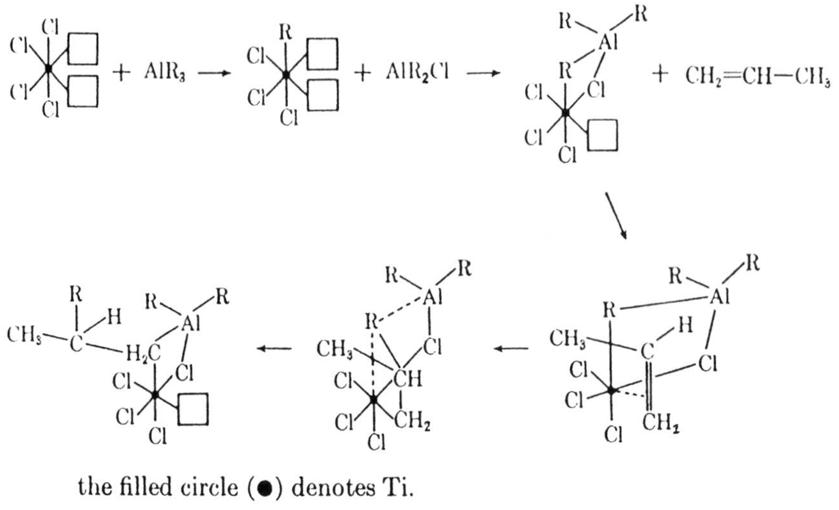

the filled circle (●) denotes Ti.

Fig. 1.49: The Rodriguez and van Looy mechanism [from reference 127, copied with permission from John Wiley & Sons]

and aluminium.

The increased scope and understanding of organometallic chemistry in recent times has led to alternative proposals concerning the mechanism of polymerization, the two most notable publications being those of Ivin et al[128] and Schrock and co-workers[129] who have both proposed mechanisms which do not involve simple insertion of an olefin into a metal-carbon sigma bond. Both of these proposals, which originate from research groups active in the field of olefin metathesis, attempt to draw parallels between the olefin polymerization and metathesis reactions: It is tempting to support this viewpoint since Ziegler catalysts are highly active in both reactions and in certain cases virtually identical catalysts can be used for the two olefin conversions.

It is now generally accepted that metal-carbene species are operative in olefin metathesis conversions, the reaction proceeding via a metallocyclobutane intermediate. The olefin polymerization mechanism proposed by Ivin et al[128] is represented in Fig. 1.50. The essence of the proposal is that the metal alkyl (active centre) first undergoes α-hydride elimination to generate a metal-carbene which in turn interacts with the incoming olefin to form the titanocyclobutane intermediate. The important difference between

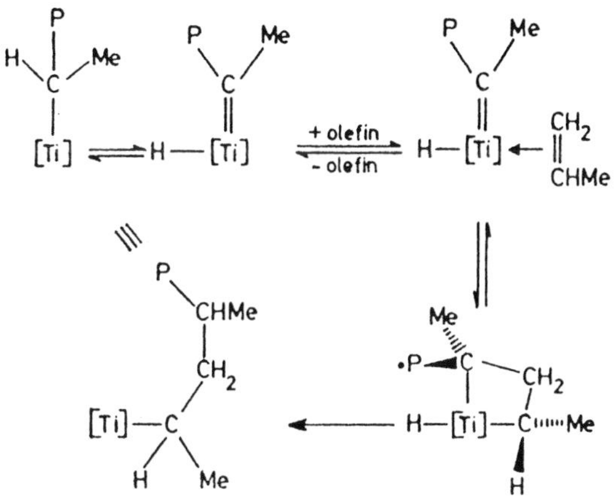

Fig. 1.50: The carbene mechanism [from reference 128, copied with permission from the Royal Society of Chemistry]

polymerization and metathesis in Ivin's scheme is that in polymerization the hydride eliminated in generating the carbene species remains attached to the metal and migrates to the sterically-hindered tertiary carbon atom in the titanocyclobutane, resulting in ring-opening and chain growth; if the hydride is lost then the only remaining reaction pathway is metathesis:

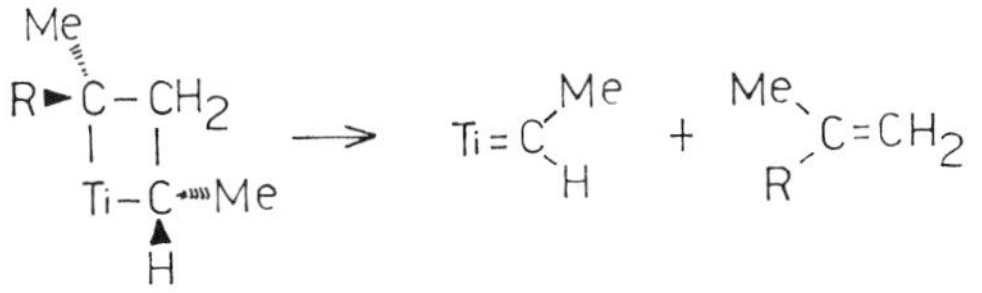

It can be argued that the work of Tebbe et al[130] in isolating the first example of a methylene-bridged (carbene) titanium-aluminium complex (Fig. 1.51) lends support to the carbene mechanism. This carbene compound was isolated from a Cp_2TiCl_2/aluminium alkyl system, well known as an ethylene polymerization catalyst. In isolating the stable carbene species the α-hydrogen (critical for the growth reaction in Ivin's mechanism) was lost, generating the "Tebbe compound" which can homologate ethylene but not polymerize it.

Schrock et al[129] proposed a mechanism based on a metallocyclo-

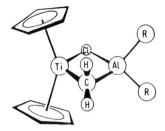

Fig. 1.51: Tebbe compound [from reference 130. Reprinted with permission from J.Am.Chem.Soc.. Copyright 1978 American Chemical Society]

pentane intermediate as illustrated in Fig. 1.52, in which two vacant coordination sites are required and in which two monomer units are incorporated simultaneously into the growing chain which results after alkyl-migration and ring-opening. Theoretical considerations by McKinney[131] lend some support to the proposal by indicating that in the case of propylene polymerization via such a mechanism isotactic, "head-to-tail" propagation would indeed occur since the most energetically favoured titanocyclopentane intermediate would be the equatorial 2,4 dimethyl substituted ring.

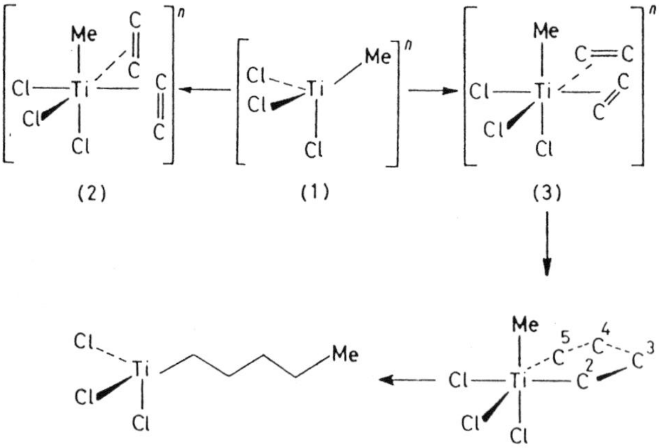

Fig. 1.52: The metallocyclopentane route [from reference 131, copied with permission from the Royal Chemical Society]

The high reactivity of the catalysts and the low concentration of active centres makes direct study to prove or disprove such mechanisms virtually impossible. Usually evidence can only be gathered via a study of the polymer generated, the structure of which forms a record of the reaction steps that have occurred. Studies of this type still point to the general mechanism of Ziegler-catalyzed polymerization being the insertion of an olefin into a transition metal carbon bond. For example a strong piece of evidence against the carbene mechanism is to be found in the work of Zambelli et al[132] who studied polypropylene generated using a labelled catalyst system, $TiCl_3$-$Al(^{13}CH_3)_2I$. Zambelli et al argued that since enriched methyls were identified in the polymer but no enriched methylene groups, which would be predicted by the carbene mechanism, that the insertion mechanism is still the favoured model.

Burfield and Tait[133] and Yermakov et al[134] have published mechanisms essentially supporting the monometallic, Cossee type proposal. The main features of their mechanisms, which are founded upon kinetic and "site-counting" or active centre concentration studies, are:

i) The active centres comprise Ti-C sigma bonds, monometallic in nature in which the alkylaluminium co-catalyst plays no direct part.

ii) The propagation reaction involves two steps; monomer coordination followed by insertion into the Ti-C bond.

iii) The role of the co-catalyst is to alkylate the pendant chlorines of the surface Ti centres affording the active propagation site
(see above). The co-catalyst also plays an important part in chain
transfer processes (section 1.6), and can influence the propagation
rate by reversibly complexing active sites which gives rise to
temporarily inactive species.

Böhm[135] published a mechanistic model in which he attempted to
include all the reaction steps involved in Ziegler-catalyzed olefin
polymerizations. In his scheme (Fig. 1.53) an isotactic site of the
Cossee type is assumed.

The question remains as to exactly how the composition and
symmetry of the active centre controls the stereoselectivity of
propylene polymerization. Broken down into the most simple terms
there appear to be three ways in which the stereoregulation of the
polymerization process might be governed:

i)   By the asymmetry of the ligand environment around the transition metal ion

ii) By the configuration of the last monomer unit incorporated (or
in other words by the asymmetry of the β-carbon atom in the growing
polypropylene chain)

iii) By the asymmetry induced by the helical conformation of the
isotactic polypropylene.

The last of the three options is perhaps the least likely
cause of the stereoregulation since the heat of polymerization is
such that in the immediate vicinity of the active centre itself the
growing polymer chain will be molten or dissolved, in which state
the helical structure, although probably present, will hardly be
sufficiently stereorigid to maintain asymmetry. Furthermore such a
method of stereoregulation could not be operative in the case of
syndiotactic polymerization since syndiotactic polypropylene has a
planar, zig-zag structure.

Regarding stereoregulation control by the last incorporated
monomer unit the situation is less clear-cut and appears to be
dependent on the catalyst applied. Zambelli and coworkers[136,137]

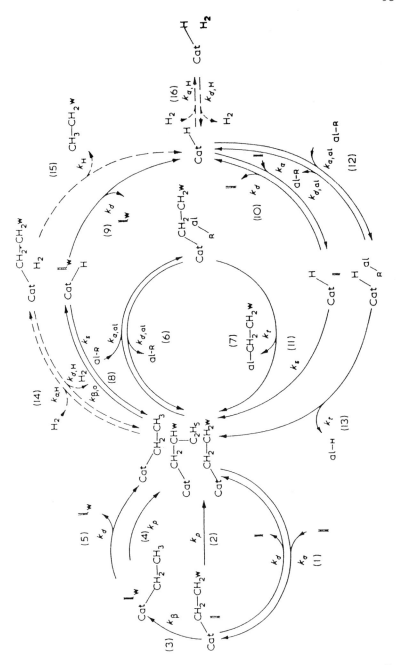

Fig. 1.53: Reaction scheme of a Ziegler-Natta polymerization process according to Böhm (cat: catalytic active centre; ∥: ethylene molecule; w: polymer chain; ∥^{w}: polymer chain with a vinyl group; al: 1/3Al; R: alkyl group; H, H$_2$: hydrogen atom or molecule) [from reference 135, by permission of the publishers Butterworth & Co Ltd]

demonstrated using ^{13}C-NMR techniques that, when an isospecific polypropylene catalyst is applied in ethylene/propylene copolymerization, the configuration of propylene insertion is retained after incorporation of ethylene. This proves that, for an isotactic polypropylene catalyst, stereoregulation is <u>not</u> controlled by the last inserted monomer unit. An exception appears to have been found in the more recent work of Ewen[44] who explains the isotactic, or perhaps more accurately the "stereoblock" nature of the polypropylene formed using a homogeneous titanium/methylaluminoxane catalyst system on the basis of a chain-end stereochemical control mechanism (section 1.4.4). In the case of syndiotactic propagation using vanadium catalysts it is largely the configuration of the last monomer unit which governs stereoregulation, a detailed mechanism having been proposed by Zambelli and Allegra[138].

In the case of isotactic polypropylene it is widely accepted that it is the asymmetry of the active transition metal centre which controls the stereoregulation. Again strong evidence for this hypothesis is to be found in ^{13}C-NMR studies of isotactic polypropylene generated with various catalysts. If stereoregulation were controlled by the last monomer unit incorporated then errors of the type symbolized in Fig. 1.54b would be expected (mrmm in the usual NMR terminology); such errors are not observed, the conformation of

a)

b)

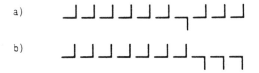

Fig. 1.54: Two possible stereostructures in polypropylene

insertion errors are almost invariably of the mrrm type symbolized in Fig. 1.54a (see also chapter 2).

Another convincing piece of evidence that stereoselectivity is achieved on the basis of chirality rather than (for example) steric grounds is to be found in studies of copolymerization of a racemic (RS) olefin and an optically active (S) olefin. In these investigations using both $TiCl_3$ catalysts[139] and, more recently, supported catalysts[140] it was observed that with both catalyst types the product comprised a mixture of two polymers:

i)   a random copolymer of the two S-monomers and

ii) the homopolymer of the R-monomer.

In this case however it could still be argued that the last inserted monomer unit is controlling chirality.

Possibly the most convincing argument that chirality is a
prerequisite for stereospecific polymerization is the elegant work
of Kaminsky et al[49]. Kaminsky developed a homogeneous zirconium
catalyst comprising bis(cyclopentadienyl) zirconiumdichloride and
methaluminoxane which exhibited unprecedented activity in the
polymerization of ethylene. However when applied in the polymerization of propylene only atactic material was produced. By using the
chiral ethylenebis-(4,5,6,7-tetrahydro-1-indenyl) zirconium dichloride Kaminsky was able to generate highly isotactic polypropylene,
with only about 1% of the product being soluble in hydrocarbons. In
these studies the racemate of the catalyst was used; the use of a
single isomer resulted in the production of optically-active
polypropylene (see section 1.4.4).

One final piece of evidence in favour of stereoregulation via
site chirality is to be found in a publication of Zambelli et al[132]
already referred to in the context of the carbene mechanism. In
this study using a $TiCl_3$ catalyst and a labelled co-catalyst,
$Al(^{13}CH_3)_2I$, it was observed (again via NMR techniques) that the
stereochemistry of the first monomer unit insertion was the same as
that of the subsequent monomer units despite the absence of asymmetry in the original alkyl group (CH_3). This finding argues
directly for control via the chirality of the active centre and
against polymer end group control.

The Cossee-Arlman model, perhaps with minor modifications as
discussed earlier in this section, still offers the most coherent
explanation in terms of stereoregulation. In this model the chirality of the surface titanium centre originates from its being fixed
rigidly within the crystal lattice, and is maintained during the
propagation/insertion step. Recent theoretical studies by Corradini's group[141] offer support for the concept of an asymmetric
centre in which the critical factor involves the orientation of the
first carbon-carbon bond of the growing polymer chain (i.e. the
bond immediately following the metal-carbon bond), this in turn
being governed by steric induction associated with the other

ligands attached to the titanium centre. Corradini's computations indicate that the exact nature of the growing polymer chain (be it polypropylene or polyethylene) has little or no effect on the mode of co-ordination and insertion of the incoming propylene monomer. This finding is in good agreement with experimental findings with ethylene-propylene copolymers such as the work of Zambelli et al[137]; in particular the fact that the retention of the configuration of propylene moieties in the growing polymer chain is preserved even after the insertion of an ethylene molecule.

(ii) Supported catalysts

Obviously in the case of supported catalysts the fundamental mechanistic steps involved in propagation and termination etc. are essentially identical to the $TiCl_3$ catalyst case. For this reason we have chosen to deal mainly with the major difference between supported and non-supported catalysts, namely the function of the electron donors and particularly the role of the co-catalyst (e.g. AlR_3 + aromatic ester) with supported catalysts. The organometallic chemistry involved is described in section 1.5, here an attempt will be made to elucidate the interactions occuring between the pro-catalyst (Ti-component) and the co-catalyst; in this respect the function of the aromatic ester is vital since in most catalyst systems the absence of ester reduces the stereoselectivity to completely unacceptable levels (50-80% isotactic index). Many publications have addressed this issue, and frequently contradictory observations and conclusions have been made.

Considering the relative "youth" of supported catalysts it is not surprising that they are less understood and that there is, as yet, no consensus regarding even some of the more basic aspects of the structure of the active site and the various chemical interactions involved. In this respect a good example is perhaps the oxidation state of the titanium in the active catalyst. Although satisfactory supported catalysts have been claimed by grinding $TiCl_3$ with $MgCl_2$[142] most supported catalysts comprise $TiCl_4$; however several authors including Chien and Wu[143], and Kashiwa and Yoshitake[144] have reported how, under polymerization conditions in the presence of co-catalyst, the titanium is reduced to Ti(III) and Ti(II) species. In contrast it is generally accepted that under normal conditions with a DEAC co-catalyst, $TiCl_3$ catalysts comprise almost exclusively Ti(III) active species; in the presence of TEA as co-catalyst, reduction to Ti(II) is observed but this is reflected in low stereoselectivity. It seems reasonable to assume that the active centre in both $TiCl_3$ and supported catalysts is a Ti(III) species, but definitive evidence has yet to emerge.

Kashiwa[145] reported studies with an $MgCl_2/TiCl_4$ catalyst (no ester in the solid component) and a $AlEt_3$ + ethyl benzoate co-catalyst. He observed that while the overall polymer yield was drastically decreased by the presence of ethyl benzoate, the yield of isotactic polymer actually <u>increased</u> (a phenomenon not observed by various other groups such as Pino et al[146] and ourselves; this difference in results may be attributable to the different pro-catalysts used). On the basis of his findings Kashiwa concluded that the addition of a suitable ester either increases the propagation rate constant by complexing isotactic sites or increases the number of isotactic sites via complexation of atactic sites. Molecular weight studies led him to conclude that there are two types of isotactic site, those capable of associating with esters and those not. The sites that become complexed with ester are characterized by a much higher molecular weight isotactic polymer than the non-complexed sites.

Pino et al[146] also favours the presence of two types of catalytic centre with different affinities for association with aromatic esters. Pino proposed a low stereospecificity site C which possesses a high Lewis acidity and a second class of catalytic centre (C_1, C_d) which are highly stereospecific, chiral and (probably for "steric reasons") less Lewis acidic. In accordance with this hypothesis the following equilibria would exist during the polymerization (where B = Lewis base, e.g. ethyl benzoate). As a first approximation it was assumed that only the non-complexed centres are catalytically active :

$$AlR_3 + B \longleftrightarrow AlR_3.B \text{ (equilibrium constant } K_1)$$

$$C_{(solid)} + AlR_3.B \longleftrightarrow C.B_{(solid)} + AlR_3 \text{ (,, \quad ,, } K_2)$$

$$C_{1(solid)} + AlR_3.B \longleftrightarrow C_1.B_{(solid)} + AlR_3 \text{ (,, \quad ,, } K_3)$$

$$C_{d(solid)} + AlR_3.B \longleftrightarrow C_d.B_{(solid)} + AlR_3 \text{ (,, \quad ,, } K_4)$$

It should be emphasized that C, C_1 and C_d represent classes of catalytically active centres explaining for example, the broad molecular weight distribution observed.

The catalytic system is further complicated by the fact that both the solid catalyst component and the co-catalyst usually contain an electron donor/Lewis base (often referred to as the "internal base" and "external base" respectively) and that there is convincing evidence that these electron donors undergo exchange reactions under polymerization conditions. Langer[147] reported a series of experiments using labelled ethyl benzoate, in which the ethyl benzoate used as an internal base was extracted by the co-catalyst and underwent exchange with ethyl benzoate used as external base. Barbé and co-workers[148] have reported that ethylbenzoate as internal base is partially extracted under polymerization conditions and that both the external base (methyl para toluate, MPT) and aluminium species become incorporated into the solid catalyst component. A significant fact is that the degree of internal base extraction and exchange is highly dependent on the molar ratio of the trialkyl aluminium to external base, as is shown in Fig. 1.55. It is particularly noteworthy that extraction of the ethylbenzoate (internal base) is complete when no external base is applied; this offers a possible explanation for the poor stereoselectivity of such catalyst systems in the absence of an external

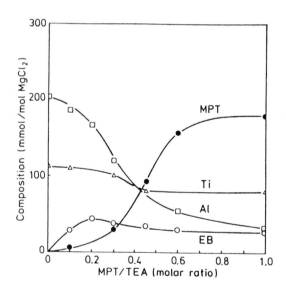

Fig. 1.55: Interaction between the catalyst $TiCl_4/EB/MgCl_2$ and TEA/MPT: effect of the MPT/TEA molar ratio on the catalyst composition (reaction conditions: temperature 50°C, TEA 50 mmol/l; Ti 7 mmol/l; time 4h) [from reference 148, copied with permission from Springer Verlag]

base. Under "normal" conditions (i.e. in the presence of an internal base) although some Lewis base exchange occurs, a considerable portion of the original internal base remains incorporated in the catalyst. Further evidence for this last observation comes from Ciardelli et al[149] who recently reported that using optically active (-)-menthyl benzoate as internal base in the polymerization of racemic 3,7 dimethyloct-1-ene gave substantial stereoselectivity affording isotactic polymer with a negative optical rotation, while using the same optically active ester as external base (and ethyl benzoate as internal base) afforded polymer with no detectable optical rotation. These experiments must surely be seen to prove that at least part of the internal donor remains attached to the catalyst and plays an important role in controlling stereoselectivity.

Concerning the interactions occuring between the co-catalyst and the catalyst, and the nature of the active polymerization centre it is possible to use a model in which it is only necessary to assume the presence of one type of surface titanium centre (which in fact represents a spectrum of similar titanium centres varying slightly in electronic and steric character). The model, described by the scheme shown in Fig. 1.56, is a "dynamic" one in which one surface titanium site gives rise to different types of

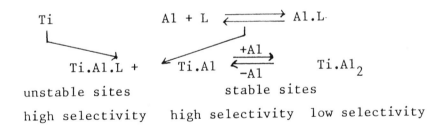

Fig. 1.56: *Interactions between pro-catalyst and co-catalyst*

active polymerization centre after exposure to the co-catalyst. In the model (Fig. 1.56) Al represents AlR_3 (the co-catalyst which exists as monomer-dimer equilibrium), L represents the external base (e.g. para ethyl-anisate) and Ti represents the surface titanium species.

As indicated in the Fig. 1.56, in the presence of the co-catalyst two types of site (stable and unstable) are generated. The concentration of each type of site is thus determined by the concentrations of the three components (Ti, Al and L) and particularly by the Al:L molar ratio. The unstable site has the form Ti.Al.L; this site decays rapidly but exhibits extremely high stereoselectivity (98-99 % isotactic index). The aromatic ester (L) is not involved with the stable sites that have the form Ti.Al and Ti.Al$_2$; these two sites (together) display a constant activity and stereoselectivity (which is poor under normal conditions) throughout the polymerization. The two stable sites are in equilibrium with each other, the position of the equilibrium being determined by the free (i.e. non-complexed) Al concentration. The Ti.Al site possesses high selectivity while Ti.Al$_2$ displays low selectivity.

The role of the aromatic ester (L) in this model is essentially twofold:

i) It converts stable (low selectivity) active centres into unstable (very high selectivity) centres. This is demonstrated by the fact that the initial polymerization rate is essentially unaltered by the presence of the aromatic ester (except at abnormally high concentrations where excess ester can directly complex the surface titanium centres giving inactive sites of the type Ti.L).

ii) It lowers the free trialkyl aluminium concentration (Al in Fig. 1.56), shifting the "stable site equilibrium" to the left (higher selectivity). This reduction of free triethyl aluminium (an efficient chain transfer agent) concentration offers a simple explanation as to why the presence of aromatic esters results in increasing molecular weight.

The "stable site equilibrium" can be used to explain why certain catalyst systems which are applied in the absence of an external base are highly selective only at low trialkyl aluminium concentrations. This equilibrium also offers an explanation for the high selectivity claimed when using sterically-hindered trialkyl aluminium co-catalysts such as ethyl di-sec-butyl aluminium (see section 1.5).

Seen in its broadest sense this model predicts that, for a given Ziegler pro-catalyst (i.e. the "solid component" of commercial catalyst systems), most aspects of catalyst performance (activity, selectivity, kinetics, etc) are determined by the equilibria in the co-catalyst system. In other words that while the asymmetry of the transition metal centre and the attached polymer chain is a major factor in controlling stereoselectivity, and the concentration of surface titanium centres a major factor in determining catalyst activity, the ligand function of the co-catalyst should not be overlooked, and the catalyst system <u>as a whole</u> must always be taken into consideration.

Even in the case of $TiCl_3$ catalysts where the "monometallic mechanism" is favoured by most researchers, certain effects such as the poisoning role of $EtAlCl_2$ and the stereoregulating influence of Et_2AlI can best be explained by considering the appropriate co-catalyst component to be ligating the titanium centre (section 1.5).

Doi[150] has proposed an active centre in which titanium is complexed by an aluminium species resulting in a species similar to that already illustrated in Fig. 1.34. The Doi bimetallic active centre is illustrated in Fig. 1.57. Although these active site models can explain many of the phenomena observed during polymeri-

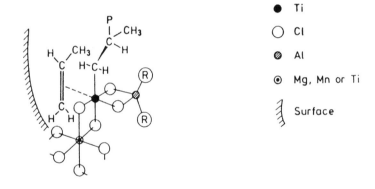

● Ti

○ Cl

◐ Al

◉ Mg, Mn or Ti

〉 Surface

Isotactic propagation

Fig. 1.57: Isotactic site model after Doi [from Transition metal catalyzed polymerizations. Alkenes and dienes, R.Quirk editor, 1983; copied with permission from Harwood Academic Publishers]

zation reactions they suffer from one major failing, namely that the all-important electron donors do not appear in the picture.

Concerning the internal base solid state ^{13}C-NMR data argues strongly that this base is bonded only to $MgCl_2$ and is not directly involved with titanium. It is very likely however that the external base <u>is</u> involved in bonding in the active centre: In the case of aromatic esters the reactivity of the ester with metal-carbon bonds seems critical for giving the desired stereoregulating effect suggesting that perhaps the ester actually exists in the active centre as an alkylated ester group, probably as an alkoxy-bridge between titanium and aluminium (see section 1.5). Another alternative is that the ester simply blocks a vacant coordination site on titanium. We favour the former argument for reasons presented in section 1.5, but it is apparent that much more knowledge of these catalysts must be generated before this important electron-donor effect can be satisfactory explained.

1.7.2 <u>Kinetic</u> <u>aspects</u>

The kinetics of propylene polymerization can be affected by a multitude of catalyst and process variables. Many of the catalyst variations (choice of catalyst, co-catalyst and third component) have already been dealt with in various sections of this chapter.

An additional catalyst variable which can have a significant beneficial influence on polymerization rate is catalyst/co-catalyst premixing and/or prepolymerization. Particularly in the case of $TiCl_3$ catalysts the premixing of the catalyst with the DEAC co-catalyst is frequently claimed to be advantageous. In addition to bestowing the catalyst with a degree of "protection" against accidental exposure to moisture, the polymerization rate is enhanced since the DEAC premix results in the solubilization of co-crystallized $AlCl_3$ (via alkylation) and adsorbed $EtAlCl_2$ (via formation of the sesquichloride) and hence a reduction in the effective concentration of these catalyst poisons. It has been claimed that these effects can be even more efficiently brought about by a concomittant prepolymerization step which offers the additional benefit of dispersing the catalyst under mild conditions allowing for more effective heat dissipation during the polymerization proper, eliminating the risk of catalyst explosion (see section 1.3). Generally speaking typical third components (esters, ethers etc.) are not involved in premixing procedures since long exposure to the, frequently reactive, Lewis bases can result in active centre complexation and deactivation.

Process variables which significantly affect the polymerization kinetics for both $TiCl_3$ and supported catalysts include the monomer concentration, temperature, diluent choice and time:

monomer concentration

        Under steady-state conditions the polymerization rate is
directly proportional to the monomer concentration. This fact was
first recognized by Natta and Pasquon[122], whose results for propylene polymerization using an α-TiCl$_3$/TEA catalyst are illustrated
in Fig. 1.58. The influence of monomer concentration on a supported

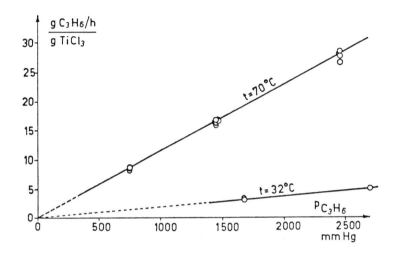

Fig. 1.58: Dependence of polymerization rate on concentration of propylene [from
reference 122, copied with permission from Academic Press]

catalyst containing 1.6 %m/m titanium is illustrated in Fig. 1.59
and again the linear dependence is apparent.

temperature

        It is perhaps to be expected that the initial polymerization
rate will increase with increasing reaction temperature but this
effect is not always observed. For TiCl$_3$ catalysts the rate does
indeed increase with increasing temperature (e.g. Fig. 1.60) but
the fact that the resulting polypropylene isotacticity decreases
with increasing temperature dictates that most commercial TiCl$_3$
catalyst-based plants are operated in the 55 to 65°C temperature
range which combines maximum rate with acceptable selectivity. For
supported catalysts it is found that the polymerization rate shows
a distinct maximum[151,152] in the range 60 to 70°C and then decrea-

ses with increasing temperature - while the polymer isotacticity usually increases. This decreasing rate at higher temperatures points to catalyst deactivation either by overreduction of the active centres or via alkylation processes involving the third component (usually aromatic esters). Some of our own data confirming the above-described temperature optimum for a $MgCl_2$-supported catalyst is represented in Fig. 1.61.

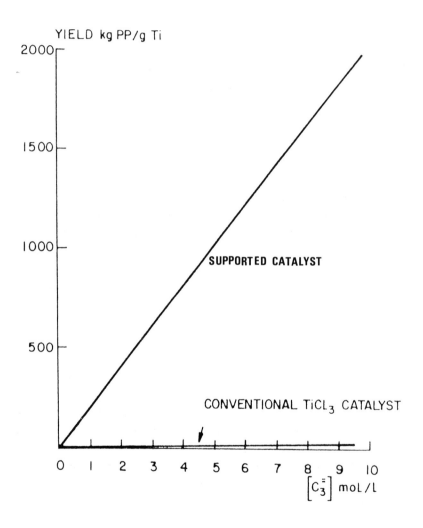

Fig. 1.59: *Relationship between polymer yield and propylene concentration*

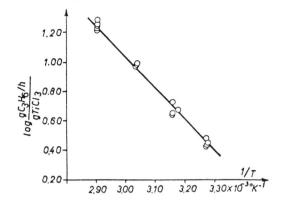

Fig. 1.60: Temperature dependence of initial rate [from reference 122, copied with permission from Academic Press]

diluent

The choice of diluent applied in propylene polymerization can have a marked effect on the polymerization rate observed, even when experiments are run at equal propylene concentration (i.e. even after correcting for the relative propylene concentration in each solvent). The effect is particularly pronounced in the case of $TiCl_3$ catalysts where aromatic solvents invariably lead to higher polymerization rates than do aliphatic ones. The most likely explanation for the rate-enhancing effect of aromatic solvents is their greater solvent power for the co-crystallized $AlCl_3$ present in these catalysts (see section 1.4).

polymerization time

All $TiCl_3$ and $MgCl_2$-supported catalysts exhibit decay-type kinetics in which the instantaneous polymerization rate decreases as a function of time. These decay or deactivation kinetics and their cause and effects are commercially of such importance that they are discussed in some depth; the large differences in behaviour betweeen $TiCl_3$ and supported catalysts results in an obvious division into two categories.

(i) $TiCl_3$-catalysts

Our discussion will be limited to the kinetics of the commer-

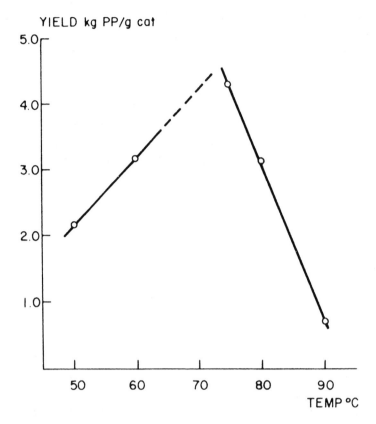

Fig. 1.61: Effect of polymerization temperature on yield (TEA/Ti=30; TEA/PEA=3.2; [TEA]=0.5 mmol/l; [propylene]=2.6 mol/l; [H_2]=1 mmol/l; 2.5h)

cially-used $TiCl_3.1/3AlCl_3$ catalysts in combination with DEAC as co-catalyst; for background information on other catalyst compositions and comparative data with (for example) TEA as co-catalyst the reader is referred to previously published reviews[121,153,154].

Despite the many detailed experimental studies that have been published over the years there is still no universally accepted kinetic model to explain the frequently complicated and diverse kinetic behaviour of these catalyst systems. This is perhaps not surprising considering the complexity of the system under study; the polymerization kinetics being affected by many process variables, by the composition, structure and physical form of the catalyst, by the choice of the co-catalyst, by the presence of (or choice of) third components to control stereoselectivity and a variety of miscellaneous factors such as whether or not the catalyst/co-catalyst are premixed or prepolymerized.

It is well known that γ-TiCl$_3$/DEAC catalyst systems exhibit decreasing polymerization rates with increasing residence time; this phenomenon is usually referred to as "catalyst decay". The decay in polymerization rate is frequently accompanied by a gradual loss in stereoselectivity. Using both first generation (TEA-reduced) TiCl$_3$ catalysts and a typical second generation (Solvay-type) catalyst with DEAC as co-catalyst, under both liquid propylene polymerization conditions and at low pressures (2 to 5 bar) in a C$_8$-fraction hydrocarbon diluent, it was found that the polymer yield at time t (hours) can be conveniently expressed as follows:

$$Y_t = Y_1 * t^p$$

where Y_1 is the yield after 1 hour and p is a measure of the catalyst decay; a value of unity corresponds to a non-decaying catalyst and the smaller the value of p the more rapid the rate of catalyst decay. Catalyst decay is very general and it is more pronounced in the case of the isotactic propagation rate than for the rate of atactic propagation, reflecting the observed decrease in catalyst stereoselectivity as a function of residence time. At least in the case of the first generation TiCl$_3$ catalyst the rate of catalyst decay decreases with increasing polymerization temperature.

It was invariably found that, while there is no direct correlation between the rate of catalyst decay and the hydrogen concentration applied (for chain transfer purposes, see section 1.6), the decay in isotactic propagation rate is greatly affected by the presence or absence of hydrogen. Indeed in the total absence of hydrogen a much slower rate of catalyst decay is observed. We attribute this hydrogen effect to one of reduction of Ti(III) species in the active centre to Ti(II). This hypothesis is supported by ESR measurements, where Ti(II) was observed, and by the fact that in "ageing" experiments in the presence of hydrogen, purple γ-TiCl$_3$ turned black, indicative of Ti(II). We visualize the hydrogen-mediated reduction process as occurring via an essentially two step process as indicated below:

a. $\quad$ Ti$-$R $+$ H$_2$ $\longrightarrow$ Ti$-$H $+$ RH

b1. $\quad$ Ti$-$R $+$ Ti$-$H $\longrightarrow$ 2 Ti $+$ RH

b2. $\quad$ 2 Ti$-$H $\longrightarrow$ 2 Ti $+$ H$_2$

The first step is simply hydrogenation and represents the mechanism of chain transfer in the presence of hydrogen. The two mechanisms indicated in the second step are bimolecular displacement or elimination reactions well-described in homogeneous titanium chemistry. The presence of titanium pairs on the surface of γ-TiCl$_3$ has been proven by our own ESR investigations.

This reaction scheme explains the catalyst deactivation since the active centres are converted into inactive ones in which no titanium-carbon bond or loosely-bound ligand remains. The difference in behaviour between isotactic and atactic polymerization sites is somewhat more difficult to explain, particularly since hydrogen actually causes the atactic polymerization rate to increase while the overall rate decreases. Generally speaking the stereoselectivity of a γ-TiCl$_3$ catalyst remains constant throughout a polymerization run carried out in the absence of hydrogen while in its presence the stereoselectivity (as reflected by the level of xylene-soluble polymer formed) begins lower and decreases with residence time.

We speculate that these effects are caused by the hydrogen-mediated reduction, and deactivation, of isotactic sites as described above while the number of atactic sites remains constant but that these sites experience an overall rate-enhancing effect. This can be rationalized by assuming that (part of) the atactic polymer is generated on sites containing two loosely bound chlorides and two vacancies. Reduction of such a site leads to an even more coordinatively-unsaturated centre which is more monomer-accessible and thus more active:

$$Cl-\overset{\overset{\displaystyle Cl}{|}}{\underset{\underset{\displaystyle Cl}{|}}{Ti}}\overset{\square}{\underset{\square}{\diagdown}} \quad \xrightarrow{\text{red.}} \quad Cl-\overset{\overset{\displaystyle \square}{|}}{\underset{\underset{\displaystyle Cl}{|}}{Ti}}\overset{\square}{\underset{\square}{\diagdown}}$$

It should be noted that there is a significant difference in the magnitude of the "hydrogen effect" in first generation and second generation $TiCl_3$ catalysts, the effect of hydrogen is much more dramatic in the case of the latter type of catalyst, a pheno-menon which we attribute to the fact that the first generation catalysts contain much higher levels of EADC which strongly comple-xes surface sites blocking other chemical interactions (see secti-ons 1.3.1 and 1.5).

The reader is also referred to the literature for further details on the effects of hydrogen on the selectivity of both $TiCl_3$ catalysts[155,156] as well as supported catalyst systems[157].

(ii) supported catalysts

Compared to conventional $TiCl_3$ catalysts the rate of propaga-tion for supported catalysts is at least an order of magnitude higher. The kinetic curves are of the decay type and show little or no sign of an induction period, as witnessed by the rapid establish-

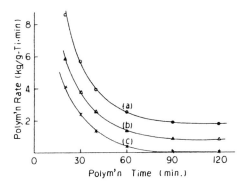

Fig. 1.62: Polymerization rate at various TEA/Ti ratio's (polymerization conditions: temperature 65°C, propylene pressure 10 bar, [Ti]=0.021 mmol/l; [donor]=0.82 mmol/l; donor is ethyl toluate): TEA/Ti is 200 (a); 150 (b); 130 (c) [from reference 159, copied with permission from Plenum Press]

112

ment of the maximum (initial) activity. In the case of the SHAC catalyst[158] initial polymerization rates of the order of 5000 kg PP/g Ti.h are typically observed in liquid propylene polymerizations. The kinetics are complicated by the choice of co-catalyst and related chemistry. In particular the electron donor used in the co-catalyst and the triethyl aluminium/electron donor molar ratio are critical in determining the polymerization kinetics. As explained in earlier sections the electron donors most frequently applied are aromatic esters and consequently these will be discussed in the most detail.

Tashiro et al[159], who investigated the kinetics and thermal stability of a typical $MgCl_2$/ethylbenzoate/$TiCl_4$ catalyst system, reported that the polymerization rate is a function of the TEA/Ti molar ratio employed. Their kinetic curves are reproduced in Fig. 1.62, where it should be noted that a constant electron donor/catalyst (para ethyl anisate/Ti) molar ratio was applied. In our opinion it is more appropriate to talk in terms of the TEA/para ethyl anisate ratio with lower ratios leading to lower initial polymerization rates. The overall effect of varying these ratios is

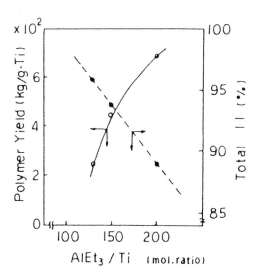

Fig. 1.63: The relationship between TEA/Ti ratio and polymer yield, and TEA/Ti ratio and stereospecificity (polymerization conditions: temperature 65°C; propylene pressure 10 bar; time 2h; [Ti]=0.021 mmol/l; [EB]=0.82 mmol/l) [from reference 159, copied with permission from Plenum Press]

shown in Fig. 1.63; the higher the TEA/Ti (or TEA/ester) ratio, the higher the polymer yield but at the expense of selectivity, or polymer isotactic index.

The cause of the rapid deactivation rate has been the subject of much study, debate and speculation in recent years. It has been proven in several publications that the rapid rate of decay is not associated with monomer diffusion through the growing polymer particle. For example Keii[151] reported that catalyst deactivation occurs when a catalyst is aged (in the presence of a co-catalyst) in the absence of monomer, and the same author with co-workers[150] studied the effect of substituting propylene with nitrogen during polymerization, followed by reintroducing propylene after about 1 hour. They found that the rate of catalyst deactivation during the monomer-less period was the same as that during the polymerization, hence ruling out any diffusion effects. These workers concluded that the decay kinetics were second order and could be attributed to instability ("overreduction" to Ti(II)) of the active centre (Ti(III)). Chien and co-workers[143,160] have also attributed decay to overreduction processes.

Although we recognise that reduction and overreduction of titanium in these supported catalysts can and does occur we do not feel that these processes play the principle role in determining polymerization kinetics. We believe that catalyst deactivation can be attributed to the electron donors used and their chemistry. Esters undergo irreversible chemical reactions with the co-catalyst (tri ethyl aluminium) under polymerization conditions. Such reactions were first studied by Pasynkiewicz et al[93,94], while we found[158] that the reactions could be most conveniently investigated using nuclear magnetic resonance techniques. Described in detail in section 1.5, the reaction between p-ethylanisate and triethyl aluminium is a double alkylation reaction resulting in the formation of a new alkoxy aluminium species, as already shown in Fig. 1.29.

We believe that it is a similar alkylation process that accounts for catalyst decay:

| active site | ester | inactive |

It is known that Ti-alkyls are even more reactive than aluminium alkyls for this type of reaction[97,98].

For aromatic esters, Pasynkiewicz et al[94] have shown that the reactivity can be steered by suitable substitution of the aromatic nucleus; electron-donating substituents such as methyl and methoxy were shown to reduce activity, which explains why alkyl anisates and toluates are preferred to (for example) ethyl benzoate.

A similar explanation for catalyst decay was proposed by Tashiro et al[159] who also suggested that any free ester in solution can complex active centres and reduce polymerization activity. Furthermore these authors showed that the reaction products of aromatic esters with triethyl aluminium (alkoxy aluminium dialkyls) have no effect on the polymerization reaction.

In studying the reaction kinetics various authors have come to differing conclusions regarding the nature of the decay kinetics. Galli et al[161] proposed first order kinetics, while Chien et al[143] described second order kinetics. Keii[151] has proposed second order kinetics but suggested that variations from second order are observed under certain conditions, such as sub-ambient temperatures, during initial polymerization period and after long (>3h) polymerization times. Support for the latter is given by Tait[154] who expressed the second order rate equation as follows:

$$(R_{p,t})^{-1} = (R_{p,0})^{-1} + k_d*t$$

where R_p is the polymerization rate at times t and zero, t is the polymerization time and k_d the second order decay rate constant.

We favour the first order kinetics first described by Galli et al (above) and suggest the following first order rate equation :

$$\ln(R_p - R_{p,\infty}) = k_i - k_d*t$$

where R_p is the polymerization rate, $R_{p,\infty}$ is the residual (usually low) polymerization rate observed after long (> 3 h) polymerization times and k_i initial polymerization rate constant and k_d the first order rate decay constant, and finally t is the polymerization time.

In studies of this type of supported catalyst two distinct phases in the polymerization behaviour can be observed, although the second phase is missed if short (<3 hour) polymerization times are applied. The first phase of the polymerization is characterized by the above-described rapid decay kinetics, while the second phase represents a constant "stationary state" with a residual, relatively low polymerization rate. This behaviour can be recognized in Fig. 1.62. Furthermore it seems that the level of this residual activity or stationary state is determined by the TEA/ester ratio applied; at very low ratios (i.e. high ester concentrations) the activity decays to zero while at higher ratios the level of the stationary state increases proportionally. This again is indicated in Fig. 1.62. Another factor which greatly affects polymerization kinetics is the temperature applied: Higher temperatures result in a higher initial polymerization rate but this is coupled with a much increased rate of decay.

When applying the above-described type of supported catalysts in a custom-designed or very flexible process version, either liquid- or gas-phase, no insurmountable problems are encountered in the manufacture of homopolymer. However, the rapid decay in catalyst activity can have significant consequences for certain types of process. Indeed when applied in a copolymer process the rapid decay becomes an issue of major importance since, as described elsewhere in this book (see chapter 5), such copolymers are invariably manufactured in a two- (or multi-) reactor series configuration; homopolymer being prepared in the first reactor, the live slurry from which is transported to the second reactor where the ethylene/propylene rubber (EPR) is prepared. The combination of the long residence times involved and the rapid decay of catalyst activity can present significant technical problems in the manufacture of copolymers of commercially acceptable impact properties due to the lack of catalyst reactivity in the copolymerization reactor. Therefore there is an incentive to modify the catalyst/co-catalyst package such that flatter kinetics result and consequently the manufacturing of such polymers becomes easier (see chapter 5 for more information about toughened polypropylene).

An example of a catalyst modification which has been succesful in reducing the rate of catalyst deactivation involves the use of alternative electron donors for both the internal and external bases. Recent patents indicate that the combination of a suitable di-ester, particularly a phthalate, and an organosilicon derivative as internal and external base respectively result in catalysts of high initial activity and excellent stereoselectivity combined with a greatly reduced rate of catalyst deactivation. The catalysts are prepared using any of the methods described elsewhere in this chapter, the only difference being that the simple aromatic ester such as EB is substituted by a di-ester such as di-isobutyl phthalate.

Soga and Shiono[33] recently published a detailed study of the polymerization performance of these catalysts, with particular reference to the electron donor effects involved. In this respect the kinetic data illustrated in Fig. 1.64 and Fig. 1.65 is especially relevant. Fig. 1.64 represents data obtained using simply TEA as co-catalyst with no external Lewis base; with the exception of the electron donor-free $TiCl_4/MgCl_2$ catalyst which produced polymer with an isotactic index of 27 %m/m all the catalysts resulted in

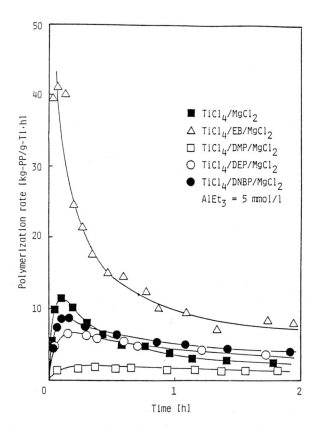

Fig. 1.64: Rate-time profile for the polymerization of propylene in the _absence_ of phenyl triethoxy silane (PTES) [from reference 33, copied with permission from Cambridge University Press]

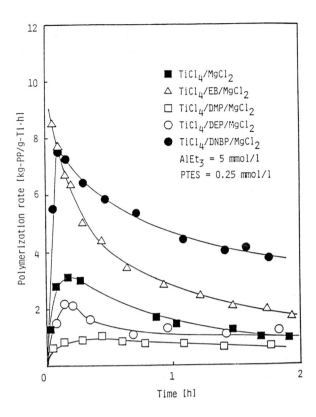

Fig. 1.65: Rate-time profile for the polymerization of propylene in the <u>presence</u> of phenyl triethoxy silane (PTES) [from reference 33, copied with permission from Cambridge University Press]

I.I.'s of around 75 %m/m. It should be noted that the "ethyl benzoate catalyst" possesses far and away the highest activity. The application of phenyl triethoxy silane (PTES) as external base (Fig. 1.65) improved isotacticity to 90% or above except in the case of the internal Lewis base-free $TiCl_4/MgCl_2$ system (I.I. 74%). Notably the polymerization kinetics of the di-n-butyl phthalate (DNBP)-modified system are essentially unaltered by the presence of PTES and are relatively flat (low rate of decay) whereas the EB-modified catalyst experiences a loss in initial rate of a factor of five and still exhibits the typical rapid decay kinetics discussed earlier. These effects are more clearly illustrated in Fig. 1.66 where log(polymerization rate) is plotted against time for both the EB and DNBP catalysts in the presence of and in the absence of the external donor PTES.

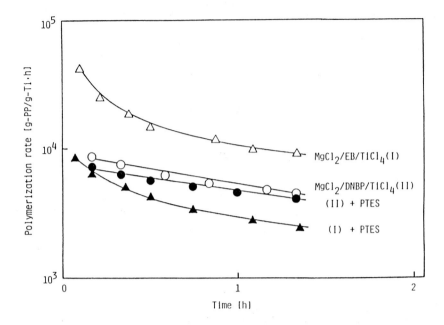

Fig. 1.66: Plots of log(rate) versus time, data ex figures 1.64 and 1.65 [from reference 33, copied with permission from Cambridge University Press]

1.8 PROCESS ASPECTS

In designing a commercial polypropylene unit there are various basic technological aspects which have to be considered:

heat removal

Propylene polymerization is a highly exothermic reaction, producing 2470 kJ of heat per kilogram of propylene polymerized. Therefore cooling of the reactor is a critical design feature even in small-scale laboratory equipment let alone todays world-scale 100+ktpa units.

polymer yield on catalyst

The yield of polypropylene per unit mass of catalyst is important since not only does it determine catalyst costs (per ton of polymer) but more significantly it determines the levels of catalyst residues ("ash", usually titanium, aluminium, chloride and some organics) which must be removed from the product before sale. With sufficiently high yields the deashing step can be obviated, resulting in a substantial saving in both capital expenditure (new plants only) and process costs. Polymer yield is a function of :

-catalyst activity,
-propylene partial pressure/concentration,
-temperature,
-residence time.

product isotacticity

For most applications homopolymer product should have a xylene solubles content of 3 to 4.5 %m/m or an isotactic index of about 95 to 97 %m/m. Indeed for many end uses it is of vital importance that the xylene solubles level is, for example 3.0 %m/m and not 4.5 %m/m (and vice-versa). The most important factor controlling isotacticity is the catalyst employed, although the co-catalyst and selected third components are often chosen to improve, control and steer the product isotacticity. Process variables which can significantly affect the isotacticity include temperature and residence time (see chapter 2). For process economics it is crucial to achieve high catalyst stereoselectivity since otherwise atactic material must be extracted from the product, resulting in increased processing costs as well as reduced propylene efficiency. In this respect it should be pointed out that in early plants 1.10 to 1.15 tons of propylene were consumed per ton polypropylene, whereas in today's monomer-efficient plants 1.01 to 1.02 tons propylene per ton product is routinely achieved, this improvement being largely a result of omitting the atactic-removal step. When it is realized that in modern plants propylene cost accounts for 60 to 70 % of the total polypropylene manufacturing cost it is obvious that propylene efficiency is a vital economic factor.

impurities in feedstreams

All commercial propylene polymerization catalysts are highly susceptible to an array of poisons such as sulphur compounds, carbon monoxide, oxygen and water. Therefore all process streams must be purified of such contaminants before exposure to the catalyst, and the reactor and catalyst vessels and lines must be designed to prevent the ingress of air and moisture.

1.8.1 Types of commercial processes

With the exception of the earliest product development units operated in the 1950's, all commercial plants are run in the continuous mode. Four distinct types of propylene polymerization process can be identified :

    -the slurry process,

    -the liquid propylene process,

    -the gas phase process, and

    -the solution process.

These four process types refer to homopolymer (and the so-called random copolymers containing a few percent ethylene, see chapters 3 and 4) and most of these types has a "block copolymer" (see chapters 5,6 and 7) variant, briefly discussed in chapter 5.

The names given to the various process types are rather trivial and not truly descriptive, for example both the slurry process and the liquid propylene process do operate with a suspension of the growing polymer particles in a liquid. In the former case it is an inert one, in the latter case this is the monomer.

Because of the huge developments in catalyst technology also process technology is changing very rapidly, processes can now be built which are much simpler than those in the past, as unit operations around atactic removal and/or catalyst removal can be left out. These economic aspects have received a lot of attention recently in the literature, a selection of which is given in references 162-166. The summary process descriptions given below are mostly based on these references. The processes which are considered to be the most economic are the liquid propylene slurry process and the gas-phase process, judging by the fact that all recent new plant announcements use one of these technologies.

i) The slurry process

The slurry or diluent process which is still common today, was the first type of commercial polypropylene process. The diluent is invariably an inert hydrocarbon, and the range applied is very wide: from butane to C_{12}'s, most common are heptane or iso octane. The partial pressure of propylene applied is typically 7 to 10 bars, and the polymerization temperature is in the range of 55 to 70 °C, usually 60 to 65 °C. The main reason that this process type was the first to be commercialized is related to the rather poor performance of early catalysts, such that both extensive deashing and atactic removal steps were required in order to give an acceptable product, and the technological difficulties involved with executing these process steps at the higher pressures involved in a liquid propylene medium. Fig. 1.67 gives a block scheme of such a

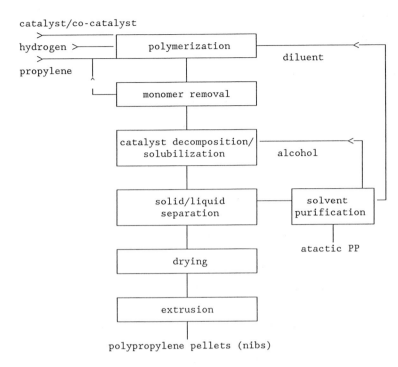

catalyst/co-catalyst

hydrogen >

propylene >

polymerization

diluent

monomer removal

catalyst decomposition/ solubilization

alcohol

solid/liquid separation

solvent purification

atactic PP

drying

extrusion

polypropylene pellets (nibs)

Fig. 1.67: Polypropylene slurry process with deashing and atactic removal

process.

The oldest versions of this process using a low-activity catalyst with mediocre selectivity, had to apply both deashing and atactic removal which makes the process quite complex and expensive. The atactic removal is effected by centrifuging the slurry after destroying and solubilizing the catalyst, i.e. in this step both catalyst remnants and atactic polymer are removed. The diluent is recycled after distillation and drying. Improved catalysts, such as the low temperature reduced ones of Shell, allowed one to skip the atactic removal, still retaining the deashing however.

An essential part of the slurry process is that the polymerization is killed on leaving the main reactor, most frequently by applying an alcohol such as propanol or butanol which also serves to solubilize the titanium and aluminium (in the form of alkoxides) residues and hence to "de-ash" the product. Water is also often applied in this deashing step. Since $TiCl_3$ forms simple Lewis acid-Lewis base adducts with alcohols rather than generating soluble alkoxy species it can be advantageous to add air or other oxidizing agents to oxidize titanium to the more reactive 4 oxidation state. From the foregoing it is apparent that purification and recycle of the reaction diluent is a costly part of these slurry processes, especially since even trace amounts of the deashing materials (alcohols, water, air, etc.) in the recycled polymerization diluent would inhibit the reaction on entering the main reactor.

Another important function of the inert diluent in slurry processes is to ensure adequate cooling via heat transfer to jackets/coils and via its latent heat of evaporation (particularly in the case of the lower boiling alkanes such as butane and hexane).

ii) the liquid propylene process

This type of process is often referred to as the bulk-phase or liquid-pool polymerization process. The process is closely related to the slurry process except that liquefied propylene is used as the reaction medium, a simple change which has a number of important consequences in both design and operation of the plant.

The major advantage of bulk processes is the high monomer concentration (typically 90 to 96 %m/m depending on propylene quality) and the consequent higher reaction rate and polymer yield on catalyst. Another benefit is the low boiling point of propylene which effectively means that in normal operation the monomer is boiling in the reactor, giving a very efficient cooling by virtue of the latent heat of evaporation. The pressure in these bulk polymerization processes varies from about 25 bar at 60 °C to 35 bar at 70 °C, necessitating skilful chemical engineering to address handling problems in those plants requiring both deashing and atactic removal.

Liquid propylene is a much poorer solvent than the higher alkanes used in most slurry processes (see chapter 2). This fact has important consequences for plants where a catalyst with mediocre stereoselectivity is employed because in a liquid propylene medium the atactic polymer formed is nearly insoluble and precipitates out along with the isotactic material; thus atactic removal in a liquid propylene process is a more complicated procedure than the relatively simple centrifugation applied in slurry processes when better solvents (hexane, heptane, etc.) are used as diluent. The removal of atactic really requires a solvent change.

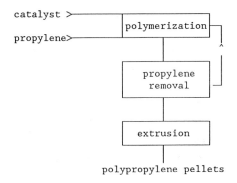

Fig. 1.68: Bulk polypropylene process, non-deashing, non-extracting

The development and implementation of the highly active supported catalysts has resulted in new, simple, energy- and monomer-efficient processes which involve neither deashing nor atactic removal. These new catalysts have already found commercial application in both new, custom-built plants as well as in older, existing plants in which the deashing and atactic extraction trains were by-passed via a so-called retrofit. The flow scheme given in Fig. 1.68 illustrates its simplicity. The Himont spheripol process takes simplification one step further and by employing spherical catalyst particles with a narrow particle size distribution, produces polypropylene spheres which (at least for captive use) require no extrusion step in the manufacturing process.

iii) <u>the gas phase process</u>

The impossibility of deashing polypropylene produced in a solvent-less medium presented a major limitation in the commercialization of gas phase processes until the fairly recent development of super-active catalysts. The first gas phase process to be commercialized was that of BASF which, because a conventional $TiCl_3$ catalyst is used, incorporates a solvent-less dechlorination step in order to achieve acceptable product quality. The BASF gas phase process involves the use of a reactor into which liquid propylene (to ensure adequate cooling) and catalyst are introduced at the bottom and co-catalyst injected at the top. Unreacted propylene exits the reactor as vapour which is condensed before returning to the reactor.

Amoco developed a gas phase process employing a rather unusual horizontal stirred reactor. Again the heat of reaction is removed via evaporation of liquid propylene (or a suitable low boiling alkane) and a conventional TiCl$_3$ catalyst is employed; the relatively low polymer yield on catalyst has serious consequences for product quality in the absence of a deashing step. The reactor is stirred using a series of paddles.

A newer version is the Unipol process (from Union Carbide and Shell Chemical US) which through the use of a highly active and highly selective catalyst results in a simple process[165-167] (see Fig. 1.69). No deashing or atactic removal is necessary with the

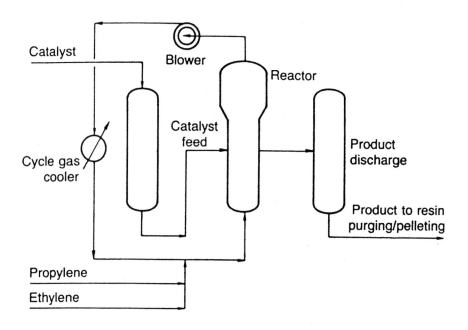

Fig. 1.69: The Unipol process, reaction line for homopolymers and random copolymers

employed catalyst - as in the other modern processes. The reactor is of the fluidized bed type. The catalyst components are fed directly into the reactor and fluidization of the polymer bed is achieved by means of a blower which circulates propylene vapour via an external cooler (to remove the heat of polymerization) and ensures a homogeneous mixing of the polymer bed and the catalyst components.

(iv) the solution process

The solution process for polypropylene is unique to Eastman Kodak. In this process the catalyst (of the $TiCl_3$ type), propylene and a diluent are fed into the reactor which is run at a high temperature (175-250 °C) so that the polypropylene formed remains in solution. The benefits of this process include the fact that the heat of polymerization at these high temperatures can be efficiently utilized elsewhere in the plant and that the problems of catalyst and polymer morphology control are overcome since the polymer is in solution. However the major problems are the limitations posed by the high viscositiy of the polymer solution. Furthermore the rapid activity decay (in particular the newer supported catalysts) at these high temperatures limits the choice of catalyst which can be utilized.

Catalyst residues are removed by filtration, although the very fine nature of the catalyst particles is thought to necessitate dilution of the polymer solution and filtration through Kieselguhr or similar filter-aids. An atactic polymer-removal step is also applied.

This process represents the most expensive of all commercial processes due to the above described complexities.

REFERENCES
1.   J.J.Eisch, Karl Ziegler: Master Advocate of the Unity of Pure and Applied Chemistry, J.Chem.Educ, 60 (1983) 1009-14
2.   F.M.McMillan, The Chain Straighteners, McMillan Press, London, 1979.
3.   P.Pino and G.Moretti, The Impact of the Discovery of the Polymerization of α-olefins on the Development of the Stereospecific Polymerization of Vinyl-monomers, Polymer, 28 (1987) 683
4.   F.A.Zenz and D.F.Othmer, Fluidization and Fluid Particle Systems, Reinhold, New York, 1960.
5.   I.G.Carson, Control of particle size of polyolefins by Ziegler-Natta catalysts, paper presented at symposium "Control of particle size", London, 23 november 1976
6.   P.Mackie, M.N.Berger, B.M.Grieveson and D.Lawson, Replication in Ziegler Polymerization, J.Polym.Sci., B, Pol.Letters, 5 (1967) 493-4
7.   P.Galli, High Yield Catalyst in Olefins Polymerization: General Outlook on Theoretical Aspects and Industrial Uses, paper presented at the 28th IUPAC Macromolecular Symposium, July 12-16 (1982), Amherst, Mass.
8.   V.W.Buls and T.L.Higgins, Uniform Site Theory, J.Polym.Sci., A1, 8 (1970) 1025-35
9.   M.Kakugo, M.Sadatoshi, J.Sakai and M.Yokoyama, Growth of polypropylene particles in heterogeneous Ziegler-Natta polymerization, Macromolecules, 22 (1989) 3172-3177
10. J.Boor,Jr, Ziegler-Natta Catalysts and Polymerizations, Academic Press, New York, 1979.
11. - P.Cossee, Tetrahedron Letters, 17 (1960) 12

- P.Cossee, Proc.Int.Congr.Coord.Chem., 6th (1961) 241
- P.Cossee, *J.Catal.*, *3* (1964) 80
- E.J.Arlman, P.Cossee, *J.Catal.*, *3* (1964) 99
- E.J.Arlman, *J.Catal.*, *3* (1964) 89
- E.J.Arlman, *J.Catal.*, *5* (1966) 178
- E.J.Arlman, Anion Polarization and Lattice Energy of Some Metal Halides MX2 and MX3 with Layer and Lattice Structure, *Rec. Trav. Chim. Pays-Bas*, *87* (1968) 1217-35
- P.Cossee, Calculation of the Chain Growth During Ziegler-Natta Polymerization Using the Molecular Orbital Method, Osn. Predvideniya Katal. Deistviya, Tr. Mezhdunar. Kongr. Katal., Meeting date 1968, Vol. 1, 184-92.
- P.Cossee, Molecular Orbital Calculations on the Propagation Step in Ziegler-Natta Polymerization Using the Molecular Orbital Method, Proc. Int. Congr. Catal., Meeting date 1968, Vol. 1, 207-17.

12. S.Mostert, Thesis, T.H.Twente, 1967.
13. Brit. Patent, 1 390 355 (1971) to Shell
14. E.G.M.Tornqvist, *Annals of N.Y.Acad.Sci.*, *155* (1969) 447
15. G.Natta, P.Corradini and G.Allegra, The different crystalline forms of TiCl3, a catalyst component for the polymerization of α-olefins, I: α-, β-, γ-TiCl3, II: δ-TiCl3, *J.Pol.Sc.*, *51* (1961) 399-410
16. E.J.Arlman, unpublished results.
17. A.D.Caunt, Role of Alkylaluminium Chlorides in Polymerization of Propylene with Titanium Chloride Catalysts, *J.Polym.Sci.*, C, *4* (1963) 49
18. U.S.Patent, 4 210 738 (1980) to Solvay.
19. R.P.Nielsen, Active Center Generation in the Solvay-Type High Performance Titanium Trichloride Catalyst - An Interpretation from the Hermans-Henrioulle Patent, in Transition Metal Catalyzed Polymerizations: Alkenes and Dienes, R.P.Quirk, Ed., Harwood Academic Publishers, New York, 1983.
20. British Patents 1 579 725 and 2 002 654 to Shell.
21. British Patent 1 502 883 to Mitsubishi Chemical.
22. Brit. Patent 841 822 (1955) to Petrochemicals Ltd.
23. Brit. Patent 904 510 (1960) to Shell.
24. Brit. Patent 1 271 411 (1968) to Mitsui Petrochemical.
25. Brit. Patent 1 335 887 (1970) to Montedison.
26. C.W.Kamienski, Ph.D.Thesis, University of Tennessee (1967).
27. Brit. Patent 955 807 (1959).
28. Brit. Patent 1 387 890 (1971) to Montedison.
29. F.W.Locher and H.M. v.Seebach, Influence of Adsorption on Industrial Grinding, *Ind.Eng.Chem.*, *Process Res.Dev.*, *11* (1972) 190
30. X.Youchang, G.Linlin, L.Wangi, B.Naiyu and T.Yougi, *Sci.Sin.*, *22* (1979) 1045
31. H.Sinn, W.Kaminsky, H.J.Vollmer and R.Woldt, Living Polymers with Ziegler Catalysts of High Productivity, *Angew.Chem.*, *Int.Ed.Engl.*, *19* (1980) 390
32. Eur.Patents 45 977 to Montedison, 125 911 to Mitsui Petrochemical, 131 359 to ICI, 171 179 to Amoco, 197 310 to Mitsubishi Petrochemical, etc.
33. K.Soga and T.Shiono, Effect of Diesters and Organosilicon Compounds on the Stability and Stereospecificity of Ziegler-Natta Catalysts, in Transition Metal Catalyzed Polymerizations: Ziegler- Natta and Metathesis Polymerizations, R. P. Quirk, Ed., Cambridge University Press, Cambridge, 1988.
34. Eur. Patent 86 288 (1982) to Mitsui Petrochemical.
35. - D.G.H.Ballard, Pi and Sigma Transition Metal Carbon Compounds as Catalysts for the Polymerization of Vinyl Monomers and Olefins, *Adv.Catal.*, *23* (1973) 263-325;
- D.G.H.Ballard and P.W. van Lienden, Polymerization of Vinyl Monomers by Transition Metal Benzyl Compounds. I. Initial Observations, *Makrom.Chem.*, *154* (1972) 177-190;
- D.G.H.Ballard, E.Jones, R.J.Wyatt, R.T.Murray and P.A.Robinson, Highly Active Polymerization Catalysts of Long Life Derived from Sigma and

Pi Bonded Metal Alkyl Compounds, *Polymer, 15 (1974) 169-174*;
- D.G.H.Ballard, Transition Metal Alkyl Compounds as Polymerization Catalysts, *J.Polym.Sci., Polym.Chem.Ed., 13 (1975) 2191-2120*;
- D.G.H.Ballard, Transition Metal Alkyl Polymerization Catalysts, in Coordination Polymerization, J.C.W. Chien, Ed., Academic Press, New York (1975).

36. - U.Giannini and U.Zucchini, Benzyltitanium Compounds, *Chem.Commun. (1968) 940*;
- U.Giannini, U.Zucchini and E.Albizzati, Polymerization of Olefins with Benzyl Derivatives of Titanium and of Zirconium, *J.Polym.Sci, Part B, 8 (1970) 405-410*;
- U.Giannini, E.Albizzati and U.Zucchini, Synthesis and Properties of Some Titanium and Zirconium Benzyl Derivatives, *J.Organometal.Chem., 26 (1971) 357-372*

37. K.Soga, K.Izumi, S.Ikeda and T.Keii, Polymerization of propylene with tetrabenzylzirconium, *Makrom.Chem., 178 (1977) 337-342*

38. A.Andresen, H.G.Cordes, J.Herwig, W.Kaminsky, A.Merk, R.Mottweiler, J.H.Sinn and H-J.Vollmer, Halogen-Free Soluble Ziegler Catalysts for Ethylene Polymerization. Control of Molecular Weight by the Choice of the Reaction Temperature, *Angew.Chem., Int.Ed.Engl., 15 (1976) 630*

39. - W.Kaminsky, Stereocontrol with New Soluble Catalysts in Transition Metal Catalyzed Polymerizations: Ziegler-Natta and Metathesis Polymerizations, R.P.Quirk, Ed., Cambridge University Press, Cambridge, 1988;
- European Patent 69951 to Hoechst.

40. US Patent 4 544 762 to Hoechst.

41. W.Kaminsky, Polymerization and Copolymerization with a Highly Active, Soluble Ziegler-Natta Catalyst in Transition metal catalyzed polymerizations: Alkenes and dienes, R.P. Quirk, Ed., Harwood Academic Publishers, New York, 1983.

42. H.Sinn and W.Kaminsky, Ziegler-Natta Catalysis, *Adv.Organometal.Chem., 18 (1980) 99*

43. P.Pino and R.Mulhaupt, Stereospecific polymerization of propylene : An outlook 25 years after its discovery, *Angew.Chem., Int.Ed.Engl., 19 (1980) 857-875*

44. J.A.Ewen, Mechanisms of Stereochemical Control in Propylene Polymerizations with Soluble Group 4B Metallocene/Methylaluminoxane Catalysts, *J.Am.Chem. Soc., 106 (1984) 6355*

45. A.Zambelli, P.Locatelli, G.Zannoni and F.A.Bovey, Stereoregulation Energies in Propene Polymerization, *Macromolecules, 11 (1978) 923*

46. A.Zambelli, P.Locatelli, A.Provasoli and D.R.Ferro, Correlation between Carbon-13 NMR Chemical Shifts and Conformation of Polymers. 3. Hexad Sequence Assignments of Methylene Spectra of Polypropylene, *Macromolecules, 13 (1980) 267*

47. A.Zambelli, G.Natta and I.Pasquon, Polymérisation du propylène à polymère syndiotactique, *J.Pol.Sci., Part C, 4 (1964) 411-426*

48. F.R.W.P.Wild, L.Zsolnai, G.Huttner and H.H.Brintzinger, ansa-Metallocene Derivatives. VII. Synthesis and Crystal Structure of a Chiral ansa-Zirconocene Derivative with Ethylene-bridged Tetrahydroindenyl Ligands, *J.Organometal.Chem., 232 (1982) 233*

49. W.Kaminsky, K.Külper, H.H.Brintzinger and F.R.W.P.Wild, Polymerization of Propene and Butene with a Chiral Zirconocene and Methylaluminoxane as Co-Catalyst, *Angew.Chem., Int.Ed.Engl., 24 (1985) 507*

50. German Patent 3 443 087 to Hoechst.

51. - W.Kaminsky, K.Külper and S.D.Niedoba, Olefin Polymerization with Highly Active Soluble Zirconium Compounds using Aluminoxane as Co-Catalyst, *Makrom.Chem., Makrom.Symp., 3 (1986) 377*;
- W.Kaminsky in Catalytic Polymerization of Olefins, K.Soga and T.Keii editors, Elsevier, Amsterdam, 1986.

52. J.C.W.Chien and B-P.Wang, Metallocene-Methaluminoxane Catalysts for Olefin Polymerization. I. Trimethylaluminum as Coactivator, *J.Polym.Sci., Part A, Polym. Chem., 26 (1988) 3089-3102*

53. - J.A.Ewen, L.Haspeslagh, J.L.Atwood and H.Zhang, "Crystal Structures and Stereospecific Polymerizations with Chiral Hafnium Catalysts", *J.Am.Chem. Soc., 109 (1987) 6544-6545*;
     - Eur. Patent 284 707 to Cosden Technology (1987).
54. Jap. Patent Applications 87/45151 (1987); CA110:135913k, 87/31927; CA110:24451p both to Mitsui Petrochemical.
55. Jap. Patent Appl. 87/114409 to Mitsui Toatsu (1987); CA110:115535k.
56. Eur. Patent EP 279 586 (1987) to Mitsui Petrochemical.
57. W.Kaminsky and M.Miri, Ethylene Propylene Diene Terpolymers Produced with a Homogeneous and Highly Active Zirconium Catalyst, *J.Polym.Sci., Polym. Chem.Ed., 23 (1985) 2151-2164*
58. W.Kaminsky and M.Schlobohm, Elastomers by Atactic Linkage of α-Olefins using Soluble Ziegler Catalysts, *Makrom.Chem., Macromol.Symp., 4 (1986) 103*
59. PCT Int. Appl. WO 88/4674 (1986), WO 88/4672 (1986), Eur. Patent EP 260 130 (1986), EP 260 999 (1986), EP 226 463 (1985), EP 206 794 (1985), EP 208 561 (1985), US 4 522 982 (1985), EP 128 045 (1983), EP 128 046 (1983) and EP 129 368 (1983) all to Exxon.
60. Jap. Patent Applications 86/240012 (1986); CA109:93842h and 86/182986 (1986); CA109:74113e both to Mitsubishi Petrochemical.
61. PCT Int. Appl. WO 88/3932 (1986), WO 88/2378 (1986), WO 88/5057 (1986), WO 88/1626 (1986), Jap. Patent Applications 86/210549 (1986); CA109:111082s, 86/201838 (1986); CA109:55476j and 86/51407 (1986); CA109:55441u all to Mitsui Petrochemical.
62. T.Tsutsui and N.Kashiwa, Kinetic Study on Ethylene Polymerization with bis(Cyclopentadienyl)Zirconium Dichloride Cp2ZrCl2/Methylaluminoxane Catalyst System, *Polym.Commun., 29 (1988) 180-3*
63. W.Kaminsky, A.Bark and R.Spiehl, Isotactic Polymerization of Cyclic Olefins with Homogeneous Ziegler-Natta Catalysts, Paper presented at New Frontiers in Ziegler-Natta Catalyst Chemistry, ACS, 44th Southwest Regional Meeting, November 30-December 2 1988, Corpus Christi, Texas.
64. Eur. Patent 210 615 (1985) and US Patent 4 680 353 (1985) both to Idemitsu Kosan.
65. Eur. Patent 268 214 (1986) to Idemitsu.
66. - J.A.Ewen, R.L.Jones, A.Razavi and J.D.Ferrara, Syndiospecific Propylene Polymerization with Group IVB Metallocenes, *J.Am.Chem.Soc., 110 (1988) 6255-6256*;
     - J.A.Ewen, Syndiotactic and Isotactic Specific Propylene Polymerizations with Group 4 Metallocene Based Catalysts, Paper presented at New Frontiers in Ziegler-Natta Catalyst Chemistry, ACS Regional Meeting, Corpus Christi, Texas, November 30 - December 2 (1988).
67. Y.Doi, S.Suzuki, G.Hizai and K.Soga, Living Coordination Polymerization of Propylene and Synthesis of Tailor-Made Polymers, in Transition Metal Catalyzed Polymerizations: Ziegler-Natta and Metathesis Polymerizations, R.P.Quirk, Ed., Cambridge University Press, Cambridge, 1988.
68. A.Antberg, L.L.Böhm, V.Dolle, H.Luker, J.Rohrman, W.Spaleck and A.Winter, Stereorigid Metallocenes: Correlation Between Structure and Behaviour in 1-Olefin Polymerization, Paper presented at New Frontiers in Ziegler-Natta Catalyst Chemistry, ACS Regional Meeting, Corpus Christi, Texas, November 30 - December 2 (1988).
69. R.F.Jordan, R.E.LaPointe, P.K.Bradley and D.F.Taylor, Chemistry of Cp2Zr(R)+ Olefin Polymerization Catalysts, Paper presented at New Frontiers in Ziegler-Natta Catalyst Chemistry, ACS Regional Meeting, Corpus Christi, Texas, November 30 - December 2 (1988).
70. - R.F.Jordan, Cationic Metal-Alkyl Olefin Polymerization Catalysts, *J. Chem.Educ., 65 (1988) 285-289*;
     - R.F.Jordan, R.E.LaPointe, C.S.Bajgur, F.S.Echols and R.Willett, Chemistry of Cationic Zirconium (IV) Benzyl Complexes. One-Electron Oxidation of d^0 Organometallics, *J.Am.Chem.Soc., 109 (1987) 4111-4113*;
     - R.F.Jordan, C.S.Bajgur, R.Willett and B.Scott, Ethylene Polymerization by a Cationic Dicyclopentadienyl Zirconium (IV) Complex, *J.Am.Chem.Soc., 108*

(1986) 7410-7411

71. J.J.Eisch, A.M.Piotrowski, F.K.Brownstein, E.J.Gabe and F.L.Lee, Direct observation of the initial insertion of an unsaturated hydrocarbon into the titanium-carbon bond of the soluble Ziegler-polymerization catalyst, Cp2TiCl2-MeAlCl2, *J.Am.Chem.Soc.*, *107 (1985) 7219*

72. - F.S.Dyachkovskii, *Vys.Mol.Soed.*, *7 (1965) 114*;
    - F.S.Dyachkovskii, A.K.Shilova and A.E.Shilov, The role of free ions in reactions of olefins with soluble complex catalysts, *J.Polym.Sci.*, *Part C., 16 (1967) 2333-2339*

73. E.A.Grigorian, F.S.Dyachkovskii, G.M.Khvostik and A.E.Shilov, *Vys.Mol.Soed. A-9 (1967) 1233*

74. G.G.Hlatky, H.W.Turner and R.R.Eckman, Ionic, Base-Free Zirconocene Catalysts for Ethylene Polymerization, *J.Am.Chem.Soc.*, *111 (1989) 2728-2729*

75. Eur. Patents 277 003 and 277 004 to Exxon (1987).

76. Research Disclosure 29 236, anonymous (August 1988).

77. G.Natta, Kinetic study of α-olefin polymerization, *J.Polym.Sci.*, *34 (1959) 21-48*

78. J.Boor, Ziegler polymerization of olefins. I. Dependence on structure of metal alkyl and the transition metal compound, *J.Polym.Sci.*, *C, 1 (1963) 237-255*

79. A.P.Firsov et al, The dependence of the stereospecific action of the complex catalyst α-TiCl3-Me(C2H5)n during the polymerization of α-olefins on the metal in the metalorganic compound, *J.Polym.Sci.*, *62 (1962) S104-S105*

80. - K.Soga, Evaluation of Olefin Reactivity Ratios over Highly Isospecific Ziegler-Natta Catalyst, *Makrom.Chem.*, *190 (1989) 37-44*;
    - K.Soga and H.Yanagihara, Synthesis of highly isotactic polystyrene by using Solvay-type TiCl3 in combination with di-cyclopentadienyl dimethyl titanium, *Makrom.Chem.*, *Rapid Commun.*, *9 (1988) 23-25*

81. K.Soga, T.Uozumi and H.Yanagihara, Propene Polymerization with MgCl2-Supported Transition Metal Catalysts Activated by Cp2MtMe2 (Mt = Ti, V, Zr, Hf), *Makrom.Chem.*, *190 (1989) 31-35*

82. K.Soga, T.Uozumi and T.Shiono, Highly isospecific catalysts for propene polymerization composed of TiCl4/MgCl2 and Cp2TiMe2 without donors, *Makrom.Chem.*, *Rapid Commun.*, *10 (1989) 293-297*

83. G.Natta, I.Pasquon, A.Zambelli and G.Gatti, Highly stereospecific catalytic systems for the polymerization of α-olefins to isotactic polymers, *J.Polym.Sci.*, *51 (1961) 387-398*

84. F.Danusso, Recent Results of Stereospecific Polymerization by Heterogeneous Catalysis, *J.Polym.Sci.*, *C, 4 (1964) 1497*

85. T.Keii in Coordination Polymerization editor J.C.W. Chien, Academic Press, New York, 1975, 263.

86. J.C.W.Chien, Kinetics of Propylene Polymerization Catalyzed by α-Titanium Trichloride - Diethylaluminium Chloride, *J.Polym.Sci.*, *A, 1 (1963) 425*

87. H.Schnecko, W.Lintz and W.Kern, Polymerization with heterogeneous metalorganic catalysts. VI. Difference in polymerization activity of α-olefins and some kinetic results on butene-1 polymerization, *J.Pol.Sci.*, *A-1, 5 (1967) 205-214*

88. L.Kollar, A.Simon and J.Osvath, Ziegler-Natta catalysts. II. The liquid phase and the polymerization activity of the catalyst, *Magy.Kem.Foly.*, *74 (1968) 284-287*; CA *69* : 52497m

89. Eur. Patent 21 478 (1979) to Shell.

90. T.Mole and E.A.Jeffrey, Organoaluminium Compounds, Elsevier, Amsterdam, 1972.

91. B.L.Goodall, Olefin Dimerization with a Homogeneous Titanium Catalyst and Polymerization Studies with Sterically Hindered Aluminum Co-Catalysts - Spin-Offs in the Development of Supported Catalysts for Propylene Polymerization, in Transition Metals and Organometallics as Catalysts for Olefin Polymerization, W.Kaminsky and H.Sinn Eds., Springer-Verlag Berlin (1988) 361-71.

130

92. Eur.Patent 107 871 (1982) to Shell.
93. K.B.Starowieyski, S.Pasynkiewicz, A.Sporzynski and K.Wisniewska, Reactions of Methyl Benzoate with Methylaluminium Compounds, *J. Organometal.Chem.*, *117* (1976) C1-C3
94. S.Pasynkiewicz, L.Kozerski and B.Grabowski, Reaction of triethylaluminium with esters, *J.Organometal.Chem.*, *8* (1967) 233-238
95. A.W.Langer, T.J.Burkhardt and J.J.Steger, Chemistry of Supported Catalysts for Polypropylene, *Polym.Sci.Technol.*, *19* (1983) 225
96. Y.Baba, The reaction of triethylaluminium with esters, *Bull.Chem.Soc.Japan*, *41* (1968) 1022-1023
97. D.G.H.Ballard, W.H.Janes and T.Medinger, Polymerization of vinyl monomers by transition metal allyl compounds, Part I, *J.Chem.Soc. B.*, *(1968) 1168-1175*
98. Brit.Patent, 858 541 (1957) to Hoechst.
99. Ger.Patent 2 355 886 (1972), Brit.Patent 1 502 886 (1974), Ger.Patent 2 605 922 (1975) all to Mitsui Petrochemical.
100. Ger.Patent 2504 036 (1974) to Mitsui Petrochemical.
101. US Patent 4 146 502 (1976) to Mitsubishi Petrochemical.
102. Ger. Patent 3 040 967 (1979) to Denki.
103. Jap.Patents 8 122 302 (1979), 8126 902 (1979), 8 170 003 (1979) and 8 170 004 (1979) all to Denki.
104. A.K.Zefirova and A.E.Shilov, *Dokl.A.N.SSSR*, *136* (1961) 599
105. K.Soga and H.Yanagihara, Stereospecific polymerization of propene using MgCl2 supported Ti-catalyst combined with various alkyltitanium compounds, *Makrom.Chem.*, *Rapid Commun.*, *8* (1987) 273
106. US Patent 4 145 313 (1977), Eur.Patent 4 739 (1978), US Patent 4 148 756 (1978) and US Patent 4 215 014 (1979) all to Exxon.
107. Ger.Patent 2 633 195 (1975) to Phillips.
108. Eur.Patent 30 742 (1979) to Toyo Stauffer.
109. - G.Henrici-Olivé and S.Olivé, *Adv.Polymer Sci.*, *6* (1969) 421; *J.Pol.Sci.*, *part C 22* (1969) 965; *Makrom.Chem. 121* (1969) 70; *Angew.Chem.*, *Int.Ed.Engl.*, *7* (1968) 821 and *J.Pol.Sci.*, *B*, *8* (1970) 271
110. J.C.W.Chien, Olefin Polymerizations and Polyolefin Molecular Weight Distribution, *J.Polym.Sci.*, *A1* (1963) 1839
111. L.Westerman, The Molecular Weight Distribution of Polypropylene, *J.Polym. Sci.*, *A1* (1963) 411
112. R.L.Combs, D.F.Slonaker, F.B.Joyner and H.W.Coover Jr, Influence of preparative conditions on molecular weight and stereoregularity distributions of polypropylene, *J.Polym.Sci.*, *A*, *5* (1967) 215-226
113. W.R.Schmeal and J.R.Street, Polymerization in catalyst particles: Calculation of molecular weight distribution, *J.Polym.Sci.*, *Polym.Phys.*, *10* (1972) 2173-2187
114. S.Tanaka and H.Morikawa, Average lifetime of growing chains in propylene polymerization, *J.Polym.Sci.*, *A*, *3* (1965) 3147-3156
115. C.Tzoganakis et al, Effect of molecular weight distribution on the rheological and mechanical properties of polypropylene, *Pol.Eng.Sc.*, *29* (1989) 390-396
116. H.J.M.A.Mieras and C.F.H.van Rijn, Influence of molecular weight distribution on the elasticity and processing properties of polypropylene melts, *J.Appl.Pol.Sc.*, *13* (1969) 309-322
117. N.Kashiwa and J.Yoshitake, The number of active centres in the propene polymerization with MgCl2/TiCl4/C6H5COOC2H5 - Al(C2H5)3/C6H5COOC2H5 catalyst, *Makrom.Chem, Rapid Commun.*, *3* (1982) 211-214
118. E.Suzuki, M.Tamura, Y.Doi and T.Keii, Molecular weight during polymerization of propene with the supported catalyst system TiCl4/MgCl2/C6H5COOC2H5/Al(C2H5)3, *Makrom.Chem. 180* (1979) 2235-2239
119. U.Zucchini and G.Cecchin, Control of molecular weight distribution in polyolefins synthesized with Ziegler-Natta catalytic systems, *Adv.Polym. Sci.*, *51* (1983) 101
120. Y.Doi, Structure and stereochemistry of atactic polypropylenes. Statistical model of chain propagation, *Makrom.Chem, Rapid Commun.*, *3* (1982) 635-641

121. T.Keii, Kinetics of Ziegler-Natta Polymerization, Kodansha, Tokyo, 1972.

122. G.Natta and I.Pasquon, The kinetics of the stereospecific polymerization of α-olefins, Adv.Catal., IX (1959) 1-66

123. P.J.T.Tait , p 85-112 in Advances in the preparation and properties of stereoregular polymers, R.W.Lenz and I.Ciardelli editors, Kluwer Academic, 1980.

124. G.Allegra, Discussion on the mechanism of polymerization of α-olefins with Ziegler-Natta catalysts, Makrom.Chem. 145 (1971) 235-246

125. P.Patat and H.Sinn, Zum Ablauf der Niederdruckpolymerization der α-olefine, Komplexpolymerization 1, Angew.Chem, 70 (1958) 496-500

126. G.Natta and G.Mazzanti, Organometallic complexes as catalysts in ionic polymerizations, Tetrahedron, 8 (1960) 86-100

127. L.A.M.Rodriguez and H.M. van Looy, Studies on Ziegler-Natta catalysts. Part V. Stereospecificity of the active center, J.Polym.Sci., A1, 4 (1966) 1971-1992

128. K.J.Ivin, J.J.Rooney, C.D.Stewart, M.L.H.Green and R.Mahtab, Mechanism for the stereospecific polymerization of olefins by Ziegler-Natta catalysts, J. Chem.Soc., Chem. Commun., 1978, 604-606

129. J.D.Fellman, G.A.Rupprecht and R.R.Schrock, Rapid selective dimerization of ethylene to 1-butene by a tantalum catalyst and a new mechanism for ethylene oligomerization, J.Am.Chem.Soc., 101 (1979) 5099-5101

130. F.N.Tebbe, G.W.Parshall and G.S.Reddy, Olefin homologation with titanium methylene compounds, J.Am.Chem.Soc., 100 (1978) 3611-3613

131. R.J.McKinney, Ziegler-Natta catalysis : an alternative mechanism involving metallocycles, J.Chem.Soc., Chem.Commun., 1980, 490-492

132. A.Zambelli, P.Locatelli, M.C.Sacchi and E.Rigamonti, Stereoregular End Groups of Isotactic Polypropylene: A Challenging Test for the Reaction Mechanism, Macromolecules, 13 (1980) 798

133. D.R.Burfield and P.J.T.Tait, Ziegler-Natta catalysis : 3. Active center determination, Polymer, 13 (1972) 315-320

134. V.A.Zakharov, G.D.Bukatov, N.B.Chumaevski and Yu.I.Yermakov, Study of the mechanism of propagation and transfer reactions in the polymerization of olefins by Ziegler-Natta catalysts. 4. The kinetic scheme for propagation and chain transfer reactions, Makrom.Chem., 178 (1977) 967-980

135. L.L.Böhm, Reaction model for Ziegler-Natta polymerization processes, Polymer, 19 (1978) 545

136. A.Zambelli, G.Gatti, C.Sacchi, W.O.Cram Jr and J.D.Roberts, Nuclear magnetic resonance spectroscopy. Steric control in α-olefin polymerization as determined by 13C spectra, Macromolecules, 4 (1971) 475-477

137. W.O.Cram Jr., J.D. Roberts and A.Zambelli, Carbon-13 magnetic resonance spectra of some polypropylenes and ethylene-propylene copolymers, Macromolecules, 4 (1971) 330-332

138. A.Zambelli and G.Allegra, Reaction Mechanism for Syndiotactic Specific Polymerization of Propene, Macromolecules, 13 (1980) 42

139. F.Ciardelli, C.Carlini and G.Montagnoli, Stereochemical aspects of the copolymerization of asymmetric α-olefins by stereospecific catalysts, Macromolecules, 2 (1969) 296-301

140. C.Carlini, A.Altomare, F.Menconi and F.Ciardelli, Stereoselective copolymerization of chiral α-olefins by high-activity Ziegler-Natta catalysts, Macromolecules, 20 (1987) 464-465

141. - P.Corradini, V.Barone, R.Fusco and G.Guerra, Analysis of models for the Ziegler-Natta stereospecific polymerization on the basis of non-bonded interactions at the catalytic site. I. The Cossée model, Europ.Pol.J., 15 (1979) 1133-1141
    - ditto, Steric control in Ziegler-Natta catalysts : An analysis of non-bonded interactions at model catalytic sites, J.Catal., 77 (1982) 32-42
    - P.Corradini, G.Guerra and V.Barone, Conformational analysis of polypropylene chains bound to model catalytic sites, Europ.Polym.J., 20 (1984) 1177-1182

142. Jap. Patent 78 118 290 to Asahi Chemical (1977); US Patent 4 130 503 to Phillips (1977); Europ. Patent 10 746 to Montedison (1978).

132

143. J.W.Chien and J-C.Wu, Magnesium-chloride-supported high-mileage catalysts for olefin polymerization. III. Electron paramagnetic resonance studies, *J.Polym.Sci., Polym.Chem.*, *29* (1982) 2461-2476

144. N.Kashiwa and J.Yoshitake, The influence of the valence state of titanium in MgCl2-supported titanium catalysts on olefin polymerization, *Makromol.Chem*, *185* (1984) 1133-1138

145. N.Kashiwa, The Role of Ester in High-Activity and High-Stereoselectivity Catalyst, in Transition metal catalyzed polymerizations :Alkenes and Dienes, R.P.Quirk, Ed., Harwood Academic Publishers, New York, 1983.

146. P.Pino and R.Mulhaupt, Stereospecific Polymerization of Propylene: An Outlook 25 Years after its Discovery, in Transition metal catalyzed polymerizations :Alkenes and Dienes, R.P. Quirk, Ed., Harwood Academic Publishers, New York, 1983.

147. A.W.Langer Jr., T.J.Burkhardt and J.J.Steger, Chemistry of supported catalysts for polypropylene, paper presented at the 28th IUPAC Macromol. Symp. Amherst, Mass., 1982.

148. P.C.Barbé, G.Cecchin and L.Noristi, The catalytic system Ti-complex/MgCl2, *Adv.Polym.Sci.*, *81* (1986) 1

149. F.Ciardelli, C.Carlini, A.Altomare and F.Menconi, The role of 'internal' and 'external' Lewis bases in controlling the stereochemistry of highly active Ziegler-Natta catalysts, *J.Chem.Soc.,Chem. Commun.*, (1987) 94-95

150. Y.Doi, M.Murata, K.Yano and T.Keii, Gas-Phase Polymerization of Propene with the Supported Catalyst: TiCl4/MgCl2/C6H5COOC2H5/ Al(C2H5)3, *Ind.Eng. Chem; Prod.Res.Dev.*, *21* (1982) 580

151. T.Keii et al, Propene polymerization with a magnesium chloride-supported Ziegler catalyst. 1. Principal kinetics, *Makrom.Chem.*, *183* (1982) 2285-2304

152. R.Spitz, J.L.Lacombe and A.Guyot, Catalysateurs Ziegler-Natta supportés sur MgCl2 pour la polymérisation stéréorégulée du propène. III. Systèmes catalytiques à hautes performances comprenant un solide ternaire MgCl2, ester aromatique, TiCl4, *J.Polym.Sci., Polym.Chem.*, *22* (1984) 2641-2650

153. W.Cooper in Comprehensive Chemical Kinetics, C.H.Bamford and C.Tipper, Eds., Elsevier, Amsterdam, 1976.

154. P.J.T.Tait, Chain reaction polymerization. Part I. Coordination complex polymerization, *Macromol.Chem. (London)*, *1* (1980) 3-21

155. E.M.J.Pijpers and B.C.Roest, The Effect of Hydrogen on the Ziegler-Natta Polymerization of 4-Methyl-1-Pentene, *Europ.Pol.J.*, *8* (1972) 1151-58

156. O.N.Pirogov and N.M.Chirkov, Reactivity of Hydrogen as a Termination Agent during Polymerization of Propylene with Catalytic System C5H5N-TiCl3-Et3Al, *Vysokomol.Soedin.*, *8* (1966) 1798-1803

157. G.Guastalla and U.Giannini, The Influence of Hydrogen on the Polymerization of Propylene and Ethylene with a MgCl2 Supported Catalyst, *Makrom.Chem. Rapid Commun.*, *4* (1983) 519-27

158. B.L.Goodall, Super High Activity Supported Catalysts for the Stereospecific Polymerization of α-Olefins: History, Development, Mechanistic Aspects, and Characterization, in Transition metal catalyzed polymerizations, R.P.Quirk, Ed., Harwood Scientific Publishers, London, 1983.

159. K.Tashiro, M.Yokoyama, T.Sugano and K.Kato, *Contemporary Topics in Polymer Science*, *4* (1984) 647

160. J.C.Wu, C.I.Kuo and J.C.W.Chien, Proc. IUPAC Symposium Amherst, Mass., 1982, p 241.

161. P.Galli, L.Luciani and G.Cecchin, Advances in the Polymerization of Olefins with Coordination Catalysts, *Angew.Makromol.Chem.*, *94* (1981) 63-89

162. - Hard Task: Choosing a Route for Making Polypropylene, *Chemical Engineering* 8 July 1985, pp. 22-27;
    - Polyolefin Routes in Bloom, *Chemical Engineering*, 15 Feb. 1988, 26-29

163. P.D.Gavens, M.Bottrill, J.W.Kelland and J.McMeeking, Ziegler Natta Catalysis, Chapter 22.5 in Comprehensive Organometallic Chemistry, Volume 3, G.Wilkinson, Ed., Pergamon Press, Oxford, 1982.

164. P.Galli, Polypropylene - a Quarter of a Century of Increased Successful Development, pp. 63-92 in Structural Order in Polymers, F.Ciardelli and P. Giusti, Eds., Pergamon Press, Oxford, 1981.
165. N.F.Brockmeier, Latest Commercial Technologies for Propylene Polymerization, p. 671 in Transition Metal Catalyzed Polymerizations. Alkenes and Dienes. Part B, R.P.Quirk, Ed., Harwood Academic Publishers, 1983.
166. K-Y.Choi and W.H.Ray, Recent Developments in Transition Metal Catalyzed Olefin Polymerization - A Survey. II. Propylene Polymerization, *J.M.S. - Rev.Macromol.Chem.Phys.*, *C25* (1985) 57-97
167. S.P.Sawin and G.W.Powers, New catalysts for propylene polymerization, p355-360 in Advances in polyolefins, R.B.Seymour and T.Cheng editors, Plenum Press, New York, 1987

Chapter 2

CHARACTERIZATION OF POLYPROPYLENE HOMOPOLYMER

2.1 INTRODUCTION

Polypropylene is a semi-crystalline polymer with a relatively high melting point, this makes it a very useful plastic. Its end-use properties are mainly determined by the level of crystallinity, the crystalline morphology, and its orientation. In their turn these aspects are fixed by both the composition of the starting material <u>and</u> the processing applied (method and conditions). These interdependencies can be pictured in the following triangle:

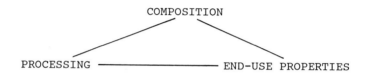

For example in modern, high speed, meltspinning operations it is known that small changes in the machine settings have a considerable impact on the resulting product and so have changes in the polymer. There are for instance rather wide differences between the products of one producer compared to that of another.

Turning now to characterization, the first question is "why is it done?", followed by "what is done?". As mentioned above polypropylene is a performance product, therefore end-use testing is the accepted way of determining the suitability of a certain type of polymer for a specific application. This is clearly a time consuming, costly affair. Moreover, the test outcome does not necessarily give a clue to the polymer scientist as to which of the basic polymer properties he has to change to get a positive answer in the next end-use test. Therefore there is a drive to basic understanding of the structure-property relationships by carrying out more "analytical" type of tests; in the end hoping to arrive at a predictive ability. These type of tests will also be considerably cheaper than end-use testing. What is it that one looks for in characterization? In principle this term can cover a wide range of techniques or aims. In this book we confine ourselves to:

-composition,

-crystallinity and morphology related aspects, and

-(in isolated cases) mechanical properties,

the main aim being to link the effect of changes in catalyst systems and/or polymerization conditions to composition of the polymer and in turn, the composition with the crystallinity, morphology, etc.

The composition is the aspect of great interest to the polymer chemist. Three aspects are of importance here in the case of homopolymer polypropylene :

-the mode of linkage of the monomer, head-to-tail, or head-to-head/tail-to-tail, i.e. its regiospecificity,

-the stereochemistry of the head-to-tail linked monomer, i.e. its tacticity,

-the molecular weight and its distribution, which are linked to chain transfer and termination processes in the polymerization.

Fig. 2.1 gives a number of structural possibilities related to the first two aspects. Once these qualities are measured the polymer

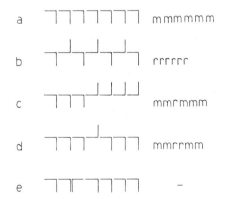

a   mmmmmm

b   rrrrrr

c   mmrmmm

d   mmrrmm

e   -

Fig. 2.1: Examples of heptad structures in polypropylene: isotactic (a); syndiotactic (b); "block"polymer (c);, "propagation error" (d); regioirregular with head-to-head and tail-to-tail (e); (a) to (d) are regiospecific structures

sample is defined in its composition, i.e. the product resulting from the combination of the catalyst system used together with the applied polymerization conditions is then described. One can then study the effects of variation in for example the catalyst type or the polymerization conditions.

The individual polymer molecules will vary in both their molecular weight (MW) and tacticity (as it turns out that the regiospecificity is very high). Thus in addition to the molecular weight distribution (MWD), a tacticity distribution is expected to be present. Hence, for a complete, molecular chemical characterization of the polymer one needs to know the tacticity distribution at each level of molecular weight and vice versa; in other words a so-called two-dimensional characterization is required. Experimentally this is very difficult and to our knowledge it has never been fully achieved. Cross-fractionation as mentioned in chapter 13 comes rather close.

In addition to composition, quite some attention will be given to crystallinity related effects and morphology, especially in the case of modified polypropylenes, such as copolymers. The reasons for this are twofold: firstly, as said above, the crystalline nature is the most important polypropylene property, and secondly the use of appropriate techniques in this area also leads to sensitive "compositional" measurements. A case in point is the measurement of the final melting point (see below and chapter 13).

Before introducing in somewhat more detail the contents of this chapter, a word about tacticity and its measurement is necessary. The most direct method of determining the tacticity of polypropylene is ^{13}C-NMR spectroscopy. Sequences of five propylene units (pentads), sometimes even longer, can be distinguished as regards their stereochemical relationships. Tacticity can thus be described for example by giving the percentage of each of the pentads present. It will be clear however that such a description is not exhaustive since those pentads can be combined in a large number of different ways to describe a polymer molecule. Nevertheless this method is the best available, especially due to the fact that the observed resonances are directly dependent on the stereochemical order in the polymer chain. Designating an identical placement of two neighbouring methyl groups along the chain by m, and the unlike arrangement by r, pentad sequences are written as mmmm, mmmr, mrrm, etc (see also Fig. 2.1). For the simplest description of tacticity one can calculate from those pentad distributions the triad or dyad distribution; the latter is in a number of cases an easy way of comparing different polypropylenes. (However, a given dyad concentration tells one very little about the real distribution of the m and r placements, much less than the pentad figures.) The NMR technique becomes difficult at very high tacticities. As the sensitivity is equal for every carbon in the chain the determination of, for instance less than 1 % non-isotactic structure, is both tedious (very many runs required in a Fourier transform mode) and not very accurate. Other methods which are sometimes claimed to measure tacticity, such as solubles level, density, melting enthalpy, wide angle X-ray diffraction, all measure a property which is not only dependent on the true tacticity but also on variables such as the thermal history and the molecular weight distribution of the polymer sample being investigated.

As an introductory example of the composition of a typical polypropylene the following fractionation result is illustrative[1]. A polymer is deposited out of a xylene solution on sea-sand at high temperature, and the slurry cooled to precipitate the polymer on the carrier material. Elution is than carried out by xylene at increasing temperature intervals, which in the sensitive area is every 1 or 2 °C. For each fraction one measures the molecular weight by GPC and the tacticity by NMR. Three different fractions are generally observed:
- material still soluble in xylene at 20 °C which is atactic in nature, the amount being around 5 % m/m for a typical polypropylene,
- material eluted between 20 and about 100 °C, amount approx. 15 to 20 % m/m, of increasing isotacticity, usually called stereoblock-polymer,
- material eluted up to 125 °C of high isotacticity and constituting the bulk of the polymer.

In the present chapter on homopolymer polypropylene characterization, from the extensive material available we shall base our discussion on the different fractions found in polypropylene: the volatiles, the solubles, and the insoluble fraction. In this way we can identify the constituents, by describing both the amount of these constituents present in homopolymer samples and their composition. Where appropriate, specific effects of catalysts or polymerization conditions will be mentioned in these sections. Finally, for true characterization studies, in order to elucidate the relationships between the independent polymerization parameters and the characteristics of the resulting polymer, the effects of catalyst type, co-catalyst type and the polymerization conditions will be systematically reviewed. The bulk of this chapter deals with atactic and isotactic polypropylene, but also the syndiotactic variant will be reviewed, as it is an interesting isomer. In addition some remarks will be made on stereoblockpolymers.

The methods applied in the characterization are described in a later part of this book (chapter 13), either concisely for the well-known standard methods (such as X-ray diffraction, gel permeation chromatography (GPC) and differential scanning calorimetry (DSC)), or more extensively in the case of own developments (as with fractionation - xylene solubles - and morphology).

2.2 CHARACTERIZATION OF THE VOLATILE FRACTION OF POLYPROPYLENE.

Measurement of the volatiles in polypropylene was done by a simple method using an extraction with a poor solvent for polypropylene and subsequently applying gas-chromatography on the extract. A typical chromatogram is given in Fig. 2.2. The peaks are clearly arranged in a repeating pattern of multiple peaks. Calibration of the GLC column with straight chain hydrocarbons indicated that the volatile peaks are equally spaced in regard to

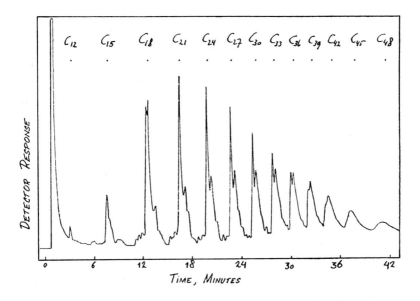

Fig. 2.2: Gas chromatogram of volatiles from a Melt Index=5 homopolymer

carbon number. This was substantiated for the lower-boiling-point part of the volatiles by a combination of mass spectrometry and GC, which showed the major peaks to be branched, saturated hydrocarbons with carbon numbers equal to multiples of three. The major part of the volatiles in polypropylene are therefore - as expected - propylene oligomers. Two complete multiplets of the chromatogram have been analyzed by mass spectrometry and shown to consist of compounds with the same carbon number, showing that these multiplets are composed of isomers. Some minor peaks have carbon numbers such as 11, 13, 16 and 22. They can be formed by copolymerization with a small amount of ethylene present, or arise from an ethyl-alkylated catalyst site or by transfer with the co-catalyst. Generally the distribution of the oligomers shows a modest maximum at C_{21} and C_{24}, which can however be caused by the extraction method applied. For a Melt Index (MI) 3 polymer each oligomer is present for roughly 200-300 ppm on total polymer. The isomer ratio within the multiplets progressively changes as the carbon number increases, as is clearly shown in Fig. 2.2.

A number of polymerization variables have been studied in relation to the amount and composition of volatiles formed:
- amount and type of the chain transfer agent
- type of catalyst
- polymerization yield

The most important of these turned out to be the amount of chain transfer agent applied. Increasing the level of hydrogen in the polymerization and thus decreasing the molecular weight of the resulting polymer, leads to an increase of the amount of volatiles present. A linear relation between the amount of volatiles and the melt index of the polymer is found and this holds for a fairly wide melt index range. This observation also supports the conclusion that the volatiles are normal low molecular weight oligomers and that cationically formed species are not present to any large extent. In line with the above, on leaving out the hydrogen completely the amount of volatiles decreases drastically. The individual oligomers are then present in amounts of around 20 to 30 ppm. The pattern within an oligomer is also dependent on the amount of hydrogen applied, one of the extremes is shown in Fig. 2.3 for a polymerization without hydrogen.

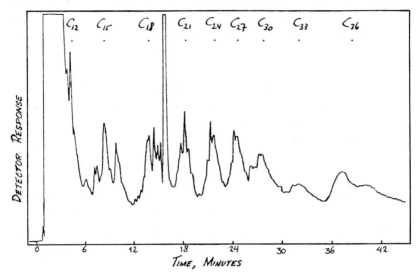

Fig. 2.3: *Gas chromatogram of volatiles from a polymer made in the* <u>*absence*</u> *of hydrogen*

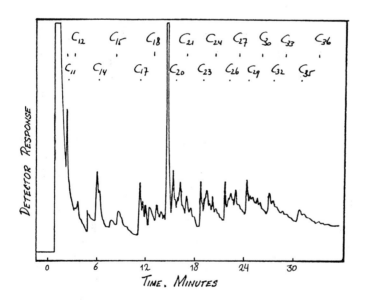

Fig. 2.4: *Gas chromatogram of volatiles from a polymer made in the presence of diethyl zinc as chain transfer agent*

Another proof that the volatiles are normal polymerization products is the effect of the use of zinc diethyl as chain transfer agent. The pattern is shifted in a manner consistent with the presence of an ethyl group in each oligomer. The chromatogram is given in Fig. 2.4.

The effect of catalyst type is found to be relatively minor. However the catalyst prepared using $AlEt_2Cl$ or sesqui for reducing $TiCl_4$ show a greater sensitivity to hydrogen with respect to the amount of volatiles than the comparable $AlEt_3$ reduced catalysts.

The effect of polymerization yield on the volatiles composition has also been investigated. The chromatograms of polymers obtained at different yields differed considerably as shown in Fig. 2.5. The main consequence of a low yield is the presence of a greater number of large peaks in each multiplet; this

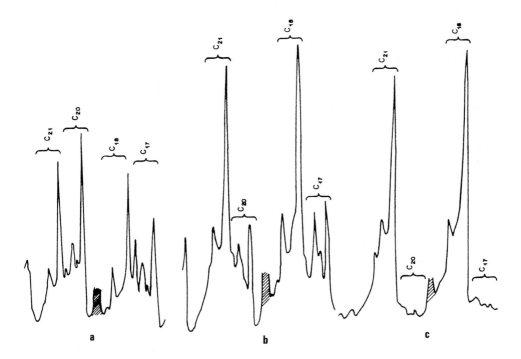

Fig. 2.5: Gas chromatogram of volatiles, effect of yield (yield expressed as g polymer/g $TiCl_3$): 340 (a); 640 (b); 5300 (c). (the shaded peak is the internal standard: $nC_{16}H_{34}$)

number is halved for the high yield polymers. The pattern of these additional peaks resembles the pattern of the preceding main peak quite closely (compare peaks marked C_{20} and C_{18} in Fig. 2.5). These pre-peaks can almost certainly be attributed to oligomers with one carbon atom less than those constituting the main, high yield peak. Thus they represent those oligomers which are terminated by an ethyl group from the co-catalyst, whereas the main peak oligomers are hydrogen terminated. If the C_{20} fraction in the multiplet $C_{20}+C_{21}$ is plotted against yield a hyperbola is found, which transforms to a straight line when the reciprocal of the yield is used as the variable. This relationship does not, however, distinguish between the case in which C_{20} is only formed initially and subsequently diluted by C_{21}, and the case in which C_{20} is formed throughout the polymerization, since both may yield a straight line. The absolute quantities of C_{20} formed are plotted against yield in Fig. 2.6, and, although the accuracy is certainly not high, it is clear that $AlCl_3$-containing catalysts behave differently from those in which this has been extracted (as in the

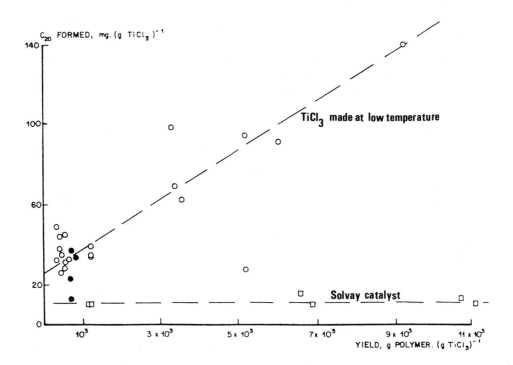

Fig. 2.6: Yield of C_{20} in the volatiles as a function of yield on catalyst (note: 150 mgC_{20}/g TiCl$_3$ represents a use of 0.08 mmol ethyl per mmol TiCl$_3$)

Solvay recipe). In the former case, chain transfer (presumably with ethyl aluminium dichloride) is an ongoing process, whereas in the latter no additional ethyl groups are consumed after the initial requirements for the alkylation of the $TiCl_3$. This is consistent with the higher hydrogen requirement of the Solvay catalyst.

It appears that yield has no profound effect on the amount of volatiles formed, one observes only an increase of around 15 ppm per 1000 increase in yield (expressed as gram polypropylene per gram of $TiCl_3$) for the individual oligomers.

Tacticity effects should also be measurable by a gas-chromatographic technique; as will be mentioned later, the use of capillary columns leads to a very high resolution (see chapter 13.5). It would be very interesting to check the tacticity, and its changes with catalyst type, etc., for these very low molecular weight oligomers.

2.3 CHARACTERIZATION OF REACTOR SOLUBLES

Reactor solubles are defined as that part of the polymer which is found in solution after the polymerization at the reaction conditions (excluding solution polymerizations). Two aspects will be dealt with in this section: the amount of the reactor solubles and their composition, with emphasis on the former. The data mentioned here are of high importance for the running of a polymerization process: the reactor solubles determine for example the slurry viscosity and thus the mixing and heat removal in the reactor, furthermore they also represent the fraction of monomer lost when a solid/liquid separation is applied in the process to remove the "atactic" fraction.

The variables that may be expected to govern the migration of polymer from inside the polymer particle to the solution - and thus determine the amount - are (when we restrict ourselves to considering one type of catalyst):
- time,
- temperature,
- type of solvent and
- average molecular weight.

Of course, with a different type of catalyst several other factors come into play as well, such as morphology and solubles level. The effects of a number of these variables will be dealt with below.

To assess the effect of time the amounts of reactor solubles were determined in freshly polymerized slurries as a function of both time and temperature. An example is given in Table 2.1. The reactor solubles level increased with time and levelled off in about 24 hours. About 2.5 hours were required with these dense catalysts to reach half of the ultimate difference in solubles level. The solubles level also increased with increasing storage temperature. Conversely, cooling the slurry resulted in a decrease in reactor solubles level, which was due to crystallization of some isotactic material, as was demonstrated by GPC data: The bulk of the material solubilized is of low molecular weight as observed in GPC. The peak molecular weight shifts to a higher value and broadens with increasing time and/or temperature. Upon cooling, the main effect is seen as a decrease at the low molecular weight end, suggesting that this is due to the precipitation of isotactic polymer which is necessarily of low molecular weight under these conditions. The crystallinities of the reactor solubles isolated at

TABLE 2.1

Level of reactor solubles at different times.
base slurry: freshly polymerized, particle size around 100μ.

Time, h	Temperature, °C	Reactor solubles, %m/m	GPC data		
			M_n*10^{-3}	M_w*10^{-4}	Q
0	60	1.5	1.03	1.1	11
0.16	60	1.5	1.07	1.0	9
1	80	1.8	1.17	1.3	11
1.5	80	2.1	1.25	1.4	11
2.16	80	2.2	1.33	2.2	16
2.5	80	2.35	1.29	2.1	16
3.3	80	2.5	1.26	2.2	17
19	80	3.1	1.44	3.2	22

90 °C are fairly high, which also points to the presence of isotactic material.

Using catalysts with different morphologies leads to differences in "washability". This is defined as the ratio of the level of the isopentane slurry wash solubles to the xylene solubles (a good measure of the total solubles level, see chapter 13 for description of the method). This represents the effectiveness of

removal of the unwanted non-isotactic polymer by a simple low
temperature slurry wash. The data described below are the result of
solubles determination on dried powders from polymerization
experiments. In this it is assumed that the reactor solubles level
is almost equal to the isopentane solubles, as the reactor solubles
will have been precipitated on the outer surface of the polymer
particles and are thus easily accessible in a solvent wash. Two
extreme examples of washability are given in Fig. 2.7 and Fig. 2.8
in which the molecular weight distributions of both the isopentane
and the xylene solubles are shown. Clearly the "Solvay" catalyst is
very dense as the amount of isopentane solubles is very low and of
low molecular weight. In the other catalyst morphologies studied,
the molecular weight distributions were generally of the same
shape. The above mentioned differences can also be observed from
the GPC parameters given in Table 2.2. A large number of data on
washability has been collected. For a specific catalyst a range of
washability is always observed, which demonstrates that this
property is affected by a great number of factors. Especially the
polymerization conditions have a large effect on the value
obtained, although the general order of washability is not changed.
From the results of laboratory polymerizations the following
sequence of increasing washability has been obtained:

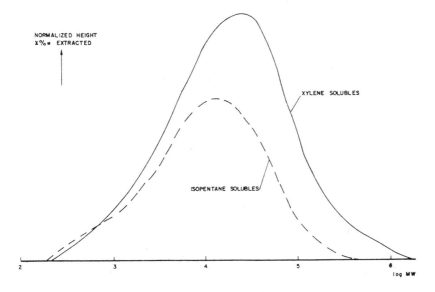

Fig. 2.7: Molecular weight distribution of isopentane and xylene solubles from a
homopolymer made with a Stauffer catalyst

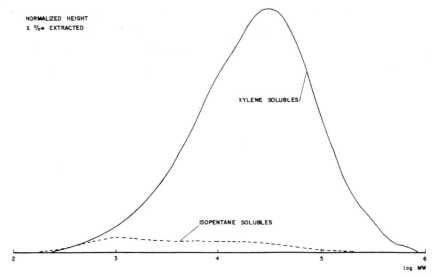

NORMALIZED HEIGHT
X %w EXTRACTED

XYLENE SOLUBLES

ISOPENTANE SOLUBLES

log MW

Fig. 2.8: Molecular weight distribution of isopentane and xylene solubles from a homopolymer made with a Solvay catalyst

Solvay < low temperature reduced $TiCl_4$ < $AlEt_2Cl$-reduced $TiCl_4$ <

Stauffer = Purechem

These catalysts cover a washability range of 0.1 to 0.6.

TABLE 2.2

Effect of catalyst morphology on MWD of isopentane and xylene solubles

Catalyst used in preparing polymer	Type of solubles	Washability[*]	GPC data			
			M_n*10^{-3}	M_w*10^{-4}	Q	R
low temperature reduced $TiCl_4$	isopentane	0.26	2.0	1.2	6.3	8.4
	xylene		4.4	3.8	8.6	4.5
Solvay	isopentane	0.11	3.5	3.8	11	6.5
	xylene		10.1	6.6	6.6	3.5
Stauffer	isopentane	0.6	3.6	2.3	6.4	3.3
	xylene		5.6	5.9	10	6.9
$AlEt_2Cl$ reduced $TiCl_4$	isopentane	0.54	2.9	2.2	7.6	4.8
	xylene		2.9	5.6	19	6.8

[*] isopentane over xylene solubles level

In a number of polymerization processes low boiling solvents such as propylene or butane are used; particularly the former is the (reactive) diluent of choice in many modern slurry plants. Because of the low critical temperature of propylene, the polymer solubility decreases with increasing temperature, the molecular

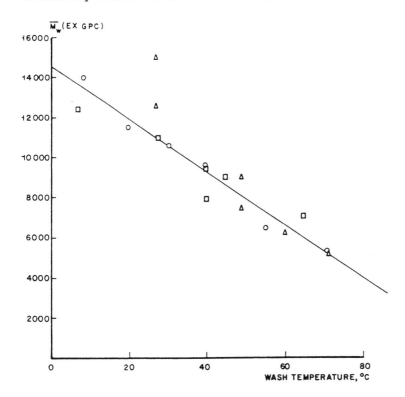

Fig. 2.9: Molecular weight of solubles obtained with liquid propylene as a function of wash temperature (MW ex GPC, calibrated with polystyrene)

weight being the determining parameter. In Fig. 2.9 a number of data are given of the MW of reactor solubles in a propylene solvent system as function of the temperature. As expected both the amount of soluble polymer and its MW decreases upon increasing the temperature.

The yield stress and the melt-flow properties have been measured for some of the residues obtained after removal of the reactor solubles. As regards the yield stress, an increase is noted upon washing, the extent of which is - as expected - proportional to the amount of <u>atactic</u> polymer removed and not to the total amount extracted. It is known that low molecular weight isotactic polymer does not affect the yield stress[2]. In the melt-flow measurements it is observed that the zero-shear viscosity increases linearly with the amount removed. Finally, as regards the molecular weight distribution a slight narrowing is found upon washing.

Most of the above experimental data fit in with a diffusion process in the extraction of reactor solubles, in that time and temperature play an important role. The effect of increasing temperature is twofold: it increases both the diffusion coefficient and the extent of swelling, thereby reducing the viscosity of the medium. The overall effect is an enhanced diffusion. Changing the temperature of a slurry also changes the total amount of potentially soluble polymer, as at higher temperatures more isotactic material is soluble. A rough idea of the temperature dependence of solubilization can be obtained by comparing the data from long duration extractions and the high-temperature extractions. In our experience, for dense polymer particles an extraction for 3 hours at 50 °C is as effective as 1000 hours at 25 °C.

The crystallinity also plays an important role especially at lower temperatures. This probably means that the slightly crystalline atactic polymer is transported from crystal to crystal, which slows down the migration process considerably. An example of this has been given by Natta et al, who studied the chromatographic separation of stereoblock polymers and found isotactic polypropylene[3] to be the most suitable absorbent due to a fairly strong interaction. This explains why, in the room temperature extraction with isopentane, the reactor solubles precipitated on the powder can never be extracted totally, as part of it is (co)crystallized and thereby becomes difficult to remove. The effect of crystallinity on migration will of course become smaller at increasing temperature.

The overall process of solubilization of both atactic and isotactic polymer is very complicated. Although the effects of the individual factors determining the migration are rather easy to describe, the integrated effect of time, morphology, temperature, solvent-type, molecular weight and crystallinity etc. cannot easily be translated into a predictive model.

2.4 CHARACTERIZATION OF THE SOLUBLES

The most common method for characterizing polypropylene with respect to its stereoregularity is the determination of its "soluble" fraction in a specific solvent and at a specific condition. Two distinct procedures exist, one an extraction, the other a re-crystallization (see chapter 13 for details). Solvents used in the first procedure are often ethylether or heptane (at their boiling points), for the latter method xylene is frequently chosen. The material obtained in the extract or remaining in solution is referred to as the "ether solubles" or the "xylene solubles" of the particular sample of polypropylene. Solubles obtained by methods referred to above have been characterized in a number of ways, both in our own laboratory as elsewhere. The most important characteristics studied are tacticity, molecular weight distribution and crystallinity. In handling the material some data on its solubility have also been collected. Additionally an elaborate fractionation of xylene solubles will be described in this section, allowing firmer conclusions on the compositional distribution of this material. Finally some data on stereoblock polymers will be reviewed, this stereoblock polymer being a special kind of solubles.

2.4.1 Tacticity

The tacticity is usually measured by ^{13}C-NMR spectroscopy. An example of a ^{13}C-NMR spectrum of polypropylene solubles is given in Fig. 2.10. Generally speaking the fractions in mm, mr and rr are roughly equal, i.e. around 30%, with the overall diad figures of [m] just below 50%. There is no difference in the tacticity of the isopentane or the xylene solubles. Comparison with the expected spectrum of pure heterotactic material is instructive. Heterotactic material is defined as a random copolymer containing equal amounts of iso- and syndiotactic placements and the NMR spectrum would show a 1:2:1 ratio for the mm, mr and rr triads, with a further split within the triads in the same ratio. Clearly the solubles isolated are by no means heterotactic and a clear preference for both the iso- and syndiotactic triads is observed. Moreover, within these triads there is a preference for the pure iso- and pure syndiotactic pentads mmmm and rrrr. This means that the solubles are to be regarded as a copolymer of m and r placements with a very blocky nature.

However both direct and indirect synthesis of true atactic

150

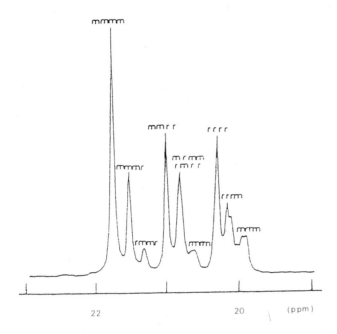

Fig. 2.10: ^{13}C-NMR spectrum of xylene solubles from a homopolymer (methyl-region) [from Y.Doi et al, p 737-749 in Transition metal catalyzed polymerizations. Alkenes and Dienes. Copied with permission from Harwood Academic Publishers]

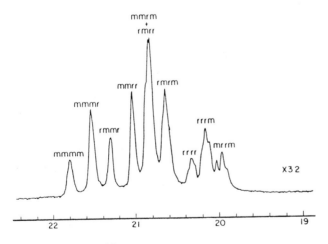

Fig. 2.11: ^{13}C-NMR spectrum of the methyl-region of hydrogenated poly(2-methyl-1,3-pentadiene) [from reference 6. Reprinted with permission from Macromolecules. Copyright 1985 American Chemical Society]

polypropylene has proved to be possible. For the direct synthesis it appeared that some homogeneous catalysts such as $Cp_2Ti(CH_3)_2/Al(CH_3)_3/H_2O$ (ref.4) or $(Me_5Cp)CpZrCl_2$/methylalumi-noxane at -30 °C as described by Ewen[5] polymerise propylene to atactic polymer. Hydrogenation of the polymer derived from 2-methyl-1,3-pentadiene also gives this type of polymer[6]. Isomerization[7] of isotactic polypropylene under hydrogen with a palladium catalyst at 270 °C is an additional approach. A ^{13}C-NMR spectrum of the methyl region is given in Fig. 2.11. Irradiation with γ-rays also brings about isomerization[8]. The extent of isomerization is larger in the liquid phase; a true equilibrium is not reached however as at high doses one reaches [mm] 31, [mr] 37, [rr] 32 %mol.

In the NMR characterization of the samples obtained in the column fractionation of the solubles (see below), end-groups can be detected in the low molecular weight fractions. A relevant part of the spectrum is shown in Fig. 2.12 and on the basis of the expected types of end-groups (ethyl from the co-catalyst and n-propyl and i-propyl from the chain transfer with hydrogen), the assignment shown in the figure has been derived. In a few resonances very considerable fine-structure is observed, for example in the i-pro-pyl end-group. The propyl and the i-propyl type end-groups are present in nearly equal amounts, as would be expected from the accepted mechanism of their formation. The isopropyl end-groups were exclusively detected by Zambelli in cases in which $Zn(CH_3)_2$ was used as the chain transfer agent[9], as would be expected. The ethyl and n-propyl end-groups are present in about a 1:6 ratio (for a polymer made with a yield of 1000 gPP/gTiCl$_3$ and a MI of 3). The stereochemistry of the isopropyl groups is roughly 50/50 threo and erythro[10], which is to be expected as they are formed from a titanium hydride species on atactic sites. A few remaining peaks would fit a head-to-head and tail-to tail structure, of which, in this particular low MW fraction, about 1 % is present. This type of structure is not detected in the polymer before extraction, indica-ting its very low overall concentration. Hayashi et al[11] measured NMR on boiling heptane solubles and on its residue. In the solubles they could discern four kinds of end-groups:
- the above mentioned propyl and isopropyl groups as the major fraction, structure I and II,
- some 4% with ethyl endgroups originating from the co-catalyst (III), and

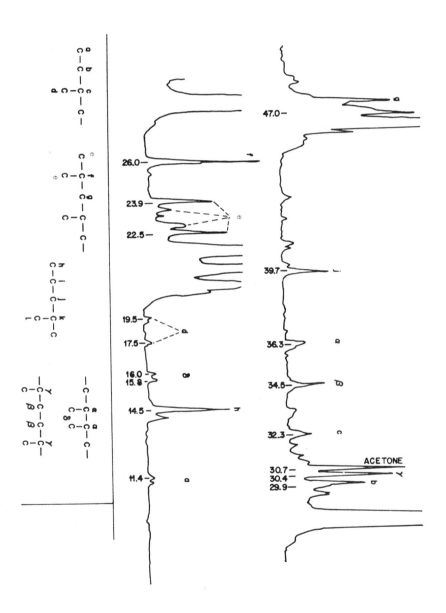

Fig. 2.12: ^{13}C-NMR spectrum of very low molecular weight polymer, with peak assignment for end-groups

- structure IV, in 10% abundance, possibly stemming from secondary insertion.

$$cat-C-C(CH_3)-C-C(CH_3)-R + H_2 \quad --->$$
$$cat-H + H-C-C(CH_3)-C-C(CH_3)-R \quad (type\ III)$$

$$cat-H + (n+1)\ C=C-C \quad --> \quad cat-(C-C(CH_3))_n-C-C-C \quad (type\ I)$$

$$cat-Cl + EtAlX_2 \quad --> \quad cat-Et + ClAlX_2$$
$$cat-Et + (n+1)\ C=C-C \quad --> \quad cat-(C-C(CH_3))_n-C-C \quad (type\ II)$$

The tactic relationships of the first units neighbouring the end-group are completely similar to that of the bulk. In the highly isotactic, higher molecular weight residue after heptane extraction, only the propyl and isopropyl groups can be observed.

2.4.2 <u>Molecular weight</u> and <u>molecular weight distribution</u>

An example of the molecular weight distributions obtained for both isopentane and xylene solubles has been given before (see Fig. 2.6 and Fig. 2.7). The peak MW is rather low at about 5,000 to 10,000, and this is about a factor of 10 lower than for the corresponding isotactic polymer. However, a broad MWD reaching up to and over the million mark is observed and the values for the distribution parameters are very high, with $Q(=M_w/M_n)$ values of around 8 and $R(=M_z/M_w)$ values of 4 to 5).

2.4.3 <u>Solubility</u>

The solubility of atactic polypropylene is generally good in solvents ranging from isopentane and isooctane up to tetrahydrofuran and chloroform. These solvents span a solubility parameter (δ) range of 13.9×10^{-3} to 19.1×10^{-3} $(J/m^3)^{0.5}$. The lower limit of the solubility region is marked by propylene, especially at higher temperatures, as already mentioned before. At the high end of the range, there is incomplete solubility in 1,2-dichloroethane(δ=20.0) and acetone(δ=20.2) at room temperature, although dichloroethane does dissolve the solubles at its boiling point.

These solubility characteristics are dependent on the starting material. For instance, it was found that freshly isolated xylene solubles are completely soluble in chloroform, tetrahydrofuran and cyclohexane, but not in isopentane. The factor responsible for the variations in solubility is the crystallinity of the material. Hence the solubility is also time-dependent, as exemplified by the observation that freshly isolated solubles are completely soluble in tetrahydrofurane, but, after a month's storage at room temperature, complete solubility is only achieved when the temperature of the solution is raised to 30-35 °C. A similar effect is noted with isooctane as the solvent.

2.4.4 Crystallinity

Both X-ray diffraction and differential scanning calorimetry(DSC) are used for measuring the crystallinity of solubles. The former not only gives information on the level of crystallinity but also on the crystal modification present.

The crystallinities derived from the DSC measurements are based on the true heat of fusion of polypropylene of 188 J/g. Careful annealing of the materials did not affect the crystallinity results, as shown in Table 2.3. Normally in DSC measurements after

TABLE 2.3

Comparison of crystallinity of annealed and non-annealed xylene solubles

sample	non-annealed crystallinity, %m/m	non-annealed melting points, °C	annealed* crystallinity, %m/m	annealed* melting points, °C
1	5	53, 61	6	50, 68
2	15	50, 82	12	53, 80, 90

* annealing: heated to 120 °C, 30 minutes hold, cooling to 40 °C at 4 °C/h, storing at room temperature

the first heating step, a cooling step and an additional heating step are applied. The solubles from polypropylene are normally cooled down to about 60 °C and no crystallization is then observed. Cooling to lower temperatures, however, gives rise to crystallization.

In the DSC trace in Fig. 2.13, a small endothermic peak from a material with a low melting point can be seen. The melting points are found around 50 °C. Generally the crystallinities observed are low, for example about 1-3 %m/m for the isopentane solubles. The

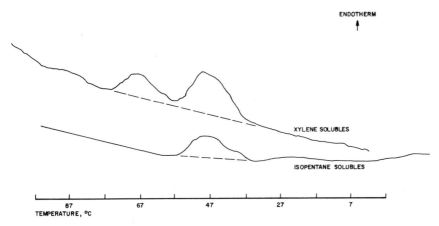

Fig. 2.13: *Example of DSC traces of typical isopentane and xylene solubles*

xylene solubles have a slightly higher crystallinity, the range being 2-10 %m/m, i.e. about three times higher than the isopentane solubles. This higher crystallinity is coupled to two endothermic peaks at around 50 and 75 °C.

The X-ray diffraction patterns for both types of solubles are shown in Fig. 2.14. In line with the DSC results only a low level

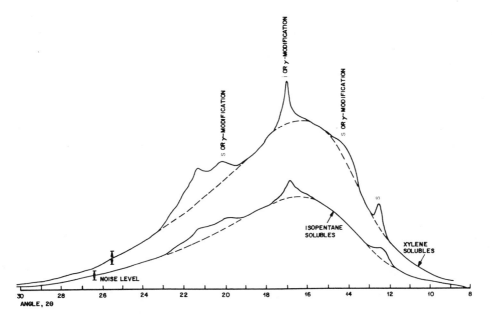

Fig. 2.14: *Example of X-ray-diffractograms of typical isopentane and xylene solubles*

of crystallinity is observed. The diffractogram also gives additional information as the patterns are specific for both the type of polymer and its crystal modification. The peaks observed can only originate from a mixture of iso- and syndiotactic crystallinities. The presence of syndiotactic material has been noted in other studies as well[12-15], even single crystals of syndiotactic polypropylene have been observed[14] in a commercial atactic polypropylene. The fact that this material is found in the solubles is due to the very high solubility of syndiotactic polypropylene[16] in the normally applied solvents in the solubles measurement. Quantitative differentiation between iso- and syndiotactic crystallinity is impossible and therefore only a qualitative assessment of both types of crystallinity can be made. The total crystallinity by X-ray is in the range 1-5 %m/m for the isopentane solubles and about double those values for the xylene solubles.

When applying solubles measurements at higher temperatures than the room temperature used for the isopentane and xylene solubles mentioned above, the level of crystallinity increases due to the larger solubility of isotactic polypropylene at higher temperatures. A few examples are given in this chapter (see e.g. Table 2.15), but this effect is mainly dealt with in chapter 13 as it is, in our opinion, strongly connected with the analytical procedure for measurement of the fraction of atactic polymer.

2.4.5 Fractionation of solubles

Further details of the structure of the solubles and firmer ideas about their formation have been obtained by fractionating the solubles. The results of a fractionation will be described below, additional data can be found in related studies[17-19]. The fractionation procedure used involved column fractionation at room temperature with Sil-O-Cel as support and cyclohexane-acetone as the solvent/non-solvent pair on a 10 gram scale. The material studied was a xylene solubles sample from a typical homopolymer grade from Shell. The relevant data obtained are given in Table 2.4.

The MWD's were found to be narrow with Q values of 2.0 ± 0.2 and R values of 1.7 ± 0.2, see the GPC curves in Fig. 2.15. An exception were the last fractions which were rather broad. This was very probably caused by the crystallinity shown by these fractions.

TABLE 2.4

Data on fractionation of a typical xylene solubles sample and characterization
of the fractions obtained

Frac-tion	Solvent composi-tion acetone/ C6H12 %vol	Weight reco-vered %m/m	X-ray data Crys-talli-nity %m/m	Relative occurence of: syndio	iso	DSC data Crys-talli-nity %m/m	melting points, °C	mm %	mr %	rr %	m %	r %
1	100/0	16.9	6.8	-	x	5.4	22,42	46	29	25	61	39
2	75/25	12.7	15.5	-	xx	14.1	45,78	49	27	24	63	37
3	60/40	19.0	12.6	-	xx	10.7	48,82	43	30	27	58	42
4	55/45	14.6	3.9	x	x	3.8	48,74	32	32	36	48	52
5	50/50	10.0	4.2	x	x	4.4	49,74	30	33	37	46	54
6	45/55	13.9	2.9	xx	x	2.1	49,74,90	25	35	40	43	57
7	40/60	10.8	2.7	xxx	?	5.3	50,74	21	36	43	39	61
8	25/75	4.3	10.3	xxxxx	?	9.3	71,96	11	24	65	23	77
9	0/100	1.3	-	-	-	15.5	70,113	-	-	-	-	-
		103.5										
starting sample						10	53,76,90					

The MW's ranged from 2700 in the first fraction to 270,000 in
fraction 8.

The X-ray results showed well developed crystallinity in the

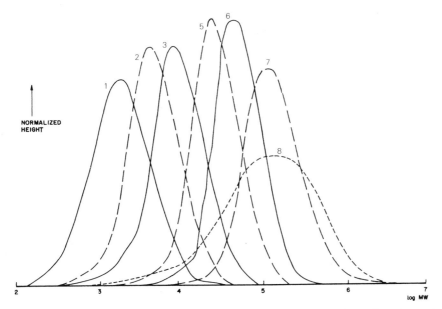

Fig. 2.15: GPC-data for the fractions obtained in the column-fractionation of a
typical xylene solubles

158

first few and last fractions only. In the first fractions the crystallinity was due to isotactic polymer. The pattern observed in fraction 3 belongs to the γ-crystal modifications of polypropylene, a form which is only observed in low molecular weight isotactic material[20,21]. The last fractions had patterns which are characteristic solely of the syndiotactic polymer. The intermediate fractions showed a low crystallinity and a rather diffuse pattern.

The crystallinities found by DSC varied in the same manner as those obtained by the X-ray measurements. The melting peaks changed with increasing molecular weight as shown graphically in Fig. 2.16. From a detailed inspection of the traces, the impression was gained that fractions 2 and 3 contained material which gave rise to rather

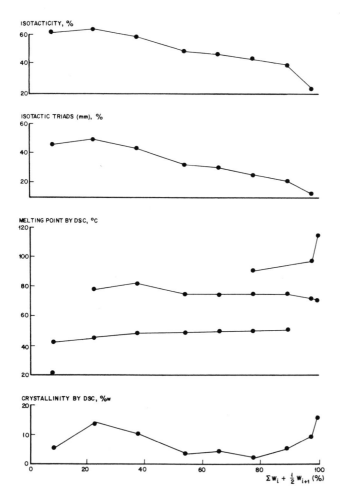

Fig. 2.16: Characterization of the fractions obtained in column-fractionation of typical xylene solubles

sharp, higher melting point, peaks which are not seen in the other fractions. This is probably low molecular weight isotactic material which is eluted later than would have been expected on the basis of its molecular weight (fraction 3 indeed showed a small low molecular weight tail in its GPC curve).

The NMR spectra, of which the methyl regions from each fraction are shown in Fig. 2.17, illustrate the change in composition of the fractions which is observed, starting with relatively isotactic-rich material, through reasonably heterotactic and finishing with fairly pure syndiotactic material.

The crystallinity and tacticity data are surveyed in Fig. 2.16. In a strict sense, apart from molecular weight, no fraction is of identical structure. However, in more general terms, the fractions fall into three distinct groups:

i)   fractions at the low molecular weight end, more crystalline and more isotactic than average; this has been observed in other fractionations as well[17],

ii)   intermediate fractions with low crystallinity and average tacticity, and

iii)   fractions at the high molecular weight end, with higher crystallinity and clearly more syndiotactic, which is contrary to

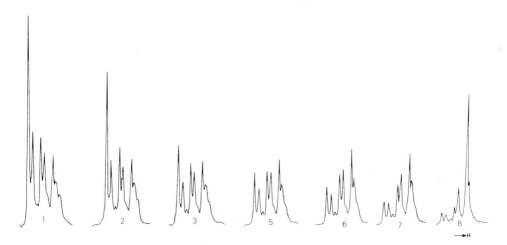

Fig. 2.17: ^{13}C-NMR spectra of some fractions obtained in column-fractionation of typical xylene solubles (methyl region only)

earlier observations using proton-NMR[19].

The first two groups of material are found in about equal amounts, whilst the last one is present in only very small amounts of about 5 %m/m on solubles, which represents roughly 0.25 %m/m on total polymer. Apparently the catalyst system used generates a small quantity of rather pure syndiotactic material, which due to its high solubility ends up in the solubles fraction of polypropylene. For fractions of the first group one is tempted to explain the differences in tacticities by assuming that the fractions contained mixtures of heterotactic material (with a composition analogous to the second group) and low molecular weight isotactic material. Since the separation of these two materials is almost impossible, this assumption can only be proved indirectly; the supporting evidence includes the following observations:

i)   the known solubility of low molecular weight isotactic polymer (see chapter 13)

ii)   the distinct crystallinity of the second and third fractions, the behaviour of which is clearly different on DSC and X-ray analysis from that of the other fractions,

iii) the observation of the γ-crystal modification which is in line with the low molecular weight of the crystalline material.

The fact that not all the low molecular weight isotactic material has been found in the first fraction is possibly due to the effect of crystallinity on the solubility (i.e. the fractionation should have been carried out at a higher temperature).

A recent study of Kakugo et al[15] also deals with syndiotactic polypropylene in the solubles. They observed in xylene solubles of polymers made by a wide variety of catalyst/co-catalyst combinations, invariably a high rr content with NMR and syndio-crystallinity with IR. Elution fractionation similar to the one described above gave also higher molecular weight syndiotactic polymer showing an rr content of 68 % m/m and syndio peaks in X-ray diffraction. The catalyst, etc. only determine the relative amount, the type of polymer is always present. Interestingly no inversions could be found in this syndiotactic polymer, which makes it a different polymer form the usual ones generated on vanadium catalysts at low temperature - they always show low regiospecificity (see section 2.8).

In conclusion, xylene solubles contain as minor components some low molecular weight isotactic material of probably a rather high steric purity ([m]> 80%) and a very small amount of high MW syndiotactic material ([r] >75%). The major component is a heterotactic polymer which is very probably not of a constant structure independent of molecular weight.

This wide range of composition, in our opinion, makes the application of propagation models on this polypropylene fraction useless, the more so when (hot!) heptane solubles are often used in these exercises. Those heptane solubles have been proved to contain an even larger fraction of isotactic material than xylene solubles - see chapter 13. Examples of attempts to model the sequence structure of atactic polypropylene can be found for instance in ref. 22 and 23. Doi[22] isolated various atactic polypropylene samples (using different fractionation methods!) from polymer made by different catalysts; all but one were found to be highly regiospecific (the exception being the $TiCl_4$/$AlEt_3$ catalyst system). Tacticity measurement by ^{13}C-NMR showed them all to be similar to the illustration given above (see Fig. 2.10), i.e. with mmmm being the largest pentad in the isotactic region, and the same holds for the rrrr pentad in the syndiotactic region. Doi can reasonably predict the measured tacticity distribution by assuming the atactic polypropylene to be a block polymer of iso- and syndiotactic runs, with the enantiomorphic-site statistics holding for the isotactic part and the Bernouillian for the syndio. Three parameters in all were used, two of which are propagation probabilities in a specific steric sequence, the third is the weight fraction of isotactic stereoblocks. The latter turns out to be the most sensitive parameter and in our opinion its value also reflects the extraction method applied (i.e. being highest with boiling heptane solubles and lowest when boiling pentane is used).

2.4.6 Stereoblock polymer

Natta[24-26] in characterizing polypropylene found a fraction which showed both a low crystallinity as well as a lower melting point. As the molecular weight of this fraction is still rather high, the low melting point could not been simply caused by the presence of low molecular weight isotactic polymer. A blocky structure was then proposed of alternating highly isotactic placements and heterotactic placements, the sequence-length of the isotactic part being small in accordance with the observed melting point. The stereoblock polymer is obtained in fractionation of polypropylene after the initial removal of the highly atactic material such as by a low temperature ether, pentane or hexane wash. The greatest fraction of this type of polymer has been found in $TiCl_4/AlR_3$-catalyzed polymerizations[24]. Increasing the polymerization temperature lead to an increase in the stereoblock fraction[27]. The rather non-specific catalyst system $VOCl_3/Al(iBu)_3$, gives next to about 60 %m/m of atactic polymer also some 20 %m/m of stereoblock polymer[28].

That indeed the fraction so recovered is not a mixture of atactic and low molecular weight isotactic polymer has been proven by Wijga and van Schooten[29] via a high temperature refractionation of this material: the melting point of around 142 °C was found to be independent of both the molecular weight of the fractions and of their crystallinity. It is however thought that the very first fraction thus obtained (with a LVN of around 0.15 and a crystallinity of about 60-80%) is indeed almost pure isotactic polymer of low molecular weight, being soluble under the extraction conditions employed. In a later publication[2] this fraction was again obtained, the melting points of the other fractions were less constant however.

It has been shown that the stereoblock polymer fraction shows a different crystal modification upon cooling from the melt[20,21]. The polymers showing this γ-modification have LVN's of 0.5-1.5 dl/g, melting-points of 125-150 °C and crystallinities of 40-60 %m/m. They were obtained by successive extractions of polypropylenes with temperature intervals of 5-10 °C in the range of 30 to 70 °C.

A different way of extraction is the initial removal of atactic polymer by the xylene solubles method, repeating this once to make sure that all the atactic is gone, and then finally extracting the resulting residue with isooctane at its boiling temperature. The amount extracted lies, for various polymers, in the range of 2-10 %m/m. The main variable affecting the amount of extract appears to be the type of catalyst used in the polymerization: polymers prepared with $AlEt_2Cl$ or sesqui-reduced $TiCl_4$ show much larger extracts than the polymers made with $AlEt_3$-reduced catalysts. The molecular weight of the extract is in the 30-80,000 range, the distribution is very broad. The isotacticity as measured by NMR gives relatively low values of 75-85 % and the spectrum shows next to the normal mmmm structure also peaks from rrrr and mrrr. DSC measurements gives melting points and crystallinities in line with the above data from Natta and Turner-Jones (i.e. around 140 °C and about 40 %m/m). These tacticity values are in the same range as those observed by Wolfsgruber et al[30], which includes polymers made with widely different catalysts.

Assuming that this polymer is indeed a block polymer consisting of isotactic and heterotactic sequences a likely way of formation is the polymerization on a bivacant site which is dynamically complexed with for example $AlEtCl_2$. In its complexed, monovacant, state this site produces isotactic polymer, whilst in its non-complexed state atactic polymer is formed. When the average complex lifetime is shorter than the polymer chain lifetime a blocky polymer will result. This can also be called chronotacticity as used by Theyssié in butadiene polymerizations with a nickel catalyst. The above idea was tested by isolating this type of fraction from polymers made at different temperatures and thus probably showing different complexation dynamics. Table 2.5 gives the results, the polymers were made using an experimental aluminium free catalyst of high stereospecificity. Both the xylene solubles and the subsequently isolated hot isooctane extract do increase

TABLE 2.5

Data on "stereoblock" polymer made at different temperatures

Polym temp. °C	LVN total dl/g	Xylene solubles, %m/m	iC8 extract on X2R* %m/m	isooctane extract				isooctane residue			
				LVN dl/g	T_{ml} DSC °C	Crystal-linity %m/m	[m] NMR %m/m	LVN dl/g	T_{ml} DSC °C	Crystal-linity %m/m	[m] NMR %m/m
50	2.90	1.9	0.9	0.26	139	34	83	2.8	164	48	97
60	2.95	2.7	0.9	0.28	143	24	81	2.9	164	48	95
80	3.00	9.0	5.2	0.59	146	17	83	3.0	162	46	96

*: boiling isooctane solubles on residue from a twice repeated xylene solubles test.

with increasing polymerization temperature. Comparing the characteristics of the <u>residue</u> after the isooctane extraction no large difference is seen in the series. The isooctane <u>extract</u> however shows changing properties: the crystallinity decreases and the melting point increases slightly and also MW increases. The NMR tacticity is constant however. The melting point can be correlated with molecular weight and values of around 1500, 2700, and 3900 apply for the respective isotactic fragments. At constant overall tacticity this means that the atactic blocks also increase proportionally in length. Thus although it would have been expected that the lifetime of the complex would be lower at higher temperature, the higher propagation rate more than compensated for this. Moreover the increase in propagation rate with temperature is roughly equal for both forms of the complex. The observed lower crystallinity of the stereoblock fraction is much more difficult to explain. The only factor which could apply is the increase in the number of "propagation errors" made in the isotactic fraction with increasing temperature (see also sections 2.5 and 2.6). The lower melting point of the residue of the highest temperature experiment also points in this direction. A complexed, originally bi-vacant, isotactic propagating site might very well have more possibilities for producing propagating errors than a normal single vacancy site - and thereby possibly be also more temperature dependent.

In conclusion on these stereoblock polymers:
- the main properties are reasonably high molecular weight, showing low crystallinity - mostly in the γ-modification, lower isotacticity and lower melting points compared to isotactic polypropylene (e.g. the xylene residue),
- the most probable composition is a multiblock polymer of isotactic and atactic or syndiotactic runs (the explanation using just propagation errors, but more of them, does not apply, as the errors normally observed are of the mmrrmm type, see next section); the distinction between these two possibilities is difficult to make, and
- at an overall tacticity [m] of 80% and an isotactic run length of 3000/42, this requires a non-isotactic run length of about 2000/42, i.e. indeed many of these runs can be present in stereoblock polymers of molecular weight 30-80,000.

A completely different type of stereoblock polymer has been described recently. Ewen[5,31], using $Cp_2Ti(C_6H_5)_2$/aluminoxane as catalyst, observed a polymer showing as main non-isotactic structure mmmrmmm. This represents a block polymer of (l) and (d) isotactic blocks (see Fig. 2.1). A NMR spectrum was already shown in the previous chapter (Fig. 1.19) an additional one is represented in Fig. 2.18. The observed melting points are low at the attainable isotacticities; at [m] = 83% the final melting point is 55 °C and at [m] = 85% this is 62 °C. A polymer structure as the above is thought to arise by a chain-end controlled mechanism, as once a different placement arises it propagates in the same manner. In a first attempt to characterize this type of polymer de Candia et al[32] have measured some crystalline and (dynamic) mechanical properties but have not given the corresponding NMR. In DSC the polymer melts at 60 °C, and a glass transition of 0°C is observed just as for other propylene polymers, as expected. The crystalli-

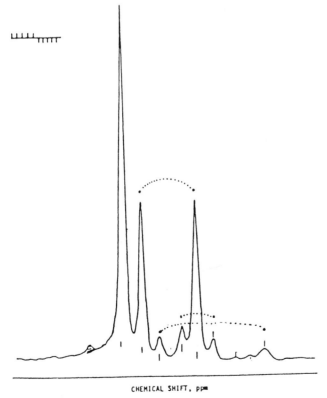

CHEMICAL SHIFT, ppm

Fig. 2.18: ^{13}C-NMR spectrum of a stereoblock polypropylene made with $Cp_2Ti(C_6H_5)_2$ at -30 °C. [from reference 31, copied with permission of Elsevier]

nity is in the γ-form, in line with the smaller isotactic blocks expected in a mmmrmmm structure. The mechanical properties do resemble those of a thermoplastic rubber, similar to the original observations of Natta's group.

From the above it is clear that many different stereoblock structures can be thought of. Firstly there are the chains of isotactic placements linked via one racemic dyad as one class and than the various combinations of atactic, syndiotactic and isotactic diblock structures as the second class. The latter, in a diblock configuration, leads to three different types: (iso + a)$_n$, (iso + syndio)$_n$, (a + syndio)$_n$. We know only a few of those in reality yet.

Using bis(arene) complexes of zerovalent titanium, zirconium and hafnium on alumina Tullock et al[33,34] prepared very high molecular weight polypropylenes of rather low isotacticity with high ethylether extracts. A similar polymer is formed[35,36] using tetraneophyl zirconium on alumina (neophyl is 2-methyl-2-phenyl-propyl). The MW's prepared can be very high and these polymers show very high ether and boiling heptane solubles, up to about 75% of the total polymer, i.e. they contain a large fraction of stereoblockpolymer. Melting points of the fractions are, as expected, low; i.e. about 50 °C for the ether extract and around 135 °C for the heptane extracts. These types of polymers show elastomeric properties, the material behaves as crosslinked through co-crystallization of the stereoblock with the isotactic fraction.

2.5 CHARACTERIZATION OF THE ISOTACTIC FRACTION

In this section on homopolymer characterization the data regarding the isotactic fraction will be given, i.e. on those polymer fractions which have been carefully freed from the solubles or the non-isotactic polymer. Characterization of the isotactic fraction uses the residues obtained after rather intensive extractions. For instance boiling heptane or octane insolubles[37,38], or in our own studies a residue after repeating the xylene solubles test twice.

The most interesting property is of course the tacticity, its measurement by ^{13}C-NMR being the obvious choice. However, due to the fact that the residues are indeed very isotactic, the measurement of small amounts of non-isotactic placements is particularly difficult, certainly in a quantitative sense. An example of the spectrum obtained is given in Fig. 2.19. The main non-isotactic peaks are mmmr, mmrr, and mrrm, i.e. those peaks stemming from a

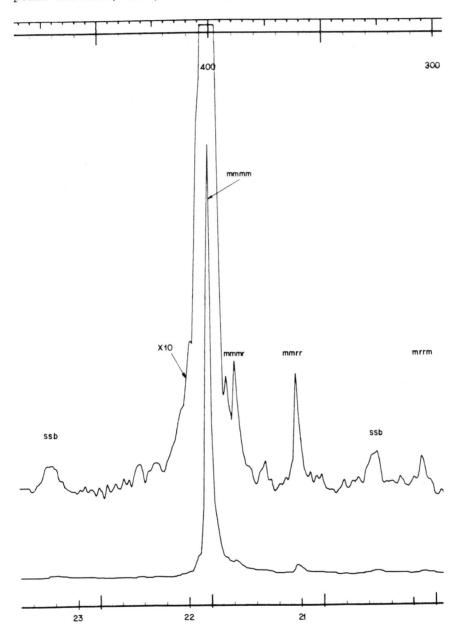

Fig. 2.19: Example of ^{13}C-NMR spectrum of re-crystallized polypropylene (methyl region only)

structure in which two inversions are present: mmrrmm. If only this structure is present than the ratios of the peaks mmmr, mmrr, mrrm should be 2:2:1. Mostly there is quite some problem to discern the first peak mmmr as this one is easily drowned in the major mmmm peak. Still, as shown in Fig. 2.19 for example, the above mentioned ratio does hold, given the large inaccuracy. This has also been observed in other residues (see below), and other laboratories have come to the same conclusion[37,38]. Not all spectra of residues are this simple however, see for example the data in ref.39. But it is thought that the use of heptane insolubles makes a dependable assessment of the quality of the residue more difficult, as a part of the atactic polymer is still present in this residue (see chapter 13). This residual atactic will obscure the analysis.

Peaks such as rmmr and rmrr are not normally observed in homopolymer residues which are carefully freed from non-isotactic polymer. Therefore the propagation errors are probably not grouped together in a "blocky" manner, but are reasonably isolated. This conclusion could be made more firm by extending the NMR measurements to longer sequences. The type of propagation error observed implies that the steric control is determined by the active polymerization complex and not by the last added monomer unit: the stereostructure of the chain does return to its original one after a "wrong" placement.

As expected, the crystallinities as measured by DSC increased on removal of the solubles and the melting point also increased slightly. There are, however, almost no differences in the final melting point, TMF, between the original and recrystallized polymers. This is quite general, a large quantity of a different material is required before a measurable change in the melting point is observed (provided of course that the interaction is almost zero, as it will be if the other material is a polymer). A quantity of 20 %m/m or more is necessary to give a change of about 1 °C.

2.6 CHARACTERIZATION OF TOTAL POLYMERS, EFFECTS OF CATALYST, CO-CATALYST AND POLYMERIZATION CONDITIONS

The effect of both catalyst/co-catalyst and polymerization parameters on the composition of the polymer will be the last and most important subject on homopolymer characterization. This is the most interesting part when looking from the viewpoint of tailoring the polymer properties by judicious choice of the reaction conditions and the catalyst/co-catalyst combination applied. The main body of the data will be on first generation catalyst derived polymers for obvious reasons, available literature on the newer catalysts is reviewed. Some data on syndiotactic polypropylene will also be given, this latter paragraph is necessarily small as only very few studies have been done on the characterization of this polymer.

2.6.1 The effect of catalyst type on the polymers produced

The importance of the type of transition metal compound for the quality of the resulting polymer was already recognized in the early days of polypropylene (see also chapter 1). Especially the effect of the different crystal forms of $TiCl_3$ has been extensively described[40,41]. For more recent reviews on the effect of catalyst transition metal compound on stereoregularity (mostly defined by the amount of boiling heptane solubles) the reader is referred to references 42 and 43.

Attempts to characterize more fully the polymers generated on different catalysts have also been made. A comparison of the residues from polymers made with the chlorides of titanium, vanadium, chromium and zirconium[30], with NMR analysis shows them to be of very similar quality. All contain as propagation error the structure mmmrrmmm, except for the VCl_3 derived polymer. The amount of soluble material varies of course widely but again not so much in tacticity: the non-random tacticity as shown by the solubles generated with commercial $TiCl_3$ catalysts is observed for all samples.

In another study of Martuscelli et al the properties of polymers made with second and third generation catalysts were compared[39]. The polymers were of equal molecular weight, the fractionation was done via successive extractions with hexane, heptane and octane. Differences in the overall tacticity([m]) of the heptane residues are rather small, they range from 96.1 to 96.7, only the tacticity of the high-solubles-containing polymer (10%) is somewhat lower at 94.6. However, the remarks made earlier about the efficiency of the heptane extraction apply even more strongly in this case. If one regards the mmmm pentad specifically, the differences are larger, ranging from 89.4 to 93.0. The heptane soluables show a slightly larger spread in their tacticity, here [m] is 62.9 to 71.5 (if commercial samples had been used in this study, this value would be affected by the normal practice of removing reactor solubles which includes an appreciable fraction of low molecular weight isotactic polymer). No outstanding difference between the various catalysts employed is observed, the only remarkable feature is the presence of the mrmr pentad in all the fractions of the polymer derived from the second generation catalyst, which is virtually absent in all the other polymers. The observed melting points in the DSC (range 4 °C) follow the same trend as the isotacticity. However, the equilibrium melting points, determined by crystallization experiments at higher temperatures lead to much higher values for the polymers made with the third generation catalysts and to a wider range of 16 °C. This points to differences in the polymer structure not discernible by NMR. Molecular weight distribution effects cannot be the sole factor involved.

In similar studies on supported catalysts[38,44] by Doi et al only mmrrmm structures were found in the octane solubles and residue, completely in line with findings on $TiCl_3$ catalysts. Interestingly $TiCl_2$ also gives an isotactic fraction with an identical structure to, for instance, the $TiCl_3$/DEAC-derived polymers[45]. This $TiCl_2$ is used as such in a polymerization without a co-catalyst being present. Clearly, this is further proof that the structure of the polymer is determined by the titanium centre.

The highest tacticity reported[46] to date stems from a catalyst made up of Solvay-$TiCl_3$ combined with $Cp_2Ti(CH_3)_2$. This catalyst is also very isospecific for other α-olefins as well as for styrene[47]. Using this catalyst at 40 °C gives an extremely pure isotactic polypropylene notable for having only 0.8 %m/m boiling heptane solubles and 99.5 %m/m isotactic pentads in the residue. A comparative example of a very good supported catalyst ($MgCl_2$/$TiCl_4$/dibutylphthalate + phenyl-triethoxysilane/TEA) gives 5 %m/m solubles and just 93.1 % mmmm pentads in the residue. The DSC melting point of the very pure polymer is 165 °C.

Recently[5,48-50] considerable success was reported in the use of <u>homogeneous</u> catalysts for the formation of predominantly isotactic polypropylene. Ewen[5] reported on the use of (ethylene)bis-(indenyl) titanium dichloride as a mixture of the meso and the racemic form, which at -60 °C with aluminoxane as co-catalyst gave a polypropylene of which the non-pentane soluble fraction shows a dyad tacticity, [m], of 80%. The non-isotactic structure, as derived from the ^{13}C NMR studies, is mmrrmm. At this low isotacticity the melting point is low as well: 94 °C. Kaminsky[48-50] reported higher isotacticities using the (ethylene)bis(4,5,6,7-tetrahydro-1-indenyl) zirconium dichloride combined with the same co-catalyst. The xylene solubles are below 1 % which is an extremely low value, normally only observed for heterogeneous catalysts at very low polymerization temperatures. The triad distribution for the polymer, derived from NMR, gives [mm] = 95, [mr] = 3.2 and [rr] = 0.9 %, a high isotacticity indeed. The observed ratio of the mr and rr triads suggests other errors than just the mmrrmm type of error from the heterogeneous catalyst. More data are required to confirm this. A more recent publication[51] gives further details of the characterization of a similar polymer as mentioned above, made on the S-enantiomer of the transition metal compound at -10 °C. The polymer shows a bimodal molecular weight distribution, suggesting two different active sites. By extraction with n-heptane a surprisingly selective separation is brought about, with the individual fractions being about equal in size and showing a narrow distribution (Q≈2.7). The surprising feature of these samples lies in their melting behaviour and NMR analysis. The heptane insoluble fraction is nearly identical to what we know of isotactic polypropylene: it melts at 160 °C and shows in NMR the expected resonances together with some small ones indicating tail-to-tail/head-to-head sequences. The heptane solubles however melt at around 149 °C, and in NMR not only some inversions are observable but also resonances stemming from the following structure

$$-C-C(CH_3)-C-C(CH_3)-C-C-C-C-C(CH_3)-C(CH_3)-C-$$

This structure arises by 1,3-hydrogen transfer polymerization of propylene as surmised by the authors. Its probability of occurrence, on the sites generating the heptane solubles is 0.068, i.e. once in 150 propagations. Apart from this the normal 1,2-polymerized sequence contains as main non-isotactic placement the mmrrmm sequence, similar to the one mentioned above for heterogeneous catalysts.

Tsutsui et al[52] characterized the low molecular weight, atactic polypropylene made on Cp_2ZrCl_2 and methylaluminoxane, this enables one to discern the end-groups easily. One observes a true atactic structure ([mm] = 22 %m/m), very regiospecific. The end-groups are isopropenyl and propyl in similar quantities:

$$CH_2=C(CH_3)-C-C(CH_3)-C-C(CH_3)- \quad \text{and} \quad CH_3-C-C-C(CH_3)-C-C(CH_3)-C-$$

These end-groups are formed by transfer with monomer and probably not by β-elimination followed by re-initiation, as in copolymerization with ethylene exactly the same end-groups are observed, whilst if β-elimination occurred one would have expected some ethyl end-groups as well. In very low yield polymer made in the first 5 seconds of reaction also methyl end-groups can be found, probably originating from alkylation of the Zr-Cl by the methylaluminoxane.

In a sequel to the above, the same group studied[53] the polymer structure generated on rac.-ethylenebis(1-indenyl)zirconium dichloride and methylaluminoxane as function of the polymerization temperature. In fractionation with boiling solvents the polymers appeared to be fairly homogeneous. Pentane solubles are very low. The molecular weights do increase with decreasing polymerization temperature, reaching a plateau at about -10 °C and below. The distribution is narrow. In NMR all fractions show a very high isotacticity, in addition to the high mmmm peak, only mmmr, mmrr and mrrm peaks are observed, i.e. the active centre determines the stereoregularity. However there is some regio-irregularity, both 2,1 and 1,3 insertion are observed. The first leads to head-to-head and tail-to-tail structures, the other one to the structure already mentioned above. The quantities of these chain irregularities increase with polymerization temperature: from about 0.1 %mol to 0.7 %mol from -30 °C to 50 °C (sum of both structures). Whether the amount of irregularity is higher in the more soluble fractions is not disclosed, but is expected. The melting points in DSC are sharp, and increase with decreasing polymerization temperature (130.6 to 151 °C when T_{pol} decreases from 50 to -30 °C). The values are still considerably lower than "standard" polypropylene. Apparently the regio-irregularities affect the melting point to a large extent, possibly to a larger extent than the other chain error (mmrrmm).

Upon using the tetrahydro-derivative of the zirconium catalyst at -15 °C Tsutsui et al[54] prepared a polypropylene with 58 %m/m boiling heptane insolubles. The differences between the two fractions were marginal in terms of melting point (151 and 153 °C) and isotacticity; the main difference is in the molecular weight whilst the regio-irregularities (only 2,1 insertion this time) are only slightly higher for the solubles (0.66 %mol) than for the insolubles (0.51 %mol). No 1,3 polymerization is observed, which could be related to lower polymerization temperature applied (although the difference with Soga and Kaminsky's study[51] is only marginal). Zambelli (see below) does observe 1,3 polymerization at his highest polymerization temperature of 25 °C.

The Zambelli group also studied the newer homogeneous catalysts[55] with their well renowned NMR techniques. Various types of homogeneous catalysts were compared in their regio- and isospecificity, in all cases of course using methylaluminoxane as the co-catalyst. Cp_2Tidiphenyl gives high regiospecificity (as expected from titanium) but low isospecificity. With the ethylene-bis(tetrahydroindenyl)$ZrCl_2$ at 0°C the opposite is observed, i.e. vic. methyl groups are observed but in very isospecific sequences, giving rise to sharp peaks in the NMR. At 25 °C 1,3 addition is observed as peaks stemming from 4 methylene groups are present. Apparently this addition mode is fairly temperature dependent.

A far less isospecific, homogeneous catalyst has been developed by Fontanille et al[56], made by reacting $TiCl_4$ with long-chain, soluble lithium compounds such as polybutadienyl lithium. (A more detailed description is given in the polyethylene synthesis chapter, see chapter 9.) A low crystalline polypropylene is formed with this catalyst, with as melting range 130 to 145 °C. In NMR mmmm is still found as main sequence, although with a low overall isotacticity ([m]≈60%). Whether this system is as regiospecific as the $TiCl_3$-based systems is not totally clear.

It also appears that only $TiCl_4$/$AlEt_3$ as catalyst gives rise to a sizable amount of non-regiospecific structures (inversions or head-to-head and tail-to tail structures)[44], mostly in the soluble material however. A low amount of these structures is present in $TiCl_3$/$AlEt_3$ catalyzed polymerizations. Using a sensitive gas-chromatographic method Tsuge et al[57] have obtained the following values for the fraction of inversions: 1.7 % in an extracted polypropylene made by $TiCl_3$ (probably Toho)/$AlEt_2Cl$ and 6.8 % in the corresponding atactic from the same original polymer. These values are much higher than similar observations made elsewhere (see above) and we would conclude that practically no inversions are present in commercial polypropylenes, i.e. the catalysts applied are very regiospecific.

We turn now to a more detailed study on the effects of different $TiCl_3$ preparation conditions on the resulting polypropylene. Various catalysts were polymerized under standard conditions at 70 °C, allowing an easier comparison of the polymers. The catalysts used were a few of Shell's low temperature $AlEt_3$ reduced ones (A and B), an $AlEt_2Cl$ reduced one (C), the aluminium reduced ones Stauffer and Toho, and finally an example of the Solvay catalyst. The relevant data on the polymerization performance are given in Table 2.6, this performance has been discussed in chapter 1. The level of xylene solubles found for polymers with these $TiCl_3$ catalysts ranges from about 3 to 6 %m/m. The highest values were found with the Stauffer and $AlEt_2Cl$ reduced catalysts. The level of isopentane solubles was, naturally, lower but showed a wider range

TABLE 2.6

Performance of different catalysts under standard polymerization conditions Standard conditions: [$TiCl_3$] about 3 mmol/l except for high activity catalysts where it was halved, $AlEt_2Cl$ co-catalyst with Al/Ti = 2, temperature 70 °C, isooctane solvent, propylene pressure 2.6 bar, time 4 hours, 0.6 %vol hydrogen in the gascap.

Catalyst type*	Activity, yield. $h^{-1}bar^{-1}$	Melt Index g/10 min	Reactor solubles %m/m	Solubles isopentane %m/m	Solubles xylene %m/m	Yield Stress original polymer MPa	Yield Stress xylene residue MPa
A	42	8.6	n.d.	1.4	3.3	39	41.5
B	46	3.9	2.1	1.1	3.6	37	40.5
B-act	73	3.2	2.9	1.1	3.9	36.5	40
C,β	52	n.d.	38	n.d.	44	10.5	34.5
C,γ	30	3.7	7	2.0	5.5	33.5	40
Stauffer AA	35	3.0	7	3.7	6.2	32	39
TOHO TYC	47	2.0	3.5	0.8	4.1	38	n.d.
Solvay	147	3.6	0.9	0.4	4.7	35.5	39.5

*: B-act is an activated form by a hot decanting in the β to γ conversion, C,β made at high concentration which leads to a mixture of β and γ forms, Solvay catalyst prepared according to the description in the German Patent Application 2,213,086 example 1.

from 0.4 to about 4 %m/m. This is a powder morphology related effect and has been mentioned before (see chapter 2.3). The reactor solubles were also largely morphology dependent; they correlate much better with the isopentane than with the xylene solubles.

The composition of the solubles is given in Table 2.7. The crystallinities as measured by DSC and X-ray crystallography were low and no striking differences between the solubles obtained with

TABLE 2.7

Characterization of xylene solubles obtained from polymers made with different catalysts

Catalyst type	DSC data		X-ray data			NMR data				GPC data		
	Crystallinity, %m/m	Melting points, °C	Crystallinity, %m/m	Relative occurence of		mm %	mr %	rr %	m %	$M_w 10^{-4}$	Q	R
				syndio	iso							
A	5.1	49,71	6.3	xx	xx	30	31	39	46	2.8	9	5
B	6.1	50	5.3	xx	xx	28	33	39	44	3.5	8	3.9
B-act	3.5	46,65	5.0	xxx	-	24	33	43	41	6.4	11	7
C,β	-	-	3.8	?	x	28	36	36	46	7.3	8	4.4
C,γ	5.9	49,71	5.2	xx	xx	-	-	-	-	5.0	10	5.7
Stauffer AA	2.9	46,65	3.9	?	?	27	34	39	44	5.9	10	6.9
Solvay	1.8	47	1.4	xxx	-	28	32	40	44	4.5	7	3.2

different catalyst were observed. In ^{13}C-NMR measurement the solubles show similar tacticities, even those generated on a β-TiCl$_3$ catalyst. Slightly lower values for the isotacticity were, however, found in the case of the activated catalyst, although the differences are small.

One frequently measured property of the isotactic fraction is the final melting point. Since the presence of solubles does not affect this value very much, the measurements are mostly made on the total polymers. In Table 2.8 the TMF's have been collected for polymers made with various catalysts. From the comparison of the various polymers obtained under standard polymerization conditions, only the Solvay-catalyzed polymer really stood out. The TMF's of the other polymers were quite close to one another, the range being 0.8 °C with the "C" polymer at the low end of the scale. The yield stresses of the recrystallizates are all very high and nearly

TABLE 2.8

Final melting points

Catalyst	TMF °C
A	166.9
B	167.3
B-act	167.4
C	166.6
Stauffer AA	166.9
Solvay	168.9

identical (see Table 2.6). Even those from polymers with a high solubles content had nearly the same high yield stress, indicating that the quality of the isotactic fraction was nearly constant.

In catalyst preparation via reduction of $TiCl_4$ with aluminium alkyls the ratio of the reactants is an important variable in determining the properties of the resulting catalyst. When the Al/Ti ratio is increased in such cases, one of the changes in the catalyst is an increase in the concentration of $AlEtCl_2$ present and a decrease in the amount of $AlCl_3$ generated (at Al/Ti=0.33 in the catalyst preparation $AlCl_3$ is the main product, increasing the ratio to 0.5 gives mostly $AlEtCl_2$). The accessible $AlCl_3$ in the catalyst is converted to $AlEtCl_2$ upon reaction with the $AlEt_2Cl$ co-catalyst. Making a number of assumptions regarding the fraction of the $AlCl_3$ in the catalyst which is available for reaction with the co-catalyst, the range of the ratio of $AlEtCl_2$ present in the polymerization to the $TiCl_3$ is about 0.1 for the Solvay catalyst up to 0.6 for the catalysts made with a large excess of reducing agent. The TMF appears to be inversely proportional to this ratio suggesting a role of $AlEtCl_2$ in the formation of propagating errors.

It is widely known that on conversion of the β to the γ form of $TiCl_3$ the level of solubles in the resulting polymer decreases greatly. This decrease is a function of heating time and temperature. More intensive heat treatments leads to a decrease in xylene solubles, in melt index, in activity and to an increase in the TMF. If indeed the presence of $AlEtCl_2$ is one of the causes of propagation errors being generated one is led to the assumption that upon prolonged heating of $TiCl_3$ at temperatures around 170 °C, $AlCl_3$ is transported into the bulk of the crystal, becoming less and less available for reaction with the co-catalyst.

Similar studies with third generation and homogeneous catalysts have still to be reported.

Effects of the catalyst type on the molecular weight and the molecular weight distribution have been discussed in chapter 1.6.

2.6.2 The effect of the co-catalyst on the polymer produced

Activity of the catalyst in the polymerization and the stereospecificity of the resulting polymer are very dependent on the co-catalyst used. Many investigations have dealt with this topic, see for example ref. 41, 42 and 58. In commercial terms, with $TiCl_3$ catalysts one commonly uses $AlEt_2Cl$ which shows a good balance of activity and specificity.

The absolute co-catalyst concentration applied in a polymerization with first generation $TiCl_3$ catalysts is not a very sensitive parameter, provided a certain minimum is used. Probably this is necessary to scavenge a number of very reactive impurities, such as oxygen and oxygenated compounds. Above the minimum a plateau exists where the polymer properties are not to any appreciable extent affected. At very high values (e.g. Al/Ti ratio's of 10 at millimolar $TiCl_3$ concentrations) adverse effects are noted resulting in somewhat higher solubles and slightly lower TMF's.

A more detailed study looked in particular into the qualitative effects of varying the co-catalyst type on the soluble and non-soluble, isotactic fractions, since the effect on the quantity was roughly known from earlier studies. The following co-catalysts were compared : AlR_3, AlR_2Cl, $AlRCl_2$, $AlRCl_2$ + HMPTA (hexamethylphosphoric acid triamide), AlR_2I and $AlR(OR)Cl$. The data obtained for the polymerization performance (activity and solubles level) and the characterization of the solubles and recrystallizates are given in Tables 2.9 and 2.10. The activity as function of the co-catalyst followed the usual pattern, i.e. AlR_3 is about 4 times as active than $AlEt_2Cl$, while the $AlEtCl_2$-co-catalyzed system is almost inactive. Both $AlEt_2I$ and the $AlEtCl_2$ + HMPTA combination show intermediate activities, while the ethoxy derivative is, especially at 80 °C, slightly more active than the standard co-catalyst. The effect of polymerization temperature is considered in more detail later.

The lowest levels of solubles have been found for polymers

TABLE 2.9

Effect of the co-catalyst on the polymerization.
Catalyst: $TiCl_3$ made at low temperature; no hydrogen present in the polymerization

Co-catalyst type	Polymerization Conditions Al/Ti	Temp, °C	Propylene partial pressure, bar	Time, h	Activity, yield. $h^{-1}bar^{-1}$	Solubles isopentane, %m/m	xylene, %m/m	LVN total polymer, dl/g	Melt Index, g/10min
trioctyl-aluminium	2.0	70	2.6	3	195	n.d.	72	2.0	n.d.
AlEt$_2$Cl	2.0	60	4.7	4	25	0.31	2.0	9.8	<0.01
	2.0	80	4.45	4	40	0.70	4.9	4.1	0.19
AlEtCl$_2$	1.6	80	4.45	7	0.3	-	27	1.9	-
	1.4	94	4.12	7	0.3	-	31	1.6	-
AlEtCl$_2$ + HMPTA	2.0	60	4.7	4	10	0.3	1.5	5.0	0.13
	2.2	80	4.45	4	13	0.66	2.8	3.3	1.2
AlEt$_2$I	2.1	60	4.7	5	10	0.28	1.2	10.8	<0.01
	2.2	80	4.45	5	6	-	4.9	5.3	0.2
AlEt(OEt)Cl	1.9	60	4.7	4	19	0.56	3.0	10.7	<0.01
	1.8	80	4.45	2.5	48	1.2	4.5	4.7	0.2

made with the AlEtCl$_2$ + HMPTA combination, although these levels were accompanied by low activities. The iodide, which has been claimed to give high stereospecificity[41,59], does so only at the lower temperature of 60 °C, in line with Natta's data. The optimal combination of activity and solubles level is obtained with AlEt$_2$Cl at the 60 °C polymerization temperature, whereas at 80 °C AlEt(OEt)Cl is slightly better than AlEt$_2$Cl. The hydrogen response of the latter was however not checked. Natta concluded from his studies that the ethoxy compound reacts with the AlCl$_3$ present leading to formation of AlEt$_2$Cl[60]. The removal of AlCl$_3$ might then account for the higher polymerization rate observed.

Characterization of the solubles showed a high crystallinity for the solubles from polymers made with AlEtCl$_2$. This may have been partly due to the rather low molecular weight of these polymers leading to a high solubility in the xylene solubles test. The same applied to the solubles of the (AlEtCl$_2$ + HMPTA)-co-catalyzed polymer. The tacticities of the solubles (except in the cases where AlEtCl$_2$ and variations in the polymerization temperature were used) were not greatly affected by the various co-catalysts, and even the solubles from a polymer made with AlR$_3$ as co-catalyst did not stand out in this respect.

The MWD's of some of the solubles were extremely broad owing to the presence of a high MW tail in low amounts. With $AlEtCl_2$ as co-catalyst, lower MW's were found for both the whole polymer as well as the solubles. This is due to the greater relative rate of chain transfer with this co-catalyst. All the other co-catalysts gave polymers with about the same molecular weights, which in the case of some of the solubles were temperature independent. Contrary to this, the MW of the isotactic residue was found to decrease strongly with increasing temperature.

As regards the influence of the co-catalyst on the isotactic residue, only its steric purity has been looked into. From both the TMF and the NMR data it is clear that the iodide gave polymers of the highest purity, followed by $AlEtCl_2$ + HMPTA as co-catalyst. The lowest tacticity value was found when $AlEtCl_2$ was used as co-catalyst, although the range in TMF-values (all polymers) was not too large, about 4 °C. In the $AlEtCl_2$-co-catalyzed polymer the bulk of the errors were syndiotactic in nature, contrary to our findings with all the other polymers studied, in which mmrrmm is the most frequently observed error.

TABLE 2.10

Effect of the co-catalyst on the polymer structure
The data are mentioned in the same order as in the previous table, i.e. in the order of the polymerization temperature. TMF is the average of at least three measurements.

| Co-cata-lyst type | Characterization of xylene solubles | | | | | | | | | | xylene residue | | | | |
| | X-ray data | | 13C-NMR data | | | | GPC-data | | | TMF | NMR-data | | | |
	Crystal-linity, %m/m	Relative occurence of syndio iso	mm %	mr %	rr %	m %	M_w*10^{-4}	Q	R	°C	mm %	mr %	rr %	m %
trioctyl-aluminium	2	x ?	29	34	37	46	16.5	10	4.4	n.d.	-	-	-	-
$AlEt_2Cl$	6	x x	24	31	45	39	15.0	26	10	166.7	96	2	2	97
	6	x x	30	30	40	45	17.0	8	2.7	165.5	95	3	2	97
$AlEtCl_2$	10	x xxxx	47	26	27	60	8.1	52	29	163.7	89	5	6	93
	-	- -	-	-	-	-	-	-	-	164.3	92	3	5	94
$AlEtCl_2$+ HMPTA	-	- -	31	31	38	46	6.2	12	6.7	167.2	96	2	2	97
	9	x xx	36	29	35	50	5.2	12	11	166.5	96	3	1	98
$AlEt_2I$	6	xx x	27	30	43	42	20.0	50	33	167.5	99	1	0	99
	6	x x	31	32	37	47	11.0	22	28	167.2	94	3	3	96
AlEt(OEt)Cl	-	- -	25	35	40	43	13.4	11	10	166.7	94	4	2	96
	-	- -	26	33	41	43	12.7	12	6.5	165.5	90	6	4	93

These results show a reasonable correlation of TMF with the NMR-isotacticity. Generally the melting point decreases with increasing fraction of errors. The slope in the plot of TMF against [m] gives a value of 0.6 °C per unit in [m] or 1.2 °C per %mmrrmm error. Martuscelli and co-workers[61-63] studied the crystallization behaviour of polymers of varying chain regularity, and the properties of the resulting crystals in considerable detail. Generally it is found that chain irregularities are not totally excluded from the folded chain crystal and consequently the enthalpy of fusion decreases with increasing chain irregularity. Moreover the fold length in the lamella appears to have no relationship with the sequence length of the regular structure. Most often the lamellar thickness increases with increasing irregularity. In their study on isotactic polypropylene[63], in which polymers showing a wide range of isotacticities were used, they observed a decrease in the melting temperature of about 1 °C per 1% decrease of the isotactic pentads. This is completely in line with our own findings, on a more narrow and homogeneous polymer series. A much steeper relationship is given by Ewen et al[64] in their studies on homogeneous catalysts, a value of 6.5 °C per %mol inversions is mentioned!

A study was carried out with mixtures of the normal co-catalyst $AlEt_2Cl$ and $AlEtCl_2$ in order to investigate the molecular weight effects of the latter. Table 2.11 gives the data obtained.

TABLE 2.11

Effect of added $AlEtCl_2$ on the molecular weight of polypropylene

Polymerization conditions: low temperature reduced catalyst, $[TiCl_3]$ = 2 mmol/l, $[AlEt_2Cl]$ = 4 mmol/l, isooctane solvent, no hydrogen.

Polymerization temperature, °C	Propylene pressure, bara	$AlEtCl_2$ added, mmol/l	Activity, yield. $h^{-1}bar^{-1}$	Xylene solubles, after 4h, %m/m	LVN after 4h, dl/g
80	4.5	0	61	12.2	5.2
80	4.5	0.67	57	8.9	4.1
80	4.5	2	25	6.2	3.6
80	4.5	5	9	10.9	2.3
60	4.6	0	30	3.9	12.2
60	5.0	1	17	2.7	5.3
60	5.0	3	9	5.0	5.2
60	5.0	5	5	5.9	3.9

As expected the MW's generated in a fixed time decreased strongly with increasing AlEtCl$_2$ concentration. The decrease was larger at the lower temperature of 60 °C. Also the polymerization activity decreased strongly. The MW was however not directly proportional to the polymerization activity. This rules out the possibility that AlEtCl$_2$ merely dynamically blocks the sites for propagation, leaving the number of active sites unaffected. Thus AlEtCl$_2$ complexes with different sites to different extents.

In order to check whether the effect is equal for sites forming iso- or atactic polymer, one series was split into iso- and atactic fractions by the xylene solubles method and subsequently the MW's of the fractions were measured. The data are shown in Table 2.12. The molecular weights of both the iso- and atactic

TABLE 2.12

Effect of AlEtCl$_2$ on the molecular weight of iso- and atactic polypropylene. The 80 °C series from Table 2.11 is the basis of this test.

AlEtCl$_2$ added, mmol/l	total AlEtCl$_2$, mmol/l	$\dfrac{[AlEt_2Cl]_o}{[AlEtCl_2]_o}$	LVN, dl/g M_w*10^{-5}	$\dfrac{M_{w,iso}}{M_{w,a}}$
0	0.8	4.0	5.23 i 7.3	
			0.89 a 0.75	9.7
0.67	1.47	2.2	4.35 i 5.7	
			0.71 a 0.71	10.2
2.0	2.8	1.1	3.26 i 4.0	
			0.49 a 0.34	11.8
5.0	5.8	0.55	2.80 i 2.80	
			0.29 a 0.29	18.3

total AlEtCl$_2$: AlCl$_3$ ex the catalyst also transformed into AlEtCl$_2$ (i.e. 0.3 mol/mol TiCl$_3$) and an additional 0.1 mol/mol TiCl$_3$ from alkylation.

polymer are lowered by the addition of AlEtCl$_2$ but not exactly to the same extent as the ratio of the two changes. Plotting the reciprocal MW against the total AlEtCl$_2$ concentration present in the system, as in Fig. 2.20, gave a 10 times greater slope for the atactic fraction. Assuming equal propagation rates for the different sites this implies a much faster chain transfer with the co-catalyst on the atactic sites. Also the intercept is different, pointing to a starting molecular weight in AlEtCl$_2$-free system at

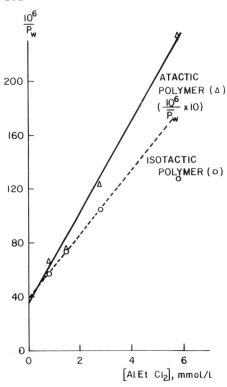

Fig. 2.20: *Effect of AlEtCl$_2$ on molecular weight of xylene solubles and the corresponding residue from polypropylene*

80 °C of one million and 120,000 respectively for iso- and atactic fractions. This could mean an eight-times faster β-hydrogen transfer for the atactic sites and/or more transfer with monomer for the atactic sites.

Using Cp$_2$Ti(CH$_3$)$_2$ with Solvay-TiCl$_3$ no chain transfer with this type of co-catalyst is observed over a wide concentration range[65]. This characteristic has been used in the preparation of blockpolymers, see chapter 5.

In the case of supported catalysts different rules apply. The co-catalyst used in these systems is more complicated by being a mixture of AlEt$_3$ and an aromatic ester, such as ethylbenzoate. These components do react with each other to form alkoxy dialkyl-aluminium species[66]. Changing the ester/AlEt$_3$ ratio affects strongly the activity as well as the selectivity, see for instance chapter 1 and references 4 and 67 to 70. Generally one observes a very strong decrease in the rate of atactic polymer production

coupled to a modest decrease for the isotactic polymer. Thus the
ester is a very selective solubles control agent, this is illustrated[70] for instance by the decreases of solubles from 40 %m/m to 3
%m/m at ester/AlEt$_3$ ratios of 0 to 0.4. This significant change in
overall polymer composition is accompanied by more modest changes
in the composition of either the atactic or the isotactic polymer.
For the atactic polymer a slight decrease in [m] is mostly
noted[68-70], and some broadening of the molecular weight distribution. The isotacticity of the regular polymer fraction increases,
illustrated by the change in TMF found in recent studies using
analogous systems. A range of 2 °C in TMF can easily be obtained.
As expected also the tensile properties of the resulting polymers
follow similar trends.

In a very detailed study Kakugo et al[71] also compared a number
of catalysts by applying the TREF (temperature rising elution
fractionation) technique to characterize the polymers. The range of
catalysts is wide with Toyo Stauffer HRA and AA, Solvay type,
β-TiCl$_3$ and supported catalysts; the co-catalyst with TiCl$_3$ is
normally DEAC, but substitution of the cloride has also been looked
into. The applied fractionation method is very good, i.e. the
xylene solubles residue is further extracted with boiling heptane;
this indeed leads in the end to residues which in NMR only show the
mmrrmm error. Similar to our own studies, the range of atactic
content is huge with such a catalyst series, however the range in
mmmm pentad in the residues is very narrow: only from 95.6 to 98.3
%m/m. The highest isotacticity in the residue is again found for
the combination of Et$_2$AlI with Solvay TiCl$_3$. A single correlation
of atactic level and fraction mmmm pentads is observed, which is
very surprising and not found in our series of tests. The interesting feature of their TREF studies of the isotactic fraction is
that with some catalysts two peaks can be distinguished in the plot
of cumulative eluted weight fraction and elution temperature.
Normal TiCl$_3$/DEAC derived polymers give for instance a small peak
at 109 °C, with the bulk at 120 °C; interestingly, the small amount
of isotactic polymer made on a β-TiCl$_3$ shows a peak at 109 °C. In
addition, when the normal TiCl$_3$-catalyst is aged in the presence of
the co-catalyst, the fractionation of the polymer subsequently made
on that catalyst shows an enhanced peak at 109 °C. This suggests an
active site population around two different types, each leading to
a distinctly different inherent tacticity. Similar effects have

been noted for supported catalysts, i.e. in the absence of a donor
in the co-catalyst a peak is seen at 111 °C, whilst in the polymer
made in the presence of methyl-p-toluate the main peak is at 121
°C, but a small one at 111 °C is present as well. Interestingly
they observe a rather good relation between the fraction mmmm
pentad and the observed melting point in DSC; the slope of the plot
is 1.9 °C per %mol [mmmm]; which is appreciably larger than either
our own observations or those of Martuscelli[63]. These observations
certainly need further study as they point in interesting direc-
tions in the explanation of the composition of polypropylenes; the
studies could possibly be extended with true cross-fractionation to
exclude molecular weight effects, or be made to include polymers
made on homogeneous catalysts (i.e. of a single homogeneous composition) to check the discriminating power of the TREF technique.

2.6.3 The effect of polymerization conditions on the composition of the polymer

In this section factors such as the polymerization temperature, catalyst concentration, monomer concentration, the concentration of the chain transfer agent and the time of polymerization
are discussed. They determine the polymerization conditions. Apart
from kinetics, surprisingly little information is found in the
literature on these aspects, only the polymerization temperature
has been studied in relation to the amount of solubles. Also when
the polymer tacticity itself is the subject of study, practically
nothing can be found in the literature. A number of these variables
will be discussed in the following.

(i) solvent/diluent

In a practical sense one can distinguish two aspects in the
effect of solvent on the polymer(ization): firstly the effect of
the solvent structure as such, and secondly the effect of the
impurities present. As regards the former the polymerization rate
is known[72] to increase in aromatic solvents compared with aliphatic
ones for first generation catalysts, due to a change in the equilibria especially between $AlEtCl_2$ and the active sites. Also the
solubles level increases. In slurry polymerizations the morphology
can be affected if the interaction of the solvent and polymer is
large. This is of course dependent on solvent quality, with the
best solvents (such as cyclohexane and toluene) the bulk density of
the polymer is already lowered at temperatures over 60 °C (the same

holds for polyethylene). In a study into the effects of aliphatic solvents of different chain length, the best selectivities were observed with both n- and i-dodecane, for reasons unknown.

Regarding the effects of impurities, those which strongly interact with either catalyst or co-catalyst should be controlled. Apart from the chemical structure, their effect is also dependent on the concentration of the catalyst and co-catalyst as this determines the ratio of impurity to active sites. Examples of impurities interacting directly only with the catalyst are CO, acetylenes, conjugated dienes and probably sulphides. A number of those are so strongly interacting that they are useful in titrating the number of active sites in the catalyst. As an example of the effect of an impurity the presence of piperylene (1,3-pentadiene) in equal concentration to a $TiCl_3$ catalyst reduces the activity by 40 %. Oxygen and oxygenated compounds such as ketones and alcohols also interact with the co-catalyst. A small amount of an oxygen-containing compound, such as an alcohol, can be advantageous for the activity due to its strong complex formation with for example $AlEtCl_2$, thereby increasing the number of sites which are effectively active[73]. The sensitivity to impurities is rather large especially for the most active catalysts (used in low concentrations), therefore great precautions are taken in each manufacturing plant to ensure a good solvent quality.

(ii) catalyst concentration

For $TiCl_3$ catalysts containing $AlCl_3$ there is a general effect of increasing activity and solubles level upon decreasing the catalyst concentration. At the same time the molecular weight also increases. Both effects are due to the shift in the equilibrium between active sites and $AlEtCl_2$ with the catalyst concentration. At low catalyst concentration more sites are active and also more atactic sites are available if one assumes that atactic sites would show a stronger interaction with $AlEtCl_2$ than isotactic sites.

(iii) monomer pressure

One example of a reduction in solubles level with an increase in the monomer pressure has been reported[41]. From our experience, monomer pressure, or rather monomer concentration, has no noticeable effect on the polymer quality. This is proved by the equality of polymers from a low pressure process and one in which the

186

polymers are made in the liquid monomer. Even the sensitive measurement of TMF gives identical values, provided the catalyst and temperature are equal in both types of polymerizations (see below).

In polymerizations without a chain transfer agent the molecular weight is in principle dependent on the monomer concentration. In the system $TiCl_3$-$AlEt_2Cl$ at 60 °C the molecular weight is linearly proportional to the monomer concentration, i.e. doubling the one leads to doubling of the other, when analyzing at identical polymerization temperature and time. An example is given in Fig. 2.21. Chain transfer with propylene is clearly unimportant under those conditions or is compensated by the increase in propagation rate.

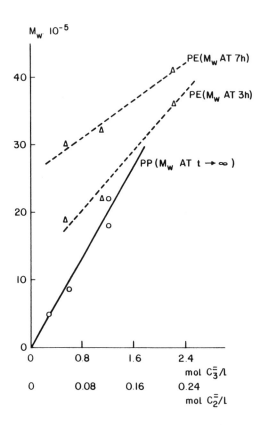

Fig. 2.21: Molecular weight dependence on monomer concentration (polymerization conditions: 60 °C, isooctane, low temperature reduced $TiCl_3$ 1mmol/l, DEAC 6 mmol/l, no hydrogen)

(iv) Chain transfer agent

Hydrogen is the only commercially applied chain transfer agent and most of this section will deal with this agent. Some interesting data with diethyl zinc will be mentioned at the end.

Hydrogen is known to influence a large number of properties in the polymerization of propylene (see also chapter 1). Since it acts as a chain transfer agent, its main effect is to lower the molecular weight of the polymer. In addition, the presence of hydrogen substantially increases the activity of the catalyst, particularly $AlEt_3$-reduced ones. No explanation is known for this effect. Also a negative effect has been noted in a lowering of the activity with time and an increase in the level of solubles by operating a cata-

TABLE 2.13

Characterization of the HOXS-extract ex homopolymers of a wide melt index range

Melt Index, g/10min	LVN, dl/g	Xylene solubles, %m/m	HOXS extract %m/m	XRD of HOXS-E Crystallinity, %m/m	DSC of HOXS-E T_m, °C	ΔH, cal/g	Fraction(a) i-PP in HOXS-E g/g	a-PP in HOXS-E %m/m
2.2	2.87	4.3	5.0	14.6	48,101	4.5	0.17	4.2
6.5	2.32	4.8	5.9	18.0	48,102	6.0	0.22	4.6
14	2.01	5.5	6.5	17.2	47,99	5.1	0.19	5.3
30	1.71	7.0	7.5	17.8	45,107	5.9	0.22	5.85
65	1.46	8.0	8.8	20.5	46,106	7.8	0.29	6.25
120	1.29	9.6	10.3	18.4	44,105	6.5	0.24	7.8
160	1.21	6.6	8.4	28.4	48,106,114	10.7	0.40	5.0
230	1.12	7.6	8.8	28.6	45,105	8.6	0.32	6.0

(a): for i-PP fraction: assume normal crystallinity is 60%, ΔH of pure crystalline PP is 45 cal/g thus divide observed ΔH by 0.6*45=27; the a-PP fraction is equal to (HOXS-E) * (1 - f_{iPP})

HOXS : follows a procedure similar to the xylene solubles measurement, equilibration is done at 50 °C instead of at room temperature (see chapter 13).

lyst in the presence of hydrogen[74,75], see also section 1.7.2.(i).

Own data on the solubles level and their characteristics from a series of polymers made with various levels of hydrogen are given in Table 2.13. A definite relation of solubles level and melt index cannot be given on the basis of these data, however from other series of polymers an increase of roughly one unit in solubles per factor 10 in melt index has been noted. The composition of the solubles has been studied by DSC, XRD, NMR and GPC. Both the DSC and XRD patterns do not change through the series, in the XRD both iso- and syndiotactic crystallinity is noted. The only change is in the level of crystallinity, which increases with melt index, in line with this the NMR isotacticity increases as well. Assuming the crystallinity to be from co-extracted low molecular weight isotactic polymer only, one can arithmetically calculate the amount of pure atactic polymer in the total extract. These data are also given in Table 2.13 and they illustrate clearly the increase in level of atactic polymer with melt index. The GPC data of a few extreme samples are given in Fig. 2.22. A low molecular weight peak at about 1500 grows strongly as the melt index increases. This is undoubtedly the co-extracted low molecular weight isotactic polymer, whose amount and molecular weight is determined by the extraction technique employed.

The same conclusion of increase in atactic polypropylene level with MI can be drawn from the following measurements. A few mate-

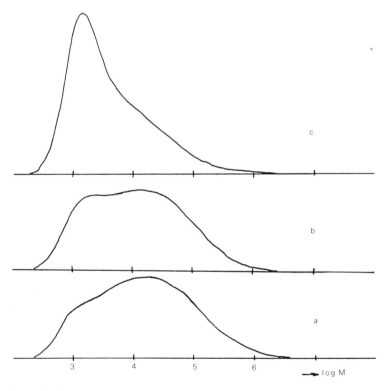

Fig. 2.22: Molecular weight distributions of hot-xylene solubles from homopolymer samples varying widely in melt index (in g/10min): 2.2 (a); 14 (b); 160 (c)

rials of widely different melt index have been extracted with a few different solvents at their boiling points. The extract is re-extracted with isopentane to split it into a largely atactic (isopentane soluble) and isotactic (isopentane insoluble) fraction. The results are given in Table 2.14. The observed increase in solubles level with melt index is caused by the increase in both atactic polymer (see increase in isopentane-soluble part) and co-extracted isotactic polymer (see increase in isopentane-insoluble part).

TABLE 2.14

Solubles from polymers with widely different melt indices

Melt Index g/10min	isopentane soluble part (absolute) of solubles obtained with			non-isopentane soluble part (absolute) of solubles obtained with		
	$n-C_6$	CyC_6	iC_8	$n-C_6$	CyC_6	iC_8
0.65	2.3	2.6	2.5	0.2	1.7	1.05
3.6	2.4	3.2	2.8	0.2	3.0	1.9
10	3.0	3.9	3.8	0.6	3.8	2.3
65	4.2	5.8	5.7	0.9	6.9	4.2

solubles measured in Soxhlet at the boiling point for 24 hours, $n-C_6$ is n-hexane, CyC_6 is cyclohexane and iC_8 is isooctane

The molecular weight distribution is found to broaden considerably upon lowering the molecular weight by hydrogen. Polymers made in the absence of hydrogen (all other conditions being equal) show the narrowest MWD. The M_z/M_w ratio increased from 1.8 to 2.8 when lowering the MW from 600,000 to 200,000, in laboratory-made polymers with an activated low temperature reduced catalyst. More data on the MWD are given in chapter 1.6. Similar effects are observed in commercial polymers.

   Not only is the MW of the isotactic polymer fraction lowered but also that of the solubles. In Fig. 2.23 the molecular weights of both fractions are plotted as function of the melt index. The lines are approximately parallel, indicating a similar response of the two types of polymer to hydrogen.

   The isotactic fraction of polymers with melt indices of 0.2 and 67 have been compared and found to be identical as regards tacticity and TMF. Additional data in smaller melt index ranges substantiates this, thus hydrogen does not affect the tacticity of polypropylene.

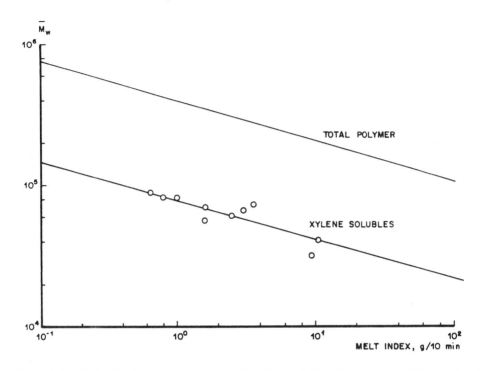

Fig. 2.23: *Relationship between the molecular weight of xylene solubles and melt index (molecular weight derived from LVN)*

Burfield[76] in a study aiming at very low molecular weight polypropylenes, preferably containing a large fraction of metal-alkyl bonds for subsequent oxidation, used large amounts of diethyl zinc as chain transfer agent. The resulting molecular weights were between 1000 and 10,000 (M_n), which was up to 10 times lower than in the absence of the chain transfer agent. Even at these immense levels of chain transfer agent the MWD remained broad, even a slight broadening was observed, both on a $TiCl_3$ and a supported catalyst.

Parsons and Al-Turki studied[77] the change in fraction of active sites upon addition of hydrogen to a catalyst system made up from supported $TiCl_4$/dibutylphthalate and TEA/diphenyl-dimethoxy-silane. In the presence of hydrogen the fraction of active sites increased measurably; for example from 0.4 to 0.5 %m/m of titanium present, the corresponding rate increase was about 50%, i.e. the increase in the number of sites only partly explains this effect.

(v) polymerization temperature

The polymerization temperature affects quite a number of dependent parameters such as activity, molecular weight, solubles level and tacticity. An overview of data extracted from the literature is given in Table 2.15. In cases in which the activation energies have been calculated, these have been based on the rate expressed as $yield.h^{-1}.[monomer]^{-1}$ in order to correct for the difference of the propylene solubility with temperature. When only the solubles data were available, we plotted the ratio of the amounts of isotactic to atactic polymer against reciprocal temperature to obtain the difference in activation energies between the isotactic and atactic propagation rates. However, since many of these plots were based on just two or three observations, their accuracy is limited.

TABLE 2.15

Effects of polymerization temperature on activity and solubles

Cocatalyst	Catalyst	Hydrogen in gascap, %v	Approx. monomer conc., mol/l	Temp. range, °C	Activation energies, kJ/mol, total polymer make	atactic polymer make	difference	type of solubles measured	Ref.
$AlEt_2Cl$	TiCl3, 170	0	0.5	50-70	-	-	27.3	boil.C6	own
	Stauffer AA	0.5	10	10-90	48.3	69.1	16.8	boil.C6	78
	Stauffer AA	0.7	10	54-71	33.6	63	29.4	xylene	79
	Stauffer AA	0.7	10	54-71	44.1	86.1	42	xylene	79
	Stauffer AA	6.3	1	50-90	47		-	boil.C7	80
	Stauffer HA	0	0.3	35-70	37.8	37.8	0	boil.C7	81
	Toho St. AA	?	?	40-85	-	80	-	xylene	15
	sesqui-reduced TiCl4	0.05	0.3	50-70	42	81.9	39.9	boil.C7	82
	TiCl3 made at low temperare	0	1	50-90	50.4	96.6	46.2	xylene	own
		0.6	1	70-85	46.2	96.6	50.4	xylene	own
		1	10	55-65	-	-	42	xylene	own
$AlEt_3$	TiCl3, 170	0	0.3	20-80	33.6	-	-		own
	α-TiCl3	0	0.5	30-70	58.8	-	-		40
	Stauffer AA	0	0.5	0-40	-	-	12.6	boil.C7	83
$AlEt_2I$	TiCl3, 200	0	1.5	70-100	-	-	84	boil.C7	41,84
	low T reduced	0	1	60-80	8.4	67.2	58.8	xylene	own
	"TiCl3"	0	0.3	15-130	-	-	44	boil.C7	85

The apparent activation energy for total polymerization activity in $AlEt_2Cl$ co-catalyzed systems is reasonably constant, varying only between 34 and 50 kJ/mol with an average of 44 kJ/mol.

The hydrogen concentration had no noticable effect on this activation energy. The activation energy found for atactic polymer make
varies widely from a low of 38 up to 97 kJ/mol. Part of this might
very well be due to the difference in the solubles measurement
applied. Still, in nearly all cases this activation energy (difference) is larger than that for the propagation itself, meaning that
the solubles level increases with increasing temperature. Of the
process conditions which were varied in this series, the monomer
and hydrogen concentration did not affect the activation energy;
the differences in the catalyst type could have been the main cause
of its wide scatter.

The data given for the other co-catalysts again underline the
profound effect of the co-catalyst on the behaviour of the system.
The largest temperature sensitivity as regards solubles production
was found with $AlEt_2I$, the activity of which was found to be only
weakly temperature dependent.

As mentioned above, polymerization temperature also affects
other properties and its effects on the molecular weight and
tacticity of both fractions, TMF and molecular weight distribution
are briefly discussed below using own data as a basis. Of these the
polymerization data are given in Table 2.16 whilst the characterization of solubles and residues is mentioned in the following Table
2.17.

TABLE 2.16

Polymerizations with low temperature reduced catalyst at different temperatures

Temp. °C	Propylene partial pressure, bar	Time, h	Hydrogen in gascap, %v/v	Activity, yield* $h^{-1}*[mon]^{-1}$	Melt Index, g/10min	LVN, dl/g	$10^6 \frac{[mon]}{Mw}$	Xylene solubles, %m/m
50	4.8	2	0	42	0.04	7.8	1.31	1.8
60	4.7	4	0	76	0.01	7.0	1.40	2.0
70	4.6	2	0	156	0.2	5.2	1.44	3.0
70	2.6	4	0	129	0.66	4.3	1.29	3.3
80	4.4	4	0	176	0.19	4.0	1.96	4.9
90	4.2	2	0	364	5.6	2.33	2.50	11.5
90	4.2	2	0	370	4.5	-	-	10.2
70	2.6	4	0.6	205	5.6	2.18	-	4.8
85	2.3	4	0.6	410	-	1.3	-	9.8
60	2.7	7	5	155	11	-	-	4.8
70	2.6	7	5	248	67	-	-	7.6

An increase in temperature is known to cause a large decrease in MW. When corrected for the change in propylene solubility our data show the difference in MW to be small between 50 and 70 °C and only to become appreciable above 70 °C. However, as no hydrogen was used, a check on the attainment of steady state should have been made to make this conclusion more firm (see below).

The MWD has been found to broaden with increasing temperature, contrary to the early findings of Chien[86]. For the atactic polymer we noted a steady _increase_ in $\bar{M}_n$ with increasing temperature, whereas M_w varied less and more irregularly. A plot of the reciprocal of M_n, corrected for monomer concentration, against reciprocal temperature gave a good straight line with a slope of -50 kJ/mol. The fact that, in this case, the molecular weight increased with increasing temperature can be attributed to the greater dissociation of the complex between the site and the aluminium alkyl, thereby reducing the possibility of chain transfer (see also the concluding discussion at the end of this chapter).

TABLE 2.17

Characterization data of polymers ex Table 2.16 (first part only)

Temp. °C	Characterization of xylene solubles								Ditto of residue			
	13C-NMR data				$\bar{M}_n 10^{-4}$	GPC data		10^6[mon]	TMF °C	13C-NMR data		
	mm %	mr %	rr %	i %		Q	R	Mn		mmrr %	mrrm %	i %
50	24	32	44	40	0.38	19	4.9	516	168.5	2.1	0.8	95
60	24	31	45	40	0.58	26	10	266	166.7	2.3	1.1	97
70	27	32	41	43	0.71	14	8	176	167.0	2.6	0.6	98
70	27	31	42	43	0.47	17	7	151	166.8	2.1	0.6	98
80	29	31	40	45	1.17	8	2.7	85	165.5	3.3	0.9	97
90	27	35	38	45	1.1	9	5.6	73	164.9	4.4	1.4	97
90	-	-	-	-	1.1	9	5.3	74	164.9	4.3	1.7	-

The variation in tacticity of the isotactic fraction with temperature was rather difficult to determine because of the inherent inaccuracies of the NMR method. The data obtained in the reported study showed the ratio of the two pentads mmrr and mrrm to be of the right order for stemming from mmrrmm structures exclusively. The quality of the isotactic polymer decreased with increasing temperature when looking at the intensity of the individual peaks, although no large change occurred between 50 and 70 °C. An inverse relationship with polymerization temperature was also found in respect of the TMF of the isotactic fraction. The TMF data of Table 2.18 together with some additional data have been plotted in Fig. 2.24, which showed a general downward trend in TMF over the whole temperature range studied. The change in TMF was about 1 °C per 10 °C in polymerization temperature.

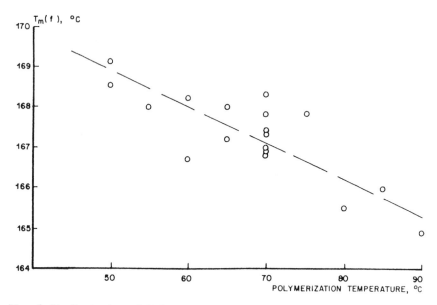

Fig. 2.24: *Variation of TMF with polymerization temperature (AlEt$_3$ reduced TiCl$_4$ as catalyst, DEAC as co-catalyst)*

A similar study was recently reported by Kakugo et al[15] but then at constant molecular weight. The activation energy observed for xylene solubles formation on the Toho Stauffer AA catalyst is completely in line with other observations (see Table 2.15), i.e. they also observe a rather steep increase in solubles with temperature. No change of isotacticity with polymerization temperature is reported by them. The reason might be that only NMR measurements have been carried out on the residues, and this effect has not been noticed due to the lower sensitivity of the NMR measurements as compared to melting temperatures. Note that the range of melting temperatures in Fig. 2.24 is only 3.5 °C.

Zambelli et al[85], in their study of the stereoregularity of polymers generated on TiCl$_3$/AlEt$_2$I, calculated the activation energy difference for syndiotactic and isotactic placements.

However, the basis for this calculation was the splitting of the polymer in a boiling heptane soluble and insoluble fraction (which is in itself not specific enough), measuring the stereoregularity of the solubles with NMR and averaging the dyad compositions taking the non-soluble fraction as totally, 100%, isotactic. In this way we think one determines the temperature effect on the rates of formation of iso- and atactic polypropylene, but not the above mentioned energy difference between a syndio- and isotactic placement on an isospecific site. One should have measured the isotacticity of the heptane insolubles (or rather the xylene recrystallizates) to arrive at a meaningful value. The authors also studied a syndiospecific catalyst system, without fractionating the polymer, therefore in this case their value is probably a true measure of the activation energy difference in syndio- and isotactic placements along one polymer chain.

Measurement of the yield stress, also a tacticity-related property, was meaningless for the polymers of Table 2.16 because the melt indices were both low and very different. However, the melt index of the polymers made at 90 °C were normal. The yield stress of these polymers were still low even after removal of the solubles (37.5 and 38 MPa resp.). This certainly reflected the lower quality of the isotactic fraction made at such a high polymerization temperature.

The tacticity of the solubles varied only slightly, the important feature being the increase in the rr triad with decreasing temperature. Since the solubles contain syndiotactic material this may have been due to the known tendency for the amount of syndiotactic polymer to increase with decreasing temperature[16]. The crystallinity of the solubles decreased with increasing temperature. The absolute amount of crystalline material in the solubles did increase, and this was undoubtedly due to the decrease in the MW at the higher temperatures leading to more isotactic polymer in the solubles.

In homogeneously catalyzed systems also increasing temperatures have a large decreasing effect on the molecular weight as reported in chapter 1.6. These polymers all show a narrow molecular weight distribution however, with Q values close to a value of 2, as expected for systems in which the lifetime of the chain is determined by propagation and transfer processes and having only one distinct type of propagating species. The distributions found with heterogeneous catalysts are always much broader. A nice illustration of the transition of these two situations has been given by Doi and Keii[87] in their study of VCl_4-catalyzed polymerizations. At low temperatures narrow molecular weight distributions and syndiotactic polypropylene is formed in a homogeneous system, whilst at higher temperatures more isotactic polypropylene is found in a heterogeneous system and showing very broad MWD's. Similar effects have been noted[88] with the catalyst system $Ti(OBu)_4/Al_2Et_3Cl_3$, although in this case the polymer produced is much more isotactic.

(vi) polymerization time

Apart from the kinetics, already discussed earlier (chapter 1), not many properties have been found to depend on the polymerization time. For δ-$TiCl_3$/DEAC Kakugo et al[71] describe a loss in isotacticity with time as measured by both the solubles level and the melting point of the polymer. The isotactic index drops from 89 to 85 %m/m and the melting point from 167.2 to 166 °C upon increasing the polymerization time from 1 to 6 hours. This suggests that the most stereospecific site is preferentially made inactive. In polymerization systems of α-$TiCl_3$/AlEt_3 time affected neither molecular weight nor the solubles level[40,41]. For the molecular weight distribution a slight decrease has been noticed in the width with time[86].

Numerous studies deal with the increase in molecular weight with time in systems not containing hydrogen[89]. This will be discussed in more detail in the chapters on copolymerization. One normally observes a very steep initial increase in molecular weight with time or yield, followed by a tapering off. Of interest is the observation that the atactic molecular weight increases with time in the same manner as the isotactic, i.e. the ratio of the two molecular weights remains the same. This suggests equal lifetimes for both types of sites.

2.7 CRYSTALLINE PROPERTIES

A few words are necessary on the crystalline nature of polypropylene. Isotactic polymer crystallizes easily, and the polymer chain structure in the crystal is a helix containing 3 monomer units in one complete turn. The elucidation of this crystal structure was another feat of Natta and his group in the early days of isotactic polyolefins. Some crystalline properties are given in table 2.18. A number of crystal modifications exist: the normally encountered one, α, is monoclinic and has a crystalline density of 0.94 g/cc and an equilibrium melting temperature around 180 °C. When one quenches the melt, a less perfect, β, modification is formed, trigonal with a lower crystalline density of 0.92 g/cc as well as a lower melting temperature of about 150 °C. In polymers which have shorter isotactic sequences (stereoblock or copolymers with other olefins) a γ-modification develops, of which more will be said in the copolymer chapter 4. Fig. 2.25 gives the X-ray diffraction patterns of the α, β and γ-modification. For an excellent, early review see reference 21.

TABLE 2.18

Crystalline properties of polypropylene (ex reference 90)

	isotactic	syndiotactic
melting point, °C	171	138
melting enthalpy, kJ/mol	8.79	2.09
density, g/cc	0.932-0.943	0.889-0.91
density atactic, g/cc	0.850-0.854	0.858

Polypropylene, as other crystalline polymers, crystallizes in lamellae which contain folded chains. As said above, a certain number of "alien" groups can be accommodated in the crystal itself, but only when there is no large mismatch (as for example in non-isotactic placements or in copolymers with ethylene).

The yield stress of a polypropylene sample also reflects the crystallinity, therefore this was measured on compression moulded plates both for total polymers as well as for the recrystallizates obtained from the xylene solubles test. A number of examples are given in Table 2.19. Clearly, considerable gains in yield stress are made when the solubles are removed. The data have been used to correlate the solubles level and the yield stress. For total polymers, in a melt index range of 2 to 5 this leads to:

$$YS = 41.7 - 1.15*XS \qquad \text{(standard deviation 1.1)}$$

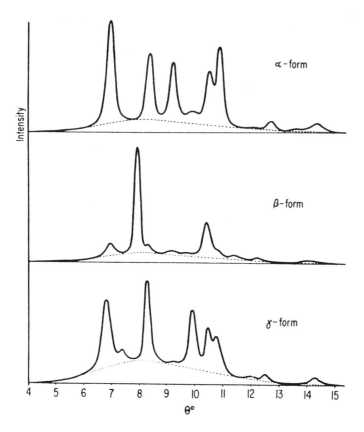

Fig. 2.25: X-ray diffractograms of α, β and γ modifications of polypropylene [from reference 21, by permission of publishers, Butterworth & Co, Ltd]

TABLE 2.19

Examples of the effect of solubles level on yield stress

Xylene solubles, %m/m	Melt Index, g/10min	Yield stress, MPa		Gain in yield stress
		original polymer	recrystalli- zate	
3.3	8.6	39	41.5	2.5
3.6	3.9	37	40.5	3.5
5.5	3.7	33.5	40	6.5
4.7	3.6	35.5	39.5	4
5.9	7.6	34.5	41	6.5
5.7	2.4	37	40.5	3.5
4.4	4.0	36	41	5
4.4	2.5	37.5	40.5	3

with the yield stress (YS) in MPa and the solubles in %m/m. Fig. 2.26 gives the experimental points. The inaccuracy is rather high, illustrating that quite a number of other factors affect this tensile property. Correlating the gain in yield stress with the amount of xylene solubles removed gives the following relationship:

$$\delta YS = -0.58 + 1.02*XS \qquad \text{(standard deviation 0.8)}$$

As a rule of thumb one can say that one unit of solubles corresponds to one unit of yield stress. The above relations are valid only for polymers made at polymerization temperatures of 60-70 °C, with a $TiCl_3$ catalyst made at low temperature and within a melt index range of 2 to 5. They are certainly not general since for example the quality of the isotactic material also plays an important role. However for every family of polymers such a relationship between fraction of solubles and the stiffness can be established.

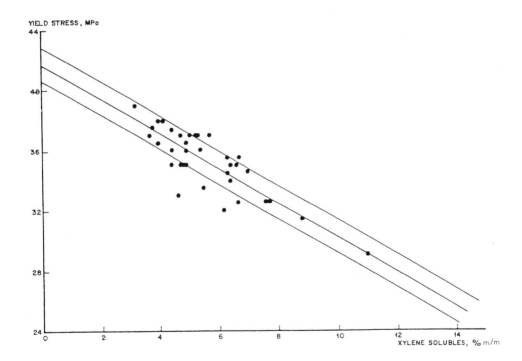

Fig. 2.26: Relation between xylene solubles and yield stress of original polymer (melt index range between 2 and 5, yield stress on compression molded samples)

The crystallization rate is of course dependent on such factors as the undercooling applied, molecular weight, its distribution and presence of nucleating agents. But effects of the catalyst type were also reported[91] in that polymers made on the third generation catalysts were much faster in crystallization than $TiCl_3$ derived products. And in line with this they show a finer spherulitic texture. It was suggested that increased tacticity was the true cause for this; it is thought however that more evidence is required to substantiate this statement.

2.8 CHARACTERIZATION OF SYNDIOTACTIC POLYPROPYLENE

Syndiotactic polypropylene has understandably not been studied in the same detail as its isotactic counterpart, since it is not a commercially attractive polymer. Its preparation has been described and reviewed by Natta[92] and Boor[16,93,94]. Starting the discussion on characterization with the composition of syndiotactic polypropylene, the earlier studies[16] with proton-NMR showed only poor resolution and no discriminating power was obtained. Using ^{13}C-NMR of course one has more possibilities, and a number of investigators report microstructures using this technique. Inspecting their data leads one quickly to the conclusion that the stereoregularity achieved is far lower compared to the isotactic case. The polymer with the highest syndiotacticity (made with VCl_4/$AlEt_2Cl$ as catalyst) has only 76% of its triads in the rr range and the remaining 24% in the mr triad[95]. Polymers made with $V(acac)_3$/$AlEt_2Cl$(ref 96) or VCl_4/anisol/iBu_2AlCl (ref 14) show a slightly lower tacticity, which is decreasing with increasing polymerization temperature, albeit only slowly. On the basis of limited data an activation energy difference of about 1.1 kJ/mol was calculated between an isotactic and a syndiotactic placement of the monomer. It would be interesting to extend the polymerization conditions to temperatures below -80 °C to enhance the steric purity of the polymer.

In a recent addition to the range of possibilities with the new homogeneous catalysts Ewen et al[97] reported a syndiospecific catalyst: isopropyl(cyclopentadienyl-1-fluorenyl) Zr or Hf dichloride with methylaluminoxane as co-catalyst. In a slurry polymerization high syndiotactic polypropylene is formed up to 70 °C. The syndiotacticity increases with decreasing polymerization temperature. The highest fraction of rrrr pentad of 0.86 (!) is observed at 25 °C. NMR shows that mm dyads bridge the syndiotactic runs in the structure rrmmrr, i.e. again a site stereochemical control is present. Almost no atactic polypropylene is present.

Mostly poorly resolved detailed NMR spectra are reported; the best spectrum found in the literature is given in Fig. 2.27. Using a simple fractionation method the syndiotacticity was shown to decrease with decreasing solubility at room temperature[98]. As mentioned above (chapter 2.4.5) the fractionation of solubles generated on an isospecific catalyst gave a very small amount of syndiotactic polymer. Its ^{13}C-NMR spectrum was given already in Fig. 2.17, and this polymer has [rr] = 65% , [mr] 24% and [mm] 11%, which is surprisingly not too far different from the quality obtained from syndiospecific catalysts applied at some 130 °C lower temperatures! Inspecting the spectrum shows the mmrm + rmrr pentad to be the second largest peak, followed by mmrr, mmmm, rmmm. No mrrm and only a very small amount of rmmr structure is found to be

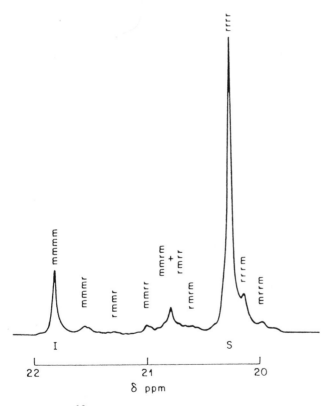

Fig. 2.27: ^{13}C-NMR of highly syndiotactic polypropylene [from reference 14. Reprinted with permission from Macromolecules. Copyright 1988 American Chemical Society]

present. This is consistent with the presence of a block polymer consisting of syndiotactic and heterotactic placements, a view also held by the groups of Zambelli and Doi.

Fractionation at increasing temperature of polymers made with VCl_4/$AlEt_2Cl$/anisole under different polymerization conditions, showed the fractions to decrease in syndiotacticity. The residue of the boiling heptane extraction was found to be pure isotactic polymer by X-ray diffraction[98].

The co-catalyst affects the stereospecificity of the $V(acac)_3$/AlR_2X catalyst-system. At -78 °C Doi et al[99] observed a decreasing syndiotacticity in the following series:

$$AlEt_2Cl > AlnPr_2Cl > AliBu_2Cl > AlEt_2Br$$

([r] decreasing from 81 to 66 and [rr] from 66 to 48). This implies a bimetallic propagation centre in which the co-catalyst actively partakes, contrary to isospecific first and second generation catalysts for which the isotacticity is far less or even independent of the alkyl and its chainlength (for the xylene residues!).

The regiospecificity of the catalyst used in syndiotactic polypropylene preparation is considerably lower than in the isospecific case. In all syndiotactic polypropylenes inversions are observed, i.e. head-to-head and tail-to-tail additions of the propene monomer. This give rise to $-CH(CH_3)-CH(CH_3)-$ and $-CH_2-CH_2-$ structures. These can be rather easily determined in IR, NMR[87] and pyrolysis-GLC[57]. The amounts observed range from a few percent up to 15 %, they are dependent on the type of catalyst used but especially on the co-catalyst applied[87,100]. The inversions are the connecting units between the syndiotactic and heterotactic blocks[100,101], which is understandable when assuming that syndiotacticity can only arise after a secondary insertion which is followed by a run of secondary insertions in a syndiotactic manner due to the steric influence of the methyl branch close to the vanadium. An inversion to primary addition removes this steric requirement leading to a more heterotactic structure. Doi has shown[100] that the fraction inversions is inversely proportional to the observed syndiotacticity, indicating that both aspects are governed by the same rules. The steric requirements of the growing chain-end are the determining factor is his conclusion. In a study[101] with ^{13}C-enriched aluminium alkyls and VCl_4, observing the stereochemistry of the end groups Zambelli et al concluded that the regiospecificity is dependent on the size of the alkyl group in the active complex. With a methyl group only primary insertion was found, whilst with an ethyl group on the vanadium secondary insertion also was inferred.

The secondary insertion was nicely proved by the group of Doi et al[102,103], using their living vanadium catalyst system (see chapter 5) to produce iodine terminated, low molecular weight polymers. They are made by adding molecular iodine to the living system, which then transforms according to

$$V-R + I_2 \quad ---> \quad V-I + R-I.$$

In proton-NMR the only structure observed in the polymer product was

$$I-CH(CH_3)-CH_2-CH(CH_3)-CH_2-$$

in its threo-form, i.e. the endgroup showed the same stereochemistry as the attached sequence.

^{13}C-NMR is more sensitive to structural detail and with this technique one observed :

-tail-to-tail units, but no head-to-head,

-ethyl end-groups in a majority, propyl endgroups being in the minority,

-the similar iodine containing species as mentioned above.

The end-groups are formed via :

V-Et + C=C-C --> V-C(CH_3)-C-Et, secondary insertion, i.e.

propyl end-group

or   --> V-C-C(CH_3)-Et, primary insertion, i.e ethyl

end-group.

Thus the observed endgroup distribution showed primary insertion to be predominant in the very first insertion, later propagation changes into a secondary one, giving rise to the tail-to-tail units. The change over is followed by almost-exclusively secondary insertion as the composition of the iodine terminated end proves.

It was subsequently shown[102] that with longer alkyl groups in the co-catalyst the erythro-placement in the iodine-terminated

polymer did increase, completely in line with the overall tacticities mentioned earlier[98].

It should be mentioned here that in the excellent studies of
Zambelli's group the mechanism of the propagation in syndiotactic
polymerization was unravelled, mostly by NMR investigations, see
for example ref 104. It can be concluded that the insertion is
normally secondary, but that upon a chance changeover to primary
insertion this is propagated as such, i.e. the last chain end
determines to a very large extent the stereochemistry. This was
also nicely proven in low-ethylene-containing copolymers with
propylene. The isolated ethylene units brought about a change from
secondary to primary insertion, and thus from syndio to isotactic
propagation.

The molecular weight distribution of these polymers can be
very narrow compared to the isotactic polypropylenes. With
$V(acac)_3/AlEt_2Cl$ at low temperature one obtains a living polymerization with its very narrow Poisson distribution[96], with Q values
in the order of 1.2. This is further described in chapter 5. With
other catalyst systems the MWD is found to be broader, but still
with Q values around 2 as is generally the case with homogeneous
catalysts with only one active propagating complex.

Some remarks on the crystallinity of these polymers conclude
this section. In DSC melting temperatures of up to 140 °C have been
observed[98], but this temperature showed no relationship with the
level of syndiotactic placements. This is in contradiction with the
findings of Boor and Youngman[16] who found a general increase of the
melting point from about 90 to 130 °C with the increase in the
crystallinity as measured by X-ray. The best value for the enthalpy
of fusion is 3150 J/mol. A kinetic study of the crystallization has
been reported[105].

Martuscelli[61] studied the morphology and properties of solution-grown crystals of syndiotactic polypropylenes of varying
tacticity. As mentioned before, upon taking polymers having lower
regularity, the thickness of the lamella in solution-grown crystals
increases and the heat of fusion decreases. The double melting
peak, regularly observed with syndiotactic polypropylene, is
interpreted by this author as an initial melting of the highly
disordered crystals, followed by the melting of the more regular
material. The relative enthalpies are a function of the syndio-

tacticity.

In the study of Lotz et al[14] crystals were grown with lamellae 6 to 10 nm thick and many μm in lateral dimension. They define also a new unit cell containing both left and right handed helices. The surprising feature of this is that such perfect crystals can form from such a poorly tactic polymer! From the lamellar thickness one can calculate that 30 to 50 monomer units have to fit in, which with a rr fraction of only 70% is rather a lot if one assumes that this can only be a pure stereosequence. Maybe the lattice is forgiving with respect to the stereostructure, this needs to be firmed up by conformational calculations.

2.9 CONCLUSION (mainly for TiCl$_3$-catalysts)

Having compiled the effect of catalyst and co-catalyst and of the polymerization conditions on the composition of the polypropylene formed, a number of conclusions can be drawn :

- in a first approximation the tacticity of the atactic and isotactic fraction (obtained by e.g. the xylene solubles method) is to a large extent independent of catalyst or co-catalyst chosen, whilst their respective amounts cover a wide range (e.g. for heterotactic 1 to 70%, and the corresponding 30 to 99 % range for isotactic). This conclusion holds for heterogeneous catalysts only.

- however, the tacticity of the isotactic fraction is found to vary in a narrow range dependent on catalyst/co-catalyst combination and the polymerization temperature (checked for first and second generation TiCl$_3$-catalysts),

- the observed non-isotactic structure is mmrrmm, which can only arise when stereochemical control is exerted by the active centre,

- similarly small differences in the molecular weight distribution are observed,

- mostly these differences can be traced to slight compositional differences in the catalyst. An example is the reactive fraction of AlCl$_3$ in TiCl$_3$ catalysts, which influences both the molecular weight and the tacticity. These smaller tacticity differences are best measured by using the TMF method which is more sensitive and quicker than the NMR method,

- heterogeneous titanium catalysts are regiospecific to a very high degree, i.e. inversions of propene addition are rare, and are predominantly found in the more heterotactic fraction,

- the greatest range in tacticity can be obtained with homogeneous catalysts, examples are true heterotactic polymer, syndiotactic

polypropylene and the mmmrmmm block structure,
- the most stereospecific $TiCl_3$-catalyst is the Solvay type, this
is due to the fact that all Al-species are removed in the catalyst
preparation.

The above conclusions can be reconciled by assuming that there
is but a limited number of different types of sites on the catalyst
surface. As a matter of fact this number is of course limited as
the degrees of freedom are small in number: a titanium atom can
reside in a plane, an edge, a corner, a ridge; it can possess one
or more vacancies; it can have different neighbours (e.g. Ti or Al)
and - in the total system - can be complexed with different co-
catalysts (mostly $AlEt_2Cl$ and $AlEtCl_2$ in first and second generation catalysts). Roughly speaking three kinds of polymer are formed
on a $TiCl_3/AlEt_2Cl$ combination: syndio- ,hetero- and isotactic. It
is assumed that the sites forming syndio and hetero have more than
one vacancy and are more open and exposed, whereas the isotactic
forming site is monovacant. The lower molecular weight of atactic
polymer fits in with this, as on a bivacant site the chain transfer
agent needs not even to compete with the monomer for complexation
as a prerequisite for the termination reaction, thus its chain
transfer rate will be larger. Also the use of a third component
(including $AlEtCl_2$) to lower the fraction of solubles can similarly
be explained. A stereoblock polymer could be formed on such bivacant sites that are still polymerization active after having
complexed an aluminium alkyl, supposedly becoming temporarily a
more isotactic site.

The very steep temperature dependence of heterotactic polymer
formation could be related to the equilibrium between the site and
aluminium alkyl (probably the dichloride). Increasing the temperature will lead to more dissociation and progressively more solubles, the overall rate expression will then be dependent on both
propagation and complexation rate, thus giving no straight line
Arrhenius plot. The explanation for the almost steady molecular
weight of the heterotactic polymer when changing the temperature
could be the compensation of the increased rates of chain transfer
at higher temperature by the increased propagation rate of the
polymerization active complex.

In the isotactic polymerization on the monovacant site the
aluminium alkyl has a minor, but distinct effect. The $AlEtCl_2$,

formed from $AlCl_3$ ex the catalyst and the $AlEt_2Cl$ co-catalyst, is preferentially complexed to the site. Thus the more $AlEtCl_2$ is present, the less active the catalyst will be. Moreover, the complex can also give rise to chain transfer by exchanging the alkyl ligand and formation of polymer-$AlCl_2$, which is removed from the site. Thus not only is the activity lowered but the molecular weight as well, upon increasing the concentration of $AlEtCl_2$. There appears also to be a relation with the fraction of mmrrmm errors, but we can not find an acceptable explanation for this.

In the work of Kakugo et al[1,71] it is suggested that in some catalysts two (classes of) isotactic sites are present as shown in the TREF/DSC analysis of the truly isotactic fraction, with both types of sites having their own extent of stereoregular control and kinetics. A relation was found between the amount of solubles and the mmmm isotactic fraction in the residue; it is not clear whether this is an universal relation. It would be very interesting to extend this type of study to a wider variety of polymers, including those made on single site, homogeneous catalysts.

At present there is no clear information on the <u>distribution</u> of propagation errors in the isotactic chain, is it random or not? Much work is required to clarify this question.

The molecular weight distribution of the polymers generated on heterogeneous catalsysts is very broad (Q >> 2), whilst using a homogeneous catalyst gives mostly the most-probable distribution (Q = 2) and sometimes even a Poisson distribution (Q = 1 + 1/n). These observations are independent of the fact whether the polymer is soluble or not under the polymerization conditions (see e.g. reference 106 for a case in point with polyethylene). The generally accepted[107] explanation for the broad distributions with heterogeneous catalysts is the spectrum of different active sites present on the catalyst surface, each with its own combination of propagation and chain transfer constants. The broadening effect of hydrogen on the molecular weight distribution remains elusive.

REFERENCES

1. M.Kakugo et al, Characteristics of ethylene-propylene and propylene-1-bu-
   tene copolymerization over TiCl3.0.33AlCl3/DEAC, *Macromolecules, 21 (1988)*
   2309-2313

2. J. van Schooten, H. van Hoorn and J.Boerma, Physical and mechanical
   properties of polypropylene fractions, *Polymer 2 (1961) 161-184*

3. G.Natta, M.Pegoraro and M.Peraldo, Chromatographic fractionation of
   stereoblock polymers, *La Ricerca Sci., 28 (1958) 1473*

4. P.Pino and R.Muhlhaupt,Die stereospezifische Polymerisation von Propylen:
   Ein Ueberblick 25 Jahre nach ihre Entdeckung, Angew. Chemie, *92* (1980)
   869-878

5. J.A.Ewen, Mechanisms of stereochemical control in propylene polymerizations
   with soluble group 4B metallocene/methylalumoxane catalysts,
   J.Am.Chem.Soc., 106 (1984) 6355-6364

6. X.Zhongde et al, Molecular characterization of poly(2-methyl-1,3-pentadi-
   ene) and its hydrogenated derivative, atactic polypropylene,
   Macromolecules, 18 (1985) 2560-2566

7. U.W.Suter and P.Neuenschwander, Epimerization of vinyl polymers to
   stereochemical equilibrium. 2. Polypropylene., *Macromolecules, 14 (1981)*
   528-532

8. P.F.Barron et al, Extensive stereoregularity changes induced in molten
   isotactic polypropylene by high energy radiation and other morphology
   effects, *J.Polym.Sc., Pol.Letters, 26 (1988) 225-228*

9. A.Zambelli, P.Locatelli and E.Rigamonti, Carbon 13 Nuclear magnetic
   resonance analysis of tail-to-tail monomeric units and of saturated end
   groups in polypropylene, *Macromolecules, 12 (1979) 156-159*

10. A.Zambelli, P.Locatelli, M.C.Sacchi and I.Tritto, Isotactic polymerization
    of propene : Stereoregularity of the insertion of the first monomer as a
    fingerprint of the active site, *Macromolecules, 15 (1982) 831-834*

11. T.Hayashi et al, Chain-end structures in polypropylene prepared with
    δ-TiCl3/DEAC catalytic system in the presence of hydrogen, *Macromolecules,*
    21 (1988) 2675-2684

12. G.Natta, I.Pasquon, P.Corradini, M.Peraldo, M.Pegoraro and A.Zambelli, *Atti*
    Accad. Naz. Lincei, Rend., Cl.Sci.Fis.Mat.Nat., 8 (1961) 539

13. I.Pasquon, G.Natta, A.Zambelli, A.Marinangeli and A.Surico, Some aspects of
    the polymerization mechanism of α-olefins to isotactic polymers, *J.Pol.Sc.,*
    C16 (1967) 2501-2516

14. B.Lotz et al, Crystal structure and morphology of syndiotactic polypropy-
    lene single crystals, *Macromolecules, 21 (1988) 2375-2382*

15. M.Kakugo et al, Microstructure of syndiotactic polypropylenes prepared with
    heterogeneous titanium-based Ziegler-Natta catalysts, *Makrom.Chemie, 190*
    (1989) 505-514

16. J.Boor and E.A.Youngman, Preparation and characterization of syndiotactic
    polypropylene, *J.Pol.Sc., A1, 4 1966) 1861-1884*

17. W.R.Moore and G.F.Boden, Heptane soluble material from atactic
    polypropylene. I. Fractionation and characterization of fractions,
    J.Appl.Pol.Sc., 9 (1965) 2019-2029

18. W.R.Moore and G.F.Boden, Heptane soluble material from atactic
    polypropylene. II. Interaction with liquids, *J.Appl.Pol.Sc., 10 (1966)*
    1121-1132

19. R.S.Porter, M.J.Cantow and J.F.Johnson, The effect of tacticity on
    polypropylene fractionation, *Makrom. Chemie, 94 (1966) 143-152*

20. E.J.Addink and J.Beintema, Polymorphism of crystalline polypropylene,
    Polymer 2 (1961) 185

21. A.Turner-Jones, J.M.Aizlewood and D.R.Becket, Crystalline forms of isotac-
    tic polypropylene, *Makrom. Chemie, 75 (1964) 134-158*

22. Y.Doi, Structure and stereochemistry of atactic polypropylenes. Statistical
    model of chain propagation, *Makrom.Chem., Rapid Commun., 3 (1982) 635-641*

23. Y.Inoue et al, Studies of the stereospecific polymerization mechanism of
    propylene by a modified Ziegler-Natta catalyst based on 125 MHz 13C NMR
    spectra, *Polymer, 25 (1984) 1640-1644*

24. G.Natta, G.Mazzanti, G.Crespi and G.Moraglio, Isotactic and stereoblock polymers of propylene, *Chim e Ind(Milan)*, _39_ *(1957) 275*

25. G.Natta, Properties of isotactic, atactic, and stereoblock homopolymers, random and blockcopolymers of α-olefins, *J.Pol.Sc.*, _34_ *(1959) 531-547*

26. G.Natta, Isotactic and stereoisomeric polymers, *Chim e Ind*, _77_ *(1957) 1009*

27. G.Natta, G.Mazzanti and P.Longi, Atactic and stereoblock polymers of α-olefins, *Chim e Ind*, _40_ *(1958) 183*

28. I.Pasquon, The coordinated anionic polymerization of propylene to atactic and stereoblock high molecular weight polymers. III : Characteristics of polypropylene obtained with the catalytic system VOCl3-Al(iBu)3, *Chim e Ind*, _41_ *(1959) 534*

29. P.W.O.Wijga, J.van Schooten and J.Boerma, The fractionation of polypropylene, *Makrom. Chemie*, _36_ *(1960) 115-132*

30. C.Wolfsgruber, G.Zannoni, E.Rigamonti and A.Zambelli, Stereoregularity of polypropylene obtained with different isospecific catalyst systems, *Makrom. Chemie*, _176_ *(1975) 2765-2769*

31. J.A.Ewen, Ligand effect on metallocene catalyzed Ziegler-Natta polymerizations; in Catalytic polymerization of olefins, T.Keii and K.Soga editors, Kodansha, Tokyo, 1986 (Studies in surface science and catalysis 25)

32. F.de Candia et al, Physical behaviour of stereoblock-isotactic polypropylene, *Makrom.Chemie*, _189_ *(1988) 815-821*

33. C.W.Tullock, R.Mulhaupt and S.D.Ittel, Elastomeric polypropylene (ELPP) from alumina-supported bis(mesitylene)titanium catalysts, *Makrom.Chemie, Rapid Commun.*, _10_ *(1989) 19-23*

34. C.W.Tullock et al, Polyethylene and elastomeric polypropylene using alumina-supported bis(arene) titanium, zirconium and hafnium catalysts, *J.Pol.Sc., Pol.Chem.*, _27_ *(1989) 3063-3081*

35. C.-K.Shih and A.C.L.Su, Poly-α-olefin based thermoplastic elastomers, Chapter 5 in "Thermoplastic Elastomers", editors N.R.Legge, G.Holden and H.E.Schroeder, Hanser Publishers, Munich, 1987

36. R.A.Setterquist, F.N.Tebbe and W.G.Peet, Tetraneophylzirconium and its use in the polymerization of olefins, p 167-192 in "Coordination Polymerization", editors C.C.Price and E.J.Vandenberg, Plenum Press, New York, 1983

37. A.Pavan, A.Provasoli, G.Moraglio and A.Zambelli, Microstructure and melting point of stereoregular polymers. A tentative experimental correlation for isotactic polypropylene, *Makrom. Chemie*, _178_ *(1977) 1099-1109*

38. Y.Doi, E.Suzuki and T.Keii, Stereoregularities of polypropylenes obtained with highly active supported Ziegler-Natta catalyst systems, *Makrom.Chemie, Rapid Commun.*, _2_ *(1981) 293-297*

39. E.Martuscelli et al, Stereochemical composition-properties relationships in isotactic polypropylenes obtained with different catalyst systems, *Polymer*, _26_ *(1985) 259-269*

40. G.Natta and I.Pasquon, The kinetics of the stereospecific polymerization of α-olefins, in Advances in Catalysis, Volume XI, Academic Press, New York, 1959

41. G.Natta, I.Pasquon, A.Zambelli and G.Gatti, Highly stereospecific catalytic systems for the polymerization of α-olefins to isotactic polymers, *J.Pol.Sc.*, _51_ *(1961) 387-398*

42. D.O.Jordan, Ziegler-Natta polymerization : catalysts, monomers, and polymerization procedures, in The Stereochemistry of Macromolecules, Volume 1, edited by A.D.Ketley, Marcel Dekker, New York, 1967

43. P.J.T.Tait, Ziegler-Natta and related catalysts, in Developments in Polymerization-2, Free radical, condensation, transition metal and template polymerizations, edited by R.N.Haward, Applied Science Publishers, London, 1979

44. Y.Doi, E.Suzuki and T.Keii, Regioselectivity and stereospecificity of titanium based catalysts for propene polymerization, p 737-749 in Transition metal catalyzed polymerizations, Alkenes and Dienes, Part B; R.P.Quirk editor, Harwood Academic Publishers, 1983

45. G.D.Bukatov, V.A.Zakharov, Yu.I.Yermakov and A.Zambelli, Comparative study

of polypropylene stereoregularity for one- and two-component catalysts based on titanium chlorides, *Makrom.Chem.*, *179* (1978) 2093-2096

46. K.Soga and H.Yanagihara, Extremely highly isospecific polymerization of olefins using Solvay-type TiCl3 and Cp2TiMe2 as catalyst, *Makrom.Chemie*, *189* (1988) 2839-2846

47. K.Soga and H.Yanagihara, Synthesis of highly isotactic polystyrene by using Solvay-type TiCl3 in combination with dicyclopentadienyldimethyltitanium, *Makrom.CHemie, Rapid Commun.*, *9* (1988) 23-25

48. W.Kaminsky, K.Kuelper, H.H.Brintzinger and F.R.W.P.Wild, Polymerization of propene and butene with a chiral zirconocene and methylalumoxane as co-catalyst, *Angew.Chemie, Int.Ed.Engl.*, *24*(1985) 507-508

49. W.Kaminsky, Preparation of special polyolefins from soluble zirconium compounds with aluminoxane as co-catalyst, p 293-304 in Catalytic polymerization of olefins, T.Keii and K.Soga editors, Kodansha, Tokyo, 1986

50. W.Kaminsky, K.Kuelper and S.Niedoba, Olefin polymerization with highly active soluble zirconium compounds using aluminoxane as co-catalyst, *Makrom.Chemie, Macromol. Symp.*, *3* (1986) 377-387

51. K.Soga, T.Shiono, S.Takemura and W.Kaminsky, Isotactic polymerization of propene with (1,1'-ethylenedi-4,5,6,7-tetrahydroindenyl)zirconium dichloride combined with methylaluminoxane, *Makrom.Chem., Rapid Commun.*, *8* (1987) 305-310

52. T.Tsutsui, A.Mizuno and N.Kashiwa, The microstructure of propylene homo-and copolymers obtained with a Cp2ZrCl2 and methylaluminoxane catalyst system, *Polymer*, *30* (1989) 428-431

53. T.Tsutsui, N.Ishimaru, A.Mizuno, A.Toyota, N.Kashiwa, Propylene homo- and copolymerization with ethylene using an ethylenebis(1-indenyl)zirconium dichloride and methylaluminoxane catalyst system, *Polymer*, *30* (1989) 1350-1356

54. T.Tsutsui, A.Mizuno and N.Kashiwa, Characterization of isotactic polypropylene obtained with ethylenebis(4,5,6,7-tetrahydro-1-indenyl) zirconium dichloride as catalyst, *Makrom.Chem.*, *190* (1989) 1177-1185

55. A.Grassi et al, Microstructure of isotactic polypropylene prepared with homogeneous catalysis : Stereoregularity, regioregularity, and 1,3 insertion, *Macromolecules*, *21* (1988) 617-622

56. F.Cansell, A.Siove and M.Fontanille, Polymerization of propene initiated with solubilized Ziegler-Natta Li/Ti(III) based systems, *Makrom.Chem.*, *186* (1985) 379-387

57. Y.Sugimara, T.Nagaya, S.Tsuge, T.Murata and T.Takeda, Microstructural characterization of polypropylene by high-resolution pyrolysis-hydrogenation glass capillary gas chromatography, *Macromolecules*, *13* (1980) 928-932

58. F.Danusso, Recent results of the stereospecific polymerization by heterogeneous catalysis, *J.Pol.Sc.*, *C*, *4* (1964) 1497-1509

59. Y.Doi et al, Role of surface halogen ligands of titanium metal sites in stereospecific polymerization with Ziegler-Natta catalyst, *Makrom. Chemie*, *176* (1975) 2159-2161

60. G.Natta, U.Giannini, E.Pellino and D.de Luca, Polymerization of propylene with TiCl$_3$ and alkyl-alkoxy aluminium chlorides, *Europ. Pol.J.*, *3* (1967) 391-398

61. A.Marchetti and E.Martuscelli, Effect of chain defects on the morphology and thermal behaviour of solution grown single crystals of syndiotactic polypropylene, *J.Pol.Sc.,Pol.Phys.*, *12* (1974) 1649-1666

62. E.Martuscelli, A review of the properties of polymer single crystals with defects within the macromolecular chain, *J.Macromol.Sci.-Phys.*, *B11* (1975) 1-20

63. E.Martuscelli, M.Pracella and A.Zambelli, Properties of solution grown crystals of fractions of isotactic polypropylene with different degrees of stereoregularity, *J.Pol.Sc., Pol.Phys.*, *18* (1980) 619-636

64. J.A.Ewen et al, Propylene polymerization with group 4 metallocene/alumoxane systems, in "Transition metals and Organometallics as catalysts for olefin polymerization", editors W.Kaminsky and H.Sinn, Springer, Berlin, 1988

65. K.Soga and H.Yanagihara, Isotactic polymerization of propene using living

catalysts originally found by Hercules Inc., Activation, stereospecific control and model of isospecific catalytic centres, p21-33 in Transition metals and organometallics as catalysts for olefin polymerizations, W.Kaminsky and H.Sinn editors, Springer, Heidelberg, 1988

66. B.L.Goodall, Super high activity supported catalysts for the stereospecific polymerization of α-olefins : History, development, mechanistic aspects and characterization, p 355 to 378 in "Transition metal catalyzed polymerizations : Alkenes and Dienes", Part A, Harwood Academic Publishers, 1983

67. Chapter 5.5 in "Catalysis by supported complexes", by Yu.I.Yermakov, B.N.Kuznetzov and V.A.Zakharov, Elsevier, Amsterdam, 1981

68. N.Kashiwa, The role of ester in high-activity and high-stereoselectivity catalyst, p 379 to 388 in "Transition metal catalyzed polymerizations, Alkenes and Dienes", Part A, R.P.Quirk editor, Harwood Academic Publishers, London, 1983

69. P.Pino, G.Guastala, B.Rotzinger and R.Muehlhaupt, Some aspects of stereoregulation by Lewis base in the polymerization of α-olefins with Ziegler-Natta catalysts supported on magnesium chloride, p 435- 463, ibid ref 68

70. V.Busico et al, Polymerization of propene in the presence of MgCl2-supported Ziegler-Natta catalysts. 2. Effects of co-catalyst composition, *Makrom.Chemie, 187 (1986) 1115-1124*

71. M.Kakugo et al, Microtacticity distribution of polypropylenes prepared with heterogeneous Ziegler-Natta catalysts, *Macromolecules, 21 (1988) 314-319*

72. T.Keii, Kinetics of Ziegler-Natta polymerization, Kodansha Ltd, Tokyo, 1972

73. T.Masuda and Y.Takami, Effect of oxygen on Et2AlCl-TiCl3 catalyst for propylene polymerization, *J.Pol.Sc., Pol.Chem.Ed., 15 (1977) 2033-2036*

74. E.M.J.Pijpers and B.C.Roest, The effect of hydrogen on the Ziegler-Natta polymerization of 4-methyl-1-pentene, *Europ.Pol.J., 8 (1972) 1151-1158*

75. D.N.Pirogov and N.M.Chirkov, The reactivity of hydrogen as a terminating agent in propylene polymerization over the catalytic system C5H5N:TiCl3/AlEt3, *Polym.Sc.USSR, 8 (1966) 1985-1991*

76. D.R.Burfield, The synthesis of low molecular weight hydroxy-tipped polyethylene and polypropylene by the intermediacy of Ziegler-Natta catalysts, *Polymer, 25 (1984) 1817-1822*

77. I.W.Parsons and T.M.Al-Turki, On the mechanism of action of hydrogen added to propene polymerization using supported titanium chloride catalysts with a phthalate ester/silane stereoregulating donor pair, *Polymer Commun., 30 (1989) 72-73*

78. German patent 2,046,020 to Chisso

79. U.S. patent 3,502,634 to Phillips

80. H.G.Yuan, T.W.Taylor, K.Y.Choi and W.H.Ray, Polymerization of olefins through heterogeneous catalysis. I. Low pressure propylene polymerizations in slurry with Ziegler-Natta catalyst, *J.Appl.Pol.Sc., 27 (1982) 1691-1706*

81. T.Keii, K.Soga, S.Go, A.Takahashi and A.Kojima, The dependence of the isotacticity of polypropylene on polymerization conditions, *J.Pol.Sc. C, 23 (1968) 453-459*

82. British patent 1,386,891 to Hoechst

83. K.Soga, Y.Takano, S.Go and T.Keii, Influence of SeOCl2 on the polymerization of propylene by TiCl3-AlEt3, *J.Pol.Sc., Al, 5 (1967) 2815-2823*

84. G.Natta, I.Pasquon, A.Zambelli and G.Gatti, Dependence of the melting point of isotactic polypropylenes on their molecular weight and degree of stereospecificity of different catalytic systems, *Makrom. Chemie 70 (1964) 191-205*

85. A.Zambelli, P.Locatelli, G.Zannoni and F.A.Bovey, Stereoregulation energies in propene polymerization, *Macromolecules, 11 (1978) 923-924*

86. J.C.W.Chien, Olefin poylmerizationa and polyolefin molecular weight distribution, *J.Pol.Sc., Al, (1963) 1839-1856*

87. Y.Doi, M.Takeda and T.Keii, Stereochemical structure and molecular weight distribution of polypropylenes prepared with vanadium-based catalyst

systems, *Makrom.Chemie*, *180* (1979) 57-64

88. Y.Doi, Y.Nishimura and T.Keii, Molecular weight distribution and stereoregularity of polypropylenes obtained with Ti(OBu)4/Al2Et3Cl3 catalyst system, *Polymer*, *22* (1981) 469-472

89. G.Natta, Kinetic studies of α-olefin polymerization, *J.Pol.Sc.*, *34* (1959) 21

90. Polymer Handbook, J.Brandrup and E.H.Immergut editors, 1975 edition, Wiley

91. P.Galli, Polypropylene : a quarter of a century of increasingly succesful development, p 63-92 in "Structural order in polymers", F.Ciardelli and P.Giusti editors, Pergamon Press, Oxford, 1981

92. A.Zambelli, G.Natta and I.Pasquon, Polymerization du propylene a polymère syndiotactique, *J.Pol.Sc.*, *C*, *4* (1963) 411-426

93. E.A.Youngman and J.Boor, Syndiotactic polypropylene, Macromolecular Reviews, Volume 2, p 33 to 69, Interscience, New York

94. J.Boor, Ziegler-Natta catalysts and polymerizations, Academic Press, New York, 1979

95. D.R.Burfield and Y.Doi, Differential scanning calorimetry characterization of polypropylene. Dependence of Tg on polymer tacticity and molecular weight, *Macromolecules*, *16* (1983) 702-704

96. Y.Doi, S.Ueki and T.Keii, "Living" coordination polymerization of propene initiated by the soluble V(acac)3-AlEt2Cl system, *Macromolecules*, *12* (1979) 814-819

97. J.A.Ewen et al, Syndiospecific propylene polymerization with group 4 metallocenes, *J.Am.Chem.Soc.*, *110* (1988) 6255-6256

98. T.Ogawa and H-G.Elias, On the structure of syndiotactic poly(propylenes), *J.Macromol.Sci.-Chem.*, *A17* (1982) 727-741

99. Y.Doi, T.Koyama, K.Soga and T.Asakura, Stereochemistry in "living" coordination polymerization of propene initiated by vanadium-based catalytic systems, *Makrom.Chem.*, *185* (1984) 1827-1833

100. Y.Doi, Correlation between regio- and syndiotactic specificity of soluble vanadium-based catalysts for propene polymerization, *Macromolecules*, *12* (1979) 1012-1013

101. P.Locatelli, M.C.Sacchi, E.Rigamonti and A.Zambelli, Syndiotactic polymerization of propene: regiospecificity of the initiation step, *Macromolecules*, *17* (1984) 123-125

102. Y.Doi, F.Nozawa, M.Murata, S.Suzuki and K.Soga, Mechanism of chain propagation in "living" polypropylene synthesis. 1H and 13C NMR analyses of chain end structures, *Makrom.Chem.*, *186* (1985) 1825-1834

103. Y.Doi, F.Nozawa and K.Soga, Stereochemical aspects on the secondary insertion of propene monomer into a vanadium-polymer bond, *Makrom.Chem.*, *186* (1985) 2529-2533

104. A.Zambelli et al, Model compounds and 13C-NMR investigation of isolated ethylene units in ethylene/propylene copolymers, *Makrom.Chemie*, *179* (1978) 1249-1259

105. R.L.Miller and E.G.Seeley, Crystallization kinetics of syndiotactic polypropylene, *J.Pol.Sc.*, *Pol.Phys.*, *20* (1982) 2297-2307

106. M.N.Berger, G.Boocock and R.N.Haward, The polymerization of olefins by Ziegler catalysts, Advances in catalysis, Volume 19, p 211 to 240, Academic Press, New York, 1969

107. U.Zucchini and G.Cecchin, Control of molecular weight distribution in polyolefins synthesized with Ziegler-Natta catalyst systems, Advances in Polymer Science, Volume 51, p 101 to 153, Springer Verlag, 1983

NB : refs 24-28 can also be found in Stereoregular polymers and stereospecific polymerization, Volume 1, G.Natta and F.Danusso editors, Pergamom Press, Oxford, 1967

Chapter 3

COPOLYMERS OF PROPYLENE AND OTHER OLEFINS ("RANDOM" COPOLYMERS AND RUBBERS). POLYMERIZATION ASPECTS.

3.1 INTRODUCTION

Polypropylene homopolymer is a highly crystalline polymer as was made clear in the previous chapters of this book. Its crystalline nature makes it a desirable polymer for those applications where a high strength/stiffness is a prerequisite, such as in fibres, films, pipes and injection moulded goods. Together with the high melting point this gives a product having attractive properties over a wide temperature range. For other applications however one needs for instance a lower melting point, as in heat-sealable films. Or one would like to have a more tough polymer for products which have to be used at lower temperatures, as the homopolymer appears to be very brittle at subzero temperatures (remember that the glass transition is around 0 °C). Finally, it would be very attractive to make a completely amorphous polymer on the basis of olefins as monomers, as in this way a saturated rubber could be made, being much more stable against oxidation compared to the usual unsaturated rubbers such as isoprene and butadiene rubber. All the above goals can be achieved by copolymerization of propylene with other monomers, and this forms the subject of the following chapters.

Many different types of copolymers can be made or thought of, an important distinction being the way the co-monomer is added. Possible modes are
    -stationary over the total polymerization, i.e. a constant feed ratio of monomers,
    -only present in a certain part of the polymerization,
    -a time dependent concentration (e.g. the so-called "tapered copolymerization")
The present chapter deals with those copolymers made in a process in which the monomers are present throughout the total polymerization, and at a constant concentration, the so-called "random copolymers". This - very trivial - name has to be contrasted to "block copolymers" which are those made with the other monomer(s) only present for part of the polymerization time. These will be the

subject of some of the following chapters. Whether the names truly describe reality will be made clear in the characterization chapters, the names are however the standard jargon in the polypropylene business. This distinction is still a useful one, as it separates different classes of end uses. The "random copolymers" contain modified chains, i.e. showing somewhat lower crystallinities and melting points than the corresponding homopolymers when low amounts of co-monomer are applied or losing nearly all crystallinity at higher levels. The latter products are rubbers, as also their glass transition temperature is low. In the rubber industry they go by the names EPR (ethylene-propylene rubber) and EPDM (ethylene-propylene-diene rubber), the latter one containing some unsaturation to assist vulcanization. The "block copolymers" are toughened polypropylenes, made to show a higher impact strength at low temperatures.

In reviewing the literature it became clear that in the early days of the Ziegler-Natta catalysis many papers were published on the topic of especially ethylene-propylene rubbers made with vanadium catalysts, including some characterization studies. Not much has been written on the properties of copolymers made on heterogeneous titanium catalysts. As the studies using vanadium catalysts have been excellently reviewed they will be only briefly mentioned in this book.

3.2 COPOLYMERIZATION ASPECTS

Regarding the preparation of copolymers, a large number of aspects found to apply in homopolymerization also apply here. For instance the rate dependence on catalyst concentration, the co-catalyst/catalyst ratio and temperature all follow similar patterns. The main difference lies in the dependence of the copolymer composition on the feed composition. Thus in this chapter the kinetics of the copolymerization, including the copolymerization parameters known as r-values, will be discussed. Moreover some aspects differing from the homopolymerization, such as molecular weight behaviour and solubles, will be mentioned as well. This discussion will be split into three parts: one for vanadium catalysts suitable for the production of amorphous or near-amorphous ethylene-propylene copolymers; the second for data on titanium systems which gives, due to the stereospecific nature of the catalyst, a different type of product; and a final section dealing, amongst others, with the new homogeneous titanium and zirconium catalysts.

The well known kinetic scheme for copolymerization in its simplest, but reasonably useful, form is given below:

$$
\begin{array}{llll}
\text{cat-M}_1 + \text{M}_1 & \dashrightarrow & \text{cat-M}_1\text{-M}_1 & \text{rate constant: } k_{11} \\
\text{cat-M}_1 + \text{M}_2 & \dashrightarrow & \text{cat-M}_1\text{-M}_2 & \text{'' } k_{12} \\
\text{cat-M}_2 + \text{M}_1 & \dashrightarrow & \text{cat-M}_2\text{-M}_1 & \text{'' } k_{21} \\
\text{cat-M}_2 + \text{M}_2 & \dashrightarrow & \text{cat-M}_2\text{-M}_2 & \text{'' } k_{22}
\end{array}
$$

These equations, together with the steady state in active species give the copolymerization equation relating the copolymer composition to the feed composition

$$
(m_1/m_2)_{pol} = ([M_1]/[M_2]) * (r_1[M_1] + [M_2]) / ([M_1] + r_2[M_2])
$$

In this, m_1 is the molar percentage of monomer 1 in the copolymer, $[M_1]$ is the monomer concentration of monomer 1 in the feed; $r_1 = k_{11}/k_{12}$ and $r_2 = k_{22}/k_{21}$, these are the reactivity ratios of the respective active species. Denoting the ratio of the monomers in the polymer by f and those in the feed by F and rearranging leads to the well known Fineman and Ross equation[1]

$$
(F/f) * (f - 1) = r_1 * F^2/f - r_2
$$

This equation is very frequently used to determine the reactivity ratios using data on composition of copolymers made at a known feed composition. In practice this is far from easy in a large number of cases, for instance in the case of rubber manufacture one has to take care that mass transfer from the gas phase to the liquid does not become rate determining due to the high viscosities sometimes encountered. Also the analytical methods used for determining the copolymer composition are not straightforward and can lead to considerable differences between different investigators. In our experience the most dependable and accurate compositional measurement can be made when one labels one of the monomers with carbon-14. The only drawback is that it is time consuming and thus expensive, as the analysis involves burning of the polymer sample, collecting the CO_2 formed and measuring its radioactivity, calibrating with a homopolymer sample, before it yields the desired composition.

The value of r_1*r_2 is rather important as it gives a clue to the expected co-monomer distribution along the copolymer chain. When its value is close to one, this means that the two monomers add to the two different active chain ends with an identical rate ratio. A <u>random</u> distribution is then expected as there is no preferential aspect in the kinetics. When however r_1*r_2 is very much smaller than 1 this implies that both cross-over rate constants (k_{12} and k_{21}) are larger than the homopolymerization rate constants, i.e. the probability of finding longer sequences of the minor monomer is low, and an <u>alternating</u> tendency is then expected. At the other end of the spectrum, a r_1*r_2 value much above 1 would mean an enhanced homopolymerization tendency over cross-over reactions. This then would lead to so-called <u>blocky</u> copolymers in which long sequences - also of the minor monomer - are present. For a discussion on the usefulness of process r_1 and r_2 data for predicting the sequence distribution, the reader is referred to the next chapter.

In a general sense the transition metal part of the Ziegler-Natta catalysts is the main determining factor in the copolymerization. This was for instance shown in a study using vanadium, vanadyl, titanium, zirconium and hafnium chlorides in combination with different co-catalysts in the copolymerization of ethylene and propylene[2,3]. The propylene reactivity increased with increasing electronegativity of the transition metal species, i.e. in the order of

$$HfCl_4 < ZrCl_4 < TiCl_4 < VOCl_3 < VCl_4.$$

3.3 VANADIUM CATALYSTS

Many vanadium compounds when reacted with a reducing agent such as aluminium alkyls give active catalysts for α-olefin polymerization. Examples are VCl_4, $VOCl_3$, vanadium-trisacetylacetonate and vanadyl esters: $VO(OR)_3$. Mostly the transition metal compound and the metal alkyl are premixed under specified conditions before a polymerization is started. Third components can also be part of the catalyst mixture, for instance in VCl_4-$AlEt_2Cl$-anisol. A fair number of these form homogeneous, or near homogeneous solutions during the polymerization. Many of these systems were extensively studied in view of the importance of the product from copolymerization: EPR rubber. Due to the homogeneous nature they are well suited for a kinetic study. Initially these studies were mainly carried out by the school of Natta and also by the Hoechst company. An extensive and excellent review is the one by Baldwin and Ver Strate (reference 4).

Reduction of higher valent vanadium species at normal temperatures is found to be rapid. Natta et al investigated the valence state in the active complex and found it to be three[5]. In many polymerizations one observes a high die-out, i.e. the polymerization activity decreases strongly with time. A reduction below the three-valent state is thought to be responsible. In line with this, reoxidation with for example perchloro compounds such as hexa-chlorocyclopentadiene, esters of pentachlorocrotonic acid or trichloroacetic acid derivatives can restore the activity[6,7]. This reoxidation reaction

$$V^{2+} + RCl \longrightarrow V^{3+}\text{-}Cl + R.$$

can be significantly accelerated by adding as a fourth component tri-butyl tin hydride[8]. This four component system shows productivity increases of up to 50-75% over tin hydride-free systems. More recent studies however observe no correlation of polymerization activity for ethylene with the concentration of V(III). Furthermore, the oxidation of V(II) with halocarbons could not be proven. Karol et al[9] favour V(II) as active valence state and observe that the role of the promotor is not clear. Additional studies are apparently required to elucidate the role of all three components in a V-compound/Al-alkyl/halocarbon system.

In a comparison of catalysts made from different vanadium compounds and various alkyls it appeared that at least one chloride substituent is necessary on either vanadium or aluminium in order to obtain an active system[10]. More than one polymerization active species is sometimes found in these catalyst systems[4] (see chapters 4 and 6).

These homogeneous catalysts have been studied extensively in a chemical and a kinetic way, to obtain insight in the detailed structure of the active species or the catalytic complex. For an excellent review by Tait on this topic the reader is referred to reference 11. The accepted view is that the active complex is a chlorine bridged complex of an alkylated vanadium species and the co-catalyst. This coordination transforms a 4-coordinated tetragonal vanadium compound into a 5-coordinated octahedral complex. The active complex contains a free coordination site on which the monomer can first form a π-complex before the true insertion into the vanadium-carbon bond takes place. This bimetallic active site has also been inferred in studies on the preparation of syndiotactic polypropylene, which uses similar catalysts (see chapter 2). In addition Natta[5] postulated that when anisole was added to the catalyst system it would form part of the active complex. Later we will see that recent evidence does not support this hypothesis (see chapter 5).

The kinetics of the copolymerization follows the normal rules, i.e. the rate is directly proportional to the concentration of the monomer and the catalyst[4,12,13]. The composition of the copolymer is independent of the die-out, the alkyl/vanadium ratio and the polymerization time. The rate does increase with temperature, the activation energy found for the system VCl_4-Al(hexyl)$_3$ is 27.7 kJ/mol for both the homopolymerizations and the copolymerization[14]. Reactivity ratios have been determined in a large number of cases. A few examples are given in Table 3.1, for more extensive data the reader is referred to the compilation in the reviews in references 4,12,13,15-17. Clearly the ethylene reactivity is much higher than that for propylene. For several catalysts the r_1 parameter is not too far different from the ratio of the respective homopolymerization rates. However examples of huge dependencies of the polymerization rate on the ethylene content of the feed were also ob-

TABLE 3.1

Examples of reactivity ratios of ethylene and propylene on vanadium catalysts

V-compound	Al-alkyl	r_1 ethylene	r_2 propylene	$r_1 {}^* r_2$
$V(acac)_3$	$AlEt_2Cl$	15	0.04	0.6
VCl_4	$Al_2Et_3Cl_3$	9.1	0.031	0.28
	$AlEt_2Cl$	5.9	0.029	0.14
	$AlEt_3$	10.3	0.025	0.25
$VO(OBu)_3$	$AlEt_2Cl$	16.8	0.019	0.32

served. In the case of VCl_4-Al(hexyl)$_3$, for example, the homopolymerization rate ratio is 1810 whilst the reactivity ratio is only
7.08. Clearly propagation rate constants alone cannot explain these
two values. A possible explanation is the increase in the number of
active species when more ethylene is present in the feed[18]. In
extreme cases homopolymerization of the second monomer is not at
all possible, for instance with internal olefins or cyclo-olefins.
Even then copolymerization with ethylene is feasible, although at a
very low rate[19]. When operating at a very low ethylene concentration in the liquid phase alternating copolymers can be made, which
also show interesting crystalline structures[13,19].

In the commercial EPDM rubbers a third monomer is added to the
system to bring side chain unsaturation in the molecules to facili-
tate crosslinking of the rubber. Often cyclopentadiene derivatives
are used, such as ethylidene norbornene. Their (ter)polymerization
apparently does not influence the relative reactivities of ethylene
and propylene[4], although it is stated that more data are required
to substantiate this statement.

The molecular weight of the copolymers is above all dependent
on the relative concentration of ethylene in the feed, upon its
increase a corresponding steep increase in the MW is noted[10,20].
Apart from this the MW decreases upon increasing the temperature,
on increasing the co-catalyst concentration, or on lowering the
monomer concentration. These factors are analogous to those for
homopolymerization. The time dependence of the molecular weight is
however low to very low[14,21], the MW reaching a steady value in 5
seconds in extreme cases[21].

In normal continuous operation in a CSTR (continuous stirred
tank reactor) one expects a molecular weight distribution characterized by a Q of around 2 or above depending on the number of
different active species present. Ver Strate et al[22] have recently
shown that very narrow MWD's (Q about 1.2 and R = 1.3) can be
obtained with VCl_4/sesqui provided that:
- the initiation is fast, in their case taken care of by premixing
the catalyst ingredients,
- a plug flow reactor is used,
- and short mixing times of catalyst with monomer apply.
The effects of residence time, co-monomer ratio (as propylene gives
transfer with monomer this ratio affects the MWD) and chain trans-

fer agent have been studied and the MWD can be well controlled. A consequence of the use of a plug flow reactor is that the copolymer composition becomes tapered, the monomer mixture becoming richer in the least reactive monomer, i.e. propylene. This inhomogeneity is mainly intermolecular as the lifetime of one chain is very short, being only a few seconds.

Other molecular weight distribution effects will be described in the corresponding characterization chapter (see chapter 4).

3.4 TITANIUM TRICHLORIDE AND OTHER ISOSPECIFIC CATALYSTS

The catalysts falling under this heading are the same ones as used for the homopolymerization of propylene. Starting with a catalyst based on $TiCl_3$ one is sure of the heterogeneity of the catalyst system.

In experiments at low ethylene/propylene ratio with $TiCl_3$-catalysts one observes an increase in the propylene polymerization rate compared to its homopolymerization. For instance at 5% ethylene in the feed the increase in rate is 35% when using catalysts made with $AlEt_3$ as reductant. This acceleration is also observed on supported catalysts for instance with chromium oxide on silica-alumina (ref 23) or a $MgCl_2$-supported one (ref 24). The rate increase is larger at higher values of ethylene in the feed, on a low temperature reduced $TiCl_4$ values of 2.1 were found independent of the polymerization temperature in the range 40-70 °C (expressed as a rate ratio, copolymerization over homopolymerization). An obvious explanation for this effect is the expected greater addition rate of propylene on a chain ending in ethylene due to the diminished steric requirements in this step as compared to propylene addition on a chain ending in propylene.

Bier observed only a slight decrease in overall polymerization rate with increasing propylene content in the feed[20]. This decrease was much faster in vanadium catalyzed systems. No reactivity ratios were determined in that study. Values of r_1 of 25.0 and r_2 of 0.10 were reported by Davison and Taylor[25]. In an own study with a low temperature reduced $TiCl_4$, the ratio of the overall ethylene and propylene polymerization rates were determined over a broad range of ethylene contents in the copolymer. The average ratio was 16 ± 2.4. For our own data on true reactivity ratios see chapter 5.

In addition to ethylene also other olefins have been copolymerized with propylene. In many studies however no kinetic data were described. The few that have been found in the literature employing heterogeneous catalysts are given in Table 3.2. These data clearly

TABLE 3.2

Reactivity ratios of olefins with propylene

catalyst used	comonomer	r_1 (propylene)	r_2	r_1*r_2	reference
$TiCl_3/AlEt_2Cl$	ethylene	0.10	25	2.5	25
ditto	,,	0.4-0.7	4.2-9.3(c		26
$TiCl_3/AlEt_3$	,,	0.15-0.18	13-14		27
Solvay-$TiCl_3$-Cp_2TiMe_2	,,	0.22	10	2.2	28
$TiCl_4/MgCl_2/PE/AlEt_3$	,,	0.09	6.1	0.55	27
$TiCl_4/MgCl_2$	,,	n.d	n.d.	4	29
Chromocene on silica	,,	n.d.	72	n.d.	30
"high yield Ti-catalyst"(a	,,	n.d.	n.d.	≈2	31
$TiCl_4/MgCl_2/EB$	,,	0.7-0.4	7.4-13.4	≈5(c	32
$TiCl_4/MgCl_2/3ROH$	,,	0.02	6	0.13(c	33
$SiO_2/TiCl_3/MgCl_2$	,,	0.14	7	1(c	34
$SiO_2/TiCl_4/MgCl_2$	,,	0.34-0.18	5-10	1.9(c	35
$TiCl_3/AlEtCl_2/0.6HMPTA/H_2$	1-butene	4.3	0.8	3.4	36
ditto, no H_2	,,	3.3	0.45	1.5	36
$TiCl_3.0.33AlCl_3/AlEt_2Cl$	,,	4.5	0.20	0.9	37
$TiCl_3/AlEt_2Cl$	,,	4.7	0.51	2.4	25
$TiCl_4/AlEt_3$	,,	2.4	0.5	1.2	38
$TiCl_3$(Stauffer AA)/$AlEt_2Cl$	4-methyl-1-pentene	6.44	0.31	2.0	39
$TiCl_3$(Stauffer AA)/$AlEt_2Cl$	1-hexene	4.18	0.16	0.67	40
$TiCl_3/AlEt_2Cl$	4-methyl-1,4-hexadiene	25(b	0.04(b	1	41

(a : probably $MgCl_2/TiCl_4$/aromatic ester and $AlEt_3$/aromatic ester as co-catalyst

(b : activity ratio, i.e. molar ratio of co-monomers in copolymer and feed

(c : all reactivity ratios determined by NMR - see chapters 4 and 13

illustrate the diminishing activity in copolymerization, the larger the co-monomer molecule. Kissin in a recent article[42], reviewing the monomer reactivity on Ziegler-Natta catalysts, elaborates this point. His conclusion is that the reactivity of olefins is determined by the steric effect of its alkyl group and to a much lower extent by the electronic changes.

Using the extremely high stereospecific catalyst Solvay-TiCl$_3$ with Cp$_2$TiMe$_2$ Soga and Yanagihara[28] determined the reactivity ratios of some α-olefins with ethylene. The values with propylene are completely on a level with other TiCl$_3$-catalysts, which is somewhat surprising as intuitively one had expected high values for r_2 because of the apparently high steric requirements of this catalyst system. In a related publication[43] three different monomer pairs were studied at a relatively low polymerization temperature of 40 °C; two pairs gave crystalline polymers also during the polymerization (ethylene and propylene, ethylene and 1-hexene respectively) whilst the last one (1-hexene and 1-octene) give a true solution polymerization. In the first two a considerable rate increase was noted (for both ethylene and propylene) when low amounts of co-monomer were copolymerized, but not so in the last case. This rate increase is a very general phenomenon, as already mentioned in this chapter, for propylene in the presence of small amounts of ethylene, whilst the converse is also true, i.e. ethylene rate increases in the presence of other α-olefins (see chapter 9 for details). The authors suggest that the rate increase in these cases is due to a greater diffusion rate of the co-monomers through the (crystalline) polymer layer around the catalyst. This is not a generally accepted view however, and more proof will be needed (see also section 9.5.8).

Kakugo and co-workers[26] and some other investigators[35] found a dependence of the kinetic parameters in ethylene-propylene copolymerization on the feed composition. The reactivity for ethylene decreasing upon increasing the ethylene concentration in the feed, for propylene reactivity the opposite trend is found (range of copolymers is rather broad, ethylene contents from 15 to 75 %mol). More information on a comparable system is given in chapter 5.

Kissin applied a different method for the determination of the reactivity ratios by following the change in the feed composition with time[27]. The values obtained in this way are comparable to the values from studies using a steady state copolymerization or using low conversions.

In the copolymerization with 1,4 dienes a very low catalyst activity is noted when using straight chain di-olefins[41,44]. This is due to rearrangement of this co-monomer to the conjugated di-olefin, mainly the 2,4 diene. These conjugated dienes are known to block the active sites by forming a strong complex. Changing the structure to a more hindered one by using 4-methyl-1,4-hexadiene gives the normal polymerization rates again[41].

When using heterogeneous catalysts in copolymerizations in which the copolymer formed is soluble under the conditions employed, the catalyst is "solubilized". No true solution is formed, but the catalyst is fragmented in very small units, probably the primary catalyst particles or even smaller. An increasing rate is observed in the first minutes of copolymerization in such situations[40].

An interesting phenomenon was observed[45] on a supported catalyst of the type γ-alumina/$TiCl_4$. The titanium can be thermally reduced to the three-valent state. Such a catalyst is able to copolymerize ethylene and propylene. Upon adding $AlEt_3$ to the system the propylene reactivity almost disappears. As it is known that $AlEt_3$ reduces the titanium to both divalent and zero valent species it was surmised that propylene cannot be polymerized on such low valent titanium species whilst ethylene can. This is just one example in which copolymerization kinetic studies are used as a probe for catalyst behaviour. Such an approach is used quite often and we will mention other examples later.

Few data are as yet published of the reactivity ratios on the newly developed supported catalysts for propylene polymerization[29,31] (see table 3.2). One observes values of the reactivity ratio product well above one, indicating a blocky nature of the copolymer, similar to $TiCl_3$-based catalyst systems. Moreover it has been reported that the $r_1 \ast r_2$ value is dependent on the feed composition, becoming smaller at propylene-richer conditions. Compared to $TiCl_3$ the supported catalyst is somewhat less reactive towards ethylene (relatively!).

The effect of the level of ethyl benzoate(EB), as external base, on the copolymerization behaviour of a supported catalyst was studied by Soga, Shiono and Doi[32]. Increasing the EB concentration led to lower propylene incorporation (coupled to a lower rate of overall polymerization). The r_1 (ethylene) value could almost be doubled - and r_2 halved as r_1*r_2 stayed almost constant. This behaviour is in line with the known effect of EB as external base in the homopolymerization of propylene: it increases the isotacticity by deactivating atactic sites, just those sites which are expected to show a less discriminating behaviour to the two monomers, i.e. sites which are expected to have low r_1 values.

Surprisingly, the use of supported catalyst variants in the copolymerization of ethylene and propylene to almost amorphous rubbers was recently described[46,59]. This is certainly a major difference with the $TiCl_3$ catalysts as the latter always give copolymers containing a crystalline fraction (see chapter 4). Use of supported catalyst in a suspension process for rubber manufacture is economically very attractive. The activity of supported catalyst is, at comparable ethylene contents in the copolymer, some 20 times that of the first generation $TiCl_3$/DEAC one[33].

Testing a non-stereospecific $MgCl_2$/$TiCl_4$/2-ethyl-hexanol catalyst in ethylene/propylene copolymerization, Kashiwa et al[33] observed a widely different kinetic behaviour in that r_1*r_2 becomes much smaller than one (0.13 as determined by NMR). This brings this catalyst into the class of the vanadium ones (except for activity!), giving rise to a less blocky type of copolymer. Also, the copolymer structurally resembled the latter type.

The relative rates of the first few insertions of ethylene and propylene on a catalyst have been studied[47,48] using a $TiCl_3$-catalyst system in which the co-catalyst was enriched in ^{13}C. This allows one to observe the endgroups separately by ^{13}C-NMR. The very first insertion on the methyl-alkylated titanium catalyst shows a rate ratio of 4 (ethylene over propylene). The second insertion gives a value of about 27 for this ratio, independently of whether the first added unit was ethylene or propylene. This last ratio is very close to the macroscopic, kinetically derived values for this type of catalyst. The reason for the low ratio in the very first insertion is undoubtedly due to the near absence of steric factors when only a methyl group is present on the catalyst. The effects of increased steric demand from the monomer were tested[48] in copolymerizations of ethylene with 1-butene or 3-methyl-1-butene using the same technique. Ratios of 5 and 7 respectively are now observed for the first insertion, subsequent insertions could not be measured.

Clearly the values are larger for these more bulky monomers. It would be interesting to use this technique to probe the catalyst for sites with different kinetic parameters, for instance by making a random copolymer with ethylene, fractionating it, followed by NMR analysis (see also chapter 4 for the non-homogeneity of copolymers generated on $TiCl_3$ catalysts). This has been attempted by the same group of investigators[49]. However a true molecular weight fractionation was carried out by dissolving an ethylene/propylene copolymer in toluene and adding progressively more methanol. The fractions all showed the same values for the various rate ratios. One would have preferred a fractionation on the basis of crystallinity, such as via an extraction at different temperatures, since in that way the split is made on the property which is of interest in this case, namely the blockiness/crystallinity. Hopefully this will be attempted in the future.

For copolymers with low ethylene contents the molecular weight is lower in copolymers compared to homopolypropylene at identical conditions, i.e. hydrogen is more effective in its transfer reaction and/or transfer with ethylene monomer is occurring. In some instances the increased formation of low molecular weight solubles can influence the molecular weight behaviour drastically. In a broader range of copolymers there is a tendency toward increasing molecular weight with increasing ethylene content[20], for other data on this see chapter 5 (see e.g. Fig. 5.1).

3.5 HOMOGENEOUS AND OTHER CATALYSTS

The kinetic parameters obtained on homogeneous zirconium and titanium catalysts are given in Table 3.3. Mostly they are drastically different from the ones given before for isospecific catalysts, especially r_1 is smaller than 0.1 and the product r_1*r_2 is mostly well below one. This indicates a much more alternating distribution of the monomers. The behaviour is however similar to the vanadium catalysts which are also homogeneous.

In the series of data given by Ewen[53] the observed reactivity ratios parallel the steric requirements of the active complex. For instance the penta-methyl-cyclopentadienyl ligand impairs the accessibility for a more bulky monomer, whilst the complex with the $Si(CH_3)_2$-bridge is of a more open structure. This leads to widely different values of the reactivity ratios. It should be remarked

TABLE 3.3
Reactivity ratios of ethylene/propylene on homogeneous catalysts

Catalyst applied	r_1 (propylene)	r_2	$r_1 * r_2$	ref
$Cp_2Ti(CH_3)_2/Al(CH_3)_3/H_2O$	0.032	18.6	0.60	50
$Cp_2Zr(CH_3)_2/[Al(CH_3)O]_n$	0.005	27	0.135	51
ditto	0.005	31.5	0.3	52
$Cp_2Ti(Ph)_2/[Al(CH_3)O]_n$	0.015	19.5	0.29	53
$Cp_2ZrCl_2/[Al(CH_3)O]_n$	0.015	48	0.72	,,
$(CH_3)_2SiCp_2ZrCl_2/[Al(CH_3)O]_n$ *	0.029	24	0.70	,,
$[(CH_3)_5Cp]_2ZrCl_2/[Al(CH_3)O]_n$	0.002	250	0.50	,,
$(CH_3Cp)_2ZrCl_2/[Al(CH_3)O]_n$	n.d.	60	n.d.	,,
$Cr(stearate)_3/MgCl_2//DEAC$	0.03**	30**	(1)	54
soluble $TiCl_3$ catalyst	0.24	18	4.3	55
$TiCl_4/Mg(hexyl)_2$	5.6	0.18***	1	56

* : this compound contains the two Cp ligands bridged by a $Si(CH_3)_2$ group
** : activity ratio
*** : 1-butene

that the data given have to be treated with care as the reactivity
ratio determination by different methods (Fineman-Ross, NMR, IR)
were not consistent. Also the way of preparation of the methylaluminoxane co-catalyst affects these ratios. Moreover the effects of
possible inversions (on the more open catalysts) have not been
taken into account. However, these types of catalysts with their
enhanced possibility of tailoring the polymers generated on them
certainly deserve a closer study.

Interestingly the homopolymerization rate ratio for ethylene
and propylene observed on the homogeneous titanium chloride catalyst, developed by Fontanille et al, is 16.5 (see reference 57) -
a value close to the r value for ethylene found in heterogeneous
systems (e.g. $TiCl_3$/DEAC in table 3.2). In a more recent study of
the same group[55] real copolymerization kinetic data were reported.
As shown in table 3.3 the r_2 value for ethylene (18) is very close
to the above mentioned ratio of the homopolymerization rates.

The data on the $Cp_2Ti(CH_3)_2$ catalyst[50] also show an immense enhancement of the propylene conversion rate in the presence of ethylene. Comparing the propylene homopolymerization rate with the first ethylene/propylene copolymerizations in Busico's paper, gives a rate increase by a factor 175! This underlines the alternating tendency of these systems.

Similarly, the chromium stearate/$MgCl_2$ catalyst described by Soga et al[54] shows only a very low homopolymerization activity for propylene, being about 500 times lower than for ethylene. Still the activity ratio in copolymerization is 30 (defined as the ratio of monomers in polymer over that in the feed), implying a considerable increase in propylene conversion rate over just its homopolymerization rate.

Kaminsky and Miri[52] reported on an elaborate study of $Cp_2Zr(CH_3)_2$/aluminoxane as a catalyst in ethylene/propylene copolymerization. The kinetics in this system are very complicated. It takes for instance some hours before the reaction reaches its steady state constant copolymerization rate, and the time required depends on both the feed ratio and the presence of a diene in the solution (ethylidene norbornene, ENB, to generate EPDM's). The higher the ethylene concentration the faster a steady state is reached and the larger the constant rate. In diene-free systems rates around 1000 kg copolymer per mol Zr.bar.h are observed, dropping to values around 100 kg in diene containing/low ethylene systems. These activities are considerably higher than in vanadium systems normally used for this type of copolymerization. The incorporation of ENB is rapid as illustrated by the rather low reactivity ratio with ethylene of 3.1 compared to 31.5 for propylene, i.e. ENB is some 10 times more reactive than propylene in this polymerization. Molecular weight - other things being equal - increases with the ethylene content in the polymer, the range being about 40,000 to 150,000 at molar ethylene contents of 63 to 87%. This holds for polymers made at 20 °C. When using lower polymerization temperatures the molecular weight goes up but much less than in ethylene homopolymerization. Addition of ENB to the system lowers the molecular weight by some 10%. These values are reasonably good from a practical point of view, but this polymerization shows less freedom in controlling molecular weight independently from other properties, particularly ethylene content. The molecular weight distributions are very narrow, with Q values around 1.7, again indicating that just one active species is operating. The normal vanadium-catalyzed EPDM's are definitely broader of distribution.

Using the isospecific rac-ethylene-bis(1,1'-indenyl)zirconium dichloride and methylaluminoxane catalyst in ethylene/propylene copolymerization[58] led to reactivity ratios (r_1=6.6, r_2= 0.06 at 50 °C) which are considerably different from the far less selective catalyst system $Cp_2Ti(CH_3)_2$/methylaluminoxane (r_1= 18.6, r_2=0.03). In both cases there is still an alternating tendency as $r_1*r_2<1$.

Surprisingly Soga et al[56] found in a modified $TiCl_4$ catalyst (using $Mg(hexyl)_2$ as co-catalyst) that, kinetically, propylene/1-butene copolymerization occurred in a random manner as r_1*r_2 is close to 1. Moreover the reactivity ratios are very close to the ratio of the homopolymerization rates, implying that k_{12} is equal to k_{21}. This is certainly different from some of the $TiCl_3$ catalyzed copolymerizations (see previous section) and could lead to interesting property differences.

3.6 COPOLYMERIZATION PROCESSES

Modified polypropylenes with small amounts of co-monomer (mainly ethylene) are manufactured in the same process as the homopolymers; the only addition required is the extra co-monomer feed. Upon using ethylene, its conversion in the process will be so high due to its much larger reactivity that almost none is left, and a co-monomer recycle is thus absent.

When manufacturing true rubbers a dedicated process is always used. Two kinds can be distinguished, one is the solution process, the other is the suspension process[4,46,59]. In the solution process the diluent is a solvent for the formed copolymer, and the process must include a recovery step for the solvent, the diene used in the terpolymerization, and also for propylene. In the older version of this process, i.e. when using low activity catalysts, in addition a deashing stage had to be included to render the copolymer non-corrosive in its subsequent processing. In fact such a process resembles of course the older polypropylene processes with the extra problem that a very viscous solution had to be handled.

In the more modern suspension process, use is made of the insolubility of EPR and EPDM in liquid propylene/ethylene mixtures. When using such a medium for the copolymerization a suspension results which can be separated from the diluent by evaporation, which requires much less energy than in the solution process. In the modern process high-activity catalysts (such as supported catalysts) are used, which obviate deashing. The space-time yield is much larger in the latter process as in a suspension the polymer concentration can be made much larger than in the solution process.

REFERENCES

1. M.Fineman and S.D.Ross, Linear method for determining monomer reactivity ratios in copolymerization, *J.Pol.Sc.*, *5* (1950) 259-265

2. F.J.Karol and W.L.Carrick, Transition metal catalysts. VII. Identification of the active site in organometallic mixed catalysts by copolymerization kinetic studies, *J.Am.Chem.Soc.*, *83* (1961) 2654-2658

3. W.L.Carrick, Mechanism studies on Ziegler-Natta catalysts using copolymerization kinetics, *Polymer Preprints*, *8* (1967) 17-21 (ACS)

4. F.P.Baldwin and G. Ver Strate, Polyolefin elastomers based on ethylene and propylene, *Rubber Chem.Techn.*, *45* (1972) 709-881

5. G.Natta, A.Zambelli et al, Polymerization of propylene to syndiotactic polymer. Part I: Valence of active vanadium in the catalytic system, *Makrom.Chemie*, *81* (1965) 161-172

6. A.Gumboldt, J.Helberg and G.Schleitzer, Über die Reaktivierung der bei der Ethylen/Propylen Copolymerisation verwendeten Vanadium-Katalysatoren, *Makrom.Chemie*, *101* (1967) 229-245

7. H.Emde, Thermoplastische Elastomere auf Polyolefinbasis mit kristalinen Segmenten, *Angew.Makrom.Chemie*, *60/61* (1977) 1-20

8. E.Giannetti, R.Mazzocchi, E.Albizzati, T.Fiorani and F.Milani, Homogeneous Ziegler-Natta catalysis: Efficiency improvement of vanadium catalyst systems by tributyltin hydride for the ethylene/propylene/diene terpolymerization, *Makrom.Chemie*, *185* (1984) 2133-2151

9. K.J.Karol, K.J.Cann and B.E.Wagner, Developments with high-activity titanium, vanadium and chromium catalysts in ethylene polymerization, p149-161 in Transition metals and organometallics as catalysts for olefin polymerization, W.Kaminsky and H.Sinn editors, Springer, Heidelberg, 1988

10. G.Natta et al, Ethylene-propylene copolymerization in the presence of catalysts prepared from vanadium triacetylacetonate, *J.Pol.Sc.*, *51* (1961) 411-427

11. P.J.T.Tait, Ziegler-Natta and related catalysts, in Developments in polymerization - 2.Free radical, condensation, transition metal and template polymerizations, R.N.Haward editor, Applied Science Publishers, London, 1979

12. G.Crespi, A.Valvasori and G.Sartori, Ethylene-propylene copolymers as rubbers, Chapter IVC in Copolymerization, G.E.Ham editor, Interscience Publishers, New York, 1964 (volume XVIII in the series High Polymers)

13. I.Pasquon, A.Valvasori and G.Sartori, The copolymerization of olefins by Ziegler-Natta catalysts, chapter 4 in The stereochemistry of macromolecules, volume 1, A.D.Ketley editor, Marcel Dekker, New York, 1967

14. G.Natta et al, Kinetics of ethylene-propylene copolymerization, *J.Pol.Sc.*, *51* (1961) 429-454

15. C.Cozewith and G.Ver Strate, Ethylene-propylene copolymers. Reactivity ratios, evaluation and significance, *Macromolecules*, *4* (1971) 482-489

16. Yu.V.Kissin, Structures of copolymers of high olefins, *Advances in Polymer Science*, *15* (1974) 91-155

17. W.Cooper, Kinetics of polymerization initiated by Ziegler-Natta and related catalysts, in Chemical Kinetics, Volume 15, Non-radical polymerization, C.H.Bamford and C.F.H.Tipper editors, Elsevier, Amsterdam, 1976

18. A.Valvasori, G.Sartori, G.Mazzanti and G.Pajaro, Kinetics of the ethylene-propylene copolymerization, *Makrom.Chemie*, *61* (1963) 46-62

19. G.Natta et al, Crystalline alternating ethylene-cyclopentene copolymers and

230

other ethylene-cycloolefin copolymers, *Makrom.Chemie, 54 (1962) 95-101*

20. G.Bier, A.Gumboldt and G.Schleitzer, Wirkungsweise verschiedener Ziegler-Katalysatoren bei der Polymerisation von α-Olefinen und Eigenschaften der Polymerisate, *Makrom.Chemie, 58 (1962) 43-64*

21. E.Junghanns, A.Gumboldt and G.Bier, Polymerisation von Ethylen und Propylen zu amorphen Copolymerisaten mit Katalysatoren aus Vanadiumoxychlorid und Aluminiumhalogenalkylen, *Makrom.Chemie, 58 (1962) 18-42*

22. G.Verstrate, C.Cozewith and S.Ju, Near monodisperse ethylene-propylene copolymers by direct Ziegler-Natta polymerization. Preparation, characterization, properties, *Macromolecules, 21 (1988) 3360-3371*

23. J.P.Hogan, Olefin copolymerization with supported metal oxide catalysts, chapter III in Copolymerization, G.E.Ham editor, Interscience Publishers, New York, 1964 (volume XVIII in the High Polymer series)

24. R.Spitz et al, Propene polymerization with MgCl2 supported Ziegler catalysts: activation by hydrogen and ethylene, *Makrom.Chemie, 189 (1988) 1043-1050*

25. S.Davison and G.L.Taylor, Sequence length and crystallinity in α-olefin terpolymers, *Br.Pol.J., 4 (1972) 65-82*

26. M.Kakugo, Y.Naito, K.Mizunuma and T.Miyatake, 13C NMR determination of monomer sequence distribution in ethylene-propylene copolymers prepared with δ-TiCl3-AlEt2Cl, *Macromolecules, 15 (1982) 1150-1152*

27. Y.V.Kissin and D.I.Beach, A kinetic method of reactivity ratio measurement in olefin copolymerizations with Ziegler-Natta catalysts, *J.Pol.Sc.,Pol.Chem., 21 (1983) 1065-1074*

28. K.Soga and H.Yanagihara, Evaluation of olefin reactivity ratios over highly isospecific Ziegler-Natta catalyst, *Makrom.Chemie, 190 (1989) 37-44*

29. V.Busico, P.Corradini, A.Ferraro and A.Proto, Polymerization of propene in the presence of MgCl2-supported Ziegler-Natta catalysts, 3, Catalyst deactivation, *Makrom. Chemie, 187 (1986) 1125-1130*

30. F.J.Karol et al, Chromocene catalysts for ethylene polymerization: Scope of the polymerization, *J.Pol.Sc., A-1, 10 (1972) 2621-2637*

31. L.Abis, G.Bacchilega and F.Milani, 13C NMR characterization of a new ethylene-propene copolymer obtained with a high yield titanium catalyst, *Makrom.Chemie 187 (1986) 1877-1886*

32. K.Soga, T.Shiono and Y.Doi, Effect of ethyl benzoate on the copolymerization of ethylene with higher α-olefins over TiCl4/MgCl2 catalytic system, *Polymer Bulletin, 10 (1983) 168-174*

33. N.Kashiwa, A.Mizuno and S.Minami, Copolymerization of ethylene with propylene by MgCl2-containing highly active Ti catalysts, *Polymer Bulletin, 12 (1984) 105-109*

34. K.Soga, R.Onishi and T.Sano, Copolymerization of ethylene and propylene over the SiO2-supported MgCl2/TiCl3 catalyst, *Polymer Bulletin, 7 (1982) 547-552*

35. Y.Doi, R.Onishi and K.Soga, Monomer sequence distribution in ethylene-propylene copolymers prepared with a silica-supported MgCl2/TiCl4 catalyst, *Makrom.Chemie, Rapid Commun., 4 (1983) 169-174*

36. H.W.Coover et al, Costereosymmetric α-olefin copolymers, *J.Pol.Sc., Al, 4 (1966) 2563-2582*

37. R.D.A.Lipman, Copolymerization kinetics of α-olefins using Natta catalysts, *Polymer Preprints, ACS, 8 (1967)396-399*

38. R.Laputte and A.Guyot, Sur les systèmes catalytique TiCl4-AlR3. II. Etude cinétique de la copolymerisation propylène-butène-1, *Makrom. Chemie, 129 (1969) 234-249*

39. V.Sh.Shteinbak et al, Polymerization of 4-methylpentene-1 and its copolymerization with propylene on the TiCl3 + (C2H5)2AlCl complex catalyst, *Europ.Pol.J., 11 (1975) 457-465*

40. A.Piloz, Q.T.Pham, J.Y.Decroix and J.Guillot, Copolymerization of olefins by Ziegler-Natta catalysts. The copolymerization of 1-hexene and propylene. Kinetic study., *J.Macromol.Sci.-Chem., A9 (1975) 517-537*

41. S.Kitagawa and I.Okada, Reactive polyolefins,1, A new crystalline copolymer of propylene with methyl-1,4-hexadiene, *Polymer Bulletin, 10 (1983) 109-113*

42. Y.V.Kissin, Monomer reactivity in stereospecific polymerization with heterogeneous Ziegler-Natta catalysts, in Transition metal catalyzed polymerizations, alkenes and dienes, part B, R.P.Quirk editor, Harwood academic publishers, New York, 1983

43. K.Soga, H.Yanagihara and D-h.Lee, Effect of the monomer diffusion in the polymerization of olefins over Ziegler-Natta catalysts, *Makrom.Chem.*, *190* (1989) 995-1006

44. J.Lal and P.H.Sandstrom, Ozone resistant diene rubbers, *Polymer Letters*, *13* (1975) 83-85

45. K.Soga, T.Sano and R.Onishi, Copolymerization of ethylene with propylene over the thermally reduced γ-alumina-supported TiCl4 catalyst, *Polymer Bulletin*, *4* (1981) 157-164

46. P.Galli, F.Milani and T.Simonazzi, New trends in the field of propylene based polymers, *Polymer Journal*, *17* (1985) 37-55

47. P.Ammendola, A.Vitagliano, L.Oliva and A.Zambelli, Ethylene-propene copolymerization in the presence of a 13C enriched catalyst: endgroup analysis and monomer reactivities in the first insertion steps, *Makrom.Chemie*, *185* (1984) 2421-2428

48. P.Ammendola and A.Zambelli, Ziegler-Natta polymerization of 1-alkenes: relative reactivities of some monomers toward insertion into metal-methyl bonds, *Makrom.Chemie* *185* (1984) 2451-2457

49. P.Ammendola, L.Olivia, G.Gianotti and A.Zambelli, Ethylene-propene copolymerization. Monomer reactivity and reaction mechanism, *Macromolecules*, *18* (1985) 1407-1409

50. V.Busico, L.Mevo, G.Palumbo, A.Zambelli and T.Tancredi, Preliminary results on ethylene/propene copolymerization in the presence of Cp2Ti(CH3)2/Al(CH3)3/H2O, *Makrom.Chemie*, *184* (1983) 2193-2198

51. W.Kaminsky and M.Schlobohm, Elastomers by atactic linkage of α-olefins using soluble Ziegler catalysts, *Makrom.Chemie, Macromol.Symp.*, *4* (1986) 103-118

52. W.Kaminsky and M.Miri, Ethylene propylene diene terpolymers produced with a homomgeneous and highly active zirconium catalyst, *J.Pol.Sc.*, *Pol.Chem.*, *23* (1985) 2151-2164

53. J.A.Ewen, Ligand effects on metallocene catalyzed Ziegler-Natta polymerizations, in Catalytic polymerization of olefins, T.Keii and K.Soga editors, Kodansha/Elsevier, Tokyo/Amsterdam, 1986

54. K.Soga, S-I.Chen, Y Doi and T.Shiono, Polymerization of ethylene and propylene with Cr(stearate)3/AlEt2Cl/metal chlorides catalysts, *Macromolecules*, *19* (1986) 2893-2895

55. F.Cansell, A.Siove and M.Fontanille, Ethylene-propylene copolymerization initiated with solubilized Ziegler-Natta macromolecular complexes. I. Determination of kinetic parameters., *J.Polym.Sc.*, *Pol.Chem.*, *25* (1987) 675-684

56. K.Soga, M.Ohtake, R.Onishi and Y.Doi, Copolymerization of propylene with 1-butene in presence of the catalytic system TiCl4/Mg(hexyl)2, *Makrom.Chemie*, *186* (1985) 1129-1134

57. F.Cansell, A.Siove and M.Fontanille, Polymerization of propene initiated with solubilized Ziegler-Natta Li/Ti(III) based systems, *Makrom.Chemie*, *186* (1985) 379-387

58. H.Drögemüller, K.Heiland and W.Kaminsky, Copolymerization of ethene and α-olefins with a chiral zirconocene/aluminoxane catalyst, p303-308 in Transition metals and organometallics as catalysts for olefin polymerization, W.Kaminsky and H.Sinn editors, Springer, Heidelberg, 1988

59. G.Foschini and P.Milani, New EP elastomers, p75-87 in Advances in polyolefins, R.B.Seymour and T.Cheng editors, Plenum Press, New York, 1987

Chapter 4

CHARACTERIZATION OF "RANDOM" COPOLYMERS AND RUBBERS

4.1 INTRODUCTION

In this chapter a number of characterization data will be described and discussed for copolymers made in a "random" procedure. Copolymer chains can be expected to differ not only in their molecular weight and the tacticity of the propylene runs as is the case in homopolymer, but also in their co-monomer content. Thus the co-monomer distribution is added to the molecular weight and the tacticity distribution. This co-monomer distribution has at least two aspects: firstly intermolecular, i.e. do separate polymer molecules possess different co-monomer contents; and secondly intramolecular, i.e. how are the co-monomer units distributed over the chain? The latter is also called the sequence distribution. The importance of the co-monomer distribution is well illustrated by experimental studies which show that copolymers made to the same co-monomer content, using catalysts giving identical r-values, can still differ appreciably in crystallinity and melting enthalpy[1]. It will be clear that a complete description of a copolymer is very difficult to achieve, certainly it is more difficult than in the homopolymer case.

The topics we will review in this chapter are the composition and chain structure, the monomer distribution (which includes fractionation and molecular weight distributions) and crystallinity effects in copolymers. Additionally the glass transition, a few examples of the mechanical properties and the sometimes dispersed nature of these copolymers will be mentioned. Finally a reconciliation is attempted between the observed characterization data and the structure and behaviour of the (TiCl$_3$-)catalyst.

The techniques used in the characterization are described in the last part of this book (chapter 13), to which the reader is referred for details on procedures, etc..

4.2 COMPOSITION, KINETIC PARAMETERS

In a macro-sense the chain structure of the copolymers described in this chapter are linear, except in some cases of EPDM's in which branching via the termonomer diene has occurred[2]. The effect

of the branching is especially noticeable in the melt-flow properties of the polymer.

The study of the <u>microstructure</u> of the polymers revealed in
the case of vanadium catalyzed ones that propylene has been added
to the chain in two ways: both a 1,2 and a 2,1 insertion is noted,
as is also observed in homopolymers made on the same type of
catalyst. In essence this means that copolymerization of ethylene
and propylene on a vanadium catalyst is a <u>terpolymerization of two</u>
"forms" of propylene with ethylene. The possibility of inverted
propylene insertion leads to even numbered ethylene sequences, as
illustrated in Fig. 4.1, of which the $(CH_2)_2$ and $(CH_2)_4$ ones can be
observed specifically in infrared[3,4,5]. A detailed study showed

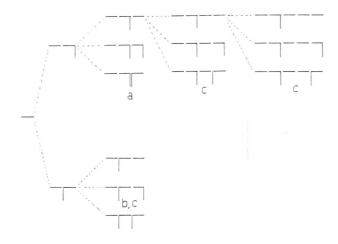

Fig. 4.1: *Possible sequences in ethylene-propylene copolymers: head-to-head
sequence (a); tail-to-tail sequence (b); $(CH_2)_n$ sequence of even number (c)*

that subsequent to an inverted propylene unit the addition of an
ethylene molecule is 4 times as probable than the addition of
propylene in a non-inverted fashion. This explains why more $(CH_2)_2$
sequences occur in copolymers of ethylene and propylene than in the
propylene homopolymers made with the same vanadium catalysts.

^{13}C-NMR is, just as in homopolymer studies, a powerful tool in
illuminating the detailed sequence structure in these copolymers.

For ethylene-propylene copolymers very extensive studies have been carried out both in determining the sequence distribution (dyads, triads, sometimes up to hexads) and in trying to relate the measured distribution to a propagation model. A non-exhaustive list of these studies has to mention those of Carman and Wilkes[6], Smith[7,8], Kakugo et al[9], Ross[10], Cozewith[11], Hayashi et al[12,13] and Cheng[14]. For instance in the last report a model is presented which includes effects of the sequence, the tacticity, inversions and even end-groups; all can in principle be unravelled and quantitatively described. One of the more easily calculated data from the NMR spectrum is the (average) value for the reactivity ratio product r_1*r_2. Carman and Wilkes[6] suggest the formula

$$r_1*r_2 = 1 + f(1 + k) - (1 + f)*(1 + k)^{0.5}$$

with f = moles ethylene over moles propylene in the copolymer and
k = fraction of propylene units in sequences of two and more over the fraction of isolated propylene units.
Or, otherwise, one can use[9]

$$r_1*r_2 = 4 [EE]*[PP]/[EP]^2$$

in which [EE] represent the dyad fraction consisting of two ethylene molecules, etc. This formula assumes that the copolymerization obeys first order Markov statistics.

For <u>vanadium-catalyzed</u> copolymers values of r_1*r_2 around one were found when not distinguishing the two different forms of the insertion of propylene. In a later study by Smith, with an enlarged resolution, the different forms of propylene could be discerned[7,8]. Conditional probabilities were obtained from the spectra. Denoting ethylene with 1, normally inserted propylene with 2, and an inverted one with 3, the following propagation probabilities apply for a copolymer containing 35 %mol propylene made on $VOCl_3$:

$$P_{11} = 0.54 \qquad P_{12} = 0.37 \qquad P_{13} = 0.09$$
$$P_{21} = 0.78 \qquad P_{22} = 0.17 \qquad P_{23} = 0.05$$
$$P_{31} = 1 \qquad\qquad\qquad\qquad\quad P_{33} = 0$$

Inversion of propylene occurs in about 20% of the propylene insertions. The conclusion from the infrared study could not be substantiated as the determination of P_{31} and P_{32} is rather inaccurate.

Reactivity ratios have also been calculated, leading to $r_{12}=18$, $r_{13}=75$, $r_{21}=0.018$ and $r_{23}=3.4$ for the above case. The $r_1 * r_2$ values are less than one and increase with increasing propylene content in the feed. The $r_1 * r_2$ values observed indicate that there is an alternating tendency in the copolymer sequences. Doi et al[15] studied the copolymer structure on his living catalyst (see chapter 5) at -78 °C. Via ^{13}C-NMR the ratio of $(CH_2)_2$ over $(CH_2)_3$ sequences was determined in low-ethylene-containing samples having no EE or longer ethylene sequences. The ratio turned out to be 5 to 6, hence the $(CH_2)_2$ sequence is the most frequently occurring one; this can only be formed when it is present in the structure

$$-CH_2-CH(CH_3)-CH_2-CH_2-CH(CH_3)-CH_2-,$$

i.e. a structure in which the secondary insertion of propylene changes into primary subsequent to an ethylene incorporation. This is well in line with Smith's data above, as P_{12}/P_{13} is about 4 and is also well known from Zambelli's studies on syndiotactic propagation.

Endgroups have been determined by infrared[16], and for this type of catalyst one observes both tr-vinylene and vinylidene, formed in transfer reactions according to

$$M-C-C(CH_3)-R + C=C-C \longrightarrow M-C-C-C + C=C(CH_3)-R \text{ (vinylidene)}$$

$$M-C(CH_3)-C-R + C=C-C \longrightarrow M-C-C-C + C-C=C-R \text{ (tr-vinylene)}$$

The latter one again resulting form a secondary inserted propylene.

Using _isospecific_ catalysts, the NMR-spectrum of an ethylene/propylene copolymer is much simpler as inversions are not at all observed[17]. An example is given in Fig. 4.2 which also shows the difference between vanadium and $TiCl_3$-catalyzed samples. Calculating the $r_1 * r_2$ value from the given triad distributions of the polymers described in reference 17 does not give a reasonably constant value. The fact that a very wide range of compositions is studied will add to the difficulties. Using only the three high ethylene containing copolymers, the $r_1 * r_2$ value is about 5, i.e. very blocky copolymers are formed. It has to be remarked that

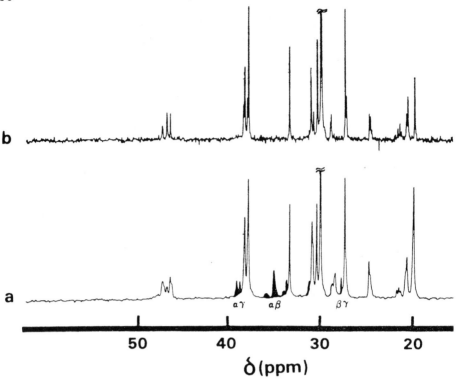

Fig. 4.2: Examples of ^{13}C-NMR spectra of ethylene-propylene copolymers made with a non-isotactic (a) and an isotactic (b) catalyst [from ref. 17. Reprinted with permission from Macromolecules. Copyright 1977 American Chemical Society]

comparison of the $r_1 * r_2$ value calculated from NMR and calculated via Fineman-Ross does not give the same value in many cases. An example is found in the study of Yechevskaya et al[16] where with TiCl$_3$ they observed 2.4 and 0.8 respectively, and for supported titanium catalysts 1.5 and 0.5. More about this will be said later. NMR studies into propylene/1-butene copolymers made on TiCl$_3$ also show relatively simple spectra[18]. The observed dyad distributions could be very well approximated by Bernouillian statistics, suggesting ideal random distribution of the propylene and 1-butene (this is very remarkable as these polymers are shown to be heterogeneous in crystallinity and fractionation studies, see below).

Catalyst effects on the sequence distribution are very large. An example for a relatively narrow range of catalysts (only hetero-geneous titanium systems) is given by Cozewith[11]. A few of his data are reproduced in Table 4.1, and clearly at the given constant overall ethylene content the individual triad populations vary considerably, such as for the PPP triad by more than a factor of 2.

TABLE 4.1
Examples of triad distributions in ethylene/propylene copolymers (from ref. 11)

Catalyst	ethylene content, % mol	triad population,		
		PPP	EEE	PEP
Mg(hexyl)$_2$-reduced TiCl$_4$//DEAC	65.1	0.060	0.294	0.091
TiCl$_4$/MgCl$_2$ Et-hexanol soluble complex//DEAC	62.4	0.089	0.298	0.0884
ditto//AliBu$_2$Cl	64.9	0.100	0.345	0.067
TiCl$_4$/Mg(2-Ethexanoate)$_2$ soluble complex//AliBu$_2$Cl	63.4	0.100	0.304	0.074
TiCl$_3$/MgCl$_2$/THF complex// AlEt$_3$	62.1	0.100	0.280	0.092
TiCl$_4$/MgCl$_2$//AlEt$_3$	62.9	0.110	0.303	0.079
TiCl$_3$ (Stauffer AA)//AlEt$_3$	62.9	0.133	0.286	0.110

The specific co-monomer chosen also affects the resulting sequence distribution. A few examples from our own work, with ethylene and 1-butene as co-monomer, are given in Table 4.2. Some corresponding spectra are given in Fig. 4.3 (ethylene as co-monomer) and Fig. 4.4 (1-butene as co-monomer). From the data in Table 4.2 a clear difference is noted for the sequence distribution at similar co-monomer contents. The butene-containing copolymers show a much lower BB dyad frequency than the EE one in the corresponding ethylene-containing copolymers. The NMR data have been used to calculate r_1*r_2 values by the method of Carman and Wilkes[6]. For polymers at the low end of the ethylene contents we find a value of 2.2, for the comparable 1-butene polymers this is 0.95. This indicates that the latter monomer copolymerizes in a less blocky fashion than ethylene, in line with data given above. In the study covering the wider range of ethylene contents (ethylene contents of 30 to 100%) the soluble material after fractionation gave rise to a value for r_1*r_2 of about 1.4 and the residues of about 3.6. Again

238

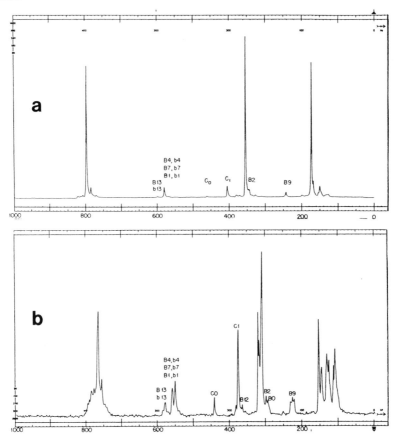

Fig. 4.3: ^{13}*C-NMR spctrum of ethylene-propylene copolymer (a) of low ethylene content (5.3 %mol) and of its corresponding xylene solubles (b)(for indexes of peaks see chapter 13)*

the values are high, illustrating the blocky character of the copolymerization. The higher value obtained with the residue is to be expected for a copolymer isolated in this way.

The endgroups of copolymers made with isospecific catalysts are only vinyl and vinylidene, i.e. those expected from chain transfer starting from a last ethylene unit or a primary inserted propylene - see also above.

Supported isospecific catalysts have also been used to prepare random copolymers. Two studies[19,20] have recently appeared. Abis et al[19] determined the copolymerization parameters, and the product of

TABLE 4.2
Examples of NMR-measurements on ethylene or 1-butene containing copolymers

Ethylene or 1-butene content %m/m	Ethylene or 1-butene in sequences of		
	1 %	2 %	>2 %
5.5	74	26	0
10.9	57	43	0
15.5	32	56	12
9.2b	87	13	0
12.6b	82	18	0
32.2b	63	37	(0)

b: 1-butene/propylene copolymer

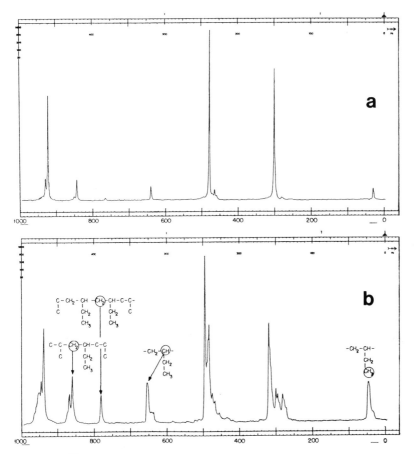

Fig. 4.4: ^{13}C-NMR spectrum of 1-butene-propylene copolymer (a) of low 1-butene content (9.8 %mol) and of its corresponding xylene solubles (b)

the reactivity ratios is well over 2, suggesting a blocky structure. In this study no inversions of propylene could be distinguished, these were however observed by Martuscelli et al[20], although at a concentration of only 0.5 %m/m or lower. Interestingly these inversions have never been reported for $TiCl_3$-catalysts.

Using a homogeneous $TiCl_4/MgCl_2/2$-ethyl hexanol catalyst with $AlEt_2Cl$ as co-catalyst Kashiwa et al[21] also found inversions to a fairly large extent, being about 1/3 of the level observed with $VOCl_3/AlEt_2Cl$. In the homogeneous $Cp_2Zr(CH_3)_2/aluminoxane$ system Kaminsky and Miri[22] describe a high regiospecificity as no inversions were detected, neither in ^{13}C-NMR nor in IR measurements.

Ross shows[10] in a recent article that calculation of reactivity parameters using either kinetic data (e.g. Fineman-Ross) or sequence distribution data (ex NMR) are bound to give different results as soon as the copolymer is not homogeneous in its composition. The triads composed of the same unit will, in an inhomogeneous copolymer, always be larger than predicted on the basis of ideal copolymerization, also other triads show differences compared to predictions. Introducing inhomogeneity in a mathematical sense these inconsistencies can nicely be removed whilst keeping r_1*r_2 equal to 1. Observed triad distribution can then be accurately described. This outcome supports the view that copolymerization kinetics can still be ideal, the catalyst forms "just" inhomogeneous copolymers. Of course this mathematical treatment does not give any insight into the true reason for the inhomogeneity; Ross suggests that monomer diffusion plays a major role, a view that we find difficult to accept. For instance, in homopolymers also there is a large compositional (in tacticity terms) distribution, and in that case it is not expected to be caused by starving of monomer but to be due to the presence of a large number of different sites. That might also well be the explanation for the observed heterogeneity in copolymers.

The sequence distribution of the monomers in the copolymer can be calculated given the kinetic constants of the polymerization system. An example can be found in reference 23. However these calculated distributions are only realistic in the case of just one type of active site, a situation which is rarely found, and then almost exclusively in vanadium systems and possibly also in some of the new homogeneous catalysts. In most practical cases however the number of types of active sites is larger and the determined copolymerization constants represent an average. A mathematical appraisal of such a situation leads to the conclusion that the observed r_1*r_2 will lie between the extremes of the individual pairs, but the calculated sequence distribution is bound to underestimate the longer sequences[1]. And these longer sequences are just the ones determining the crystallization properties of the copolymer, which are among the most important properties in view of its application as a rubber.

One can also attempt to use more elaborate models such as those by Ross[10] and Cozewith[11]. An example will be given from the latter study. Apart from the single site model (which he refers to as ideal copolymerization) he studies a multiple site model for which for each individual site r_1*r_2 equals 1. As a special case of the multiple site model a dual site case is also tested. Expressions can be derived which link the model parameters (either reactivity ratios or propagation probabilities and relative weight of each of the sites) with the measured sequence distribution. In attempting to fit measured triad distributions with a statistical computer fit, a good fit is defined as one in which the residual sum of squares is reasonably small (10^{-3}). These models are applied on a number of copolymer series, both from the literature as well as a specific series made with different catalysts (reported in Table 4.1). For instance, using the data from Kakugo et al[9], a good description is obtained with the two-site model; one of the sites is giving very high-ethylene-containing copolymers regardless of the feed composition, its fraction increasing with increasing ethylene in the feed (or copolymer). The same holds for the copolymer series reported by Ray et al[17]. (These results fit in rather well with simple fractionation tests and melting point determinations reported in the next sections.) However no good fit could be obtained in the series described by Doi et al[24] (made on silica supported $TiCl_4/MgCl_2//AlEt_3$) nor on his own series. Thus the proposed models have at best a limited applicability. (Will the situation be improved when we can measure pentads rather than triads, as then much more detail of the structure is known?)

Cozewith also observes that reactivity ratios determined from process data (rates, conversions, composition) differ sometimes appreciably from the ones derived from NMR measurements. Only in the single site case is there good correspondence. Thus for sequence distributions one should stay with measurement by NMR rather than calculations from process data.

4.3 FRACTIONATION, COMONOMER DISTRIBUTION

Fractionation of the copolymers is a very powerful tool in assessing the homogeneity of the distribution of the monomers. The solubility behaviour of ethylene-propylene and other propylene copolymers resembles to a very large extent the behaviour of the homopolymer polypropylene, when taking into account potential crystallinity differences. Thus good solvents for amorphous copolymers at room temperature are for instance heptane, cyclohexane and xylene. Experimental determinations of the solubility parameter are infrequent. On the basis of maximum swelling of copolymer films in blends of n-decane and benzene of various compositions, Kirkham[25] arrives at a value of $17.4*10^{-3}$ $(J/m^3)^{0.5}$ for copolymers containing about 70 %m/m ethylene, which is close to the calculated value of $17.2*10^{-3}$ $(J/m^3)^{0.5}$ using Small's method. Using the method of "inverse gas chromatography" a solubility parameter value (at infinite dilution) of $15.8*10^{-3}$ $(J/m^3)^{0.5}$ was obtained[26] (this type of measurement is expected to lead to values somewhat below the values derived from measurements at practical concentrations).

Elaborate fractionations have shown that the catalyst systems $VOCl_3/Al_2Et_3Cl_3$ and $VCl_4/Al_2Et_3Cl_3$ give narrow distributions (Q=2) with no noticeable change in co-monomer content with the molecular weight of the fractions[1]. The same was observed with the $VCl_4/(C_6H_5)_4Sn/AlBr_3$ catalyzed ethylene/propylene copolymers with low propylene contents[27]. These are rather exceptional as the majority of the other vanadium systems, for example the above mentioned ones with $AlEt_2Cl$ as co-catalyst instead of "sesqui", and all titanium derived heterogeneous catalysts, show broad co-monomer and molecular weight distributions. This is in line with the observations on homopolymerization of propylene with such catalysts. For instance[28] using $VCl_4/AlEt_2Cl$ narrow molecular weight distributions are only obtained at temperatures below -20 °C, whilst very broad distributions are observed at the higher temperatures normally applied in ethylene/propylene copolymerization. In GPC the chromatograms show a number of humps pointing to polymer species formed on different sites with their own specific rate constants. In all the cases of a broad distribution in a molecular weight sense the co-monomer distribution is also far from homogeneous.

Not only ethylene/propylene copolymers are inhomogeneous but also propylene/1-butene as shown by Locatelli et al[29] for TiCl$_3$(HRA) and TEA. This shows that even for these two monomers, for which it is expected that they are polymerized on truly the same sites (ethylene is small and could probably be "using" more sites than larger α-olefins), inhomogeneity is observed, and thus the catalyst must be composed of different sites. Note however that the polymer fractions obtained in their fractionations all obey first order Markov statistics, i.e. the last-added monomer unit is not having any effect on the subsequent insertion.

Another example of the heterogeneity of copolymers is given by the studies of the group of Kakugo[30]. Using copolymers with very low ethylene or 1-butene content made on TiCl$_3$(AA) and DEAC and applying the TREF fractionation technique, ethylene contents differing a factor of five are observed, with the most soluble material showing the highest co-monomer content. In the butene case the ratio of butene contents is only two. Also in the stereoblock region the co-monomer content decreases. As observed in the propylene homopolymer case (see chapter 2) here also there is the impression that two types of isotactic sites are active.

In the newly developed supported catalysts, when applied for ethylene/propylene copolymerization, a broad molecular weight distribution is again observed[31]. The distribution is however not much broader than the one for the commercially applied vanadium catalyst (Q values of 6 and 8 respectively). The same broad distributions were noted in propylene/1-butene copolymers made on the heterogeneous TiCl$_3$ catalyst, independent of the type of co-catalyst used[32-34]. In an extensive investigation on so-called "co-stereosymmetric" copolymers, made on TiCl$_3$/AlEtCl$_2$/hexamethylphosphoric triamide, Coover et al[32] showed, by a fractionation based on the dissolution temperature, a very broad distribution of monomers indeed; for instance a 40% m/m 1-butene containing copolymer gave fractions ranging from 15 to 70 %m/m 1-butene. And flash pyrolysis gas chromatography studies[35] on the same type of copolymers observe for example more trimers of the separate monomers (PPP and BBB) than expected, indicating blockiness as well. Even in the copolymerization of the closely related monomers styrene and p-t-butyl styrene with TiCl$_3$/AliBu$_3$ the copolymer composition is molecular weight dependent[36]. An extreme case of blockiness has been described by Segre et al[37,38]. In the copolymerization of (S)-4-methyl-1-hexene and styrene with $TiCl_4/AliBu_3$ as catalyst, both the nearly pure homopolymers and a copolymer of intermediate composition can be isolated by extraction with different solvents, from a copolymer made from a 50/50 feed. Interestingly the observed overall reactivity ratios are 1.65 and 0.76 respectively; their product, being 1.25, suggests relatively random copolymerization!

Copolymerization of propylene and styrene with the regular catalyst is difficult as next to copolymer also the homopolymers are formed. However, by using Solvay $TiCl_3$ with $Cp_2Ti(CH_3)_2$ as co-catalyst copolymer is formed almost exclusively[39], illustrating the much smaller range of catalytically active sites on this catalyst.

Regrettably no compositional fractionation has been caried out by Kaminsky and Miri[22], thereby still leaving the homogeneity of their copolymers - made on homogeneous zirconium catalysts - formally unproven, although this is certainly expected from other evidence (molecular weight distribution).

Random copolymers made on a supported $MgCl_2$-catalyst are also heterogeneous in their composition as a simple fractionation with

TABLE 4.3
Solubles and their composition in copolymers ($TiCl_3$-catalyzed)

Ethylene content of copolymer %mol	Xylene solubles %m/m	Ethylene content of solubles %mol	a*	1-butene content of copolymer %mol	Xylene solubles %m/m	1-butene content of solubles %mol	a*
0.33	7.2	0.36	1.0	1.5	5.5	2.8	1.9
0.61	7.6	1.13	2.0	3.6	7.1	7.7	2.3
0.69	9.5	1.9	3.3	4.2	7.7	12.0	3.4
0.96	5.8	3.4	4.2	8.6	9.4	17.1	2.2
1.85	6.8	7.0	4.7	13.5	16.7	26.6	2.4
3.2	7.2	12.3	4.9	16.6	28.2	29.9	2.7
4.5	9.8	16.3	5.1				
5.5	11.4	16.2	3.9				
8.2	14.7	21.6	3.7				

a* : the ratio of the ethylene (or 1-butene) content of the solubles fraction to that in the crystalline fraction

ether and hexane gave a 10 %m/m range in ethylene content[19]. Thus, there too, one can expect more than one type of active site.

In a simpler sense the xylene or hot xylene solubles test, coupled to a compositional measurement on the fractions, also gives information on the distribution of the monomers. For low ethylene- or 1-butene-containing copolymers made on a $TiCl_3$ catalyst, a number of our own data are given in Table 4.3. Both co-monomers do give rise to an increased solubles level as expected, as a decrease in the chain regularity is bound to give rise to a greater propor- tion of soluble material. The absolute level is however much lower for the 1-butene than for the ethylene containing copolymers. In line with other observations the co-monomer content of these xylene solubles is considerably higher than the average value for the total polymer. Xylene solubles are about four times richer in ethylene and 2.5 times in 1-butene compared to the xylene residue.

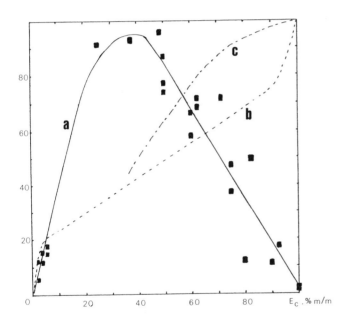

Fig. 4.5: Amount, %m/m (a), and composition, E_c (b), of hot xylene solubles (HOXS) observed in ethylene-propylene copolymers as function of their co-mono- mer composition, in addition E_c of the HOXS-residue is given (c) ($TiCl_3$ reduced at low temperature as catalyst)

This ratio is almost independent of the overall ethylene content (in the range mentioned) but must decrease at higher ethylene contents (see below). The crystallinity of the solubles, in the case of ethylene, is not much different from homopolymer solubles; in the 1-butene case it appears to be somewhat higher.

Extending the ethylene contents to the full range gives the solubility data as shown in Fig. 4.5. A maximum solubility is observed around the 30 to 50 %mol composition, but in no case is total solubility reached for copolymers made on a $TiCl_3$ catalyst. This has been described already in the early literature on copolymerization, by Bier[40]. Again when measuring the co-monomer content of the fractions one must conclude that these copolymers are inhomogeneous, the solubles are lower in ethylene content than the overall polymer, and the residues are high-ethylene containing copolymers. As in the very low ethylene content range the solubles are richer in ethylene, whilst at the higher range this is just the opposite; at a certain ethylene level the composition of solubles and residues should be nearly equal. Not surprisingly this happens around 40 %m/m ethylene, i.e. in the region of maximum solubility. It appears that the percentage of xylene(or HOXS) residue is nicely proportional to the ethylene content of the gascap, as illustrated by the data in Table 4.4. The average value of HOXS-r/E_g is 0.90 ± 0.23. The significance of this observation will be discussed later.

TABLE 4.4
HOXS-residue of ethylene/propylene copolymers, made on $TiCl_3$, related to the gascap composition

Ethylene content of copolymer %m/m	Gascap composition %V/V C_2 (E_g)	HOXS residue %m/m	HOXS-r $\dfrac{}{E_g}$	Ethylene content of copolymer %m/m	Gascap composition %V/V C_2 (E_g)	HOXS residue %m/m	HOXS-r $\dfrac{}{E_g}$
25	13	9	0.69	63	38	28.3	0.74
50	31	25.9	0.84	71	44	28.1	0.64
50	31	23.3	0.75	82	57	50.8	0.90
50	31	13.4	0.43	92.5	73	82.8	1.13
60	41	33.9	0.83	100	100	98.4	0.98
60	41	42.4	1.03				
75	58	62.7	1.08				
80	64	87.7	1.37				
90	81	88.6	1.09				
100	100	99.3	0.99				

4.4 MELTING BEHAVIOUR, (CO)CRYSTALLIZATION

We now leave the compositional characteristics of the copolymers and turn to the crystallization behaviour. In crystalline polymers one expects upon copolymerization a decrease in the melting point as the chain regularity is lowered and thus the sequence length of the crystallizing units is decreased. At the same time the crystallinity should drop as well, together with the density. In DSC the melting or crystallization peak broadens as crystals with varying perfection are formed.

Early observations of this, using differential thermal analysis, were made by Ke[41]. In a general sense this is indeed observed, but the extent of these changes differs widely for different systems and moreover deviates considerably from theory. In Flory's theory of polymer melting it is assumed that the chain irregularity cannot be accommodated in the crystalline lattice. The theory predicts a rapid lowering of the melting point with co-monomer content. For instance in the case of polypropylene disturbed by low levels of ethylene a decrease of about 1.5 °C per %mol of built-in ethylene is predicted[42]. In practice we observed in that system (for $TiCl_3$ catalyst) an even more rapid decrease of 4.6 °C per %mol ethylene over a limited range of 0 to 2% when measuring the final melting point (TMF)! Admittedly one observes here the melting of the most perfect crystal that the conditions allow to form. In normal DSC measurements, using the peak maxima, the decrease is far less and moreover quickly levels off at slightly higher co-monomer contents (for $TiCl_3$ catalyzed copolymers). The same level of initial decrease of 5 °C per %mol ethylene was observed in the random copolymers made on the high yield catalyst[20], only the equilibrium melting temperature was measured in that case. Contrary to this Greco et al[43] found no change in the equilibrium melting temperature and a value equal to that of the homopolymer in a series of ethylene/propylene copolymers made on δ-$TiCl_3$ and ranging in composition from 8 to 19 %m/m. This points to a large heterogeneity in their copolymers.

Some DSC-data on own copolymers over a broad composition range are given in Table 4.5. Clearly the melting temperature is nearly independent of the ethylene content of the copolymer over a wide range, and about 10 °C below the pure polyethylene value. Crystallization from the melt is most sensitive to the co-monomer level, as shown by the dependence of T_x on ethylene content in Fig. 4.6.

TABLE 4.5
DSC measurements on ethylene/propylene copolymers made on $TiCl_3$

Ethylene content of copolymer %m/m	DSC-data			Ethylene content of copolymer %m/m	DSC-data		
	T_x °C	T_{m2} °C	ΔH_m cal/g		T_x °C	T_{m2} °C	ΔH_m cal/g
25	92,73	117	0.4	75	108	127	9.1
50	105	124	2.4	75	104	121	10.9
50	115,103	123	3.4	75	106	124	16.1
50	102	122	2.2	90	108	125	18.8
50	102	122	1.8	100	118	134	27.7
60	105	125	2.4				

The ethylene contents are approximate

The crystallinity is lowered strongly upon copolymerization, Fig. 4.7 gives some data using crystallinity based on X-ray diffraction. The same conclusion can be drawn from the data in Table 4.5. Around the minimum crystallinity (at about 30 %m/m ethylene) both poly-

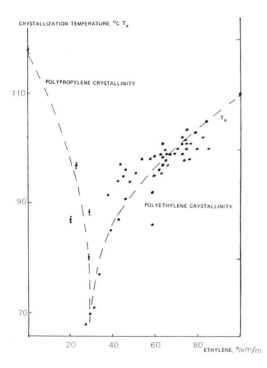

Fig. 4.6: Crystallization temperatures of ethylene-propylene copolymers as function of their composition (from DSC)

propylene and polyethylene crystallinity are observed. Some investigators[1,44-46] have tried to relate the calculated sequence
distribution to the measured crystallinity by estimating a minimum
ethylene sequence length above which all ethylene sequences parti-
cipate in the crystallinity. Values of 8 to 10 ethylene units in
the minimum sequence are a rather common result from such calculations. Davison and Taylor[45] used nitric acid digestion of terpolymers of ethylene, propylene and 1-butene to determine the amount of
crystalline material present. This was again correlated to the

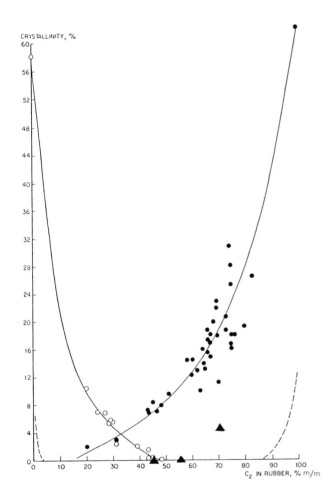

Fig. 4.7: Crystallinity of ethylene-propylene copolymers as function of their
composition, copolymers made on TiCl$_3$ prepared at low temperature (data from
X-ray diffraction); o: PP-crystallinity, •: PE-crystallinity, ▲: vanadium
catalyzed commercial EPDM's, --- predicted for completely random copolymers.

calculated sequence distribution and the crystallinity derived therefrom. It is a pity that the remaining material after the oxidation was not analyzed for its molecular weight (distribution) as a direct answer could than probably have been obtained.

Martuscelli et al[47] attempt to correlate crystallization properties such as crystallinity, rate of spherulite growth, melting points, etc. with the observed NMR triad distribution. The greatest sensitivity is observed with the EPE triad. In our opinion it is rather dangerous to correlate a fairly "long range" property such as crystallization with only a triad distribution, which in effect spans only a six-carbon chain segment. More meaningful results might be expected using pentads or even longer sequences.

Fig. 4.8 depicts a typical melting endotherm for an ethylene/propylene copolymer containing 70% of ethylene which has

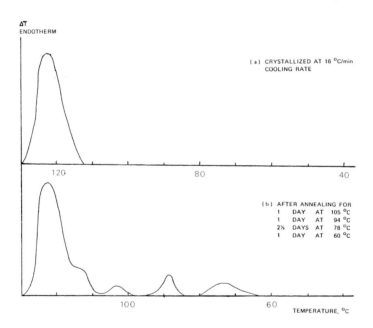

Fig. 4.8: DSC melting endotherms of a 70 %m/m ethylene containing ethylene-propylene copolymer after various annealing steps at progressively lower temperatures

been subjected to four annealing steps at progressively lower temperatures. For each temperature a corresponding melting peak was produced in the DSC, having a value about 10 °C higher than the annealing temperature. All copolymer samples produced this effect regardless of ethylene content. Polypropylene and polyethylene homopolymers do not show such an effect in the same temperature region, confirming our belief that this is a true copolymer effect. One must postulate that at such supercoolings, smaller crystallites resulting from shorter block lengths can crystallize. Since the total crystallinity increases at each annealing step, one can ignore any effect due to redistribution of crystallite sizes. An identical situation has been observed in ethylene/1-butene copolymers[48].

In vanadium-catalyzed copolymerization of ethylene and propylene the levels of crystallinity are generally lower than for $TiCl_3$-catalyzed ones (see e.g. the data given in Fig. 4.7). However, the exact level is very much dependent on the catalyst system chosen, and a general trend of increasing crystallinity with decreasing $r_1 * r_2$ is observed[1,44,45]. Of the copolymers made on the homogeneous $Cp_2Zr(CH_3)_2$/aluminoxane catalyst no levels of crystallinity have been reported[22]; the few DSC-data given show, however, low crystallinity at rather high E_c's, with low melting points (e.g. at an E_c of 80 %mol the melting point was 49 °C).

Whilst with ethylene the effect of its copolymerization on the crystallinity is large, in the case of 1-butene as co-monomer this is very much lower. Even at the composition of minimum crystallinity this is still about 50% of the homopolymer values[32-34]. Moreover the X-ray diffraction pattern of both homopolymers can be discerned in copolymers with compositions from 12.5 up to 75.5 % mol 1-butene in propylene, albeit that in the extremes the level of the minor one is only very modest. The two homopolymer patterns are not only seen in melt crystallized samples but also in solution-grown crystals[47], although only in the 50/50 composition.

The disturbance of the chain regularity is also observed in the occurrence of the γ-modification of polypropylene when small amounts of co-monomer are applied[49]. At about 6% mol ethylene up to 50% of the crystallinity is found in the γ-form depending on the thermal treatment. The fraction of this modification of course increases with increasing co-monomer level. These phenomena were studied in more detail later. Guidetti[50] studies a range of commercial and laboratory prepared samples, both ethylene-propylene copolymers as well as ethylene/propylene/1-butene terpolymers. The overall crystallinity of the samples does not correlate with the fraction of γ-crystallinity present, but a good, inverse, correlation is observed with the crystallinity remaining at 131 °C. However the relation is not unique but differs for the co- and the terpolymers with the latter forming more of the γ-crystal modification. This division also holds for the relation of the "γ-crystallinity" with the co-monomer content. Apparently the γ-form does not survive over 131 °C due to its lower melting point, which in turn is caused by the relatively short chains in the crystal. The exact reason for the more random behaviour of butene-containing copolymers was not given in this study. It can however be shown that introduction of a third monomer will distinctly lower the propylene sequence length, especially of the longer ones. This is for instance calculated by Davison and Taylor[45], admittedly for copolymers with higher ethylene content, but the principle stays the same. In a similar study[51] Busico et al investigated the crystalline fractions of random ethylene/propylene copolymers (using only the crystalline fraction means that especially at the higher ethylene levels a large fraction of the polymer is discarded). The γ-fraction appears to be very sensitive to the crystallization conditions applied. The optimum crystallization conditions to maximize the fraction in the γ-form were determined, these were: annealing a few degrees under the final melting point and subsequently cool fairly rapidly (< 10 °C/min). In this way up to 80% of the total crystallinity could be found in the γ-form. This fraction increases with the co-monomer content. They observe also a different form of the α-modification ("α-2"), its melting point in DSC is very sharp and dependent on the ethylene content. For pure polypropylene its maximum lies at about 180 °C, decreasing to 142 °C for copolymers with 20 %mol ethylene. Also the extent of α-2-crystallinity decreases sharply with the ethylene content, indicating that only polymers with a limited number of disturbances along the chain can be accommodated in the lattice of this modification.

There are numerous studies of characterization of ethylene/propylene copolymers generated on different catalysts. They do not really contribute much new knowledge to what has been described above, but they do illustrate the importance of the co-monomer sequence distribution on the properties, especially the crystallinity. An example of this is the tetraneophylzirconium/alumina catalyst used by Starkweather et al[52], which appears to give very heterogeneous ethylene/propylene copolymers, showing at E_c 36 %m/m both poylethylene and polypropylene crystallinity (like the $TiCl_3$ catalyzed copolymers of similar composition). Another example is the use of a chromium catalyst[46] ($Cr(acetate)_3$/ acetic anhydride/$AlEt_2Cl$) giving rise to a more random type of copolymer in which the crystallinity drops off rather rapidly with composition. Still, at E_c of 80 %m/m (or 20 %m/m), an appreciable crystallinity is present (in the order of 10-20 % of the respective homopolymer values). Much more true random products appear to be generated on the $Cp_2Zr(CH_3)_2$/aluminoxane catalyst.

We now consider the structure of the crystal itself. Crystallization of polymers is not a very selective process, as for instance the heat of melting does decrease[53] with decreasing crystallinity (as measured e.g. by density or X-ray diffraction). This implies an incorporation of chain irregularities in the crystal lattice, which is observed in a great number of cases: for example in homopolymer propylene (as function of propagation errors, see chapter 2) and in propylene/1-butene copolymers[47,54]. This is contrary to the treatment given by Flory of the effect of co-monomers on melting. Incorporation of the other constituent leads also to expansion of the crystal lattice for the propylene lattice in copolymerization with 1-butene, or conversely contraction in the 1-butene lattice upon incorporation of propylene[32-34,47]. For the extensively studied modified polyethylenes a similar expansion of the lattice along one of the crystallographic axis is observed, see for instance ref. 55. No changes in lattice parameters are however observed[47] for ethylene-propylene random copolymers with propylene crystallinity. The incorporation of ethylene units in the polypropylene crystal also explains the large initial drop in the final melting point, as the lattice will become less stable.

In a theoretical study Starkweather et al[52] calculates that an isolated ethylene unit in a propylene sequence can be accommodated in the polypropylene helix at no large extra expenditure of energy (about 3.6 kJ/mol). Blocks of ethylene lead to much higher energies. Surprisingly compositions close to the alternating one are fairly easily fitted into a helix, the true alternating sequence is even 6 kJ/mol more stable in the helix than in the zigzag configuration. This effect could be used to affect co-crystallization of polypropylene and the copolymer, see chapter 7.

The extent of co-crystallization varies widely. Whereas in ethylene/propylene copolymers it is rather limited - as shown by the great reduction in crystallinity observed - in the case of propylene/1-butene it is much more prominent. And almost complete co-crystallization is observed in copolymers of 1-butene and 3-methyl-1-butene or 1-pentene[33,34] (see also chapter 11 and 12). It is worth mentioning already here that in polyethylene it has been found that methyl side groups (i.e. propylene units) are incorporated in the polyethylene crystal on an equilibrium basis[56]. An interstitial crystalline model has been recently proposed for this type of co-crystallization[57]. From further studies[56,58] it is very clear that side groups larger than methyl, such as ethyl or hexyl, are strictly excluded from the polyethylene crystal. One of the effects of this behaviour is the independence of the melting point of polyethylene on the nature of the side group. Of course when side groups become much larger, co-crystallization becomes again possible as for instance in LDPE[56]. For a more extensive discussion of this see chapter 10.

4.5 GLASS TRANSITION

The β-transition of polypropylene is lowered by copolymerization with α-olefins, and the glass transition of the copolymer is ofcourse a function of the co-monomer content[2,59,60]. Measurement by dynamic mechanical analysis (DMA) using a torsion pendulum, lead to data as given in Fig. 4.9 and Fig. 4.10 for ethylene and 1-butene copolymers respectively. These data fit in very well with the data given by others who used either DMA or DSC. The glass-rubber transition temperature (T_g) of a random copolymer is often intermediate between that of the homopolymers of the constituents, in a way expressed by the Gordon-Taylor relation:

GLASS - TRANSITION TEMPERATURE

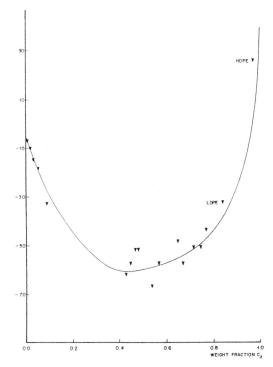

Fig. 4.9: Glass transition temperatures of ethylene-propylene copolymers

$$T_g = (w_1*T_{g1} + k*w_2*T_{g2})/(w_1 + k*w_2)$$

in which T_g is the copolymer glass-rubber transition temperature

T_{g1}, T_{g2} is the T_g of the respective homopolymers

w_1, w_2 is the weight fraction of the components

k is an adjustable constant.

For the propylene/1-butene copolymers Fig. 4.10 gives the calculated relation on the basis of the above formula using 273 and 248 K as the T_g values of polypropylene and polybutene respectively. The constant k is 0.21, and the curve represents rather well the measured values.

Fig. 4.9 shows that the T_g values of the ethylene/propylene copolymers display a minimum at intermediate compositions. This, of course, cannot be reconciled with the above equation unless we

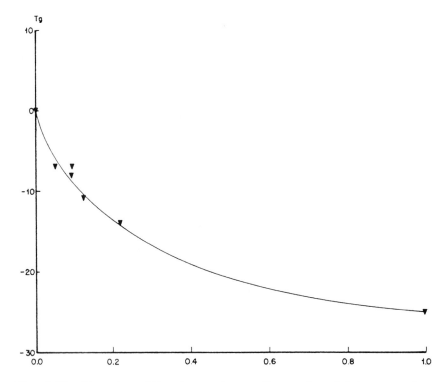

Fig. 4.10: Glass transition temperatures of propylene-1-butene copolymers as a function of the 1-butene fraction

suppose a third "homopolymer" to dominate the T_g value at these intermediate compositions. For this we suggest the alternating ethylene/propylene copolymer. This polymer results on hydrogenation of isoprene rubber. For this "copolymer" with 40 %m/m ethylene content we measured a T_g value of -61 °C (212 K). The T_g value of another regular copolymer is known[61]: poly-1-methyl-octamer, alternating three ethylene units with one propylene unit, has a value of -57 °C. This value is very close to the alternating 1:1 copolymer. Hence it might be a good approximation to apply the above equation in the range of 0 to 40% m/m ethylene using:

$$T_{g1} = 212 \text{ K} \quad w_1 = 2.5*(\text{\%m/m ethylene})$$
$$T_{g2} = 273 \text{ K} \quad w_2 = 1 - w_1.$$

We found that a k of 0.42 fitted the data best.

For the high ethylene concentration range the results of the measurements seem to refer to the polyethylene α-transition (40 °C), rather than to the T_g value of polyethylene, which is found typically in the range of -125 to -80 °C. We then find that the results can be described by using

$$
\begin{aligned}
T_{g1} &= 212 \text{ K} \qquad w_1 = 1.67*(\%m/m \text{ propylene}) \\
T_{g2} &= 313 \text{ K} \qquad w_2 = 1 - w_1 .
\end{aligned}
$$

In this case the value of k is 0.10. The above model has no true scientific basis, nevertheless it is effective in describing the measured data. Even the short chain branching in low density polyethylene fits into the scheme when equating one methyl per 1000 C atoms to 0.5 %m/m propylene.

4.6 YIELD STRESS

Due to the less-perfect crystallinity and the decreasing fraction of crystalline material upon copolymerization, the mechanical properties are greatly affected. For instance the yield

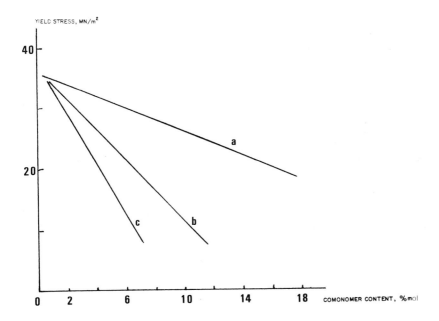

Fig. 4.11: Yield stress of copolymers as function of composition and co-monomer type: 1-butene (a); ethylene (b); 1-octene (c)

stress decreases roughly with about 2 to 3 MN/m^2 per %mol of incorporated co-monomer[62]. The effect is largest for 1-octene as co-monomer, followed by ethylene, with 1-butene as the co-monomer least affecting the yield stress. This is clearly seen in Fig. 4.11 based on own measurements. This is, for ethylene and 1-butene, in line with the co-crystallization behaviour mentioned above. To a large extent the yield stress of propylene copolymers at low co-mo-

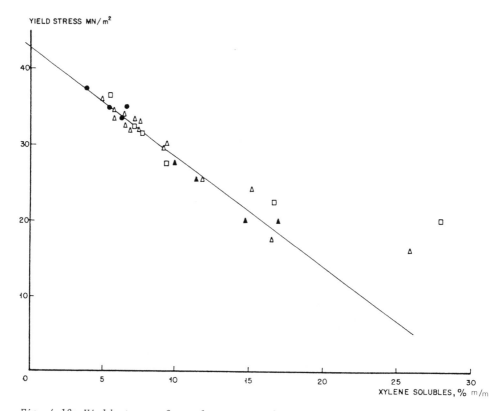

Fig. 4.12: Yield stress of copolymers as a function of xylene solubles

nomer levels is determined by the level of xylene solubles, as illustrated in Fig. 4.12. The slope of this relation is 1.4 unit in yield stress per unit in xylene solubles, larger than observed in the homopolymer case where it was found to be 1.15. This indicates the effect of the decreasing perfection of the crystalline lattice. But it also implies that the (im)perfection is reasonably constant in the range studied, which could mean that there is a limit to the number of disturbing ethylene units that can be accommodated in the polypropylene lattice.

4.7 FORMATION OF DISPERSED SYSTEMS

The last mentioned characterization aspect deals with the macroscopic homogeneity of the system. Turner-Jones[33,34] observed in the propylene/1-butene copolymers of around 50/50 composition a heterogeneous, disperse system under the microscope. Different regions showed different melting points as well. Similar disperse situations were also observed by us. Already in a 5% ethylene/95% propylene copolymer there is a dispersion visible in phase-contrast microscopy, albeit of very small particle size. This is not really surprising, keeping in mind that the polymer contains about 10 %m/m

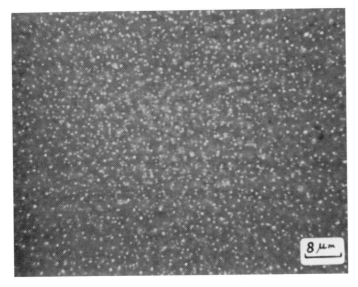

Fig. 4.13: Phase contrast micrograph of hot xylene solubles from an ethylene-propylene random copolymer

of a fraction with an average ethylene content of around 16 % mol, which is already very different from the remainder of the polymer. Moreover the xylene solubles fraction as such of this type of copolymer is also very heterogeneous, as shown in Fig. 4.13.

4.8 CONCLUDING REMARKS ON CHARACTERIZATION

Copolymers are very difficult to characterize and - with a view to the original objective of characterization - predictions of final properties are just as difficult. The main problem lies in the wide variety of characterization methods required to obtain a more or less complete picture of the behaviour of the sample. The minimum required comprises a compositional check (preferably NMR), a MWD determination and a DSC measurement on melting enthalpy and melting temperatures. This allows feedback to the polymerization proper and gives the first impression of some of the properties of the material. As the ultimate properties are almost all hinged around the crystalline nature of those polymers, the data related to this property are very important. Therefore, as an example, knowledge of which groups are excluded from the polymer crystal, and how the melting points relate to the amount and distribution of co-monomer(s) present are of crucial importance in predicting behaviour in end use. Also, the effect of the composition on the extent of the formation of the γ-crystal modification is very valuable information in view of the use of random copolymers in, for example, heat-sealable films. The studies done so-far need certainly to be extended. Thus, a diligent use of the characterization data is expected to allow reasonable prediction to be made in specific and limited applications as for instance stiffness or heat sealibility in low-co-monomer containing randoms. But it is still not possible to derive, from the above suggested measurements, all the relevant properties such as stiffness, elongation, impact (versus temperature), creep, and interaction with environment (corrosion) for a given copolymer.

Clearly, variation in the catalyst used in the copolymer preparation affects to a large extent the properties of the latter. Variations of very blocky to nearly alternating sequence distributions have been described above. The true tailoring of copolymer sequence distribution (and thus the properties) still requires more catalyst development in our opinion. It is thought that the new homogeneous catalysts might bring a breakthrough as they appear to be very amenable to changing their properties via for instance manipulating the ligands on the transition metal and thereby arriving at a situation where, at will, one could prepare alternating to random copolymers and their mixtures. More is mentioned on this topic in the polyethylene chapters, as for that polymer more studies have been done on the relation of sequence distribution and properties.

4.9 RELATION OF COPOLYMER COMPOSITION AND COPOLYMERIZATION KINETICS WITH CATALYST PROPERTIES

As was done at the end of the chapter on the characterization of homopolymers, here again, we will try to relate the observed polymer characteristics to possible catalyst properties. The types of catalysts considered are limited to $TiCl_3$ and supported ones. The facts that need explanation are:

- the reactivity ratio of propylene versus ethylene, with r_1(propylene) in the order of 0.1 and r_2(ethylene) of 15 to 25 (see Table 3.2).

- the increase in <u>propylene</u> polymerization rate when some ethylene is copolymerized.

- the sharp increase in solubles when ethylene is copolymerized, and the much smaller increases with butene as co-monomer.

- the much higher concentration of co-monomer in the solubles compared to the bulk composition in random copolymers, which holds for both ethylene and butene.

- the lower molecular weight of the solubles, similar to the homopolymer case.

- the dispersed nature of the xylene solubles of ethylene/propylene random copolymers (in optical microscopy), the same holds for the total copolymer.

- copolymers with higher ethylene contents ("rubbers") can be fractionated into a low-crystalline, mainly amorphous copolymer and a distinctly crystalline fraction which is also of an appreciably higher co-monomer content.

As a further piece of information the number of active sites and the propagation rate constants for the different monomers would bring much enlightenment. However, determination of the former still gives conflicting data, although more agreement exists on the latter. On the number of active sites Jung and Schnecko[63] report a decreasing series of values when comparing ethylene, propylene and 1-butene. More recent determinations come to values which are more equal for ethylene and propylene (supported catalysts[64,65]), or even more sites active in propylene polymerizations[66,67]. However, at maximum rates the difference is small[66,67]. The propagation rate constants as determined by the group of Yermakov and Zakharov are 12,000 l/mol.s for ethylene and 90 l/mol.s for propylene on $TiCl_3$-type catalysts, whilst on $MgCl_2$-supported catalysts the rate for ethylene stays the same, while that for propylene increases more than 10-fold to 1000 l/mol.s. This gives as rate ratios 133 and 10 (ethylene over propylene) on the respective catalysts, quite different from the observed reactivity ratios, especially in the $TiCl_3$-case, which is unexplained at this moment.

For a tentative explanation of the above observations we assume that in ethylene polymerization more sites are participating than in the corresponding propylene case. This is thought also to be the case when comparing propylene with 1-butene. The underlying reasoning (having no recourse to unequivocal facts) is the increase in steric demand of those monomers. The active centers are also thought to span a range of steric requirements, see also the homopolymer discussion. The range extends, in our opinion, to sites which are only able to polymerize ethylene. If this is the case, then the fraction of crystalline polymer in the total copolymer (being of a high ethylene content, and for this discussion equated with polyethylene) should be proportional to the fraction of ethylene in the feed. As shown in Table 4.3 this is roughly the case.

In the kinetics, the main effect of steric hindrance will be exerted by the CH_3-group of the last inserted propylene. This will hamper access for the following monomer, the more so for propylene than for ethylene of course. Calling the homopolymerization rate constants k_{22} and k_{33} respectively and then logically the crossover constants k_{23} and k_{32}, a specific kind of site can best be characterized by its ratio of k_{33}/k_{23} - propylene homopolymerization over the rate of propylene addition on a chain ending in ethylene. For explaining the rate increase of propylene when some ethylene is copolymerized, we assume there is a range of sites, with their k_{33}/k_{23} ranging from zero (no homopolymerization of propylene possible) to nearly one (no hindrance at all). Atactic sites will show the highest k_{33}/k_{23} ratio because of their open nature. Thus part of the copolymer is formed on sites not active in propylene homopolymerization, which leads directly to a rate increase. Moreover the k_{23} may well be higher than k_{33} resulting in an additional rate increase.

The above hypothesis can also be used to rationalize the

increase in solubles, as on the extra sites a copolymer is produced higher in ethylene content than the bulk because of the very low k_{33}. The solubility of this copolymer will undoubtedly be higher and it will end in the solubles fraction. The atactic sites will produce a copolymer more in line with the bulk composition or even lower than this as the sites are least discriminating. Thus the xylene solubles would indeed contain two fractions of widely different ethylene content, giving a dispersed system as these different polymers are not miscible.

The molecular weight of the random copolymer solubles is low (about 5 times lower than for the isotactic ones). This means that the ratio of propagation to chain transfer is much lower for those sites. One is tempted to reason that especially the propagation rate will suffer on hindered sites, as the chain transfer agent, hydrogen, is a very small molecule and will probably not be restricted in its action to a large extent.

For the 1-butene/propylene copolymers, using analogous nomenclature as above, it might well be that due to the much greater steric requirements of 1-butene, there are sites not able to polymerise 1-butene but still active with propylene. Then the increase in solubles can be explained by assuming a rate lowering on the isotactic sites. Then relatively more polymer will be formed on the atactic sites, being at the same time richer in the co-monomer due to their open structure.

In Fig. 4.14 an illustration is given of the above ideas.

Clearly, the number of measurements on copolymer fractions could be extended to obtain more proof for the above statements. Examples would be the measurement of the homogeneity of solubles from propylene/1-butene copolymers (according to the above hypothesis they are expected to be rather homogeneous, being only formed on the atactic sites), glass transition measurements on solubles from various copolymers (multiple, or very broad, transitions are expected in solubles from ethylene/propylene copolymers, only one transition in the 1-butene case). Also, in the copolymerization reaction one could use for instance complexing agents which would show different constants with different sites (small molecules are probably the best) and the composition of the copolymer is expected to show some kind of variation with either the type of agent or its applied concentration.

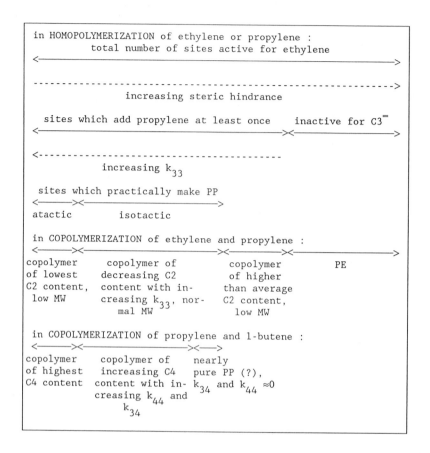

Fig. 4.14: Schematic description of the behaviour of a $TiCl_3$-catalyst in (co)polymerization

REFERENCES
1. C.Cozewith and G.VerStrate, Ethylene-propylene copolymers. Reactivity ratios, evaluation and significance, *Macromolecules, 4 (1971) 482-489*
2. F.P.Baldwin and G.VerStrate, Polyolefin elastomers based on ethylene and propylene, *Rubber Chemistry and Technology, 45 (1972) 709-881*
3. J.van Schooten, E.W.Duck and R.Berkenbosch, The constitution of ethylene-propylene copolymers, *Polymer, 2 (1961) 357-363*
4. J.van Schooten and S.Mostert, The constitution of polypropylenes and ethylene-propylene copolymers, prepared with vanadyl-based catalysts, *Polymer, 4 (1963) 135-138*
5. C.Tosi, A.Valvassori and F.Ciampelli, A study of inversions in ethylene-propylene copolymers, *Europ. Pol. J., 5 (1969) 575-585*
6. C.E.Wilkes, C.J.Carman and R.A.Harrington, Monomer sequence distribution in ethylene-propylene terpolymers measured by 13C nuclear magnetic resonance, *J.Pol.Sc., C43 (1973) 237-250*

7. W.V.Smith, Sequence distribution in ethylene-propylene copolymers. I. Relations between multads and between multads and the 13C NMR spectrum, *J.Pol.Sc., Pol.Phys., 18 (1980) 1573-1585*

8. ditto, II. The 13C NMR spectra, multad populations, and kinetic parameters, *ibid 1587-1597*

9. M.Kakugo et al, 13C-NMR determination of monomer sequence distribution in ethylene-propylene copolymers prepared with δ-TiCl3-AlEt2Cl, *Macromolecules, 15 (1982) 1150-1152*

10. J.F.Ross, Copolymerization kinetic constants and their prediction from dyad/triad distributions, *J.Macromol.Sc., Chem., A21 (1984) 453-472*

11. C.Cozewith, Interpretation of 13C NMR sequence distribution for ethylene-propylene copolymers made with heterogeneous catalysts, *Macromolecules, 20 (1987) 1237-1244*

12. T.Hayashi et al, 13C-NMR spectral assignments and hexad co-monomer sequence determination in stereoregular ethylene-propylene copolymer, *Polymer, 29 (1988) 1848-1857*

13. T.Hayashi et al, 13C-NMR chemical shifts calculation for model compounds of ethylene-propylene copolymer with a low ethylene content, *Polymer J., 20 (1988) 107-118*

14. H.N.Cheng and M.A.Bennett, Spectral simulation and characterization of polymers from ethene and propene by 13C-NMR, *Makrom.Chemie, 188 (1987) 2665-2677*

15. Y.Doi, T.Koyama, K.Soga and T.Asakura, Stereochemistry in "living" coordination polymerization of propene initiated by vanadium-based catalytic systems, *Makrom.Chem., 185 (1984) 1827-1833*

16. L.G.Yechevskaya et al, Study of the molecular structure of ethylene-propylene copolymers obtained with catalysts of different composition, *Makrom.Chemie, 188 (1987) 2573-2583*

17. G.J.Ray, P.E.Johnson and J.R.Knox, Carbon 13 NMR determination of monomer composition and sequence distributions in ethylene-propylene copolymers prepared with a stereoregular catalyst system, *Macromolecules, 10 (1977) 773-778*

18. J.C.Randall, A 13C NMR determination of the co-monomer sequence distributions in propylene-1-butene copolymers, *Macromolecules, 11 (1978) 592-597*

19. L.Abis, G.Bacchilega and F.Milani, 13C NMR characterization of a new ethylene-propylene copolymer obtained with a high yield titanium catalyst, *Makrom.Chemie, 187 (1986) 1877-1886*

20. M.Avella, E.Martuscelli, G.Della Volpe, A.Segre, E.Rossi and T.Simonazzi, Composition-properties relationships in propene-ethene random copolymers obtained with high-yield Ziegler-Natta supported catalysts, *Makrom.Chemie, 187 (1986) 1927-1943*

21. N.Kashiwa, A.Mizuno and S.Minami, Copolymerization of ethylene and propylene by MgCl2-containing highly active Ti catalysts, *Polymer Bulletin, 12 (1984) 105-109*

22. W.Kaminsky and M.Miri, Ethylene propylene diene terpolymers produced with a homogeneous and highly active zirconium catalyst, *J.Pol.Sc., Pol.Chem., 23 (1985) 2151-2164*

23. G.Crespi, A.Valvassori and G.Sartori, Ethylene-propylene copolymers as rubbers, chapter IVC in Copolymerization, G.E.Ham editor, Volume XVIII in the High Polymer Series, Interscience, New York, 1964

24. Y.Doi, R.Onishi and K.Soga, Monomer sequence distribution in ethylene-propylene copolymers prepared with a silica-supported MgCl2/TiCl4 catalyst, *Makrom.Chem., Rapid Commun., 4 (1983) 169-174*

25. M.C.Kirkham, Properties and microstructure of ethylene-propylene terpolymers, *J.Appl.Pol.Sc., 17 (1973) 101-111*

26. K.Ito and J.E.Guillet, Estimation of solubility parameters for some olefin polymers and copolymers by inverse gas chromatography, *Macromolecules, 12 (1979) 1163-1167*

27. G.W.Phillips and W.L.Carrick, Transition metal catalysts. IX. Random ethylene-propylene copolymers with a low pressure polymerization catalyst,

J.Am.Chem.Soc., *84* (1962) 920-925

28. Y.Doi, J.Kinoshita, A.Morinaga and T.Keii, Molecular weight distribution
     and stereoregularity of polypropylenes produced with VCl4-AlEt2Cl
     catalyst, *J.Pol.Sc., Pol.Chem.*, 13, (1975) 2491-2497

29. P.Locatelli et al, Propene/1-butene copolymerization with a heterogeneous
     Ziegler-Natta catalyst: Inhomogeneity of isotactic active sites,
     Makrom.Chemie, Rapid Commun., *9* (1988) 575-580

30. M.Kakugo et al, Characteristics of ethylene-propylene and
     propylene-1-butene copolymerization over TiCl3.0.33AlCl3-DEAC,
     Macromolecules, *21* (1988) 2309-2313

31. P.Galli, F.Milani and T.Simonazzi, New trends in the field of propylene
     based polymers, *Polymer Journal*, *17* (1985) 37-55

32. H.W.Coover, R.L.McConnell, F.B.Joyner, D.F.Slonaker and R.L.Combs,
     Costereosymmetric α-olefin copolymers, *J.Pol.Sc.*, *A1* (1966) 2563-2582

33. A.Turner-Jones, Cocrystallization in copolymers of α-olefins. II. Butene-1
     copolymers and polybutene type II/I crystal phase transition, *Polymer*, *7*
     (1966) 23-59

34. A.Turner-Jones, Copolymers of butene with α-olefins. Cocrystallizing
     behaviour and polybutene type II to type I crystal phase transition,
     J.Pol.Sc., *C16* (1967) 393-404

35. J.C.Verdier and A.Guyot, Microstructure of propene-butene copolymers
     studied by flash-pyrolysis G.L.Chromatography, *Makrom. Chemie*, *175* (1974)
     1543-1559

36. C.G.Overberger and S.Nozakura, Copolymerization of styrene and styrene
     derivative with an aluminium alklyl-titanium trichloride catalyst,
     J.Pol.Sc., *A 1* (1963) 1439-1451

37. E.Chiniellini, A.M.Raspolli-Galletti and R.Solaro, Optically active
     hydrocarbon polymers with aromatic side chains. 12. Synthesis and
     characterization of coisotactic copolymers of (S)-4-methyl-1-hexene with
     styrene, *Macromolecules*, *17* (1984) 2212-2217

38. A.L.Segre, M.Delfini, M.Paci, A.M.Raspolli-Galletti and R.Solaro,
     Optically active hydrocarbon polymers with aromatic side chains. 13.
     Structural analysis of (S)-4-methyl-1-hexene/styrene copolymers by 13C NMR
     spectroscopy, *Macromolecules*, *18* (1985) 44-48

39. K.Soga and H.Yanagihara, Copolymerization of propylene with styrene using
     the catalyst system composed of Solvay type TiCl3 and Cp2TiMe2,
     Macromolecules, *22* (1989) 2875-2878

40. G.Bier, Hochmolekulare Olefin-Mischpolymerisate hergestellt unter
     verwendung von Ziegler-mischkatalysatoren, *Angew.Chemie*, *73* (1961) 186-197

41. B.Ke, Characterization of polyolefins by differential thermal analysis,
     J.Pol.Sc., *52* (1960) 15-23

42. Yu.V.Kissin, Structures of copolymers of high olefins, *Advances in polymer
     science*, *15* (1974) 91-155

43. R.Greco et al, Crystallization, morphology, and thermal behaviour of
     ethylene/propylene copolymers, *Makrom.Chemie*, *188* (1987) 2231-2239

44. J.F.Jackson, Crystallinity in ethylene-propylene copolymers, *J.Pol.Sc.*, *A1*
     (1963) 2119-2126

45. S.Davison and G.L.Taylor, Sequence length and crystallinityy in α-olefin
     terpolymers, *Br.Pol.J.*, *4* (1972) 65-82

46. S-N.Gan, D.R.Burfield and K.Soga, Differential scanning calorimetry
     studies of ethylene-propylene copolymers, *Macromolecules*, *18* (1985)
     2684-2688

47. P.Cavallo, E.Martuscelli and M.Pracella, Properties of solution grown
     crystals of isotactic propylene/butene-1 copolymers, *Polymer*, *18* (1977)
     42-48

48. A.P.Gray and K.Casey, Thermal analysis and the influence of thermal
     history on polymer fusion curves, *Pol.Letters*, *2* (1964) 381-388

49. A.Turner-Jones, Development of the γ-crystal form in random copolymers of
     propylene and their analysis by DSC and X-ray methods, *Polymer*, *12* (1971)
     487-507

50. G.P.Guidetti, P.Busi, I.Giulianelli and R.Zanetti, Structure-properties relationships in some random copolymers of propylene, *Eur.Pol.J.*, *19* (1983) 757-759

51. V.Busico, P.Corradini, C.De Rosa and E.Di Benedetto, Physico-chemical and structural characterization of ethylene-propene copolymers with low ethylene content from isotactic-specific Ziegler-Natta catalysts, *Eur.Pol.J.*, *21* (1985) 239-244

52. H.W.Starkweather, F.A.Van-Catledge and R.N.MacDonald, Crystalline order in copolymers of ethylene and propylene, *Macromolecules*, *15* (1982) 1600-1604

53. G.Ver Strate and Z.W.Wilchinsky, Ethylene-propylene copolymers: Degree of crystallinity and composition, *J.Pol.Sc.*, *A2* (1971) 127-141

54. S.Cimmino, E.Martuscelli, L.Nicolais and C.Silvestre, Thermal and mechanical properties of isotactic random propylene-butene-1 copolymers, *Polymer*, *19* (1978) 1222-1223

55. P.R.Swan, Polyethylene unit cell variations with branching, *J.Pol.Sc.*, *56* (1962) 409-416

56. R.Alamo, R.Domszy and L.Mandelkern, Thermodynamic and structural properties of copolymers of ethylene, *J.Phys.Chem.*, *88* (1984) 6587-6595

57. R.Seguela and F.Rietsch, On the isomorphism of ethylene/α-olefin copolymers, *J.Pol.Sc.*, *Pol.Letters*, *24* (1986) 29-33

58. D.J.Cutler, P.J.Hendra, M.E.A.Cudby and H.A.Willis, Chain branching in high pressure polymerized polyethylene, *Polymer*, *18* (1977) 1005-1008

59. C.A.F.Tuynman, Dynamic mechanical analysis of olefin polymers, *J.Pol.Sc.*, *C16* (1967) 2379-2392

60. L.F.Byrne and D.J.Hourston, Aspects of the thermal and dynamic mechanical behaviour of EPDM rubbers, *J.Appl.Pol.Sc.*, *23* (1979) 1607-1617

61. G.Gianotti, G.Dall'Asta, A.Valvassori and V.Zamboni, Physical properties of poly-1-methyloctamer, a model of the ethylene/propylene copolymer, *Makrom.Chemie*, *149* (1971) 117-125

62. T.Huff, C.J.Buchman and J.V.Cavender, A study of the effect of branching on certain physical and mechanical properties of stereoregular polypropylene, *J.Appl.Pol.Sc.*, *8* (1964) 825-837

63. K.A.Jung and H.Schnecko, Vergleich der Zahl aktiver Zentren und einiger kinetischer Konstanten bei der Polymerisation van Aethylen, Propylen and Buten-1 mit Ziegler-Natta Katalysatoren, *Makrom.Chemie*, *154* (1972) 227-240

64. G.D.Bukatov, S.H.Shepelev, V.A.Zakharov, S.A.Sergeev and Y.I.Yermakov, Propylene polymerization on titanium-magnesium catalysts, Determination of the number of active centers and propagation rate constants, *Makrom.Chemie*, *183* (1982) 2657-2665

65. V.A.Zakharov, G.D.Bukatov and Y.I.Yermakov, Mechanism of Ziegler-Natta polymerization on the basis of data on the number of active centers and their reactivity, in Coordination Polymerization, C.C.Price and E.J.Vandenberg editors, Plenum Press, New York, 1983

66. N.B.Chumaevskii, V.A.Zakharov, G.D.Bukatov, G.I.Kuznetzova and Y.I.Yermakov, Study of the mechanism of propagation and transfer reactions in the polymerization of olefins by Ziegler-Natta catalysts, I. Determination of the number of propagation centers and the rate constant, *Makrom.Chemie*, *177* (1976) 747-761

67. V.A.Zakharov, G.D.Bukatov and Y.I.Yermakov, On the mechanism of olefin polymerization by Ziegler-Natta catalysts, p 61-100 in Advances in Polymer Science, Volume 51, Springer Verlag, Berlin, 1983

Chapter 5

IN-SITU PREPARED TOUGHENED POLYPROPYLENE ("BLOCK" COPOLYMERS),
POLYMERIZATION ASPECTS

5.1 INTRODUCTION

Toughened polypropylenes are an important member of the
polypropylene family. The incentive in preparing them lies in the
rather poor low temperature impact properties of the homopolymer.
An obvious way to alleviate this drawback is to create a dispersed
system containing an elastomer next to the homopolymer, as has been
shown to be effective in a large number of other brittle polymers,
such as polystyrene, polyvinylchloride and epoxy resins[1].

This preparation can be done in two very distinctly different
ways, the first one is by mechanically blending an elastomer into
polypropylene homopolymer. The second way is the in-situ preparation of the toughened product by polymerizing first propylene only,
followed by a copolymerization of ethylene and propylene for
example to form the elastomeric part of the product. One and the
same catalyst is used in both steps. This has led people to believe
that true block copolymers would form in this type of process,
which gave the name "block copolymers" or "copolymers" to this
class of products. As will be shown later this is certainly a
misnomer. In the following we will refer to these products as
toughened polypropylenes to distinguish them from the copolymers
described in the previous chapters. This chapter deals with the
preparation of in-situ toughened polypropylenes; the preparation of
blends will, very concisely, be dealt with in chapter 8. Additionally the attempts at preparing true block copolymers will be
reviewed.

A few words about the nomenclature used in this chapter are in
order. Regarding the toughened polypropylene (TPP) as a mixture of
homopolymer polypropylene and an ethylene-propylene copolymer, the
percentages of the fractions of each are denoted by F_{pp} and F_c. The
ethylene content of the copolymer part, as a percentage, is called
E_c, and the ethylene content of the total product is E_t. Naturally
some of these entities are related, for example $F_c = 100*E_t/E_c$ when
weight precentages are used. Further distinctions within the

fractions can of course be made, for instance between atactic and isotactic polypropylene, and they will be introduced when appropriate in the course of this chapter.

In preparing the TPP's one can use a large number of different processes and procedures. For instance one can have one or more copolymerization stages, or one can vary the feed composition of the co-monomers during the copolymerization. A large number of these routes have been patented and a few are also used commercially. However not much is known about the detailed manufacturing process of the various TPP producers, they are closely kept secrets and the only source of information lies in the patents! In contrast to toughened polystyrene manufacture for instance, where there is an abundant literature, for the polyolefins case there is almost none. This is clear when checking a number of review articles[2,3] or books, such as the excellent one of Bucknall[1]. The blending of polypropylene and ethylene-propylene copolymers however has been described quite extensively, but here the production of the toughened polypropylene is generally done by polypropylene manufacturers which do not prepare their own ethylene/propylene rubbers. Therefore one producer has to sell to the other, and must highlight the useful properties of his elastomeric product.

For the in-situ TPP we will restrict ourselves in the following to the simplest one: a homopolymer polypropylene followed by a steady state copolymerization of ethylene and propylene on the same catalyst. Using a steady state copolymerization implies that the concentration of both monomers (and thus also their ratio) is kept constant, which - assuming no change of the kinetic parameters of the catalyst with time - will lead to a copolymer composition not varying in time. The patent literature will not be reviewed to any extent as that information is regarded neither of good quality nor of sufficient detail to allow discussion on the polymerization aspects of different catalysts or monomers, nor on their characterization.

5.2 POLYMERIZATION ASPECTS OF IN-SITU PREPARED TOUGHENED POLYPROPYLENE

As already said before the literature on this topic is scarce and most of what follows stems from own studies. Obviously the polymerization of the homopolymer polypropylene part of the product follows exactly the behaviour described earlier in chapter 1. By the same token the copolymerization in general also follows the rules previously given in chapter 3, although of course exclusively heterogeneous catalysts can be used in the in-situ preparation of toughened polypropylene in order to arrive at the required high steric purity and thus high crystallinity for the polypropylene fraction.

In this section some kinetic data will be mentioned in addition to that from chapters 1 and 3. The powder morphology for toughened polypropylenes will also be briefly mentioned. The preparation of true block copolymers of propylene and ethylene will be discussed, which turns out to be very problematic. Finally a special case of sequential polyethylene and polypropylene polymerization, which leads to some unexpected kinetic effects, is shortly described.

5.2.1 Kinetics

Not many data on kinetics of copolymerization subsequent to a homopolymerization will be given, but the few that are mentioned do shed new light on the catalyst behaviour (restricting ourselves mainly to first generation $TiCl_3$ catalysts), additional to the data from the homopolymerization.

Copolymerization parameters were determined in a semi-steady-state type of operation at constant total monomer pressure. The gas cap composition could be held constant by proportionate feeding of the two monomers. The individual rate constants in the copolymerization stage, for a fixed homopolymerization time, show an increase from E_c 30 to about 60 %m/m and a levelling off thereafter. The ratio of the rate constants of the two monomers is however constant. Calculation of the copolymerization parameters using the Fineman-Ross method leads to values for r_1 = 16.7 and for r_2 = 0.036, and thus r_1*r_2 = 0.60 (correlation coefficient in the least squares fit is 0.92). These numbers apply to a $TiCl_3$ catalyst made at low temperature and a polymerization temperature of 60 °C, and are similar in value to those in straight copolymerization of ethylene and propylene. That is, prior use or "ageing" of the catalyst in the homopolymerization had no effect upon the reactivity ratios.

In a series in which the copolymerization temperature was varied between 50 and 80 °C the Arrhenius plots for the ethylene and propylene rate constants were reasonably good. The apparent activation energy was calculated to be 35 and 68 kJ/mol for ethylene and propylene respectively. The inequality of the two activation energies means an increasing r_1 value at lower temperatures, and rather strongly increasing as well. This observation fits in nicely with the much lower reactivity ratios found at high temperature (around 250 °C), see chapter 9.

As mentioned before, lowering the catalyst concentration in the homopolymerization leads to increasing rate constants. The same holds true in copolymerization, where the effect is even larger, as shown by the data in Table 5.1 (top half). Moreover ethylene polymerization rate is more sensitive to catalyst concentration than propylene, which means also some change in the copolymerization parameters.

Of even more importance is the effect of the preceding homopolymerization on the kinetics in the subsequent copolymerization and on the molecular weight of the copolymer formed. A number of experiments have been done in which a copolymerization under fixed

TABLE 5.1
Kinetic parameters in copolymerization; effect of catalyst concentration and homopolymerization time. Polymerizations at 60 °C, hydrogen only in homopolymerization, copolymerizations at around E_c 60 to 70 %m/m.

	homopolymerization			copolymerization				
cat conc, mmol/l	time h	yield gPP per gTiCl$_3$	activity yield/h.bar	halflives of C_3, h	C_2, h	ratio of halflives	LVN of copolymer dl/g	decay in % from 10 to 30 min
0.26	3.9	2220	79	0.90	0.040	23		
0.46	4.0	2160	75	0.94	0.044	21		
0.75	3.3	1570	66	1.06	0.049	22		
1.32	1.7	920	75	1.17	0.067	18		
1.66	1.7	720	59	1.49	0.11	14		
2.0*	0	<2		0.69	0.067	10	1.7	40
2.0	0.25	12		1.16	0.092	13	3.9	25
2.0	1	70		1.30	0.109	12	4.1	8
2.0	4	790		1.40	0.116	12	5.4	0
2.0	4	190		1.90	0.103	18	6.7	-

* : different catalyst batch, slightly different copolymerization conditions

conditions was preceded by a homopolymerization in which either the yield of homopolymer or the time of homopolymerization was varied. The data are given in Table 5.1 (bottom half). In nearly all the copolymerizations a fast decay is observed, which diminishes in a regular fashion with increasing homopolymer yield (this yield increase was obtained by increasing the homopolymerization time). The copolymerization rates decrease upon making more homopolymer in the preceding step, but the ratio of the individual rates stays however reasonably constant. Whether or not the homopolymer yield is the real prime variable was studied in an experiment in which, for a fixed homopolymerization time, two widely different yields were generated. No significant difference in molecular weight could be found but there is an effect on the kinetics. No definite conclusion can hence be drawn from the latter experiments.

The LVN of the copolymer part of a toughened polypropylene can be calculated from the LVN's of the homopolymer part and of the total, using the F_c in the calculation. Inspecting the data on the molecular weight of the copolymer in these experiments shows a strongly increasing trend with increasing homopolymerization yield, or rather homopolymerization time. The MW on naked, fresh catalyst is fairly low, for instance at 60 °C and an E_c of 60 %m/m the LVN is around 2 dl/g. This increases to 7 dl/g when a preceding homopolymerization of 4 hours is applied. No clear mathematical relationship could be derived between the MW and either yield or time in the homopolymerization. This behaviour can be described by the assumption of two classes of active sites widely differing in propagation rate constants and in decay by over-reduction. The so-called "hot" sites[4] are those with high propagation and decay rates, and they are relatively short lived, which is in line with the copolymerization rate data. For the observed molecular weight effects one must further assume a difference in the chain transfer rate constants as well between "hot" and "cold" sites. A different explanation could be the consumption of a chain transfer agent present in the system ($AlEtCl_2$?).

The dependence of the molecular weight on the ethylene content of the copolymer(E_c) is very strong especially at the upper E_c-range. This fits in with the very high polyethylene MW's one always encounters when no hydrogen is applied in the polymerization. Fig. 5.1 gives a general impression of the behaviour of the system. It also illustrates a large temperature sensitivity.

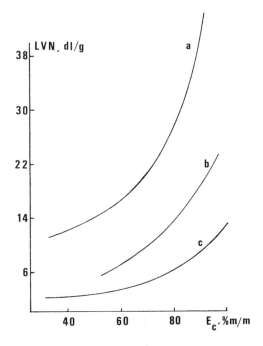

Fig. 5.1: LVN of copolymers of ethylene and propylene as function of their composition and polymerization temperature: 50 °C (a); 60 °C (b,c); (a) and (b) for copolymers with a preceeding homopolymerization of propylene, (c) for copolymerization direct on "naked" catalyst

5.2.2 <u>Morphology</u> <u>of</u> <u>ex</u> <u>reactor</u> <u>products</u>

In preparing a TPP the final stage entails the formation of a copolymer, which, even with the normally applied heterogeneous catalyst, is to a certain extent soluble in the polymerization solvent. Of course, the extent of solubilization will be very dependent on the polymerization conditions (temperature, time) and on the base morphology of the catalyst, as was described before in the discussion concerning homopolymer reactor solubles (see chapter 2.3). In most cases some of the copolymer will dissolve, and on driving off the polymerization solvent by evaporation, this dissolved polymer will precipitate on the surface of the particles, rendering them sticky and leading easily to agglomeration of the polymer particles. In the normally applied separation of the solid polymer particles and the solvent by centrifuging or other means, this of course does not happen. However, for obvious economic reasons, one tends to minimize the amount of dissolved polymer as this constitutes a direct monomer loss.

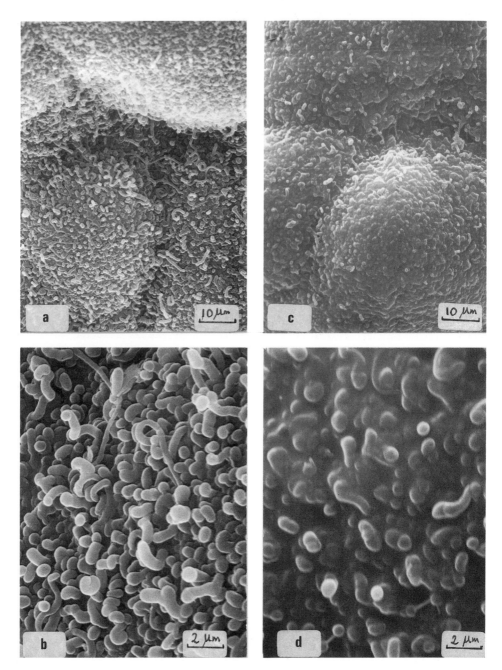

Fig. 5.2: Scanning electron micrographs of polymer powders: homopolymer (a and b); toughened polypropylene with F_c of 12 %m/m (c and d)

The use of polymerization solvents with a very low polymer solubility, such as propylene/ethylene mixtures, would overcome this problem totally. Moreover one could expect that in such a case the morphology of the polymer particles would be identical to the one arising in a homopolymerization. This is however not generally the case. Using TiCl$_3$'s made at low temperature, we did observe a clear copolymer layer on the surface of the polymer particles. Fig. 5.2 gives a few examples. This apparent transport of the copolymer to the surface decreases upon polymerizing at lower temperatures, lowering the fraction copolymer, generating higher E$_c$'s i.e. more crystalline copolymers, or preparing higher copolymer molecular weights. All these effects point to the viscosity of the copolymer phase as a determining factor. It is thought that the copolymer is squeezed out of the rather rigid homopolymer matrix once the free space in the particle is filled by the copolymer. Therefore the relative amount of copolymer phase generated in the preparation is of course also very important. In our experience complete prevention of migration of the copolymer to the surface is difficult to achieve, however the situation improves considerably when taking a catalyst with a very high density generating a homopolymer of high density, such as the "Solvay" catalyst. Probably the pores in that case are so small that transport of copolymer is very hampered but that of monomer is not. Conversely, an open catalyst structure might also lead to TPP powders not having a copolymer layer on their surface, at least up to a level of copolymer equal to or less than that required for filling the void volume.

5.2.3 Block-copolymerization

Block copolymers are polymers consisting of molecules containing two or more different sequences connected via a chemical bond. The sequences can be either homopolymer or copolymer. Within the α-olefins group examples would be: ethylene-propylene block copolymer E(E)$_n$E-P(P)$_m$P or propylene-ethylene/propylene block copolymer containing a copolymer sequence as part of its structure. Generally speaking, preparation of such polymer molecules can be done either in a sequential polymerization or by an after-reaction such as coupling of two different polymer molecules. In the former method the catalyst must be of the "living" type, which means that the active site must not deactivate during the polymerization, and for high efficiencies of block copolymer formation no chain transfer should occur and addition of the first unit (both in the first and second step) must be fast compared to propagation. Only then would each individual site forms just one block copolymer molecule. A true block copolymer can be obtained in this way, as is for instance the case in alkyl-lithium initiated polymerizations of dienes or styrene (see for instance ref. 5).

In Ziegler-Natta catalysis, block copolymers have been mentioned very many times[3,6], in nearly all cases very loosely however as for instance mentioned by Boor in his book[7]. We have already seen in the previous chapters that not only deactivation of sites does occur, but especially chain transfer reactions are very common. These latter are the main cause of the rather unsuccesful block copolymerizations, as chain transfer can happen with the standard ingredients of the polymerizing system, i.e. the co-catalyst and the monomer(s). Were it only for the active-site die-out lowering of the temperature might still allow a reasonable performance we expect.

The main parameter governing a successful block copolymerization is the lifetime of an active site. This can be measured in a number of different ways. An example is the measurement of the molecular weight of the polymer formed and its distribution as a function with time, which leads to the number of polymer molecules present (e.g.ref 8), and thus to the number of active sites (an example is given below for the $V(acac)_3$ system). Attempts have also been made to label the polymer molecules containing a titanium or aluminium bond by reacting them with tritiated alcohols[9,10]. As in most cases aluminium alkyl is a chain transfer agent one has to extrapolate to time=zero to obtain the number of active sites at the start of the polymerization. This technique does not allow the measurement during a polymerization. Finally use can be made of a very selective and potent poison of the active site to "titrate" them, a technique developed to a great extent by Zakharov, Bukatov and Yermakov[11] and Tait. For instance CO is a useful agent. For extensive reviews on those measuring techniques see Tait[12,13].

Using those techniques the answers appear to vary a great deal. The range given for heterogeneous catalysts is 1 minute to over 10 hours, see for instance the reviews of Pasquon[14], Cooper[15] and Berger[16]. Although the differences between various measurements are large, nearly all investigators will agree that the

$TiCl_3/AlEt_2Cl$ catalyst has at least a lifetime of the order of minutes at 70 °C. The lifetime does strongly increase at lower temperature, whilst the use of ethylene lowers it. The latter result is probably due to the increased transfer with monomer. If we take the straight line portion of the plot of molecular weight against time as an indication of the lifetime of a site, then in our own experience a low temperature reduced $TiCl_3$ shows a 3 hours "lifetime" at 40 °C with propylene, but only about 25 minutes with ethylene.

With supported catalyst[17] very low lifetimes were obtained of less than 5 seconds as the molecular weight generated on this catalyst appeared to be constant when varying the polymerization time from 5 seconds to 3 hours.

In homogeneous systems the observed lifetimes are normally very low. An example is found in reference 18 with $VOCl_3/Al_2Cl_3Et_3$ where the molecular weight is already at a constant value in the very short polymerization time of 20 seconds. The same behaviour was observed by Natta[19]. An exeption will be mentioned below.

It will be clear from the foregoing that the polymerization system itself is not behaving in such a way that one can be sure of true block copolymer formation under certain conditions. But proving that a block copolymer has been obtained by analytical methods is also very difficult. As one normally aims for fairly high molecular weights, the linking unit of the blocks is very diluted, and not discernible by NMR or any other technique sensitive to composition. Therefore one looks for evidence in the increase of molecular weight with each of the process steps, or for more circumstantial type of evidence in the properties of the product, for instance in its mechanical properties, its dispersion characteristics or in its solubility behaviour. As an example of the latter case for a test on the success of an attempted preparation of a block copolymer of propylene and ethylene/propylene, the product is treated with boiling heptane. When, after such a treatment, ethylene/propylene copolymer is still observed in the residue, this is regarded as evidence for block copolymer formation, as non-block-copolymerized ethylene/propylene copolymer would be effectively removed in this test.

We now consider the more recent reports in the literature dealing with block copolymers. For a good review of earlier literature see Heggs[3]. For <u>heterogeneous</u> catalyst systems block copolymerization was already mentioned very early by Natta[20] and Bier[6]. The latter used a $TiCl_3$ made by reducing $TiCl_4$ at low temperature with sesqui, and polymerized at low temperature. A steadily increasing molecular weight with time or yield was observed, as reported before (see chapter 2). The main drawback in all these systems is the fact that only a minute fraction of the $TiCl_3$ catalyst is really active in the polymerization, which means that at reasonable yields the molecular weights are very high, for instance at yields of around 50 g polymer/g catalyst the molecular weight is in the order of a million. This makes the processing of these polymers very difficult and also the comparison of their properties with the normal polymer blends doubtful. A technique used to overcome this problem involves a cyclic process which uses short polymerization times alternated with a hydrogen treatment to generate a block copolymer molecule and a hydrided active site. Bier showed his polymers to possess better impact properties, higher elongations at break and much lower solubility in the normal solvents as compared to "random" copolymers. At this moment in time it is thought more likely that toughened polypropylenes were generated in that study, due to the difficulty of removing non-reacted monomers completely, and to the still-present chain transfer with monomers. Some block copolymer could however been present.

Changing from a solvent to a gas phase polymerization is attractive, as the monomer removal between stages is made easier, and probably chain transfer with aluminium alkyl is hampered by its relative immobility in such a system. A number of groups have used this set-up[21,22] with Stauffer AA as catalyst component, and at relatively low temperatures of 40-55 °C. No molecular weights were reported at all. The proof for the composition of the polymer is very marginal, in a cyclic preparation with 200 steps alternating ethylene and propylene, ethylene-propylene links are observed in ^{13}C-NMR. The melting point of the polyethylene part (126 to 128 °C) is lower than expected, pointing to some copolymerization with propylene. Upon attempting the preparation of propylene-ethylene/propylene block copolymers by this route, the mechanical stress--strain test showed reinforcing behaviour with high moduli and tensile strength. This could however also be caused by the probably very high molecular weight of these products. Busico[23,24] et al have shown that the conditions applied give rise to a chain life-time of about 6 seconds when using a monomer pressure of 1 bar. As the transfer with monomer will be the main process determining the lifetime, one expects an increase when lowering the monomer pres-sure. Indeed a value of 60 seconds is found at 0.1 bar (derived from the relation of molecular weight with time, taking the straight line portion as an indication for the lifetime). At the latter monomer pressure, a cyclic polymerization with 10 seconds ethylene, 10 seconds nitrogen, 20 seconds ethylene plus propylene and again 10 seconds nitrogen, repeated some 200 times, leads to a polymer which after a boiling heptane extraction still contains ethylene/propylene copolymer. Even the copolymer fraction is nearly equal to the fraction before the extraction. Performing a similar experiment at 1 bar monomer pressure gives a product from which all the copolymer is extracted with boiling heptane and the residue is nearly pure polyethylene. This is regarded as fairly strong circum-stantial evidence of block copolymer formation in the low monomer pressure case. One is then of course very curious about the mecha-nical and dispersion properties of such polymers. This will be fairly difficult to study on a reasonably practical scale as the succesful polymer mentioned above was made in a yield of 0.027 g polymer/g $TiCl_3$ per hour!

Very recently a similar experimental set-up was used for the determination of propagation rate constants for propylene and ethylene[25]. With a yield of 0.05 g/gTiCl$_3$ the product was split into three fractions by extraction with boiling hexane and heptane. ^{13}C-NMR on these fractions showed about 4, 1 and 0.5 %mol of EPE+PEP groups in the respective fractions, i.e. groups not belon-ging to the expected block structure. They may originate from incomplete removal of the preceeding monomer. Thus, although the polymerization conditions might have favoured block copolymer formation, definite proof from only the compositional measurement is not convincing.

A completely different approach has been used by researchers of Hercules[26,27] by changing the co-catalyst from an aluminium-con-taining one to $Cp_2Ti(CH_3)_2$ (or similar ones with substitution on the cyclopentadienyl moiety). Using δ-TiCl$_3$ and this co-catalyst at 45 °C for example leads to polymerizations in which the yield and the molecular weight increase linearly with time. The molecular weight distribution is rather broad (Q≈4.5) but stays constant.

This means that chain transfer both with co-catalyst as well as with monomer is very low under these conditions. Or conversely it indicates that the aluminium alkyl co-catalyst normally applied is responsible for most of the observed chain transfer in the absence of hydrogen. The molecular weight is indeed independent of co-catalyst concentration over a wide range spanning one order of magnitude[28]. This catalyst system has been used to make ABA block copolymers with small polypropylene A-blocks and a ethylene/propylene copolymer as centre block. In these polymerizations also the molecular weight (or LVN) increases with every step. Fractionation results also point to a large fraction of true blockcopolymer being present in the mixture. The yields of blockcopolymer on the catalyst are, as expected, very low, even at the rather high molecular weights required to obtain interesting mechanical properties (50,000-100,000 for the A-block and 300,000-400,000 for the B-block).

Surprisingly, in a recent patent[29], a method is described for preparation of block copolymers of styrene and for instance propylene by using $MgCl_2$-supported catalysts! The process, following the description of example 1 of the patent, is sequential in that first styrene is polymerized mixed with toluene for 1 minute at room temperature, propylene being added, the mixture left for 1 hour before raising the temperature to 50 °C and allowing 3 hour polymerization. The desired product has to be isolated by fractionation, i.e. homopolymers are also present. A more extensive purification is described later[30] in an article by Del Giudice, Cohen et al which also tests the compatibilizing effect of this block copolymer. The latter is positive, proving almost beyond doubt its block copolymer character (see chapter 6.4).

A different approach was later used by Drzewinski and Cohen[31] in preparing a living polybutadienyllithium for the formation of one part of the block structure, the second part being polypropylene which is formed after reacting the polybutadienyl-lithium with $TiCl_4$ and propylene. A more extensive description of this method of polymerization is given in the polyethylene chapter (see chapter 9). The resulting polymerization product was extracted successively with 2-pentanone, hexane and xylene. It is assumed that the expected blockcopolymer is in the xylene soluble fraction, the characterization and especially the dispersion data are positive in this respect.

Also in <u>homogeneous</u> systems there is an early Natta reference[32] to block-copolymer formation using VCl_4/anisole/$AliBu_2Cl$,
although the extraction data given show it to be at least a mixture
of polymers. Later extensive studies have been made by Doi and
co-workers and they came up with a real living polymerization
system[33-37], for a review by Doi and Keii see reference 38. Doi and
his team use as catalyst $V(acac)_3$/$AlEt_2Cl$ at -78 °C, which with
propylene as monomer gives a polymer with a very narrow molecular
weight distribution, and a number-average molecular weight which is
linear in polymerization time. About 4 % of the vanadium in the
system is active in the polymerization. Using other co-catalysts,
such as $Al_2Et_3Cl_3$ or $AlEtCl_2$, gives no true living system as the
distributions are broader, with Q values tending to 2, and the
molecular weight is not linear with time. The same holds for the
use of $Al(CH_3)_2Cl$ which shows however a considerable increase in
activity[39]. The catalyst system $V(acac)_3$/AlR_2Cl has subsequently
been studied extensively, especially its kinetics[34,40] and tacticity; the latter aspect has been mentioned in chapter 2. For the
present discussion it is important to note that transfer to alkyl
does not occur at -78 °C, but at temperatures above -65 °C it
starts to happen. The molecular weight of the polymer is not linear
in monomer concentration, due to the two step reaction: a complexation of monomer with the active site is followed by the insertion
in the vanadium-carbon bond itself. The productivity of the catalyst system can be increased substantially by adding anisole[35].
This additive appears only to increase the fraction of active
vanadium species as the rate constants all stay the same, nor is
there any effect on the stereoregularity. One must conclude that
the anisole is not present in the active complex. The above data
give rise to high hopes for enabling one to make true block copolymers. However it so happens that the same catalyst system with
ethylene as monomer gives broader MWD's with Q values of 2. In line
with this the investigators found no evidence for block copolymer
formation in sequential polymerizations of ethylene and propylene.
A different method probably led to slightly better results: a
steady propylene monomer concentration was maintained and ethylene
was added to the system at a certain moment. As ethylene is rapidly
consumed one could obtain a propylene-ethylene/propylene-taper-pro-
pylene three block copolymer. The GPC-s given show a progressive
increase in molecular weight.

By reacting the living system with functional reactants one

can prepare polypropylenes carrying one specific end group; examples are iodine, amine, hydroxy or vinyl, see for instance reference 41. Some reactions are given below:

$$V-P + I_2 \longrightarrow V-I + P-I$$

$$V-P + C=C-C=C \longrightarrow (after\ hydrolysis)\ C=C-C-C-P(75\%) + C-C=C-C-P(25\%)$$

$$V-P + CO \longrightarrow V-C(=O)-P$$

$$V-C(=O)-P + HCl \longrightarrow V-Cl + H-C(=O)-P$$

$$H-C(=O)-P + LiAlH_4 \longrightarrow HO-CH_2-P$$

The results of Doi's group have been largely substantiated by Evens and Pijpers[42] who studied similar systems. At the low temperature of -78 °C the living character was proven, only some differences in the kinetics were observed resulting in a figure of about 12 % active vanadium centres as opposed to about 4 to 5 % in Doi's investigations. Also in their hands ethylene gives polymers with Q-values of around 2 which they ascribe to increased transfer with the co-catalyst, as no terminal unsaturation could be detected. The best chance of making a block copolymer is to follow a procedure identical to the one mentioned above: carry out a propylene polymerization and spike some ethylene into the system. From a product point of view, the presence of syndiotactic instead of isotactic crystallinity in the propylene blocks makes the above product a less attractive study material.

A further improvement of this catalyst system was recently announced by Doi et al[43-45]. Using 2-methyl-1,3-butanedionato complexes instead of the previous acetylacetonato ones, gives a catalytic system in which all of the vanadium atoms are active in the propagation. This discovery brings Ziegler-Natta catalysts to the same efficiency as the lithium alkyls which are so frequently used in block copolymerization. A trial experiment in block copolymer formation to propylene-ethylene/propylene was reported to be succesful. Later it was shown[46,47] that - by using MW and MWD as a criterion - a living ethylene/propylene and propylene/1,5-hexadiene copolymers can be generated using DEAC as co-catalyst.

On soluble titanium catalysts of the type $Cp_2TiCl_2/AlEt_3$ in dichloroethane it has also been claimed[48] that by sequential polymerization of ethylene followed by propylene, block copolymers could be prepared. Fractionation was required to isolate the block-copolymer. Since 1976 no further information has been reported however, so the product might have turned out to be less interesting than initially thought.

5.2.4 Sequential polymerization of ethylene and propylene

The polymerization of propylene on catalysts which initially polymerized some ethylene was also studied. The surprising feature is the rate of propylene polymerization: it starts quite low, but finally reaches the rate of a blank - propylene only - experiment. This effect is much more pronounced with larger catalysts. An induction time can be defined as the time required to reach the undisturbed rate. This induction period appears to be proportional to the yield of polyethylene on the catalyst, with different relations for catalysts of different size. The morphologies of the polymer powders show, at relatively high polyethylene fractions, the well known cobweb structure[49] (see also chapter 9 on polyethylene). An example is shown in Fig. 5.3. The explanation of the above observations is linked to the ethylene polymerization. This appears to happen on the surface of the catalyst particle only, in contrast to the accepted "expanding universe" manner of propylene polymerization. In other words the catalyst is so active for ethylene that the concentration of ethylene in the inner part of the catalyst is very low. In line with this we have found the ethylene polymerization rate to be approximately inversely proportional to the gross catalyst particle size. This also means that at identical yields a thicker layer of polyethylene covers the surface on a bigger catalyst compared to a smaller one. When propylene is now polymerized on such a composite catalyst, the monomer transport to the active sites is hampered by the polyethylene, which explains the observed slow starting rate. On reaching the catalyst, the reactivity of propylene is low enough to permit diffusion throughout the total catalyst particle. This eventually leads to the break-up of the catalyst as is normally the case. The observed morphology of the polymer supports this hypothesis, as in a polymerization in which the ethylene was polymerized at every primary $TiCl_3$ particle (as in the "expanding universe") the subsequent

Fig. 5.3: Scanning electron micrographs of polymer powders made in a sequential ethylene-propylene polymerization: homopolymer PP at yield 1545 g/g (a); sequential polymer, about 50 %m/m PE at yield 400 g/g (b); ditto, about 10 %m/m PE at yield 1120 g/g (c and d)

polymerization of propylene would not lead to fibrous structures, as the distances that the primary particles would grow apart are then much smaller.

5.3 TOUGHENED POLYPROPYLENE PROCESSES

As said above the data published on toughened polypropylene are relatively small in number and this holds also for its process technology. Patents, however, do sometimes describe a process scheme in some detail. This section will describe "copolymer" processes in very general terms - more information, particularly on batch versus continuous operation, is given in chapter 7.

Toughened polypropylene can be made in a batch mode and this requires in principle a very simple reactor system, one reactor being sufficient. One works mostly in an inert diluent. After a set homopolymerization time (yield), ethylene and propylene are fed to the reactor in a predefined ratio and a certain yield of ethylene-propylene copolymer is generated in line with the desired composition. Monomers have to be removed from the system before work-up, and so does the catalyst. The diluent is removed (by flashing or solid/liquid separation) and the product pelletized. This operation is very flexible with respect to composition and also allows one to change the copolymer composition during its formation by playing around with the ethylene/propylene ratio; for example tapered copolymers can be made in this way.

Economically speaking however, one favours continuous processes. The technology used for continuous toughened polypropylene manufacture follows closely the ones described for homopolymer in chapter 1 (see section 1.8). In a continuous process more reactors are needed, minimally two. The first, homopolymerization reactor is then followed by a second reactor for the copolymerization, and sometimes an intermediate propylene removal stage (totally or partly) is required. The live slurry from the first reactor is fed to the second, and ethylene and propylene are metered in such as to generate a desired copolymer composition. The residence time ratio is the variable governing the F_c of the product (at given (co)polymerization rates). Obviously the catalyst should still be active in the second stage. Toughened polypropylene processes exist for both slurry (inert hydrocarbon and propylene) and gas-phase technologies. As an example the flow scheme of a Unipol gas-phase process is given in Fig. 5.4. The solubility of the copolymer in the

286

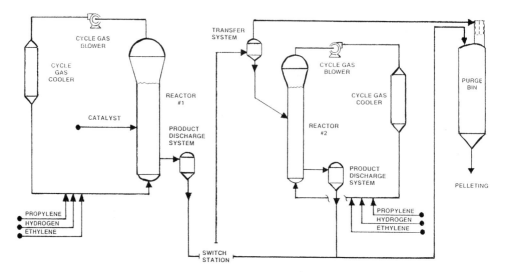

Fig. 5.4: The Unipol toughened polypropylene process

usually applied inert diluents in slurry processes is a drawback as it gives rise to increased viscosity of the slurry and, after diluent removal by evaporation, to sticky polymer particles. One could of course apply a solid/liquid separation, but this lowers the yields on monomer very considerably. Obviously, these aspects are not present in either the liquid propylene slurry nor in the gas-phase processes.

REFERENCES
1. C.B.Bucknall, Toughened Plastics, Applied Science Publishers, London, 1977
2. A.Noshay and J.E.McGrath, Block copolymers, overview and critical survey, Academic Press, New York, 1977
3. T.H.Heggs, Block copolymers made with Ziegler catalysts, chapter 4 in Block copolymers, D.C.Allport and W.H.James editors, Applied Science Publishers, London, 1973
4. T.Keii, Kinetics of Ziegler-Natta polymerizations, London, 1972
5. M.Szwarc. Carbanions, living polymers and electron-transfer processes, Wiley, New York, 1968
6. G.Bier, Hochmolekulare Olefin-Mischpolymerisate, *Angew.Chemie, 73 (1961) 186-197*
7. J.Boor, Ziegler-Natta catalysts and polymerizations, Academic Press, 1979
8. S.Tanaka and H.Morikawa, Average lifetime of growing chains in propylene polymerization, *J.Pol.Sc., A, 3 (1965) 3147-3156*
9. G.Bier, W.Hoffmann, G.Lehmann and G.Seydel, Zahl der aktiven Zentren bei der Polymerisation von Propylen an Ziegler-Misch Katalysatoren, *Makrom.Chemie, 58 (1962) 1-17*
10. C.F.Feldman and E.Perry, Active centres in the polymerization of ethylene using TiCl4-alkylaluminium catalysts, *J.Pol.Sc., 46 (1960) 217-231*

11. N.B.Chumaevskii, V.A.Zakharov, G.D.Bukatov, G.I.Kuznetzova and
    Y.I.Yermakov, Study of the mechanism of propagation and transfer reactions
    in the polymerization of olefins by Ziegler-Natta catalysts.1.
    Determination of the number of propagation centers and the rate constant,
    Makrom. Chemie, *177* (1976) 747-761

12. P.J.T.Tait, Ziegler-Natta and related catalysts, chapter 3 in Developments
    in polymerization-2, Free radical, condensation, transition metal and
    template polymerizations, R.N.Haward editor, Applied Science Publishers,
    London, 1979

13. P.J.T.Tait, Studies on active centre determination in Ziegler-Natta
    polymerization, p115-147 in Transition metal catalyzed polymerizations,
    Part A, R.P.Quirk editor, Harwood Academic Publishers, London, 1983

14. I.Pasquon, A.Valvassori and G.Sartori, The copolymerization of olefins by
    Ziegler-Natta catalysts, chapter 4 in The stereochemistry of
    macromolecules, A.D.Ketley editor, Marcel Dekker, New York, 1967

15. W.Cooper, Kinetics of polymerization initiated by Ziegler-Natta and related
    catalysts, chapter 3 in Chemical Kinetics, C.H.Bamford and C.F.H.Tipper
    editors, volume 15, Non radical polymerization, Elsevier, Amsterdam, 1976

16. M.N.Berger, G.Boocock and R.N.Haward, The polymerization of olefins by
    Ziegler catalyst, *Advances in Catalysis*, *19* (1969) 211-241

17. E.Suzuki, M.Tamura, Y.Doi and T.Keii, Molecular weight during
    polymerization of propylene with the supported catalyst system
    $TiCl4/MgCl2/C6H5COOC2H5/AlEt3$, *Makrom. Chemie*, *180* (1979) 2235-2239

18. E.Junghanns, A.Gumboldt and G.Bier, Polymerisation von Ethylen und Propylen
    zu amorphen Copolymerisaten mit Katalysatoren aus Vanadiumoxychlorid und
    Aluminiumhalogenalkylen, *Makrom. Chemie*, *58* (1962) 18-42

19. G.Natta et al, Kinetics of ethylene-propylene copolymerization, *J.Pol.Sc.*,
    51 (1961) 429-454

20. G.Natta and I.Pasquon, The kinetics of the stereospecific polymerization of
    α-olefins, Advances in Catalysis, volume XI, 1959, 1-66

21. P.Prabhu, A.Schindler, M.H.Theil and R.D.Gilbert, Evidence for
    ethylene-propylene block copolymer formation, *J.Pol.Sc.,Pol.Letters*, *18*
    (1980) 389-394

22. P.Prabhu, A.Schindler, M.H.Theil and R.D.Gilbert, Synthesis and properties
    of ABA propylene-ethylene block copolymers, *J.Pol.Sc.,Pol.Chem.*, *19* (1981)
    523-537

23. V.Busico, P.Corradini, P.Fontana and V.Savino, On the formation of block
    copolymers in the presence of isotactic-specific Ziegler-Natta catalysts,
    Makrom.Chemie,Rapid Commun., *5* (1984) 737-743

24. ibid, part 2, *Makrom.Chemie,Rapid Commun.*, *6* (1985) 743-747

25. P.Ammendola, A.Zambelli, L.Oliva and T.Tancredi, Polymerization of
    α-olefins in the presence of δ-TiCl3/Al(CH3)3: chain propagation rate
    constants for ethylene and propylene, *Makrom. Chemie*, *187* (1986) 1175-1188

26. G.A.Lock, Thermoplastic elastomers based on block copolymers of ethylene
    and propylene, p59-74 in Advances in polyolefins, R.B.Seymour and T.Cheng
    editors, Plenum Press, New York, 1987

27. Eur.Patent, 0 041 361, published 31 july 1985, to Hercules Incorporated

28. K.Soga and H.Yanagihara, Isotactic polymerization of propene using living
    catalysts originally found by Hercules Inc.. Activation, Stereospecific
    control and model of isospecific catalyst centers, p21-32 in Transition
    metals and organometallics as catalysts for olefin polymerization,
    W.Kaminsky and H.Sinn editors, Springer, Heidelberg, 1988

29. European Patent Application 0 099 271, to Montedison, filed 14.07.86

30. L.Del Giudice, R.E.Cohen, G.Attala and F.Bertinotti, Compatibilizing effect
    of a diblock copolymer of isotactic polystyrene and isotactic polypropylene
    in blends of the corresponding homopolymers, *J.Appl.Pol.Sc.*, *30* (1985)
    4305-4318

31. M.A.Drzewinski and R.E.Cohen, Block copolymers of isotactic polypropylene
    and 1,4-polybutadiene, *J.Pol.Sc.*, *Pol.Chem.*, *24* (1986) 2457-2466

32. A.Zambelli, G.Natta and I.Pasquon, Polymerisation du propylene a polymère
    syndiotactique, *J.Pol.Sc.*, *C*, *4* (1963) 411-426

33. Y.Doi, S.Ueki and T.Keii, Preparation of "living" polypropylenes by a soluble vanadium-based Ziegler catalyst, *Makrom.Chemie, 180 (1979) 1359-1361*

34. Y.Doi, S.Ueki and T.Keii, "Living" coordination polymerization of propene initiated by the soluble V(acac)3-AlEt2Cl system, *Macromolecules, 12 (1979) 814-819*

35. S.Ueki and Y.Doi, Activation effect of anisole on the "living" polymerization of propene with the soluble V(acac)3/AlEt2Cl system, *Makrom.Chemie,Rapid Commun., 2 (1981) 403-406*

36. Y.Doi, S.Ueki and T.Keii, Block copolymerization of propylene and ethylene with the "living" coordination catalyst V(acac)3/AlEt2Cl/anisole, *Makrom.Chemie,Rapid Commun., 3 (1982) 225-229*

37. Y.Doi, S.Ueki and T.Keii, "Living" coordination polymerization of propylene and its application to block copolymer synthesis, in Coordination Polymerization, C.C.Price and E.J.Vandenberg editors, Volume 19 in Polymer science and technology, Plenum Press, New York, 1983

38. Y.Doi and T.Keii, Synthesis of "living" polyolefins with soluble Ziegler-Natta catalysts and application to block copolymerization, *Advances in Polymer Science, 73/74 (1986) 201-248*

39. A.Giarusso, G.Amari, S.Verona and L.Porri, The soluble tris(acetylacetonato)vanadium/dialkylaluminium chloride system for alkene polymerization. Enhancement of activity by the use of dimethylaluminium chloride, *Makrom.Chem., Rapid Commun., 8 (1987) 315-319*

40. Y.Doi etal, A kinetic study of "living" coordination polymerization of propene with the soluble V(acac)3-AliBu2Cl system, *Polymer, 23 (1982) 258-262*

41. Y.Doi, G.Hizal and K.Soga, Synthesis and characterization of terminally functionalized polypropylenes, *Makrom.Chem., 188 (1987) 1273-1279*

42. G.G.Evens and E.M.J.Pijpers, "Living" coordination polymerization, in Transition metal catalyzed polymerizations, alkenes and dienes, Part A, R.P.Quirk editor, Harwood Academic Publishers, New York, 1983

43. Y.Doi, S.Suzuki and K.Soga, A perfect initiator for "living" coordination polymerization of propene: tris(2-methyl-1,3-butanedionato)vanadium/AlEt2Cl system, *Makrom.Chemie, Rapid Commun., 6 (1985) 639-642*

44. Y.Doi, S.Suzuki, F.Nozawa and K.Soga, Structure and reactivity of "living" polypropylene, in Catalytic polymerization of olefins, T.Keii and K.Soga editors, Kodansha/Elsevier, Tokyo, 1986

45. Y.Doi, S.Suzuki and K.Soga, "Living" coordination polymerization of propene with a highly active vanadium-based catalyst, *Macromolecules, 19 (1986) 2896-2900*

46. Y.Doi, N.Tokuhiro, S.Suzuki and K.Soga, Synthesis and characterization of living copolymer of ethylene and propene, *Makrom.Chem., Rapid Commun., 8 (1987) 285-290*

47. Y.Doi et al, Living polymerization of olefins with highly active vanadium catalysts, p379-388 in Transition metals and organometallics as catalysts for olefin polymerization, W.Kaminsky and H.Sinn editors, Springer, Heidelberg, 1988

48. J.N.Hay and R.M.S.Obaid, The polymerization of ethylene with a soluble titanium-aluminium complex, *European Pol.J., 14 (1978) 965-969*

49. R.J.L.Graff, G.Kortleve and C.G.Vonk, On the size of the primary particles in Ziegler catalysts, *Polymer Letters, 8 (1970) 735-739*

Chapter 6

CHARACTERIZATION OF IN-SITU MADE TOUGHENED POLYPROPYLENES

6.1   INTRODUCTION

The characterization of in-situ made toughened polypropylenes
(TPP's) is the subject matter of this chapter. The aim for the
characterization is fairly easy to define as one would like to know
the relation between the process parameters applied and the mechanical properties, and thus be able to control the latter. This, in
contrast, turns out to be very difficult. Most of the studies done
in the area of TPP's deal with ethylene/propylene systems. Our
discussion also centres around those types of polymers, although
many of the techniques described are equally suitable for other
combinations. We assume that the polymers under consideration are
at least build up from the following constituents:

- polypropylene (PP) in both isotactic and atactic forms,
- a copolymer of ethylene and propylene (EPC) - which mostly
  means that a broad co-monomer distribution is present,
- frequently polyethylene (PE) is present, whether
  deliberately added or "just" formed in the polymerization
  process, and finally
- a possible fraction of real block copolymer.

For this complicated system, in an ideal characterization case,
from the compositional point of view one would like to know the
amounts of the individual components, and their relevant parameters
such as the distributions of their molecular weight, tacticity and
co-monomer. Moreover, as these systems are all disperse in nature,
one needs to know the distribution of the various constituents over
the phases, the dispersed phase size, shape, interaction with the
matrix, the effect of the one phase on the crystallization (rate)
of the other, etc.. And last but not least the relation of all the
above with the mechanical, or rather end-use properties, has to be
pursued. In this chapter we will describe some of the mechanical
properties, but as it turns out the linking of all available data
is not truly possible yet.

In the case of TPP's the same restriction as mentioned in
chapter 5 (dealing with the polymerization) apply, i.e. only those

polymers made in two stages and with a steady state in co-monomer feed ratios will be discussed when referring to own work; this restriction does not apply to commercial samples simply because their preparative conditions are not known.

The results obtained from the individual characterization methods will be treated first separately. This will be followed by a discussion drawing a number of conclusions both on the suitability of the methods as such and on the structure of toughened polypropylenes. In a final discussion an attempt is made to correlate structure to performance, including the relationship of morphology with composition. Aspects such as particle size effects on impact behaviour and the extent of compatability of the ingredients on this will be highlighted.

6.2 CHARACTERIZATION OF IN-SITU PREPARED TOUGHENED POLYPROPYLENES

In this section we discuss generally the same aspects as in previous characterization chapters, i.e. with the main emphasis on composition and crystallinity. Due to the special character of TPP's a few extra items are very relevant to the characterization, especially their morphology and the mechanical properties.

6.2.1 Composition

In the open literature almost no references to compositional analysis are found, exceptions being those dealing with the method per se such as NMR or IR; the application of these methods is mostly not given.

An exception to this is a recent publication[1] describing the fractionation of a TPP (containing 15.5 %m/m ethylene) and characterization of the fractions obtained. The fractionation was based on the separation of a polymer solution in two phases near the lower critical solution temperature (LCST), which was followed by turbidity measurements. Using this technique three peaks are observed in a plot of the turbidity with temperature; the location of the peaks corresponds with those of polyethylene, ethylene/propylene copolymer and polypropylene. These fractions were separated and their composition measured. The "polyethylene" fraction contains indeed 92 to 98 m/m ethylene, and constituted 12 %m/m of the total sample; the second copolymer fraction is 23 %m/m of the total but contained only 7 %m/m ethylene; the main fraction (65 %m/m) still contained 4 %m/m ethylene. The melting point of the first fraction is 120 °C and of the last one 160 °C, the same melting peaks were visible in the original sample. The effectiveness of the separation is not extreme, although the ability to separate out the PE fraction is very positive. The question as to whether true blockcopolymer is present cannot be answered by this technique.

As mentioned in the chapter dealing with the different characterization methods (see chapter 13), a combination of a simple separation into a crystallizable and a non-crystallizable fraction, coupled to a compositional method such as NMR or IR applied on the fractions, is in our opinion an attractive method to obtain insight into the composition of TPP's. The best way of illustrating this is the application of this combined technique to a number of commercial products made with ethylene and propylene. In Table 6.1 a few examples are given, detailing the overall ethylene content, the amount of the non-crystallizable fraction, its composition as derived from NMR measurements, and the same data on the crystallizable fraction. From these primary measured data the composition of the fractions is derived in terms of atactic polypropylene

TABLE 6.1
Primary characterization data of some selected commercial TPP's.

Manufac- turer	Melt Index dg/min	E_t ex IR %m/m	HOXS extract %m/m	^{13}C NMR data							
				HOXS extract				HOXS residue			
				E_t,%m/m	N_E	E_1 %	E_2 %	E_t,%m/m	N_E	E_1 %	E_2 %
Shell	1	8.1	12.2	37.4	2.8	16	11	2.5	13	<1	<1
ICI	1	18.6	11.8	28.6	2.9	16	12	15.2	20	<1	<1
DSM	1	12.7	11.2	43.8	2.5	21	9	8.5	8.7	5	5
Montedison	10	9.5	10.2	44	2.9	13	9	-	-	-	-
Mitsui	1	20.3	7.0	15.7	1.9	36	20	-	-	-	-
BASF	3	10.4	13.3	44	3.0	14	11	-	-	-	-
Hoechst	10	6.0	14.0	26	2.5	20	13	-	-	-	-
OSW	1	22.2	6.0	19	2.1	29	17	-	-	-	-

(F_{aPP}), non-crystallizable ethylene/propylene copolymer (or amorphous copolymer, F_{cr}), isotactic polypropylene (F_{iPP}) and crystalline ethylene/propylene copolymer (F_{cx}) using the method as described in chapter 13. Those data are given in Table 6.2. One surprising outcome from these data is the very large differences in the ratio of the two ethylene/propylene copolymer fractions and their respective ethylene content.

TABLE 6.2

Calculated composition of some commercial toughened polypropylenes, using the data from table 6.1

Manufacturer	F_{cr} %m/m	E_{cr} %m/m	F_{aPP} %m/m	F_{cx} %m/m	F_{iPP} %m/m	fraction of ethylene in soluble components	approx. peak mol.wt. (a)
Shell	8.4	54	3.8	2.4	85.4	0.56	5
ICI	5.9	57	5.9	14.1	74.1	0.18	3
DSM	9.8	50	1.4	8.5	80.3	0.39	14
Montedison	7.9	57	2.3	5.6(b)	84.2	0.47	3
Mitsui	3.0	40	4.0	21.3(b)	71.7	0.05	4
BASF	10.1	58	3.2	5.0(b)	81.7	0.56	2
Hoechst	7.3	50	6.7	2.6(b)	83.4	0.61	-
OSW	2.7	42	3.3	23.4(b)	70.6	0.05	-

(a): ex GPC on the HOXS extract, $\times 10^{-5}$

(b): calculated assuming E_{cx} equal to 90%m/m

Some commercial polymers are almost completely made with non-crystallizable copolymer, whilst in others the main copolymer fraction is in the crystalline residue. As we will see later these differences are also seen in the morphology of those TPP's. Apart from the monomer composition, the molecular weights of the copolymer part also differ, as shown by the GPC-data on the non-crystallizable fraction, given in Table 6.2.

From the GPC measurement on the soluble fraction one can derive a value for the width of the MWD of the soluble ethylene/propylene copolymer. It is not the true Q or R-value, as the non-crystallizable fraction is composed of at least two parts, namely the atactic PP and the non-crystallizable EPC. The MW of the atactic PP is always rather low however (see chapter 2), so by taking the ratio of M_z over the peak molecular weight there will be only a minor effect of the atactic PP on this ratio, if at all. Fig. 6.1 shows a plot of this ratio against the peak molecular weight for a large number of commercial and laboratory-made TPP's. As is clear they all fall on a straight line, and the relation found is very similar to the one found for homopolymers in the sense that upon lowering the molecular weight (by increasing the polymerization temperature or by applying hydrogen) the width of the distribution increases.

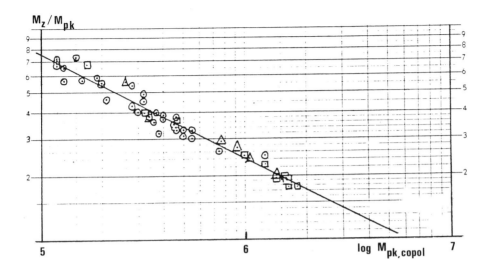

Fig. 6.1: The width of the molecular weight distribution of copolymers, expressed as M_z/M_{peak} calculated from MWD's of hot xylene solubles, as function of the peak molecular weight

6.2.2 Morphology

A fair number of studies related to the morphology of toughened polypropylenes are found in the open literature, although the preparative details of those polymers are mostly lacking. This makes correlation of the observed morphology with the composition or impact properties well nigh impossible. Still a number of worthwhile conclusions can be derived from such studies, therefore we will mention them in the following. Subsequently some of our own studies, made on polymers of better known composition will be mentioned and discussed.

Fracture surfaces were studied by Kojima and Magill[2] on two commercial TPP's of which only the melt index and the ethylene content were mentioned. The most important conclusion from the observations is the disperse nature of these polymers as the fracture surfaces made at liquid nitrogen temperature showed globular micron-sized particles within a matrix. The size of the particles was sensitive to annealing treatments at temperatures of around 140 °C, they become larger and more uniform in size. Interesting morphologies were observed by Levij and Maurer[3] in a commercial TPP; using a special staining technique irregular particles are observed with a skin of rubbery material and a core of polyethylene-like polymer. In blending this type of polymer with LLDPE the size of the core increases whilst the skin decreases in thickness, this substantiates the identification of the core phase. In a study using wide and small angle X-ray techniques, Wenig[4] showed the two different crystalline constituents PE and PP to be present in separate crystals and not to be mixed. Moreover the two polymers crystallize undisturbed from the melt. In a later study of Kojima[5] on stress whitening, again the dispersed nature is clear from both fracture surface studies as well as from the surface of injection moulded sheet when properly etched. Sizes of the particles that are visible range from 0.5 to 3 μm. The conclusion from this study is that <u>three</u> phases are present in his samples namely crystalline polypropylene, polyethylene and amorphous ethylene/propylene copolymer. A number of the statements made leave the impression that they are based on experimental evidence not given in the paper, a situation which is more often found in those areas of polymer studies where companies are (partly) secretive about their way of making and designing certain products. In an optical microscopic study[6] using two different TPP's of non-described origin, Prentice grows large spherulites and infers, from the low magnification photographs, differences especially in size and shape of the dispersed particles. He suggests that more true block copolymer is formed in a gas-phase toughened polypropylene preparation than in a diluent process. More evidence for proving this suggestion is certainly required. Ramsteiner et al[7] produced very informative micrographs of sections of TPP using a transmission electron microscope technique coupled to a selective etching/staining of the amorphous regions. Lamellae of polyethylene are clearly visible in the dispersed particles as well as amorphous (co)polymer. In deforming the material they showed voids to form partly inside the dispersed phase – at the outer rim – and partly in the matrix just outside the particle. This is a very different behaviour when compared with the much studied high impact polystyrene polymers, where the voids will only be generated in the matrix starting from the interface. Making solution-cast films of TPP's and blends[8] gives great similarity in the observed morphology. Treating those films with a solvent for the copolymer leaves a film full of holes, indicating that most of the ethylene/propylene copolymer in in-situ toughened polypropylene is not chemically bound to the crystalline polypropylene. Of course the same proof is obtained in the fractionation experiments described in chapter 6.2.1, as the rubber is isolated in that case. Finally in an optical phase contrast study by Karger-Kocsis et al[9] of thin sections of compression moulded samples the dispersed nature was very clear, moreover it was observed that the size and shape did not change appreciably during long crystallization times. The authors suggested a network structure which would stabilize the dispersed phase, this network could be originating from the block copolymer present in the mixture.

Concluding from the above, the dispersed nature is well documented, in many cases polyethylene is clearly determined (of course its presence is expected to depend on the preparative conditions), but in general no direct link with the impact or other mechanical properties is made. This appears to be very difficult in any case, but to us the necessary condition for finding these correlations is the study of the morphology on the same specimen as used for the impact testing. In the following section dealing with our own findings on the morphology of in-situ toughened polypropylenes, care was taken to use only the impact bars as samples. Moreover the interpretation is helped substantially when one knows more about the detailed composition.

In our own studies a few series of TPP's were prepared batchwise on a first-generation $TiCl_3$ catalyst. In the first series the ethylene content (E_c) of the ethylene/propylene copolymer phase was varied at constant copolymer fraction (F_c), at constant molecular weight (expressed as LVN, being about 6 dl/g) and constant melt flow. In the second series the molecular weight of the ethylene/propylene copolymer was varied at constant copolymer level and composition. In this section the morphological data obtained will be mentioned mainly using anoptral contrast microscopy on thin microtomed sections of injection moulded Izod bars.

The first set of micrographs are given in Fig. 6.2 of the first (E_c) series. A number of features are very clearly visible:
- the products are dispersed systems, the contrast between the phases changing from E_c=70 %m/m onwards, contrast is almost absent in the E_c=80 %m/m product and it reverses in the higher ethylene containing copolymers,
- a distinct increase in the dispersed particle size with ethylene content, being about 1 micron at E_c of 30 %m/m up to

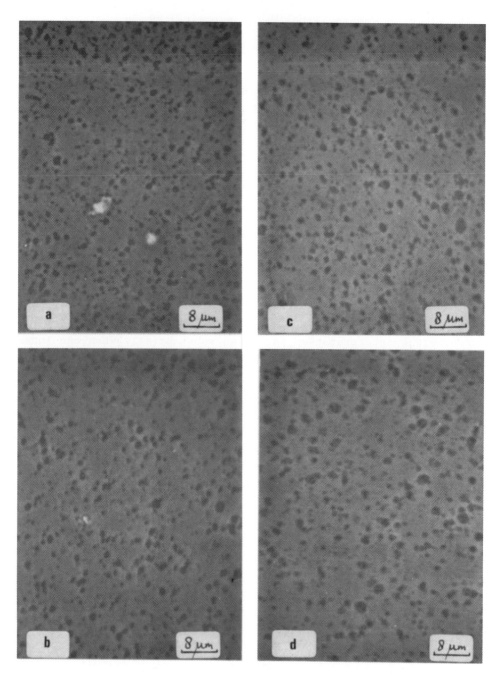

Fig. 6.2: Anoptral phase contrast micrographs of toughened polypropylenes as function of the ethylene content (E_c, in %m/m) of the copolymer part: 30 (a); 40 (b); 50 (c); 60 (d)

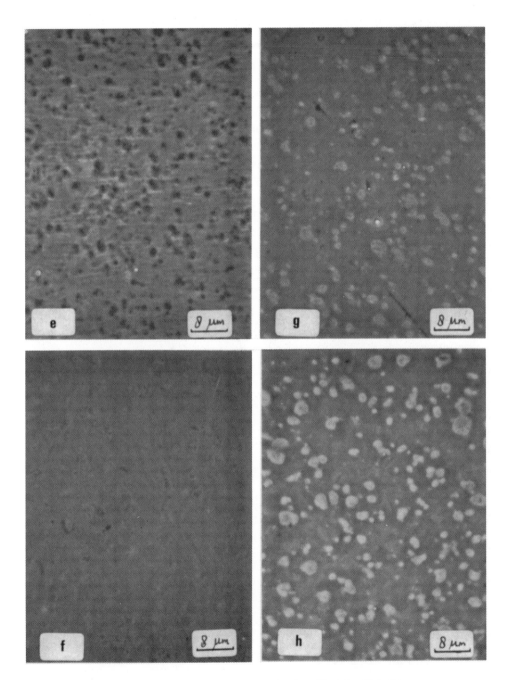

Fig. 6.2 continued: E_c in %m/m of 70 (e); 80 (f); 90 (g); 100 (h)

(MI≈1g/10min; F_c≈15%m/m; LVN_{copol}≈6 dl/g)

roughly 5 micron at pure polyethylene,
- a distinct change-over from internally homogeneous rubber-
like particles (at E_c=30-60 %m/m) via particles with a
substructure to polyethylene-like particles, although also in
the latter some substructure remains visible.

On a similar series at higher molecular weight of the ethylene/pro-
pylene copolymer (LVN's around 15 dl/g), a few samples were studied
with TEM. The micrographs obtained are depicted in Fig. 6.3 together with the corresponding optical micrographs. The size effects
are identical to the ones mentioned above, the distribution of
particle sizes is very large. The shape of the particles can be
more clearly judged in these large magnification photographs and
it is obvious that only the E_c=33 %m/m polymer shows nearly perfect
spheres, whereas the other two show more irregular shapes. A
substructure is only evident in the high E_c sample (80 %m/m)
however vague, in the E_c 62 case there is an indication of lamellae
but not in distinct segregated areas.

The size differences can be caused by a number of different
factors, of which the increasing difference in solubility parame-

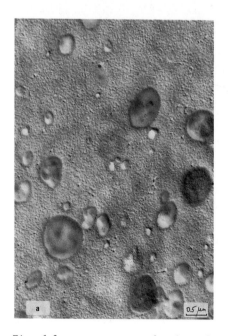
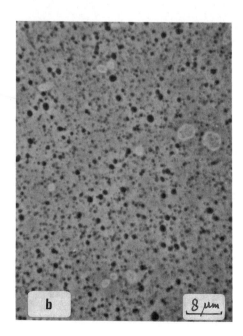

Fig. 6.3: see next page for legend

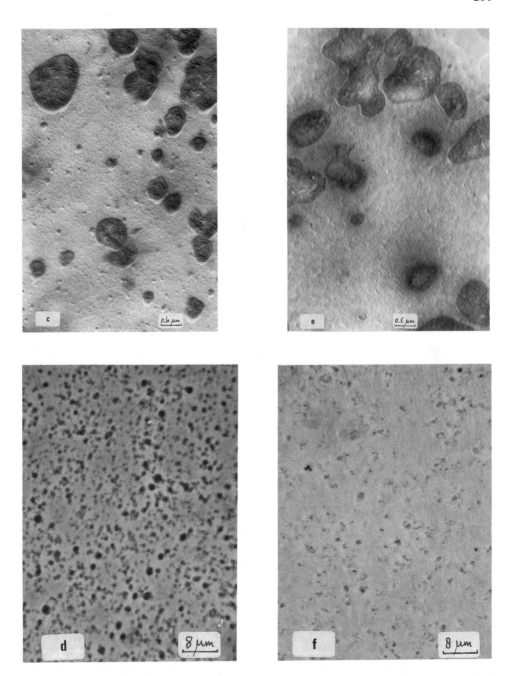

Fig. 6.3 continued: Morphology of three TPP's varying in E_c (%m/m): 33 (a,b); 62 (c,d); 80 (e,f). (a), (c) and (e) are transmission electron micrographs (microtomed Izod bar, plasma etched, Pt shadowed); (b), (d) and (f) are anoptral phase contrast micrographs

300

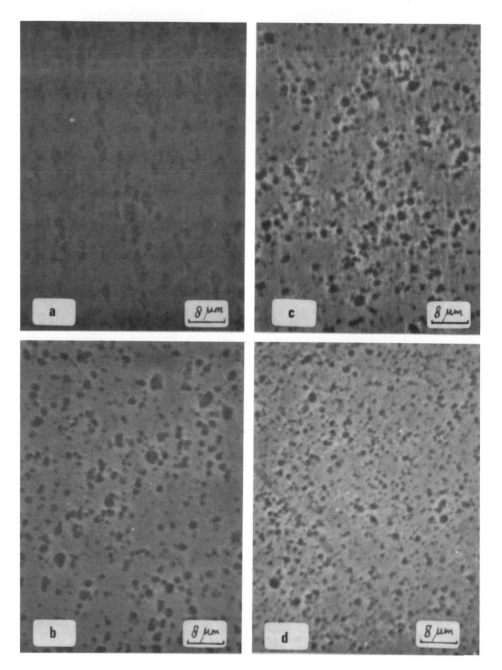

Fig. 6.4: Anoptral phase contrast micrographs of TPP's ($E_c \approx 60\%$ m/m, $F_c \approx 15\%$ m/m, MI≈1g/10min) as function of MW of the copolymer phase (as LVN, dl/g): 2.4 (a); 7.2 (b); 14.4 (c); 27.5 (d)

ter, i.e. the increasing tendency to be non-compatible, seems the most likely. Still other elements such as the small variation which can exist in the true viscosity ratio between the matrix and the ethylene/propylene copolymer could also play a role (although the LVN's were kept fairly constant, the melt viscosity increases with increasing E_c, see for instance reference 10). From blending studies (see chapter 8) it has been shown that increasing viscosity ratios lead to larger particles in the final blends.

On the second series, in which only the molecular weight of the ethylene/propylene copolymer phase was varied in a range of LVN from about 2 to about 25 dl/g, optical microscopy studies were also carried out. The results are shown in Figures 6.4 to 6.7. The general effects of increasing the molecular weight are:

- the size of the dispersed phase reaches an maximum value at some intermediate value of the molecular weight, this is especially clear in Figures 6.4 and 6.7, sometimes agglomerates are visible (e.g. Fig.6.4.a)

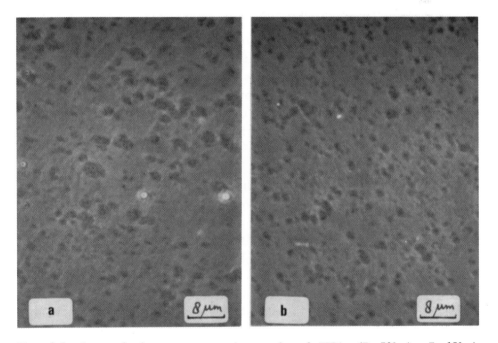

Fig. 6.5: Anoptral phase contrast micrographs of TPP's (E_c=70%m/m, F_c≈15%m/m, MI≈1g/10min) as function of MW (as LVN, dl/g) of the copolymer: 3.8 (a); 6.1 (b)

- there is more substructure visible in the lower molecular weight samples (one should realize the danger in this type of statements: as the particle size decreases substructures will automatically be more difficult to discern).

To explain the effects of the molecular weight on the dispersed particle size another observation is very important, relating to the base morphology of the TPP right after the polymerization. In the following, a number of experiments will be described which define this base morphology: When using a diene in the copolymerization stage of the preparation of these polymers one can crosslink the copolymer part completely. It is expected that subsequent working of the product will not change the base morphology or, more prudently, the morphology observed will reflect a maximum in the particle size. Fig. 6.8 gives an optical and a STEM micrograph of such a product. The optical micorgraph is "empty" and in the STEM micrograph the dispersed phase size is about 0.05 micron, and very regularly at that. In another attempt to observe the base morphology of as-polymerized products, a composition of 90 % polypropylene and 10 % polyethylene was made and studied with TEM after plasma etching. In that case no dispersion could be found at all. This fits in nicely with the picture one has about the composition of a first generation catalyst: being an agglomerate of a multitude of very small particles. Using the data given in chapter 1 on the sizes of, for example the first generation $TiCl_3$-catalysts (say 15 nm) one can calculate that at a fraction copolymer of 20 %m/m the base size of the copolymer particle formed around the primary crystallite is about $0.1\mu m$ (this size depends on the yield of product made on the catalyst - in the calculation this is assumed to be 200 g copolymer per g of catalyst). The above mentioned sizes of the in-situ formed rubber tie in quite well with the most recent data on homopolymer morphology as given by Kakugo et al[11]. These investigators observe a minimum size of the globules in a growing polymer particle of 0.1 to 0.2 μm.

Using this information in the explanation of the observed changes in particle size with dispersed phase molecular weight, the base morphology from the polymer powder after the polymerization will be influenced by the work done on the system in, for example the extrusion step from powder to nib and subsequently from nib to injection moulded product. Thermodynamically the system would

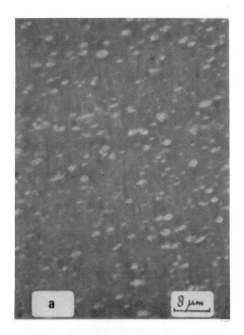

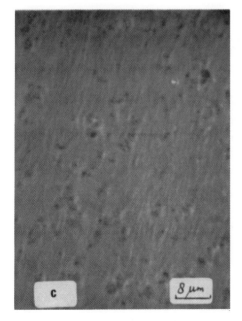

Fig. 6.6: Anoptral phase contrast micrographs of TPP's ($E_c=80\%m/m$, $F_c\approx15\%m/m$,
$MI\approx1g/10min$) as function of MW (as LVN,
dl/g) of the copolymer part: 3.8 (a),
7.6 (b), 11(c)

304

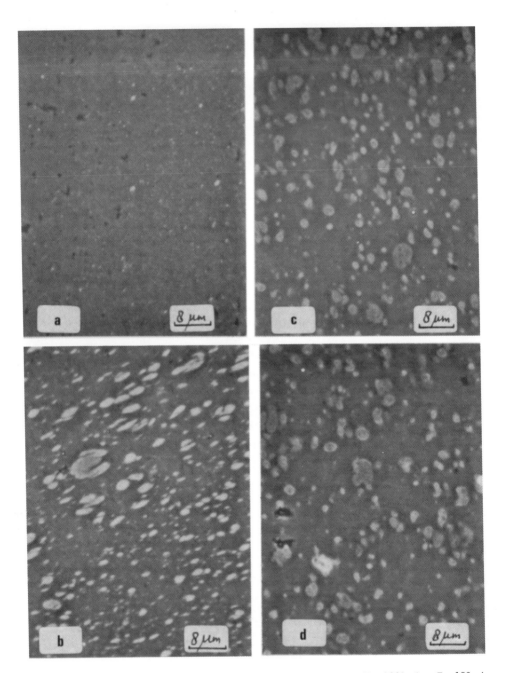

Fig. 6.7: Anoptral phase-contrast micrographs of TPP's (E_c=100%m/m, $F_c \approx$15%m/m, MI$\approx$1g/10min) as function of MW (as LVN, dl/g) of the copolymer: 1.8 (a); 3.9 (b); 8.2 (c) 13.8 (d)

prefer a state of complete separation of the phases, but this is counteracted by the work done on the system and moreover the rate of phase separation by coalescence will be very low due to the high viscosities of the components. The final size is determined, at constant E_c, both by the chosen processing method and by the viscosity ratio of the constituents. The latter can most easily be illustrated by describing two extremes:

- if the dispersed phase viscosity is far lower than that of the matrix the dispersed phase will flow very easily, in for example injection moulding it is easily elongated and then forms small droplets, finally resulting in a very fine dispersion. In practice one indeed observes small (e.g. < 0.5 micron) sizes after injection moulding, in compression moulding (i.e. almost no work done on the system) phase separation will be relatively undisturbed and very large particles can result. In Fig. 6.4a one observes a very broad particle size distribution which is due to the presence of agglomerates not completely broken up by the mechanical working applied in this case. This situation is also encountered in blends (see

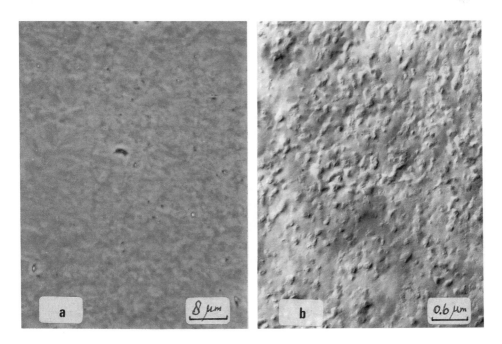

Fig. 6.8: Micrographs of a TPP sample containing a heavily crosslinked copolymer phase: anoptral phase-contrast (a); transmission electron microscopy (sample etched and shadowed) (b)

chapter 8),
- if the dispersed phase viscosity is much larger than that of the matrix, the high viscosity and thus the difficult flow of the dispersed phase will lead to a much lower rate of agglomeration in the processing, as on collision of particles coalescence will occur only with a low frequency, thus the dispersion will stay fine. This is in the extreme case seen in the crosslinked sample, but also in the high molecular weight dispersed phase samples in Fig. 6.4 for example. In the intermediate range larger particles will result, leading to a (probably flat) maximum in the dispersed phase size as a function of the molecular weight. The precise relationship cannot be given as it will depend not only on the processing method but also on the composition of the ethylene/propylene copolymer. As regards the stability of the dispersion with mechanical working, the low viscosity ones are probably fairly reproducible on repeated processing, whilst one could expect some slight growing of the dispersed phase size in the very high molecular weight cases. The shape of the particles will at the same time be influenced by the same parameter, being spherical at high flowability and becoming more irregular at lower flows because of incomplete coalescence.

Regarding the dispersed phase structure as observed under the microscope, one observes changes with its molecular weight as well. This is especially distinct in the E_c 80 toughened polypropylenes in Fig. 6.6, where at low molecular weight only a white dispersed phase is visible, interpreted as "totally" crystalline, at an intermediate value of the molecular weight a small dark nucleus is surrounded by a whiter phase, i.e. some amorphous copolymer is present, whilst at the highest value studied the fraction of darker phase definitely increases. These effects are very probably related to the extent of crystallization of the copolymer phase as will be discussed in the following paragraph, although it can not be excluded that the higher levels of hydrogen used to obtain the lower molecular weights not only changed this but also changed the ratio of the amorphous to crystalline copolymer. We have some evidence for this but more data are required to confirm it.

In Figures 6.9 and 6.10 examples are given of the morphology of various commercial TPP's. Using the knowledge obtained from the foregoing studies we can try to interpret these pictures. Clearly - from the optical microscope studies - these commercial products span a range of E_c's with grades of Shell, Montedison and BASF

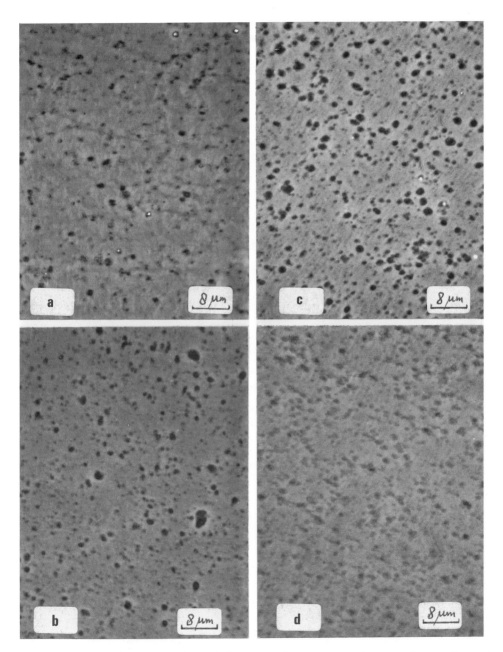

Fig. 6.9: Optical micrographs, with anoptral contrast, of typical examples of various commercial TPP's: Moplen EPC40R (a); Hoechst PPU 1062 (b); Shell GMT (c); Hercules Profax 7731 (d)

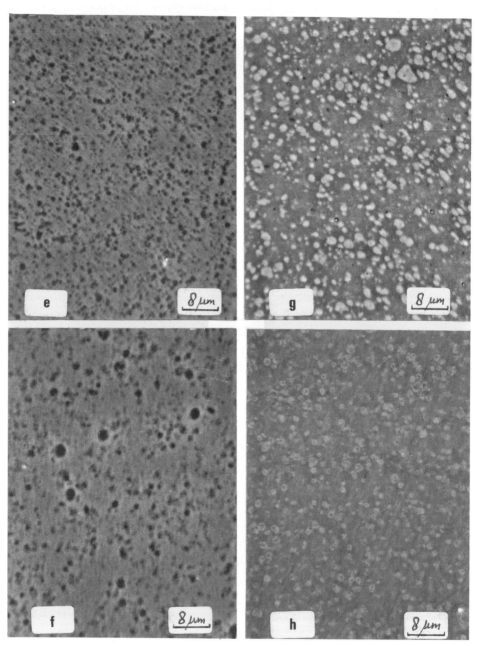

Fig. 6.9 continued: Saga(low E_c!) (e); DSM 83M10 (f); Mitsui (g); ICI GWM213 (h)

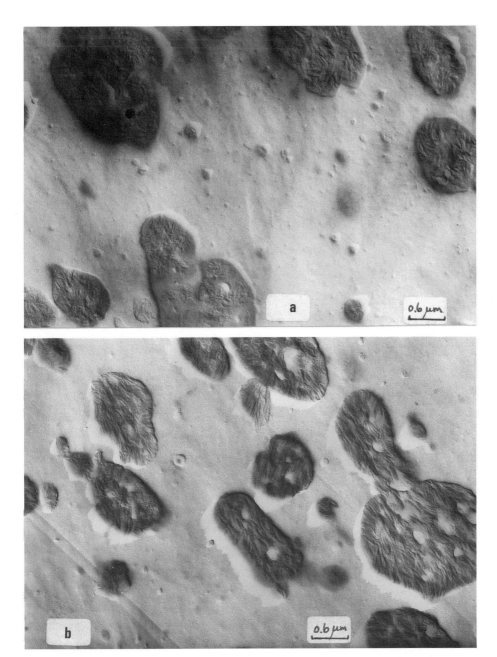

Fig. 6.10: Transmission electron micrographs of two examples of commercial TPP's
(corresponding with two samples in figure 6.9): DSM 83M10 (a); ICI GWM213 (b)

being at the lower E_c side and ICI and Mitsui at the high end, near-polyethylene like. This fits in nicely with the behaviour of these samples in the HOXS extraction test. From the TEM studies one also finds sometimes a broad particle size distribution (DSM, BASF), or very irregular particles (DSM), whereas in others the particles are more spherical (e.g. ICI). Substructure in the dispersed phase is sometimes noted as well, but mostly only in optical studies, not in TEM.

An example of fracture surfaces on some commercial Shell toughened polypropylenes are given in Fig. 6.11. The dispersed nature is again very clear, with sizes in the order of 1 to 2 microns.

6.2.3 Crystallization

There are not many studies on crystallization in TPP's, certainly not on well-described series of polymers. In differential thermal analysis or DSC studies all researchers observe upon melting two endotherms in most TPP's, the first at a temperature slightly below the normal polyethylene melting and the second at the expected polypropylene value. Barrall et al[12] found a fairly good relationship of the respective enthalpies of melting with the composition of his TPP's. Upon cooling normally one exotherm is noted as PP and PE, when supercooled, crystallize nearly at the same temperature. However Ke[13] observed a double exotherm in what was called blockcopolymers and only one in blends. (This block copolymer was a sample he obtained from Bier, see chapter 5). Barrall finds a similar double peak in crystallization but only if the ethylene content is over 10 %m/m. More recent studies[1,2,5,8,14] on commercial polymers again find polyethylene-like crystallinity in most of them, always to a low extent and the melting points observed are all in the range of 120 to 130 °C, i.e. substantially lower than the melting point of pure HDPE. The reason for this is probably the copolymerized propylene and/or its very high MW. The observation of this polyethylene-like phase is independent proof that four constituents are present in TPP's, i.e. the atactic and isotactic polypropylene, the amorphous, extractable ethylene/propylene copolymer and a crystalline polyethylene of relatively low melting point.

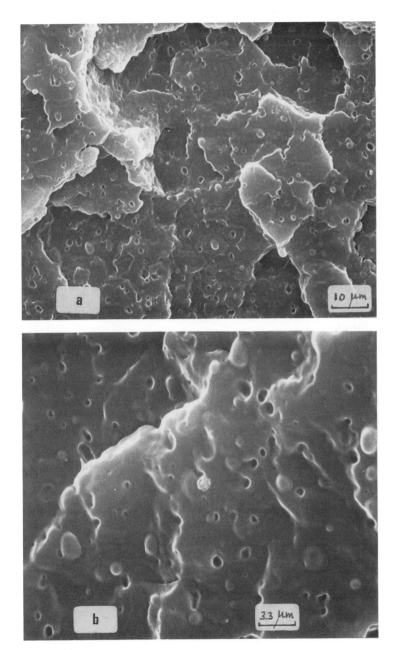

Fig. 6.11: Scanning electron micrographs of a fraction surface (typical Shell grade)

312

DSC-measurements have been done on the same series mentioned above in the section on morphology. Figures 6.12a and b give the melting and crystallization runs of TPP's with all parameters constant except for the E_c. All but the lowest E_c sample studied show a melting peak in the low temperature range, stressing the conclusion that a polyethylene-like phase is present in almost every toughened polypropylene made with an isospecific first generation catalyst. The observed melting points are all rather close, ranging from 117 to 122 °C, with no clear relation to the E_c of the copolymer fraction. Even the "100 %m/m" E_c product shows a

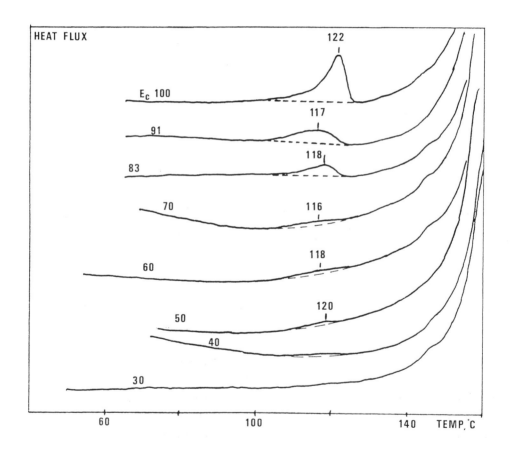

Fig. 6.12a: Normalized DSC melting <u>endo</u>therms of the polyethylene crystallinity of TPP's varying in E_c (E_c nominal, $F_c \approx 15\%m/m$, $LVN_{copol} \approx 6$ dl/g, DSC scanrate 20K/min)

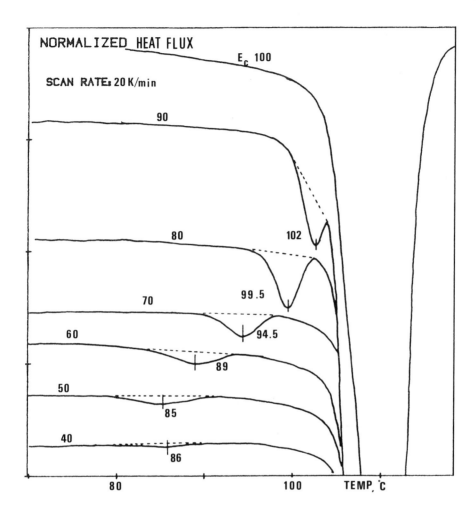

Fig. 6.12b: Normalized DSC crystallization _exotherms_ of the polyethylene crystallinity of TPP's varying in E_c (E_c nominal, $LVN_{copol} \approx 6$ dl/g, $F_c \approx 15$ %m/m)

melting point of only 122 °C, indicating the copolymerization of some propylene, despite the fact that the polymerization slurry after the homopolymerization was extensively flushed with nitrogen to remove unreacted propylene. The melting enthalpy uniformly increases with the E_c as expected. The crystallization exotherms in Fig. 6.12b show a clear dependence on the ethylene content both in their position and size. The temperature range is fairly large, from 85 to 102 °C. Clearly the presence of longer ethylene sequences in higher E_c copolymers results in a higher crystallization

314

temperature and a larger exotherm.

Also on the series in which, at constant fraction of copolymer, the E_c and independently therefrom the molecular weight were varied, DSC-measurements have been done. The results are given in Figures 6.13 to 6.15. In the crystallization the E_c 60 series (given in Fig. 6.13) illustrates nicely the increasing crystallization temperature with decreasing MW as well as an incrasing crystallization enthalpy. The range of crystallization temperatures is 86 to 96 °C. The same general behaviour is seen in all other

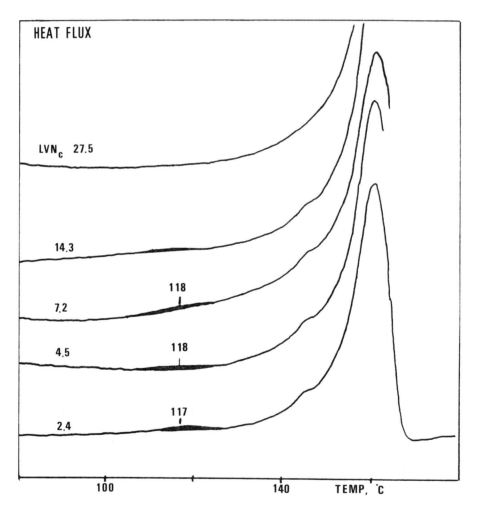

Fig. 6.13a: Normalized DSC melting <u>endo</u>therms of the polyethylene crystallinity of TPP's varying in LVN_{copol} (E_c=60%m/m, F_c≈15%m/m)

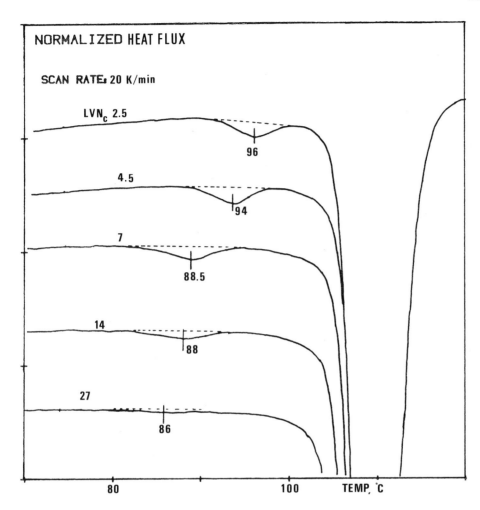

Fig. 6.13b: *Normalized DSC crystallization exotherms of the polyethylene crystallinity of TPP's varying in LVN*$_{copol}$ *(E$_c$=60%m/m, F$_c$≈15%m/m)*

cases, only the temperature range shifts to higher levels at increasing E_c, which at the compositions near the 100% mark means that the polyethylene crystallization peak is buried under the polypropylene crystallization peak (with a crystallization temperature of 110 °C). But in all other cases two crystallization peaks are thus noted, and as we have good reasons to believe that almost no true blockcopolymer is present in these compositions (see chapter 5), this contradicts the observations of Ke and Barrall mentioned above. Studying the melting of these products shows an

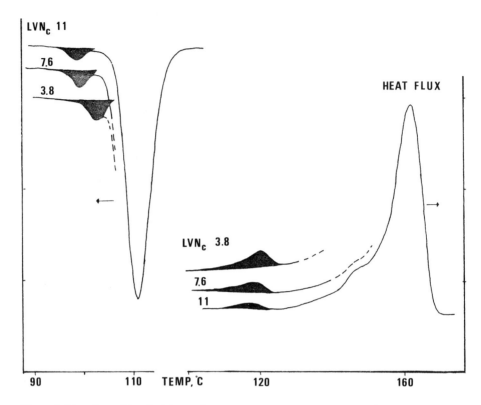

Fig. 6.14: Normalized DSC melting and crystallization of TPP's varying in LVN$_{copol}$ (E_c=80%m/m, $F_c \approx 15$%m/m)

almost constant melting temperature independent of the MW, and even nearly independent of the E_c of the copolymer in line with the observations mentioned above. At the very high MW's, even in the E_c 60 case no distinct melting is observed. The melting enthalpy does again increase with decreasing MW, analogous to the crystallization behaviour. The trends observed are completely in line with the general effect of MW upon crystallization, i.e. at higher molecular weight and hence melt viscosity the ease of crystallization becomes less and thus also the final extent of crystallization is also less. Note that the range of MW's studied is huge. Annealing will of course have a large effect on the higher MW samples, this however was not studied. The observed differences in the optical micrographs of these products are in line with the DSC results in that at lower molecular weight more crystallinity and thus more "light" phase is present.

As a last item a few final melting points are mentioned for total TPP's and their homopolymer component (as it is sampled from the polymerization, just before the second stage copolymerization is started). At the copolymer fraction level of 20 to 25 %m/m the difference between the two TMF's is 1.0 ± 0.1 °C with the homopolymer being the largest. This difference is small, but as the TMF is a very sensitive measurement (see chapter 2.5), this difference precludes the use of TMF on toughened polypropylenes as a measure of the <u>homopolymer</u> quality. Note that the peak melting point in DSC is not sensitive to the presence of the copolymer fraction.

6.2.4 <u>Dynamic</u> <u>mechanical</u> <u>analysis</u>, <u>glass</u> <u>transition</u>

Using torsion pendulum type measurements on TPP's, one finds in addition to the normally observed polypropylene-related transition around 0 °C further transitions in the region of -55°C and -120 °C. The former is certainly related to the presence of ethyle-

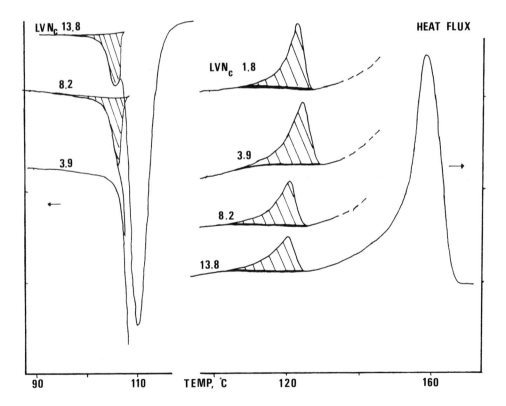

Fig. 6.15: Normalized DSC melting and crystallization of TPP's varying in LVN$_{copol}$ (E$_c$=100%m/m, F$_c$≈15%m/m)

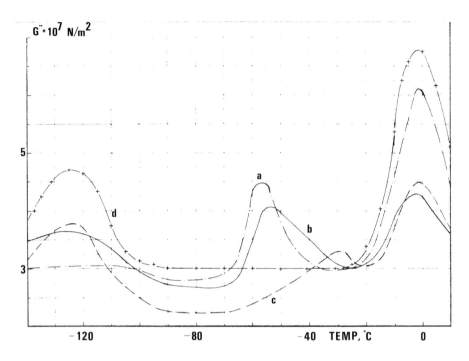

Fig. 6.16: The loss shear modulus, G", of some commercial TPP's as a function of temperature: Shell GMT (a); DSM 83M10 (b); Mitsui SJ310 (c); Hüls V4700 (d)

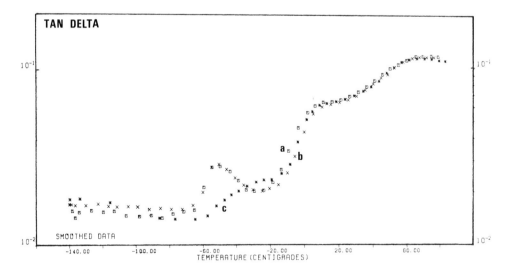

Fig. 6.17: The loss factor, tanδ, of three TPP's with F_c=20%m/m varying in E_c: 43 (a); 70 (b); 88 (c)

ne/propylene copolymer, and the latter to polyethylene. The same general behaviour has been reported in the literature[7,14-16]. In measurements on commercial TPP's almost invariably the rubber glass transition has been found in the range of -50 to -60 °C. The loss modulus as a function of temperature is shown in Fig. 6.16 for a number of commercial samples widely differing in their dynamic response. One TPP shows only a rubber transition, another only the PE-one, whilst the others show both transitions.

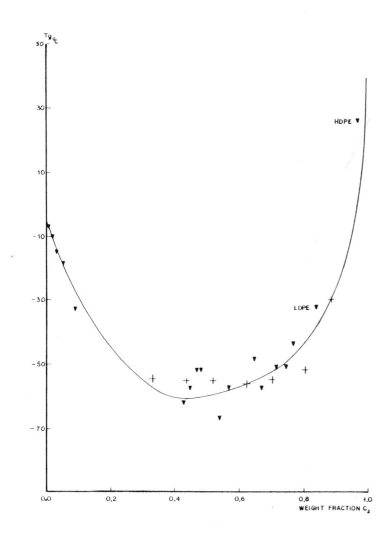

Fig. 6.18: Glass transition temperature of the copolymer part of TPP's compared with neat ethylene-propylene copolymers (+ represent TPP-samples)

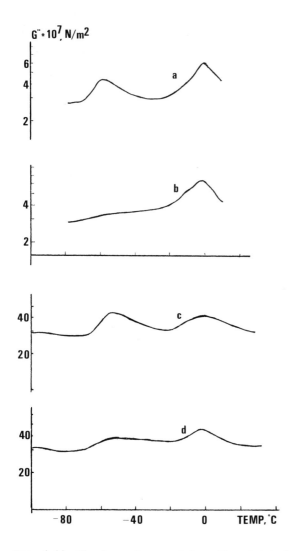

Fig. 6.19: The loss shear modulus, G", of total polymer and "hot xylene residue" of two commercial TPP's: Shell GMT (a); ditto HOXS-residue (b); MYKK BC8 (c); ditto HOXS-residue (d)

Measurements on a series of own model TPP's with varying E_c gives the glass transition temperatures and the maximum value of the loss modulus as given in Table 6.3. Clearly the values of the glass transition are all identical except for the highest E_c sample, see the illustration of this in the plot of loss modulus versus temperature in Fig. 6.17. Plotting these data in the figure given before of glass transitions of random copolymers (compare chapter 4) fits reasonably as shown in Fig. 6.18, although of course the minimum in the glass transition curve as a function of E_c becomes very flat and broad, and a band of about 10 to 15 °C exists for fixed ethylene contents. The loss modulus is more sensitive to the E_c and shows a maximum value in the range of 50 to 60 %m/m ethylene.

TABLE 6.3

Glass transition and maximum loss modulus in TPP's with varying E_c and constant fraction copolymer (15 %m/m) and molecular weights (LVN of the copolymer being 15 dl/g)

Ethylene content %m/m	Rubber phase		Homopolymer phase	
	glass transition °C	maximum loss modulus MNm^{-2}	glass transition °C	maximum loss modulus MNm^{-2}
33	-54	31	2	47
43	-55	40	0	41
52	-55	42	1	43
62	-56	41	-3	52
70	-55	38	5	40
80	-52	33	4	40
88	-30	27	3	43

Finally, when measuring the dynamic mechanical response on the residues after HOXS extraction of TPP's, indeed the loss modulus maximum around the -50 °C is no longer present to the same extent although mostly a baseline shift is still discernible (see Fig. 6.19).

6.2.5 Rheology

The normal rheological measurements at higher shear rates will not be dealt with. Only very low shear rate behaviour of a few TPP's is mentioned as this has a possible bearing on the overall structure of these systems. For a number of simple TPP's, varying only in the molecular weight of the dispersed phase, the viscosity-shear rate curves are given in Fig. 6.20. The measurements were done by starting at the lowest shear rate (the open symbols in the figure), and after reaching the highest shear rate the measurements were directly repeated but now in decreasing order of the shear rate (points labelled 1 to 3 in the figure). Finally the

322

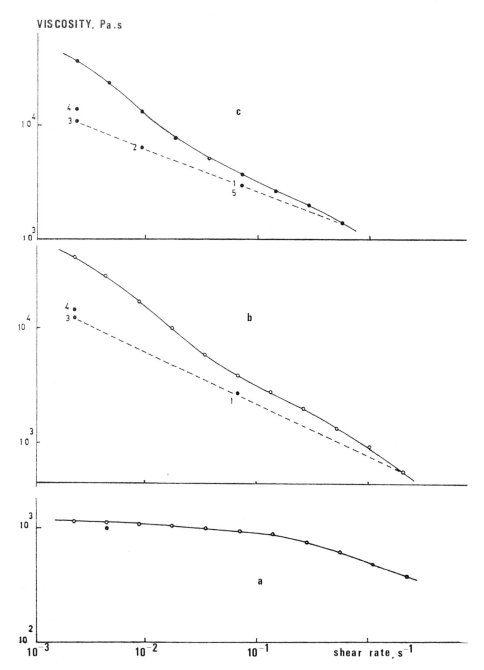

Fig. 6.20: Viscosity at very low shear rate of TPP's differing in LVN$_{copol}$: 5.2
(a); 19 (b); 24 (c) (the melt indices are resp. 4.5; 1.4 and 0.7 g/10min)

sample was left undisturbed for 30 minutes and points 4 and 5 measured. Samples with the lowest MW of the dispersed phase behave as a Newtonian liquid at low shear rates, and the order of measurement is immaterial. However the higher MW samples do show an ever decreasing viscosity with decreasing shear rate. Moreover, the subsequent measurement at decreasing shear rates shows a much lower value for the viscosity; the viscosity after 30 minutes rest is higher at the lower shear rates. These findings indicate a shear-induced disrupture of some "structure", which , although slowly, is restored at rest.

This conclusion is in line with the results of the shear stress development during the measurements. Whereas with Newtonian fluids the shear stress quickly reaches a constant value without any overshoot, the samples with the high MW dispersed phase show a maximum shear stress ("yield stress") that has to be passed before a state of equilibrium is reached. When the stress is measured after a preceding shear at a higher shear rate, this yield stress has disappeared. After a rest period of 30 minutes it is observed again.

One finds in the literature numerous studies[17-20] on "filled" polymer systems for which, in a number of cases, behaviour similar to the high molecular weight polymers described above is found. The viscosity at low shear rate increases with the volume fraction of the filler, but especially with its decreasing size. One way of interpretation of the results is the assumption that particles immobilize a polymer layer of a certain thickness.

Although the above data point to a kind of network structure, it is not at present known which constituents are connected, to what extent, the range of interaction, etc.

6.2.6 Mechanical properties

Data on impact and stiffness properties of in-situ prepared TPP's of well described composition are very rare. Excepting data from patents, the only good set of data has been discussed by Heggs[15]. The TPP's described and tested in that reference were prepared along the same lines as the ones from our own studies mentioned before, except for using a different catalyst and the removal of the reactor solubles before drying of the polymers. A series was made having constant fraction of copolymer and varying

E_c (no molecular weights or melt flows are mentioned). These products show a broad, flat minimum in flexural modulus in the range of E_c 30 to 70 %m/m. The brittle point (determining the minimum temperature still showing a fixed impact level) shows a slightly sharper minimum at E_c 55 to 80 %m/m when measured on compression moulded samples. For injection moulded samples the notched Charpy impact is much more sensitive to E_c, and it peaks at 65 %m/m. As expected the Charpy impact is very temperature dependent, being considerably lower at lower temperature, but the maximum stays at the same E_c. The stiffness is very strongly related to the fraction of copolymer present.

In the same reference similar relationships are quoted for polymers described in Exxon patents. In addition to this, Exxon mentions the effect of adding hydrogen in the copolymer stage (i.e. lowering the molecular weight of that phase) as lowering the film haze but at the same time the Izod impact is lower.
Heggs concludes that

- toughness increases in polypropylene compositions containing ethylene/propylene copolymer provided the E_c is neither very low nor very high,

- the toughest product is one having E_c's between 50 and 80 %m/m,

- the modulus is reduced by those factors which reduce the overall crystallinity of these polymers.

In the following, a number of data from own measurements will be given which in broad terms corroborate the above conclusions. On a series of TPP's, having constant F_c, LVN of the copolymer phase and melt flow (15 %m/m, 6 and 1 resp.) and only varying in the E_c, both the falling weight impact strength (FWIS) and Izod impact were measured for testing the toughness, and the flexural modulus (FM) for the stiffness. Some of the data obtained are given in Fig. 6.21. Clearly the FWIS is very sensitive to the E_c, showing an optimum at 50 to 60 %m/m. A similar situation is observed for the Izod impact, this coming very close to the optimum mentioned above[15]. The FM shows a fairly broad minimum in the E_c-range of 30 to 70 %m/m. Due to the more rigid character of the dispersed phase at higher E_c, the modulus increases above an E_c level of 70 %m/m.

For a different series with all other things equal except for the molecular weight of the dispersed phase (same melt flow and F_c), the most relevant data are given in Table 6.4. The FWIS

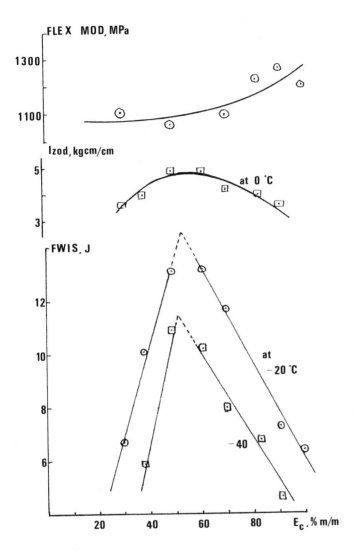

Fig. 6.21: *Mechanical properties of TPP's with* F_c *of 15%m/m as function of* E_c
(*LVN*$_{copol}$ ≈ 6dl/g, MI ≈ 1g/10min)

increases with increasing MW up to a certain level of the LVN, after which it stays constant. This behaviour is independent of E_c in the range studied, i.e. 60 to 100 %m/m. For an illustration see Fig. 6.22. It is interesting to note that in a E_c=100 %m/m composition a FWIS level of 5 J at -30 °C can be reached provided the LVN of the dispersed phase is over 8 dl/g. Izod impact as function of the LVN is also shown in Fig. 6.22, from which it is

TABLE 6.4

Mechanical properties of in-situ made toughened polypropylenes with E_c and the molecular weight of the copolymer as a variable. Overall melt index constant at around 1.

E_c %m/m	F_c %m/m	LVN copol dl/g	FWIS(J) at -20 °C	-30 °C	-40 °C	-50 °C	ISO Izod at 0 °C kgcm/cm	-20 °C	Flex properties modulus MPa	strength
62	14.2	2.4	10.8	9.5	5.6	a	3.4	2.9	1130	34.3
61	14.4	4.4	12.0	11.4	9.6	6.4	4.2	3.8	1120	33.9
61	14.9	7.2	13.2	12.5	10.2	5.9	4.9	4.2	1110	34.1
59	15.4	14.4	13.5	12.4	10.2	6.7	6.2	4.5	1160	35.4
60	16.8	27.5	-	-	11.1	7.0	10.2	5.4	1060	32.8
72	13.3	3.8	10.2	8.6	5.9	3.1	3.8	3.1	1210	37.1
71	13.2	6.1	11.7	9.7	8.0	4.1	4.2	3.9	1100	33.8
84	11.7	3.8	7.7	5.5	a	a	b	b	1310	39.2
83	11.8	7.6	-	8.5	6.8	-	4.0	3.5	1230	37.6
82	12.0	11.0	9.6	8.3	6.6	-	4.2	3.8	1220	37.5
100	13.4	1.8	a	a	a	a	-	-	1280	40.0
100	12.6	3.9	a	a	a	a	-	-	1250	39.4
100	13.9	8.2	6.4	4.5	-	a	-	-	1200	37.6
100	13.4	13.8	6.5	5.5	-	a	-	-	1160	36.4

a: <2.7J ; b: < 2.5 kgcm/cm

clear that the Izod increases steadily with increasing MW in contrast to the behaviour of the FWIS. The difference in response between FWIS and Izod demonstrates that these impact tests measure different aspects of energy absorption/fraction behaviour. In the E_c 60 series the FM is not dependent on the molecular weight, it is essentially constant in this series. At the higher E_c's and especially at E_c=100 %m/m a decrease in rigidity is found upon increasing the molecular weight, in line with the decreased crystallinity (see 6.2.3).

6.2.7 Discussion

Can we relate any of the observed characterization data with the mechanical performance properties found? The following is an attempt using some of the accepted theories of impact improvement[21-26].

(i) theory of energy absorption

The generally accepted theory about toughening runs as follows: the normal fracturing in glassy polymers starts by forming crazes roughly perpendicular to the applied stress. A craze is

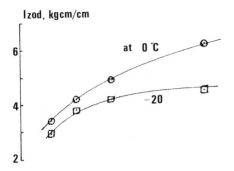

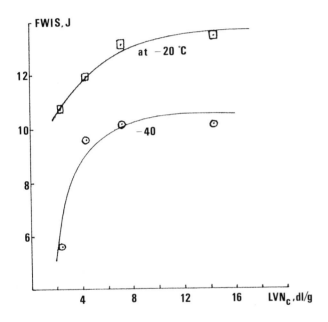

Fig. 6.22: Mechanical properties of TPP's with F_c of 15%m/m as function of LVN$_{copol}$ (Melt Index about 1g/10min)

filled with extended polymer, the density in a craze is low. Crazes "heal" when the stress is relaxed. A craze can grow, and this costs energy in forming the new surface with its extended chains. Above a certain size and width the craze can transform into a crack (in which tie material is no longer present) after which catastrophic failure of the specimen can occur. For toughening of brittle polymers a dispersed phase is necessary, of a rather small size. Around the dispersed phase - with a lower modulus than the matrix - a stress concentration is present when a stress is applied and this can initiate craze formation. If the dispersion is indeed able to initiate the formation of a multitude of small crazes, a high energy absorption can result without any large craze or crack formation, i.e. the impact is increased. The dispersed phase particles can also stop or delay the growth of a craze. Thus the particles serve two roles: initiation of crazes, assisting energy absorption and secondly stopping growth of crazes. To function properly, the dispersed phase particles should adhere fairly strongly to the matrix as otherwise a crack would easily form at the interface. This adhesion is sometimes obtained via grafting of matrix and rubber (as in high impact polystyrene and ABS-resins) thereby forming a "surfactant" molecule which will connect the two phases over the interface, or in other cases one plays with the compatibility of the ingredients (as in PVC) - one needs a fairly compatible rubber but not too good as otherwise no dispersion is formed. The crazes are also responsible for the "blushing" frequently observed in these polymers, also in toughened polypropylene. In addition to crazes, shear bands can also play a role in the energy absorbing mechanism, in polypropylene it is believed to be a minor role however, and then only at higher temperatures.

(ii) <u>effect</u> <u>of</u> E_C

For the effect of E_C on the properties one finds, as main aspects, the optimum in impact at E_C = 55 %m/m and a broad minimum in the stiffness in the E_C range of 30 - 70 %m/m.

We start by mentioning a non-relation: morphological features as discussed in section 6.2.2 do not relate with the impact, as for instance the TPP's at E_C 40 and 80 %m/m have a widely different morphology whilst they have an equivalent impact performance. Assuming a constant rubber particle size, then a possible explanation for poor impact at high E_C's may be the polyethylene crystallinity of the dispersed phase and thus a reduced interfacial adhesion between matrix and "rubber". Namely, an increasing rubber crystallinity leads to a decreasing rigidity-difference between matrix and rubber and this may result in a decreasing interfacial stress concentration which may hinder the possibility of craze formation. At high E_C's this is demonstrated by a decreased tendency to stress-whitening, which is a reflection of craze formation. Secondly, the interfacial adhesion which is required for good impact may be reduced at high E_C due to the inherent greater difference in solubility parameter. This description cannot explain the poor impact at low E_c's because, in contrast to high E_c's there is no crystallinity at E_c of 30 %m/m (and moreover no significant reduction in stress-whitening). Furthermore the interfacial adhesion may be higher due to greater similarity in solubility parameter of matrix and dispersed phase. Also the glass transition is rather insensitive to E_c in this region and thus cannot be a reason for the lower performance. A possible explanation may be in the smaller dispersed particle size at these lower E_c values (see also below). The one characterization property which also is a maximum at E_c of about 55 %m/m is the loss modulus in dynamic mechanical measurements. Such a correlation has also been mentioned in the literature[27].

Finally, the broad minimum in rigidity can relatively easily be explained by the broad minimum in rubber crystallinity at the same E_c range.

(iii) effect of molecular weight, effect of particle size

In trying to explain the effects of the molecular weights of the dispersed phase on the impact properties, one of the crucial elements is the effect of the particle size of the dispersed phase on the impact. It has been observed in other polymer systems that the particle size has an optimum when testing for impact[21,22,24,25,28]. The optimum value is smaller in more ductile matrices (e.g. about 1 μm in polystyrene and 0.2 μm in polyvinylchloride)[26]. Very small particles are probably ineffective in terminating crazes, i.e. are ineffective in stopping a craze from transforming into a crack. On the other side of the spectrum, large particles have a low effective interfacial surface to generate crazes, i.e. the energy absorbing capacity decreases. Also the interparticle distance increases at a constant dispersed-phase loading so that a craze can more easily develop into a crack. The above implies that an optimum particle size could exist, however the studies by Jang[29] imply that in TPP the transition of crazing to shear banding occurs at about 0.5 μm dispersed phase size, but impact still increases at particle size below that value. This is not in line with our findings at the low E_c products. Still assuming that impact shows a maximum at some specific particle size, this reasoning can be used to assist in explaining a number of observations. For instance the lower impact at low E_c's: due to the greater compatibility of the matrix and the rubber, smaller sizes of the dispersed phase are present, and this could explain the impact difference with higher E_c materials. Also the general trend of impact with molecular weight of the dispersed phase fits in: at very low MW the dispersed phase will be small, and the same will happen at very high MW's (see section 6.2.2). Hence a maximum in impact at some intermediate MW is expected when studying a wide range of dispersed phase MW's. This, in the mentioned cases is not observed, only the initial rise in impact is seen. However we have limited evidence that in higher flow TPP's a maximum is found (this is due to the fact that at constant MW of the dispersed phase the resulting particle size will become smaller the lower the MW of the matrix is, as the mechanical work done on the dispersed phase will then decrease). An extreme case was obtained in crosslinking the polymerized copolymer in-situ which leads to very small particle sizes (see section 6.2.2), indeed these products have a very low impact.

Recently some examples were given of beneficial effects of having a bimodal distribution of particle sizes of the low-modulus phase[30]. Some theoretical rationale is available for this, and it was speculated to be worthwhile in polypropylene[31].

In our experience the Izod impact is far less sensitive to the particle size of the dispersed phase. In contrast to a FWIS test, the sharply notched ISO Izod test specimen requires almost no energy for crack initiation. Practically all the energy in the Izod test is therefore absorbed in crack propagation. It may be that the amount of energy required to propagate a crack increases with increasing rubber strength/elasticity. This in turn may then explain the ever-increasing Izod impact with increasing MW of the dispersed phase.

Just as was the case in the E_c series, also in the MW series no consistent correlation of the morphological features as revealed in the anoptral contrast microscopy and the impact performance is found. Identical morphologies can be found giving very different impact values, a case in point is E_c 60 %m/m at high MW and E_c 30 %m/m at intermediate MW. On the other hand in the E_c 80 %m/m series, the improving impact preformance with increasing MW is parallelled with the change to a darker (more rubbery) dispersed phase.

Effects of particle shape may very well be present, especially at the higher molecular weight polymers, but can in no way be determined from the above experiments.

The non-influence of the MW of the dispersed phase on the stiffness was to be expected from the known theories on composite materials. They predict the overall rigidity to be dependent on the rigidity and volume fraction of the component phases. Although the modulus of the rubber will increase with increasing MW, the values will stay still small compared to the modulus of the matrix. Therefore no effect is expected on the overall rigidity.

(iv) blends versus in-situ prepared toughened polypropylenes

The commercial TPP's are generally of two classes: the blends and the in-situ prepared products. Although blends show a considerable improved impact performance and have the advantage of freedom in the choice of at least one of the blend components, in our opinion the in-situ prepared compositions can better use all the possibilities in tailoring the properties. Indeed the best toughened products presently available are made in-situ. The main advantage of in-situ generation of the product is the intimate mixing right at the start, independent of MW and composition (see our discussion of the morphology above). Furthermore, for in-situ prepared compositions, one is free to choose the MW (within the bounds of the desired melt index of course), whilst for blends the conditions for good mixing and thus dispersion requires near-identical melt viscosities for the components. All this together will probably also allow one to affect the particle shape or size distribution in in-situ TPP's, but whether this is already practiced is not known to us.

(v) rubber-matrix compatibility

A recurring theme in all discussions on toughening is the compatibility of the ingredients. Particularly in the class of polymers which form the subject matter of this book this is an interesting topic as poylethylene, polypropylene and - even more strongly - their copolymers are rather close in structure and thus in compatibility. The data described in the chapter on blends (see chapter 8), especially on their crystallization effects, show however only a minor compatibility. Only when the polymers are very

close in composition, such as HDPE and LLDPE, is a complete miscibility observed. A small-angle neutron scattering study[32] of blends
of polypropylene and ethylene/propylene copolymer showed all of
them being incompatible, even with a blend of polypropylene and a
copolymer containing only 8 %m/m ethylene. Another illustrative
example was reported on copolymers of ethylene and vinylchloride[33],
in which compatibility was only observed (as measured by DSC) when
both constituents were amorphous and did not differ by more than 15
%m/m in their composition. This result fits in with the theory of
polymer compatability[34,35]. Just as illustrative are the following
examples[36]: fully chlorinated polybutadiene does not molecularly
mix with polyvinylchloride (PVC), despite the equal overall composition, the only difference being the head-to-head versus the
head-to-tail structures; even chlorinated polyethylene with a
chlorine content equal to PVC does not mix with PVC. This clearly
shows that the chain structure is a very important determining
factor in compatibility.

Compatibility is not uniquely to be defined, except when also
naming the method to be employed in testing it. It has frequently
been observed that different techniques give different answers. A
case in point is the study of Fujioka et al[37]: with visible light
spectroscopy a much wider range of compatibility was noted than
with DSC, an even smaller range was observed using dielectric
relaxation spectroscopy.

Using the data as given in van Krevelen's book[38] the solubility parameters were calculated for polyethylene, polypropylene,
their copolymers and a few other polyolefins. Note that the numbers
are only valid for the amorphous polymers, in cases where crystallinity is present the numbers apply to the non-crystalline part
only. The results are given in Table 6.5. The ethylene/propylene
copolymers have, due to the equal densities of the parent polymers,
a linear dependence of the solubility parameter with E_c.

Of importance is the difference of the solubility parameters
of polypropylene and the copolymers. For a few common E_c's this is
given in Table 6.6. These values have to be compared to the condition of total miscibility. Taking the $(\chi_{AB})_{cr}$ approach as described
by Krause[34] gives as $|\delta_{PP} - \delta_{EP}|_{cr} < 0.06$ for polymerization grades
of 10,000. This corresponds to some 13 %m/m range in ethylene
content. Similar values are obtained using Paul and Barlow's
treatment[35].

TABLE 6.5

Calculated solubility parameters of amorphous polymers

Polymer	Density g/cc at		Solubility parameter $(cal/cc)^{0.5}$ at	
	25 °C	200 °C	25 °C	200 °C
polyethylene	0.854	0.755	8.02	7.09
polypropylene	0.851	0.744	7.37	6.64
EPC, E_c = 35			7.60	6.80
,, = 50			7.70	6.87
,, = 65			7.79	6.93
,, = 80			7.89	7.00
polyisobutylene	0.840	0.769	6.86	6.28
poly-1-butene	0.860	0.760	7.61	6.72

The first conclusion is: there is only limited miscibility in this polymer series, which fits in with the observation as even E_c 30 copolymer mixed with polypropylene gives a visible dispersion. The polymer coming close to polypropylene is polybutene, of which one could expect miscibility in the melt (at equal molecular weights). Polyisobutylene shows a $|\delta_{PP} - \delta_{PIB}|$ of about 0.4, i.e. equal to an E_c 60 copolymer. This does not explain the greater interaction found in crystallization experiments by Martuscelli[39].

TABLE 6.6

Calculated differences in solubility parameter

E_c , %m/m	30	50	65	80	100
$\delta_{EP} - \delta_{PP}$ (25 °C)	0.20	0.33	0.42	0.52	0.65
,, (200 °C)	0.14	0.22	0.29	0.36	0.45

The next question is: do the numbers of Table 6.5 relate in any way to the observed properties? Or put differently, is there an optimum $|\delta_{EP} - \delta_{PP}|$ for impact and what is its value? Not much is known, or investigated along these lines. Generally speaking there must be an optimum as with very small solubility parameter differences no dispersion will result and thus no toughening obtained, whilst at high values (e.g.> 1) the adhesion is expected to be poor, so despite the presence of a dispersion the impact improvement is poor. Stehling[40] gave in his paper on polypropylene-ethylene/propylene copolymer blends an example of blends of polyvinylchloride with acrylonitrile/butadiene copolymers, for which a $|\delta_{PVC}-\delta_{ANB}|$ of 0.7 gave much better impact than either the values 0.3 or 0.1. The range of $|\delta_{EP}-\delta_{PP}|$ considered in our case is 0.2 (at $E_c=30$) to 0.4 (at $E_c=65$), with much better impact at the latter condition. In view of the above the poorer impact at E_c 30 might not only be due to the smaller particle size but also to a too good compatibility. (A complicating factor might be the increased compatibility at processing temperatures, i.e. more mutual solubility could be present at higher temperatures - whilst demixing is not fast enough on cooling.) One would like to test the behaviour of systems with $|\delta_{EP}-\delta_{PP}|>0.4$, but unfortunately this is not readily possible as the higher E_c materials do crystallize to a larger extent and give a too-high modulus for adequate toughening. A different way to decrease compatibility is to increase the MW of the dispersed phase, and this was found effective by different groups. This lends some support to the hypothesis that the compatibility is too good instead of too poor in this polymer series. By the same token, a true blockcopolymer of polypropylene-ethylene/propylene would not drastically improve the impact performance. An experimental test of this would increase our understanding considerably.

(vi) <u>approach to impact grade development</u>

Finally, what are the best methods for characterizing a toughened polypropylene, especially with a view to developing improved impact grades? As shown above, not much primary knowledge exists on which to base a sound characterization nor development. Thus the first recourse in impact grade development is to a practical "trial and error" method. But, one could assist this very much in our opinion by

- measuring the morphology of the system in great detail, i.e. measuring both the size, size distribution and the shape of the dispersed phase. One will need to develop some sound analytical methods to do this, especially in the contrast enhancement necessary for distinguishing the different phases, the measurement itself can be automated,

- having a thorough knowledge of the catalyst system used in preparing the polymers, especially as regards its molecular weight behaviour and the composition of copolymers (and its spread), and

- preparing a polymer series in which preferably only one parameter is changed at a time, to assist the interpretation of the results.

In characterizing an unknown sample the method of simple fractionation and preforming NMR and GPC on the fractions is - in our opinion - one of the best around.

REFERENCES

1. M.Besombes, J-F.Menguel and G.Delmas, Composition analysis of an ethylene-propylene block copolymer by fractionation and turbidity measurements at the lower critical solubility gap, *J.Polym.Sc., Pol.Phys., 26 (1988) 1881-1896*

2. M.Kojima and J.H.Magill, Fracture surface morphology of some crystalline block copolymers - propylene/ethylene and tetramethyl-p-silphenylenesiloxane dimethylsiloxane, *J.Appl.Phys., 45 (1974) 4159-4166*

3. M.Levij and F.H.J.Maurer, Morphology and rheological properties of propylene-linear low density polyethylene blends, *Polym.Eng.Sc., 28 (1988) 670-678*

4. W.Wenig, Zur Phasenstruktur von schlagzähen Polypropylen, *Angew. Makrom. Chemie, 74 (1978) 147-164*

5. M.Kojima, Stress whitening in crystalline propylene-ethylene block copolymers, *J.Macromol.Sci., Phys., B19 (1981) 523-541*

6. P.Prentice, Morphology of ethylene-propylene copolymers, *Polymer, 23 (1982) 1189-1192*

7. F.Ramsteiner, G.Kanig, W.Heckmann and W.Gruber, Improved low temperature impact strength of polypropylene by modification with polyethylene, *Polymer, 24 (1983) 365-370*

8. T.Takahashi, H.Mizuno and E.L.Thomas, Morphology of solution-cast films of polypropylene homopolymer/ethylene-propylene random copolymer blends and polypropylene/ethylene-propylene sequential copolymer, *J. Macromol. Sci.,Phys., B22 (1983) 425-436*

9. J.Karger-Kocsis, L.Kiss and V.N.Kuleznev, Optical microscopic study on the phase separation of impact-modified PP blends and PP block copolymers, *Polymer Communications, 25 (1984) 122-126*

10. S.Danesi and R.S.Porter, Blends of isotactic polypropylene and ethylene-propylene rubbers; rheology, morphology and mechanics, *Polymer, 19 1978) 448-457*

11. M.Kakugo et al, Transmission electron microscopic observation of nascent polypropylene particles using a new staining method, *Macromolecules, 22 (1989) 547-551*

12. E.M.Barrall, H.Roger, S.Porter and J.F.Johnson, Characterization of block and random ethylene-propylene copolymers by differential thermal analysis, *J.Appl.Pol.Sc., 9 (1965) 3061-3069*

13. B.Ke, Differential thermal analysis of high polymers. V Ethylene copolymers, *J.Pol.Sc., 61 (1962) 47-59*

14. P-L.Yeh, A.W.Birley and D.A.Hemsley, The structure of propylene-ethylene sequential copolymers, *Polymer, 26 (1985) 1155-1161*

15. T.G.Heggs, Physical properties of block copolymers made with Ziegler-Natta catalysts, in Block copolymers, edited by D.C.Allport and W.H.James, Applied Science Publishers, London, 1973

16. J.Karger-Kocsis and L.Kiss, Dynamic mechanical properties and morphology of propylene block copolymers and polypropylene/elastomer blends, *Pol.Eng.Sc., 27 (1987) 254-262*

17. H.Münstedt, Similarities in the rheological behaviour of polymer melts with various fillers and melts of copolymers, Proceedings of the VIIth International congress on Rheology, Gothenborg, 1976

336

18. N.Minagawa and J.L.White, The influence of titanium dioxide on the rheological and extrusion properties of polymer melts, *J.Appl.Pol.Sc.*, *20* (1976) 501-523

19. T.Kataoka, T.Kitano, M.Sasahara and K.Nishijima, Viscosity of particle filled polymer melts, *Rheol.Acta, 17* (1978) 149-155

20. T.Kataoka, T.Kitano, Y.Oyanagi and M.Sasahara, Viscous properties of calcium carbonate filled polymer melts, *Rheol.Acta, 18* (1979) 635-639

21. C.B.Bucknall, Toughened plastics, Applied Science Publishers, London, 1977

22. R.P.Kambour, A review of crazing and fracture in thermoplastics, Macromolecular Reviews, volume 7, 1973, p. 1-154

23. M.Baer, Studies on heterogeneous polymeric systems. I. Influence of morphology on mechanical properties, *J.Appl.Pol.Sc.*, *16* (1972) 1109-1123

24. C.G.Bragaw, The theory of rubber toughening, paper 7 in "Multicomponent polymer systems", Advances in chemistry series, 99, ACS, Washington, 1971

25. J.Silberberg and C.D.Han, The effect of rubber particle size on the mechanical properties of high impact polystyrene, *J.Appl.Pol.Sc.*, *22* (1978) 599-609

26. C.B.Bucknall, Fracture phenomena in polymer blends, chapter 14 in volume 2 of Polymer blends, D.R.Paul and S.Newman editors, Academic Press, New York, 1978

27. J.Karger-Kocsis and V.N.Kuleznev, Dynamic mechanical and impact properties of polypropylene/EPDM blends, *Polymer 23* (1982) 699

28. C.B.Bucknall, P.Davies and I.K.Partridge, Rubber toughening of plastics, part 11. Effects of rubber particle size and structure on yield behaviour in HIPS, *J.Mater.Sc.*, *22* (1987) 1341-1346

29. B.Z.Jang, Rubber toughening in polypropylene, *J.Appl.Pol.Sc.*, *30* (1985) 2485-2504

30. M.E.Fowles, H.Keskkula and D.R.Paul, Synergistic toughening in rubber modified blends, *Polymer, 28* (1987) 1703-1711

31. C.B.Bucknall, The micromechanics of rubber toughening, *Makrom.Chemie, Macromol.Symp.*, *20/21* (1988) 425-439

32. D.J.Lohse, The melt compatibility of blends of polypropylene and ethylene-propylene copolymers, *Pol.Eng.Sc.*, *26* (1986) 1500-1509

33. T.N.Bowmer and A.E.Tonelli, Blends of homo- and copolymers of ethylene and vinylchloride: A compatibility study, *Macromolecules, 19* (1986) 498-500

34. S.Krause, Polymer-polymer compatibility, chapter 2 in volume 1 of Polymer Blends, D.R.Paul and S.Newman editors, Academic Press, New York, 1978

35. D.R.Paul and J.W.Barlow, A binary interaction model for miscibility of copolymers in blends, *Polymer, 25* (1984) 487-494

36. M.A.Masse, H.Ueda and K.E.Karasz, Miscibility in chlorinated polybutadiene/chlorinated polyethylene blends: effect of chain microstructure, *Macromolecules, 21* (1988) 3438-3442

37. K.Fujioka, N.Noethiger, C.L.Beatty, Y.Baba and A.Kagemoto, Compatibility of a random copolymer of varying composition with each homopolymer, paper 10 in "Polymer blends and composites in multiphase systems", Advances in chemistry series, 206, ACS, Washington, 1984

38. D.W.van Krevelen, Properties of polymers, Their estimation and correlation with chemical structure, Elsevier, Amsterdam, 1976

39. E.Martuscelli, C.Silvestre and G.Abate, Morphology, crystallization and melting behaviour of films of isotactic polypropylene blended with ethylene-propylene copolymer and polyisobutylene, *Polymer 23* (1982) 229-237

40. F.C.Stehling, T.Huff, C.S.Speed and G.Wissler, Structure and properties of rubber-modified polypropylene impact blends, *J.Appl.Pol.Sc.*, *26* (1981) 2693-2711

Chapter 7

CONTINUOUS VERSUS BATCH PREPARED TOUGHENED POLYPROPYLENES

7.1 INTRODUCTION

Continuous polymerization is the preferred manufacturing method for large scale production of commodity type products. Not surprisingly this is also practiced in the manufacture of toughened polypropylenes (TPP's) by most of the interested companies. However, the literature, especially the patent literature, describes at length the problems of obtaining the same good, desired properties of the product in a continuous process mode as compared to batchwise preparation.

An instructive example is given in a patent by the Sumitomo company[1], in which a TPP composition consisting of three different (co)polymers is described and compared as to their properties after a continuous and a batch mode preparation. The composition is made up of 7 %m/m copolymer with E_c of 50 %m/m and a LVN of 8 dl/g, a second copolymer - 17.3 %m/m - with E_c = 96 %m/m and a LVN of 14 dl/g, the balance being polypropylene. For almost identical overall composition the continuously-made product showed distinctly poorer properties: the brittle temperature was 7 °C higher, the haze 12 %, Izod impact at around 60 % of the batch-product value, and a large number of "fish eyes" (i.e. inhomogeneities) are present in thin films made from this product whilst none are present in the batch-made polymer.

This observation of poorer properties in a continuous preparation is common knowledge in patents and many inventions are claimed to overcome this problem (see below). Especially the Phillips company was early in recognizing this negative aspect and was prolific in finding many solutions.

In our own experience, also the above problem has occasionally been encountered, mostly in cases where a drastic change in the polymerization procedure was tested out, for instance in introducing new catalysts. Most of the problems arose when trying larger sized catalysts. In common with observations by others, when decreased impact was noted, inhomogeneities were also found in injection moulded articles. They can more clearly be measured and assessed when making thin films, in which the inhomogeneities appear as "fish eyes", a name in common use by polyethylene-film manufacturers.

The problem can hence be redefined as "the observation of lower impact in continuously-made TPP's, which coincides with the visual appearance of inhomogeneities in thin films". All patents are clear over the fact that in batch-made products no inhomogeneities are present.

An obvious way to get rid of inhomogeneities in a polymer, is to (re)compound it. In our own experience only in a few cases has this been succesful, and only then when a very high intensity mixing was applied by using for instance a twin-screw extruder.

In the following sections an explanation is sought for the above problem, starting with characterization data on the inhomogeneities themselves.

7.2 SOME CHARACTERIZATION DATA ON INHOMOGENEITIES

The film-inhomogeneities, being the only compositional difference between batch and continuously made products, were studied rather more closely.

The sizes observed vary considerably, values up to a few 100 microns being fairly common. This directly implies that film thickness will vary considerably at the spot of the "fish eye". These sizes are huge compared to the normal final dispersed phase size of a few microns at the highest.

On melting on a hot stage, these inhomogeneities are unchanged and no crystallinity develops (for compositions of intermediate E_c). The fact that no flow at all is observed illustrates its higher viscosity/higher molecular weight character. The total film is completely soluble in the normal solvents, which rules out the presence of crosslinked material.

In optical microscopy the "fish eyes" are easily made visible, their overall form resembles the catalyst form, and moreover some substructure is found to be present, see Fig. 7.1. The detailed structure is much better observable in TEM as is shown in Fig. 7.1 and 7.2 which gives a low magnification photograph through a "fish eye" particle bordering a normal part of the film and the high magnification photograph showing a view on the interface of the two

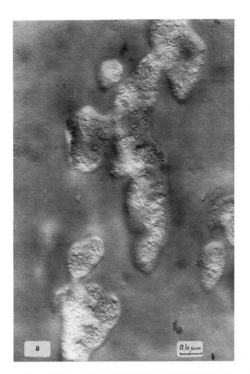

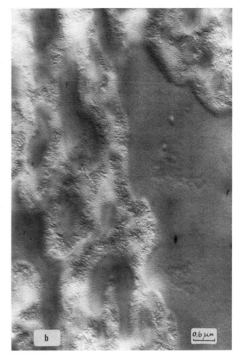

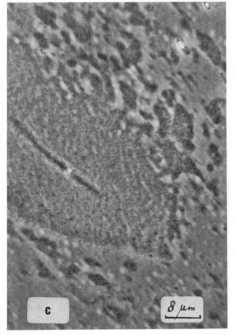

Fig. 7.1: *Transmission electron micrographs (bulk film (a) and rim "gel" (b)) and optical micrographs (anoptral contrast (c)) of an inhomogeneity in a toughened polypropylene film*

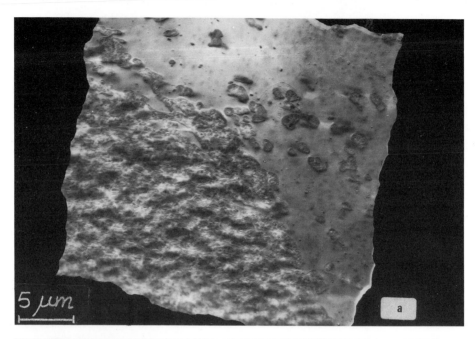

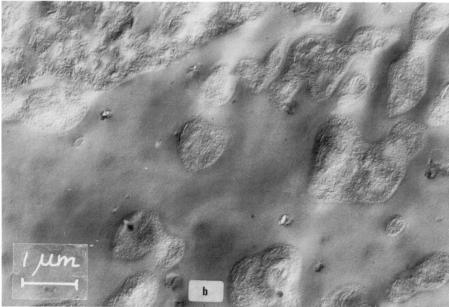

Fig. 7.2: Transmission electron micrographs of an inhomogeneity in a toughened polypropylene film

regions. At both sides of the interface a finely dispersed system
(with particle sizes of about 0.2 to 0.5 μm) is clearly discernible, the continuous phase however is different in both regions (the
featureless phase is the drawn polypropylene). Another very illustrative TEM example, using a reactive staining technique which
especially highlights phase boundaries, is given in Fig. 7.3.
Clearly the "fish eye" particle again contain a large number of
smaller particles which, given their contrast, has to be polypropylene dispersed in copolymer. Thus the reasonable explanation is
that the inhomogeneity consists of a polypropylene in copolymer
dispersion. As normally phase inversion in this type of polymers
occurs at around the 50/50 composition mark, the inhomogeneity
should contain at least some 50 %m/m of copolymer. This is of
course far more than the bulk of the product was intended to show.

To check on the composition in a "fish eye", Fourier transform
infrared spectroscopy was used to study these specific regions in
the film. Although this is far from easy, and the accuracy is low
due to presence of normal film material in the beam and inaccuracy
in the thickness measurement, nevertheless the outcome shows
unequivocally a much higher overall ethylene content translating to
a value for copolymer content in line with the microscopic observations, i.e. of the order of 50 %m/m.

It is hence concluded that the inhomogeneities are made up of
regions of very high copolymer content, and no evidence is found
for differences in composition (in terms of E_c and molecular
weight) from the bulk product.

7.3 EXPLANATION OF THE EFFECTS OF CONTINUOUS POLYMERIZATION, THE EFFECT OF PROCESS VARIABLES

The accepted explanation of the above mentioned observations
is the presence of a residence time distribution in continuously
operated reactors, as mentioned by Heggs[2]: considering individual
particles their ratio of the constituents will vary depending on
the real time spent in each of the reactors present in the process.
An F_c distribution will be present over the particles. For instance, in a two-reactor system, those particles which leave the
homopolymer reactor at a low yield (low residence time) make (on
average) high F_c particles, whilst vice versa those which stay very
long in the homopolymer reactor make a low F_c. Using the expanding
universe model for particle growth, there will be a F_c-level for

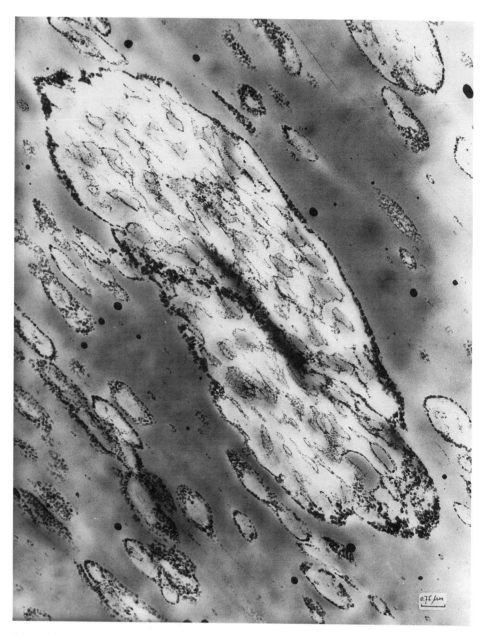

Fig. 7.3: Transmission electron micrograph of an inhomogeneity in a toughened
polypropylene film (selectively stained)

which phase inversion occurs, i.e. that level at which the ethyle-
ne/propylene copolymer becomes the continuous phase. This will be
around the 50/50 composition. For such a particle to become an
inhomogeneity a further condition has to be met: in the normally
applied compounding these particles have to stay more or less
intact. This will be the case when the copolymer molecular weight
is considerably higher than that of the homopolymer polypropylene
present in the composition (compare with the blend studies reported
in chapter 8). For lowering the impact one has to assume that the
presence of very large particles has a negative effect on the
impact. This is generally observed, for instance in high impact
polystyrene one observes a large loss in falling weight impact
strength when larger particles are present (mostly no distinct
effect on the Izod impact strength is noted in line with our own
observations).

For ideal continuous stirred tank reactors (CSTR's) the
fraction of particles which have a certain individual copolymer
content can easily be calculated. Taking particles with F_c's over
50 %m/m as an example their number of course increases with the
overall increase in rubber content. At for instance an average 15
%m/m rubber content, already 7.5 % of the particles show an F_c over
50 %m/m; for the case of 20% m/m rubber overall this figure increases rapidly to 12 %.

The above implies a density distribution over the particles as
homopolymer PP and the EPC have different densities. Indeed, in a
density gradient column a distribution is observed. However, it
appeared to be very difficult to arrive at accurate measurements,
due to an agglomeration tendency of the particles, voids in the
particles and difficult wettability for the medium chosen (alcohol-
containing). Nevertheless, in addition to the characterization data
given above, a floating fraction was taken and analyzed by IR for
its rubber content and after extraction by xylene for its molecular
weight distribution. The rubber content appeared to be distinctly
higher than for the overall product (factor of two), but the MWD of
the xylene-soluble part of the product was identical to the bulk
polymer. DSC-measurement on the floating fraction showed the
expected lower crystallinity, and no other effects. This is additional proof for the above explanation.

Using the given explanation one can derive a number of sensitivities to some process variables. For instance the catalyst size will have a large bearing on the size of the inhomogeneities as this, together with the individually experienced residence time, determines to a large extent the size of the resulting particle. Additionally it is to be expected that larger inhomogeneities have a more negative effect on the impact, which amplifies the effect of catalyst size. Only very, very small catalysts would give a guaranteed inhomogeneities-free product, namely ones which would give final particle sizes in the neighbourhood of the desired dispersed phase size, i.e. 1 to 3 μm. The effect of large catalyst size is demonstrated in an example in a Mitsui Petrochemical patent[3]. Using a 58 μm sized catalyst instead of the preferred 30 μm one gives a product with a much lower FWIS and Izod impact strength at identical stiffness (although both catalysts are in our opinion very large).

The second effect relates to the molecular weight of the ethylene/propylene copolymer generated. When inhomogeneities are present, higher molecular weight will make the situation worse as compounding in the processing machinery will have less effect. An example of this is given in a Hoechst patent[4], where a three component product is described, made in a continuous way in 2 and 2 and 1 reactor per step respectively. Applying the claimed conditions in each of the reactors leads to products with good impact and homogeneity. However, increasing the molecular weight in steps two and three by lowering the fraction of hydrogen, leads to lower impact and a much larger number of "fish eyes". Interestingly and in line with expectation, increasing the molecular weight in the second step is more harmful than in the third, and last step. Also in our own experience an increase in inhomogeneities with the molecular weight of the copolymer phase was observed, the increase was even exponential with the ratio of the LVN's of copolymer and matrix as variable.

7.4 DESCRIBED WAYS OF OVERCOMING THE PROBLEM

All the process steps have to be taken into account when discussing the problem of continuous polymerization in the manufacture of toughened polypropylenes. For instance the desired composition, both in F_c and in molecular weight, is very important but also the type of extruder present at the end of the train plays an important role as its compounding effect can remedy (some of) the inhomogeneities present.

In order to optimally use the compositional freedom inherently present in an in-situ preparation of TPP's, many companies have looked for means to overcome the problems with continuous polymerization. Apart from the relatively simple solution of carrying out each step in more reactors[4,5] and thereby narrowing the residence time distribution, other routes have also been mentioned. Amongst these are the specification of molecular weights for each individual step, whilst keeping them fairly close to that of the matrix[4], or the use of staging in reactors by building in compartments[6] or using plug flow type reactors as loop or pipe reactors (Phillips[7] and Solvay), or the use of a hydrocyclone in the outlet of the first homopolymerizer which only allows the larger particles to continue to the second reactor, whilst the finer material is recycled[8].

7.5 CONCLUDING REMARKS

The facts mentioned above on the problems encountered in producing toughened polypropylene compositions in a continuous process, hold in principle for every polymer product consisting of relatively widely different constituents and made continuously. Both monomer composition and molecular weight can lead to differences in the constituents. Thus, for instance, the manufacture of wide molecular weight distribution polyethylenes will encounter the same problems. Indeed from our own experience this is true, including the same dependencies on process variables; also in the literature on molecular weight distributions the residence time distribution is signalled as an important variable[9]. A survey of a number of patents for this application fails to reveal the remedies mentioned above, one still claims the use of only two ordinary reactors albeit with mostly well-described catalyst compositions. It may well be that the real compositions made have roughly 50 %m/m of the high molecular weight component, making the dispersion of the lower molecular weight part relatively easy.

REFERENCES
1. Belgian patent, Belg 877 781, to Sumitomo Chemical Company, published 16.11.79
2. T.G.Heggs, Block copolymers made with Ziegler catalysts, chapter 4 in Block copolymers, D.C.Allport and W.H.James editors, Applied Science Publishers, London, 1973
3. Japanese Patent Application Laid-Open No. 8094/52, application number 83 583/75, to Mitsui Petrochemical Industries Co., filing date 9.7.75

4.   German patent, OLS 28 49 114, to Hoechst, filed 11.11 78
5.   German patent, OLS 26 46 189, to Sumitomo, filed 13.10.76
6.   United States patent, US 34 54 675, to Phillips, filed 21.03.66
7.   Belgian patent, Belg 721 579, priority USA 29.9 1967
8.   German patent, OLS 26 16 699, to ICI, filed 15.04.76
9.   U.Zucchini and G.Cecchin, Control of molecular weight distribution in
     polyolefins synthesized with Ziegler-Natta catalytic systems, p 103 to 153,
     esp. p 146; in Advances in Polymer Science, Volume 51: Industrial
     developments, Springer, Berlin, 1983

Chapter 8

PREPARATION AND CHARACTERIZATION OF HIGH-IMPACT POLYPROPYLENE
BLENDS

8.1 INTRODUCTION

In the previous chapters 5 to 7 the subject was in-situ made
toughened polypropylene. However, high impact grades are made not
only in this way but also by blending a polypropylene homopolymer
with a rubber. The latter represents a commercial important class
of polypropylene products. These high-impact blends form the
subject of our present chapter.

Both the preparation of the blends and some of the characterization aspects will be discussed, although fairly concisely. Also
the effect of true block-copolymers as a blend component will be
reviewed.

There are numerous studies related to blends of polyolefins,
and a good general overview is given in the book of Paul and
Newman[1]. We will restrict ourselves to blends containing various
combinations and proportions of polypropylene, polyethylene and
ethylene/propylene copolymers. The topics to be dealt with are:
- the manufacture of blends (which is mainly related to the
effect of the melt viscosity of the ingredients),
- the morphology of the blends (in order to compare this with
in-situ products),
- the crystalline properties (especially as regards the
possible compatibility of the constituents), and finally
- the mechanical properties.
The topics that are of highest interest are morphology and compatibility.

8.2 PREPARATION OF BLENDS

Most polymers are incompatible, i.e. in a thermodynamic sense
they prefer their own environment to that of an intimate mixture.
Therefore, when wishing to prepare blends one has to put in energy
and rely on the slow rate of demixing in the highly viscous polymer
systems. The success of a preparative method can be judged from the
resulting dispersion, which should preferably show dispersed
particle sizes in the micron range.

In blending two or more different polymers, various mixing methods can be chosen. As the preparation of an intimate mixture is however rather difficult, mostly devices are used which are able to put a high amount of mechanical work into the mixture.

The main variable governing the mixing process is the _melt viscosity ratio_ of the ingredients[2-8]. Let us call the ratio of the viscosity of the main constituent to that of the minor one α. Invariably it is observed that blends made with a major component having an equal or slightly higher viscosity (i.e. $\alpha \geq 1$) give good dispersions, whilst the reverse with $\alpha < 1$ gives poorer results. The most worked out example is given by Karger-Kocsis et al[5,7] where in one PP matrix various EPDM-rubbers are blended, differing mainly in their melt viscosity. A clear relationship of the measured dispersed phase particle size with the melt viscosity ratio is found. The relation is very sensitive as an increase in the particle size is already noted when a factor two difference in the melt viscosity is present. This represents only a small difference in molecular weight, realizing the power 3.4 of the MW in the relation of melt viscosity with MW. The above studies have not been extended to situations in which α is much larger than one, theory[7] predicting, however, minimum particle size at $\alpha = 1$. The latter is contrary to the observations made in the previous chapters for in-situ made compositions (see e.g. section 6.2.2).

Of course, a number of other factors besides α influence the dispersion as well, but all are of considerable less importance. Among these factors are the shear rate in the mixer and the mixing time[6], whilst parameters such as the blending temperature have almost no effect.

The tendency to demix is shown by various studies in which particle growth by coalescence in the melt is observed when leaving a blend under static conditions at higher temperature[3,9].

Using a completely different mixing technique Blackadder et al[10] cast films at 130 °C from xylene solutions of the ingredients. The casting procedure takes up to 40 minutes, and is followed by drying and compression moulding. In this way they are able to make polypropylene/polyethylene blends which are composed of dispersions of about 20 μm. Crystallization of the co-dissolved polymers from xylene at room temperature followed by pressing gives larger sizes of around 100 μm. Still, those systems have almost not been mechanically worked, and the time given for demixing in especially the xylene solution method was rather high; so it is rather surprising that a relatively fine dispersion results. This probably results from the not too large difference in solubility parameter (see below and also chapter 6).

8.3 CHARACTERIZATION OF BLENDS

As the composition in blends can be totally inferred from the composition of the ingredients (treated in the previous chapters 2 and 4) we will deal only with the morphology and the crystalline properties of the blends as main characterization aspects. The dispersed phase morphology is discussed separately, whilst the crystalline, spherulite morphology is dealt with in the section on crystalline properties. At the end some mechanical properties will also be mentioned.

8.3.1. Morphology in blends

Generally speaking blends form dispersed systems, with the major component as matrix and the minor constituent as dispersed phase. Upon increasing the level of one of the components a situation of phase inversion is always observed, mostly in the composition range 40/60 to 60/40 of the two components. Around the phase inversion point the two constituents may both form a continuous phase[3]. The sizes that are observed in blends of for instance a 10/90 or 20/80 composition, range from a few tenths of a μm in good blends to over 10 μm in poor ones[4,6] (in which good and poor are obviously related to the impact performance shown by these blends).

One group[4] was able to measure the phase ratios of the dispersion from microscope studies, and found that all of the rubber added in the blend was found in the dispersed phase, within the errors inherent in the measurement. This indicates a high degree of incompatibility.

In a study[11] of heptane treatment of microtome sections of compression-molded sheets of melt-blended PP with EPC's, smooth holes are observed in SEM when the rubber has a 50% ethylene content. At lower ethylene contents the holes become smaller in size as well as in number, whilst at the lowest E_c studied (around

8 %m/m) almost none could be detected. This, according to the authors, points to compatibility effects at the lower E_c levels.

Three-component systems have also been widely studied, mostly using next to EPC's also PE as an extra polymer. Upon mixing PP + PE + EPC one does not observe a PP matrix in which the other two polymers are individually visible in the dispersion, but the PE and the copolymer have combined in one type of dispersed phase[4]. Even more interestingly, a sub-structure is formed in these compositions, the PE residing in the centre of the dispersed phase particle and the rubbery copolymer being found at the rim of the particle. This has been observed by many groups[4,6,12-14]. An example is given in Fig. 8.1 in which also the crystalline PE can

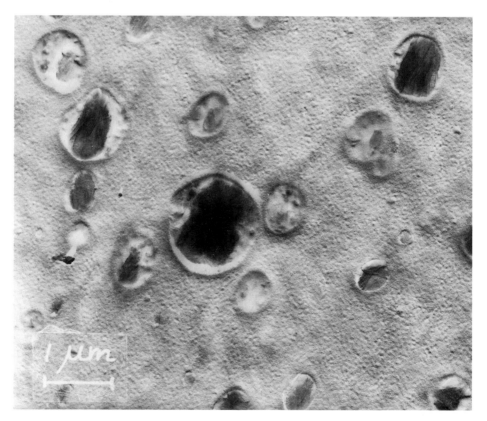

Fig. 8.1: *Transmission electron micrograph of blend of polypropylene with ethylene/propylene copolymer and polyethylene (section etched and Pt shadowed)*

be clearly observed in the centre of the dispersed phase. This substructure is also stable in molten systems under static conditions, although the overall particle size increases[6,15]. This ordering of the constituents, in principle, is in line with the solubility parameter difference, but factors of relative melt viscosity differences may play a role as well.

In a study on the effects of annealing of blends Ito et al[16] have interestingly observed that less rubber could be extracted after annealing at 140 °C. This was attributed to the formation of a relatively thick transition layer between polypropylene and the rubber upon annealing. Also observed was a considerable improvement in the impact level of the annealed blend.

Starkweather[17], on the basis of conformational analysis, concluded that certain ethylene/propylene copolymers should be able to co-crystallize with both polypropylene and polyethylene. Using an EPC containing 36 %mol ethylene, (made on a special catalyst, tetra neophyl zirconium on alumina, and showing on its own two melting points at about 121 and 147 °C), in a blend with PP and LDPE led to a much finer dispersion of 0.2 - 0.3 μm than in the absence of this copolymer (dispersion size was then 1-1.5 μm). Moreover, in blend compositions of 70/30/0 and 70/20/10 (PP, LDPE, copolymer) the impact at low temperature of the latter was much improved. Note that the EPC is very heterogeneous as in a random copolymer no crystallinity at all is expected, let alone two different crystalline phases.

8.3.2 Crystalline properties of blends

Many different blends have been studied for their crystallization and melting behaviour using for example the optical microscope for the study of the spherulite growth or the DSC-technique for the study of crystallization and melting temperatures and the corresponding enthalpies. We will bring some sort of order into this by discussing the different polymer combinations separately. In this compilation aspects of compatibility will form our main interest.

PP/(HD)PE: Wenig and Meyer[9] studied the crystallization of PP in the presence of liquid PE as a minor component, using either mill or solution blends. The results were independent of the blending method employed. The phases are already segregated in the melt, and the PE dispersion is not expelled from the growing PP spherulite. The radial growth rate of the PP spherulites is not influenced under these conditions, indicating - according to the authors - complete incompatibility.

Martuscelli et al[18] studied the total composition range. Under conditions that PE can not crystallize the crystallization rate of PP is delayed by PE, a minimum is observed just before the phase inversion. Similarly, the crystallization of PE is slowed down by PP especially at low fractions of PP present and high crystallization temperatures. The equilibrium melting temperature is not influenced by the blend composition however. Lamellar thickness in polypropylene is slightly lowered (10%) when crystallizing in the presence of molten polyethylene. The authors suggest a limited interaction between the components.

Blackadder et al[10], using solution-blended mixtures, show only one non-additive effect in 50/50 blends in that increased PP crystallization rates are noted in DSC measurements due to nucleation of PP on PE crystallites. A more extensive study was carried out by Bartzcak et al[19], who suggest that migration of heterogeneous nuclei from the PP phase to the PE phase occurs upon blending, leading to lowered crystallization rates for PP in the presence of molten PE, and giving at the same time larger PP spherulites of course. However, at crystallization temperatures low enough to allow also PE to crystallize, the PE crystallizes faster due to the presence of the nuclei and the PP nucleates in its turn on the PE. In practice, where low crystallization temperatures are always used, this would lead to smaller PP and PE spherulites.

In blends of PP with LLDPE (linear low density polyethylene) one observes considerable effects on the crystallinity at the corners of the composition diagram[20]. For instance at 95/5 PP/LLDPE the crystallinity is much lower than expected on the basis of additivity, and this effect is very dependent on the type of LLDPE used, the crystallinity being lowest with the highest molecular weight LLDPE. At the other end of the composition scale, 5/95 PP/LLDPE, the crystallinity is much higher than expected with the lower molecular weight LLDPE being the most affected.

PP/LDPE: Martuscelli et al[21] also studied melt blended PP and LDPE in the total range of compositions. The outcome of their study is very similar to that of their PP/HDPE study mentioned above. Again the spherulite radial growth rate is unaffected by molten LDPE, the rate of crystallization of PP is lowered and there is some lowering of the equilibrium melting temperature, the main effect being felt at low levels of LDPE and not much change thereafter.

PP/EPC or PP/EPDM : In injection moulded samples (i.e. made at low crystallization temperatures) of blends of PP and EPC's, Karger-Kocsis et al[22] observed a halving of the spherulite size, this size reduction being almost independent of the fraction of copolymer present. Crystallization studies by Martuscelli et al[23-25] showed the spherulite growth rate to be not affected by the presence of the amorphous rubber. The equilibrium melting temperature, however, in blends with various rubbers is sometimes lowered and sometimes increased. For the latter case it is speculated that the rubber dissolves some PP molecules, namely those which are least perfect, leaving the more perfect ones and leading to a higher equilibrium melting point. This increase in melting point has also been observed by Greco et al[11] in compression molded samples of blends of laboratory synthesized rubber with a commercial homopolymer. Even a 10 °C higher melting was found. The crystallinity as determined by DSC is considerably lower, being however fairly independent of the ethylene content of the rubber as the examples in table 8.1 show.

TABLE 8.1
Crystallinity (by DSC) of PP/rubber (80/20) blends (from ref 11)

E_c, %mm	0	8	14	19	26	78
Crystallinity, %	59	43	40	38	37	38

Using polyisobutylene as the elastomer in PP blends a slightly different behaviour was noted[26]: the elastomer is not ejected from the growing spherulite but is accomodated between the lamellae; this rubber lowers the growth rate of spherulites and lowers the melting point of PP as well. This strongly suggests a higher degree of compatibility compared with ethylene/propylene copolymers.

Generally identical conclusions were reached in a study by Jang et al[27] using injection moulded specimens made from blends. Spherulites became smaller and more irregular upon addition of rubber to PP, the melting and crystallization enthalpies changed just in proportionality to the volume of the modifier. Under these practical crystallization conditions it is found that the dispersed phase is occluded in the spherulite. The melting temperatures - as measured by DSC - are lowered and the crystallization temperatures are increased a few degrees by the addition of a rubber.

To investigate the interaction a few copolymers made on TiCl$_3$-catalysts (which varied in E$_c$, i.e. their ethylene content) were melt blended with PP (the molecular weight of both compounds being nearly equal to assist blending) and both the crystallization and melting temperatures as well as the enthalpies were measured by DSC. The results are given in Table 8.2.

TABLE 8.2

DSC-data of PP-copolymer blends with varying E$_c$ and fraction copolymer.

Sample	T_x °C	T_{m2} PE °C	T_{m2} PP °C	ΔH_x (PE+PP) cal/g	ΔH_{m2} PE cal/g	ΔH_{m2} PP cal/g
base polymer PP	119	-	162	21.7	-	21.7
+10%m/m E$_c$ 37.8	111	-	164	20.9	<0.1	20.6
+20%m/m ,,	112	122	161	19.7	0.1	17.7
+30%m/m ,,	111	122	162	17.5	0.1	16.1
+10%m/m E$_c$ 71.5	104/114	126	165	20.4	0.3	20.0
+20%m/m ,,	104/114	125	163	18.5	0.7	17.1
+30%m/m ,,	104/114	124	162	17.0	1.0	14.6
+10%m/m E$_c$ 100	113	136	161	24.6	1.7	23.0
+20%m/m ,,	116	136	163	26.6	7.5	18.0
+30%m/m ,,	115	138	162	28.3	10.6	16.3

The crystallization temperatures of the blends are all lower than that of the pure PP, although the values increase with E$_c$. Neither the crystallization nor the melting temperatures are consistently affected by the fraction of copolymer present, and the melting temperature of the blend is almost never larger than that of the homopolymer as such. The crystallization and melting enthalpies of the blends do of course decrease, except for PE as blend component. The decrease is linear with the fraction present, the values for the E$_c$ 71.5 case are however consistently lower than the corresponding values for the E_c 37.8 copolymer. This indicates a stronger effect of the former on polypropylene crystallization, which in view of possible compatibility is contrary to the expected trend as E_c 37.8 has a solubility parameter closer to PP (see chapter 6). The enthalpy data in the case of pure PE cannot be used in the same comparison as the individual contributions cannot be accurately assigned. The fact that the decrease in melting enthalpy is linear with the fraction of copolymer allows one to use DSC as a method to determine this fraction, provided that both the base polymer and the copolymer phase are of constant composition. This situation will be rarely met, but may hold for instance in a manufacturing plant.

PP/propylene-1-hexene copolymers: Blending polypropylene with propylene/1-hexene copolymers (containing 34 %mol of 1-hexene) gives rise to a system showing a melting point which is intermediate between the constituents[28,29]. Similar behaviour is observed at crystallization. The melting temperature and crystallization temperature change gradually between the two extreme values when varying the blend composition. Shih, who discovered this type of blend, concludes that there is compatibility. Even repeated melting in the DSC equipment of a powder blend leads eventually to single melting and crystallization peaks. Furthermore, annealing above the melting temperature of the copolymer does not lead to segregation. Other copolymers of propylene with higher α-olefins showed similar behaviour. The observed compatibility in this case is probably the result of co-crystallization of short isotactic polypropylene runs in the copolymer with homopolymer polypropylene. The melting enthalpy observed in blends is higher than that calculated on the basis of additivity.

PP/stereoblock-PP: Again on the basis of increased melting enthalpy in blends of PP and stereoblock PP it has been concluded[29] that co-crystallization occurred. At the same time the stereoblock polymer appears to be less extractable.

LDPE/EPDM: In blends of high E_c EPDM with LDPE, Starkweather[30] observed that the melting point of the mixture is lowered below that of the LDPE, and the unit cell lattice parameters change, especially in the a-direction. Using EPDM's with lower ethylene contents reduces these effects. Co-crystallization is the explanation of the observed phenomena, i.e. with these closely related polymers some, but not total, miscibility is present. Additional evidence comes from dynamic mechanical analysis where only one glass transition temperature is observed, intermediate between those of the constituents. In DSC however the individual glass transition temperatures stay constant over the total composition range, with no sign of an intermediate value.

In a related study, Lindsay and co-workers[31] obtained evidence of co-crystallization of LDPE and crystalline EPDM using DSC and SEM techniques. In this investigation also the crystallization temperature of LDPE is lowered, the decrease being higher with copolymers showing higher crystallinity. In DSC one can also notice the nucleation of the crystallization of the rubber by the LDPE-crystallization. This leads to copolymer crystallizing on LDPE-crystallites, forming a network with some similarity to the carbon-black-rubber network in diene rubbers. It has been shown that low MW linear PE does truly co-crystallize with a LLDPE, which contained up to 1.3 %mol co-monomer[32].

HDPE/LLDPE: This is another example of closely related polymers. A study by Edward[33] showed a HDPE of density 0.96 g/cc and a LLDPE of density 0.92 g/cc to be compatible, based on DSC and XRD evidence. The melting point is very dependent on the composition, for instance. Similar results, based on a wider range of diagnostic tools, were reported by Stein et al[34]. Even no broadening of peaks in DSC or SAXS for example is observed in these blends. Interestingly neither LDPE and HDPE nor LDPE and LLDPE[35] are compatible, they show their own melting regions in DSC irrespective of the blend composition. The above mentioned compatibility/co-crystallization is present in a limited composition range only and depends heavily on the preparative conditions (see chapter 10).

In conclusion, the interaction of the various polymers in the propylene and ethylene family is fairly slight. The main effect is the change in rate of crystallization and in spherulite size for certain combinations; for practical crystallization conditions smaller polypropylene spherulites are always observed to be formed at higher crystallization rates. As regards observations concerning a possible compatibility it is clear that generally there is none, except for polymer pairs which are very close in composition such as HDPE and LLDPE (see also chapter 10).

8.3.3 <u>Mechanical</u> <u>properties</u> <u>of</u> <u>blends</u>

Most studies which blend a rubber into polypropylene lead to the conclusion that the impact properties do improve with a concomitant decrease in the rigidity. The impact level increases with the amount of rubber applied[2]. Optimum toughening is certainly related to the degree of dispersion of the rubber, as many examples show[4,36]. Stehling illustrates this with a number of examples of blends with identical composition but blended with varying intensity, the blend with the smaller particles showing the best impact performance[4]. Jang[36] states that the optimum particle size for PP toughening is probably 0.4 μm. Views on the compatibility of the components in the blend vary strongly; whilst some consider EPC's as a "natural" toughening agent which has enough interfacial adhesion[2,4], others state that no adhesion is believed to be present[36] and suggest that the use of true block copolymers might substantially improve the performance (for a more detailed discussion on this matter, see chapter 6).

In a more practical approach, effects of the molecular weight of the rubber and its ethylene content on the impact have also been studied. Galli[37] in his review on PP developments since its discovery, mentions that higher molecular weight rubbers toughen PP to a higher extent than lower molecular weight ones. Lowering the E_c helps as well, but no values for these parameters were given. This improved performance can only be realized when a high intensity mixing is practised.

Another route to good dispersions is a masterbatch technique in which first the rubber is mill-blended with some PP, and subsequently PP is blended with the masterbatch. Using this technique, copolymers made with $TiCl_3$ were blended in PP with E_c and molecular weight as variables. This system worked quite well even for high MW products. Optimum impact, whether measured as FWIS or Izod, was obtained at E_c values in the range 50 to 60 %m/m ethylene as shown in Fig. 8.2. Generally speaking the better impact is found at higher molecular weights, in line with the data given above.

8.4 CHARACTERIZATION OF SURMISED BLOCK COPOLYMERS AND THEIR EFFECTS
IN BLENDS

In the chapter dealing with the preparation of block copolymers (see chapter 5.2.3), some characterization aspects which were used in validating claims that indeed block copolymers had been

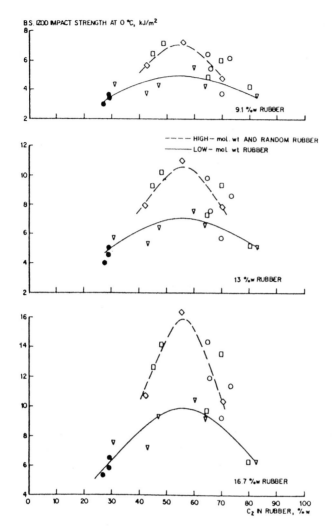

Fig. 8.2: *Izod impact strength (BS) at 0°C of PP/EPC blends (Melt Index about 1 dg/min) as function of E_c of the copolymer applied in the blending (different types of copolymer were used)*

prepared have already been mentioned. In all cases, at best limited evidence is produced, mostly based on fractionation. Prabhu et al[38] describe evidence from WAXS studies of films of surmised block copolymers before and after stretching. From the observed behaviour - high stretchability, relative low orientation - they conclude that the samples were true blockcopolymers or contain a significant fraction of these. We would like to see confirmation from morphological studies (as such, or in blends - see below), and moreover the molecular weight data should be determined.

It is a pity that for the probably one true polypropylene/rubber blockcopolymer ever made, by Hercules researchers[39], only mechanical data are reported in the literature. One would have liked to have seen the morphology of the blockcopolymer as such and its effect in blends of polypropylene and EPC.

Recently a new study appeared on the preparation of a block copolymer of propylene and styrene using the new supported catalysts[40]. In the following we will briefly discuss the characterization of these polymers and will also mention their effect in blends of polypropylene and polystyrene (PS). From the elemental analysis and IR on the blockcopolymer a composition of around 50 %m/m styrene is deduced, and the molecular weight was estimated from LVN to be in the range 300,000 to 600,000 g/mol. In DSC a broad endotherm was found around 161 °C and a sharp one at 227 °C, so that both constituent parts are therefore crystalline and highly isotactic. The NMR-data given showed no signs of copolymerization of styrene with propylene. All these data do not prove the presence or absence of a blockcopolymer. More importantly, the film morphology of this product was very fine and very different from the coarse dispersion of a blend of the two polymers. In TEM also, only very small features were visible, ranging in size from some hundreds of Angstroms to several microns. Finally in dynamic mechanical analysis the transitions were found to be less pronounced compared to the blend, the respective glass transitions were however discernible. Using the above data (which are not in contradiction with a block copolymer structure and certainly exclude just a mixture of homopolymers), coupled to the applied purification of the product (crystallization at 95 °C, at which temperature isotactic PS does not yet crystallize whilst PP would do so already at the higher temperature of 110 °C) leads to the tentative conclusion that this product is a block copolymer (or at least to a large extent). The above illustrates clearly the difficulties in ascertaining a specific structure when the polymerization behaviour is not completely understood. The most telling characterization methods appear to be the fractionation/solubility behaviour and the morphology.

Further proof of the block copolymer structure was obtained in using this product in blends of PP and PS. A true block copolymer would act as a surface-active agent and thus, with small fractions added, would lead to finer dispersions. This was indeed observed, furthermore the impact properties of the PP/PS-blends were much improved by the block copolymer addition. Similar effects have been observed before, for instance in the system polyethylene/polystyrene[41].

REFERENCES
1. Polymer blends, D.R.Paul and S.Newman editors, 2 volumes, Academic Press, New York, 1978
2. W.M.Speri and G.R.Patrick, Fiber reinforced rubber modified polypropylene, Pol.Eng.Sc., 15 (1975) 668-672
3. S.Danesi and R.S.Porter, Blends of isotactic polypropylene and ethylene-propylene rubbers; rheology, morphology and mechanics, Polymer, 19 (1978) 448-457
4. F.C.Stehling, T.Huff, C.S.Speed and G.Wissler, Structure and properties of rubber-modified polypropylene impact blends, J.Appl.Pol.Sc., 26 (1981) 2693-2711
5. J.Karger-Kocsis, A.Kallo and V.N.Kuleznev, Scanning electron microscopic investigations of particle size and particle size distribution of EPDM impact modifier in PP/EPDM blends, Acta Polymerica, 32 (1981) 578-581
6. A.Jevanoff, E.N.Kresge and L.L.Ban, Morphology of thermoplastic polyolefin blends by scanning electron microscopy, paper 23 in Polymer blends, Symposium 4 at Plasticon 81, 14 to 16 september 1981, University of Warwick, The Plastics and Rubber Institute, 1981.
7. J.Karger-Kocsis, A.Kallo and V.N.Kuleznev, Phase structure of impact-modified polypropylene blends, Polymer 25 (1984) 279-286
8. B.Z.Jang, D.R.Uhlmann and J.B.Vander Sande, The rubber particle size dependence of crazing in polypropylene, Pol.Eng.Sc., 25 (1985) 643-651
9. W.Wenig and K.Meyer, Investigation of the crystallization behaviour of polypropylene-polyethylene blends by optical microscopy, Colloid & Polymer Sc., 258 (1980) 1009-1014
10. D.A.Blackadder, M.J.Richardson and N.G.Savill, Characterization of blends of high density polyethylene with isotactic polypropylene, Makrom.Chemie, 182 (1981) 1271-1282
11. R.Greco et al, Polyolefin blends: 2. Effect of EPR composition on structure, morphology and mechanical properties of iPP/EPR alloys, Polymer, 28 (1987) 1929-1936
12. J.Karger-Kocsis, L.Kiss and V.N.Kuleznev, Optical microscopic study on the phase separation of impact-modified PP blends and PP block copolymers, Polymer Communications, 25 (1984) 122-126
13. L.D'Orazio et al, p 111-126 in Polymer Blends, Processing, morphology and properties, Volume 2, M.Kryszewski, A.Gateski and E.Martuscelli editors, Plenum Press, New York, 1984
14. L.D'Orazio, R.Greco, E.Martuscelli and G.Ragosta, Effect of the addition of EPM copolymers on the properties of high density polyethylene/isotactic polypropylene blends: II Morphology and mechanical properties of extruded samples, Pol.Eng.Sc., 23 (1983) 489-497
15. J.Kesari and R.Salovey, Mechanical behaviour of polyolefin blends, in Polymer blends and composites in multiphase systems, C.D.Han editor, American Chemical Society, Washington, 1984
16. J.Ito, K.Mitani and Y.Mizutani, Annealing of polypropylene-poly(ethylene-co-propylene) blends. I. Thermal and physical properties of blends, J.Appl.Pol.Sc., 29 (1984) 75-87
17. H.W.Starkweather, F.A.VanCatledge and R.N.MacDonald, Crystalline order in

copolymers of ethylene and propylene, *Macromolecules, 15 (1982) 1600-1604*

18. E.Martuscelli et al, Properties of polyethylene-polypropylene blends: crystallization behaviour, *Makrom. Chemie, 181 (1980) 957-967*

19. Z.Bartzcak, A.Galeski and M.Pracella, Spherulite nucleation in blends of isotactic polypropylene with high-density polyethylene, *Polymer, 27 (1986) 537-543*

20. M.M.Dumoulin, P.J.Carreau and L.A.Utracki, Rheological properties of linear low density polyethylene/polypropylene blends, part 2: solid state behaviour, *Polym.Eng.Sc., 27 (1987) 1627-1633*

21. E.Martuscelli, M.Pracella, G.Della Volpe and P.Greco, Morphology, crystallization, and thermal behaviour of isotactic polypropylene/low-density polyethylene blends, *Makrom. Chemie, 185 (1984) 1041-1061*

22. J.Karger-Kocsis et al, Morphological study on the effect of elastomeric impact modifiers in polypropylene systems, *Polymer, 20 (1979) 37-43*

23. E.Martuscelli, C.Silvestre and L.Bianchi, Morphology, crystallization and thermal behaviour of isotactic polypropylene/elastomer blends, paper 28 in Plasticon 81- Polymer blends, symposium, University of Warwick, The Plastics and Rubber Institute, 1981

24. E.Martuscelli, C.Silvestre and L.Bianchi, Morphology, crystallization and melting behaviour of thin films of isotactic polypropylene blended with several rubbers, p 57-71 in Polymer Blends, Processing, morphology and properties, Volume 2, Plenum Press, New York, 1984

25. E.Martuscelli, C.Silvestre and L.Bianchi, Properties of thin films of isotactic polypropylene blended with polyisobutylene and ethylene-propylene-diene terpolymer rubbers, *Polymer 24 (1983) 1458-1468*

26. E.Martuscelli, C.Silvestre and G.Abate, Morphology, crystallization and melting behaviour of films of isotactic polypropylene blended with ethylene-propylene copolymer and polyisobutylene, *Polymer 23 (1982) 229-237*

27. B.Z.Jang, D.R.Uhlman and J.B. vander Sande, Crystalline morphology of polypropylene and rubber-modified polypropylene, *J.Appl.Pol.Sc., 29 (1984) 4377-4393*

28. C-K.Shih, Melting and crystallization behaviour of an unusual compatible blend based on propylene copolymers, *Polym.Eng.Sc., 27 (1987) 458-462*

29. C-K.Shih and A.C.L.Su, Poly-α-olefins based thermoplastic elastomers, chapter 5 in "Thermoplastic Elastomers", N.R.Legge, G.Holden and H.E.Schroeder editors, Hanser Publishers, Munich, 1987

30. H.Starkweather, Co-crystallization and polymer miscibility, *J.Appl.Pol.Sc., 25 (1980) 139-147*

31. G.A.Lindsay, C.J.Singleton, C.J.Carman and R.W.Smith, Morphology of low density polyethylene/EPDM blends having tensile strength synergism, paper 19 in "Multiphase polymers", Advances in chemistry series, 176, ACS, Washington, 1979

32. M.T.Conde-Brana et al, Morphology of binary blends of linear and branched polyethylene: transmission electron microscopy, *Polymer, 30 (1989) 410-415*

33. G.H.Edward, Crystallinity of linear low density polyethylene and of blends with high density polyethylene, *British Polymer J., 18 (1986) 88-93*

34. S-R.Hu, T.Kyu and R.S.Stein, Characterization and properties of polyethylene blends. I. Linear low-density polyethylene with high-density polyethylene, *J.Pol.Sc., Pol.Phys., 25 (1987) 71-87*

35. T.Kyu, S-R.Hu and R.S.Stein, Characterization and properties of polyethylene blends II. Linear low-density with conventional low-density polyethylene, *J.Pol.Sc.,Pol.Phys., 25 (1987) 89-103*

36. B.Z.Jang, Rubber toughening in polypropylene, *J.Appl.Pol.Sc., 30 (1985) 2485-2504*

37. P.Galli, Polypropylene: a quarter of a century of increasingly successful development, p 63-92 in Structural order in polymers, F.Ciardelli and P.Giusti editors, Pergamon Press, Oxford, 1981

38. P.Prabhu, R.E.Fornes and R.D.Gilbert, X-ray diffraction studies of ABA propylene-ethylene block copolymers, *J.Appl.Pol.Sc., 25 (1980) 2589-2595*

39. G.A.Lock, Thermoplastic elastomers based on block copolymers of ethylene

and propylene, p59-74 in Advances in polyolefins, R.B.Seymour and T.Cheng editors, Plenum Press, New York, 1987

40. L.Del Giudice, R.E.Cohen, G.Attala and F.Bertinotti, Compatibilizing effect of a diblock copolymer of isotactic polystyrene and isotactic polypropylene in blends of the corresponding homopolymers, *J.Appl.Pol.Sc.*, *30* (1985) *4305-4318*

41. D.Heikens and W.Barentsen, Particle dimensions in polystyrene/polyethylene blends as a function of their melt viscosity and of the concentration of added graft copolymer, *Polymer, 18 (1977) 69-72*

Chapter 9

POLYETHYLENE, CATALYSTS AND (CO-)POLYMERIZATION

9.1 INTRODUCTION

This, and the following chapter, will deal with polyethylene in a similar fashion to the way polypropylene was handled in the previous chapters. Thus subjects such as the polymerization with an emphasis on the catalysts applied and the characterization of the polymers will be reviewed and discussed. The present chapter will deal with the catalysts used in the production of polyethylene and their polymerization behaviour as well as the preparation of copolymers. A short description of processes used in polyethylene manufacture will also be presented. The characterization proper is dealt with in the subsequent chapter.

Polyethylenes come in different types, the main categorizing feature is the density of the polymer, although other important polymer properties vary as well, either coupled with the density or completely independently. The ASTM organization distinguishes four classes of polyethylenes:

low density(high pressure process)(LDPE)	0.91 - 0.925	g/cc
low density(low pressure process)(LLDPE)	0.926- 0.940	,,
high density copolymer(LLDPE)	0.941- 0.959	,,
high density homopolymer(HDPE)	>0.960	,,

[The above abbreviations have the following meaning: LDPE low density polyethylene, LLDPE linear low density polyethylene and HDPE high density polyethylene.]
As will become more clear later, these density distinctions are associated with the process applied. The difference between the two low density classes is indeed one of process, in which the lowest density one is made in the oldest version of technology: the high pressure process based on a radical polymerization giving rise to a very branched polymer architecture; whereas the second originates from a low pressure copolymerization process giving a linear structure with only short sidechains stemming from the co-monomers applied (1-butene, 1-hexene and 1-octene).

The density range is wide, and these polymers find applications in very different types of end uses. A large fraction of the lower density materials is used in film. The intermediate density ones are used in injection- and blow-moulding in the packaging market, for which not only the rheological properties are tailored but also their corrosion properties, with emphasis on stress-corrosion. As the rheology is to a large extent governed by the molecular weight distribution, this property has attracted much attention, as will become clear when reviewing the catalysts applied in the polymerization. Other important end uses of these polymers are in wire and cable insulation and in pipes.

The annual production of polyethylenes is huge, the total world figure being 29 million tons per annum (for 1985). This is subdivided into 14 Mtpa LDPE, 5 Mtpa LLDPE and 10 Mtpa HDPE, with LLDPE being the fast grower. These quantities are roughly a factor 3.5 larger than the production figures of polypropylene.

The history of polyethylenes is already quite extensive, especially when the laboratory studies on diazomethane and related compounds are included. Referring to the historical reviews of Sailor and Hogan[1] and McMillan[2] it can be learnt that as early as 1897 polyethylene (and the linear form at that) was made by decomposing diazomethane. Of necessity with such an intermediate this remained a laboratory curiosity. In 1933, at the ICI laboratories, reactions at high pressure (up to 3000 bar) and high temperatures (up to 300 °C) were studied and with ethylene a polymer was isolated with a density of 0.91 and a melting point of 115 °C, reference 3 giving a more detailed history of this discovery and its further development. Its potential use was recognized as an ideal electrical insulation material, and first production was started in 1939. The branched structure of the high pressure polyethylene became apparent from infrared studies on these polymers. In 1938, in studies on the conversion of CO/H_2 mixtures on cobalt and ruthenium catalysts, a high density polyethylene was found, with a density of 0.98 and 132-134 °C as melting temperature. Laboratory studies were made of copolymers of ethylene and higher α-olefins by co-reacting diazomethane and homologues of this chemical, and the properties of the obtained copolymers were related to the composition (around 1950).

The early fifties saw a very wide range of discoveries in the field of olefin polymerization. Firstly, in 1951, Zletz of Standard Oil of Indiana, discovered a new catalyst for the polymerization of ethylene to linear polymer: molybdenum oxide supported on aluminium oxide. Then in 1953, Hogan and Banks of the Phillips company, patented the chromium oxide on silica catalyst, a catalyst which in some modified form is still used today in a large number of plants. In the same year, only a few months later, a third new catalyst was announced by Ziegler and co-workers: the reaction product of titanium tetrachloride and triethyl aluminium was found to be very active in ethylene polymerization. These new catalysts revolutionized the technology as their activity enabled one to perform the polymerization at relatively low ethylene pressures. Development to commercial production was rapid, especially for the latter two catalysts (Phillips, first production in 1957 - and Ziegler, in 1956). From then onwards not only the catalyst activity was the important study aspect, but the emphasis shifted also to tailoring the properties of the polymer. Two important dates in this respect were the introduction in 1958 of ethylene/1-butene copolymers by Phillips, which were followed 10 years later by ethylene/1-hexene copolymers of densities of 0.925 and 0.935 - the start of what is now called the "linear low density polyethylenes"(LLDPE).

As mentioned above, this chapter will contain sections on the polymerization (catalysts, polymerization, processes used) and the copolymerization of ethylene. Especially when discussing the catalysts described in the (mainly patent) literature we feel that an understanding of many of the aspects is lacking, or if it is known it might very well be kept secret; the situation in polyethylene is very competitive. For this reason our treatment of this matter will be more a compilation than a critical review.

Generally speaking ethylene is more readily converted to a polymer than the higher α-olefins. It is the smallest olefin, so the steric demand is minimal; moreover the aspect of stereo-regularity cannot play a role. Three types of polymerization mechanism have been found leading to polymer:
- radical polymerization
- true anionic polymerization, and
- "coordinated-anionic" polymerization.

The following discussion will focus on the latter type, but for the sake of completeness, and to assist the reader in the later characterization chapter, a short description of the other two mechanisms will be given.

9.2 RADICAL AND ANIONIC POLYMERIZATION

In the high pressure polymerization of ethylene a small amount of a radical initiator is admixed with the feed, the simplest one being oxygen. About the true initiation reaction not much can be said, but the subsequent reactions are better known. The important ones are given in Fig. 9.1. The propagation is the direct addition of an ethylene molecule to the growing macro-radical. Of the greatest importance are the reactions of this growing macro-radical with itself or with other molecules, as these reactions alter the

initiation
$$CH_2=CH_2 + O_2 \longrightarrow R\cdot + \;?$$

propagation
$$R\cdot + CH_2=CH_2 \longrightarrow R\,CH_2CH_2\cdot$$

chain transfer
  intramolecular
$$R\,CH_2CH_2CH_2CH_2CH_2CH_2\cdot \longrightarrow$$

$$R\,CH_2\overset{\cdot}{C}HCH_2CH_2CH_2CH_3$$

  intermolecular
$$R\cdot + \text{\small\raise2pt\hbox{$\sim\!\!\sim$}}CH_2\text{\small\raise2pt\hbox{$\sim\!\!\sim$}} \longrightarrow RH + \text{\small\raise2pt\hbox{$\sim\!\!\sim$}}\overset{\cdot}{C}H\text{\small\raise2pt\hbox{$\sim\!\!\sim$}}$$

termination
$$R\cdot + R'\cdot \longrightarrow R\,R'$$

$$R\cdot + RCH_2CH_2\cdot \longrightarrow RH + RCH=CH_2$$

Fig. 9.1: The main reactions in the radical polymerization of ethylene

macrostructure of the polymer considerably. Reacting with itself usually involves hydrogen atom abstraction from the fourth or fifth carbon from the radical end, via a five or six membered cyclic intermediate. Not surprisingly such a reaction is called "backbiting". The newly created radical adds further ethylene and in this way a short sidechain (propyl, butyl or amyl) is formed. When the growing radical abstracts a hydrogen from a different chain the result is a terminated chain and one which continues adding monomer but from some arbitrary site in the chain. This creates polymer molecules with long side-branches, a characteristic feature of low density polyethylene (LDPE). The overall molecular weight is governed by the ratio of propagation and termination processes. As the latter are fast in radical ethylene polymerization, one has to assist the propagation with high monomer concentrations and high temperatures in order to arrive at reasonable molecular weights. Furthermore, in order to lower molecular weights chain transfer agents can be added. Compounds with a more reactive hydrogen, such as isobutane serve this purpose. For reviews on the radical polymerization see for instance the somewhat old books mentioned in reference 4 and 5, for more modern reviews the reader is referred to references 6 to 8.

The anionic polymerization of ethylene is more of a laboratory curiosity. It has been found that butyl-lithium - better known for its ability to polymerize aromatic olefins, dienes and acrylates - was also able to polymerize ethylene especially in the presence of powerful chelating agents such as tetramethyl ethylene diamine. Rates are slow and to obtain reasonably high molecular weights an appreciable pressure is required. Mechanistically this polymerization is interesting and one still discusses the presence of the chelating agent in the propagating complex, see for instance reference 9.

9.3 COORDINATION CATALYSTS FOR ETHYLENE POLYMERIZATION

In the category of the coordinated type of polymerization mechanism there exists an abundant patent literature, which makes it difficult to see the wood for the trees especially since no logical classification method has yet been developed. One finds catalysts which only oligomerize, others make high molecular weight polymers; some are soluble, others supported or insoluble in reacting media; some are very simple, others are of baffling complexity. The range of transition metals claimed is also very wide. In practice catalysts on the basis of Ti, Zr, Hf (group 4b), V, Nb, Ta (group 5b), Cr, Mo, W (group 6b), Co and Ni (group 8) have all been reported. One could treat each of these metals separately in their catalytic performance. A more practical approach will be taken here, focusing on the more commercially important catalysts forming high molecular weight polyethylene (such as the titanium and the chromium containing ones) and trying to highlight the newer developments (especially the homogeneous zirconium species). Therefore the following categories will be

described:
- Ziegler type of catalysts (such as first generation $TiCl_3$),
- supported type of catalysts, treating chromium containing, titanium on magnesium compounds and supported organometallic compounds separately,
and
- homogeneous catalysts.

The review is not exhaustive, the literature taken into account was selected in order to give more weight to fundamental studies and less to patent claims or compilations. Where possible, attempts are made to generalize findings. In a number of cases patent examples on catalyst preparation are given to illustrate the complexity of this subject. As some of the catalysts used for polyethylene are closely related to the ones already described for polypropylene these are treated only summarily. This holds especially for the $TiCl_3$-type of first generation catalyst and the $MgCl_2$-supported titanium ones. More attention will be given to chromium-on-silica type catalysts as these are not dealt with elsewhere in this book.

Polymerization data on kinetics and molecular weight are treated separately, and in doing so inevitably some of the relationships between the catalyst preparation and composition and its polymerization behaviour will be mentioned there. Thus to get a more complete picture of the catalysts the reader should consider both sections. A number of review articles have recently appeared on the production of polyethylene, some of which contain catalyst data as well. See references 10 to 16.

The requirements that an ideal catalyst should meet are fairly easily defined:
It should combine high activity (making deashing superfluous and giving a non-corrosive polymer ex-reactor), with
- minimal low molecular weight polymer formation (wax),
- molecular weight control over a wide range without undue high hydrogen pressure requirements and thus loss of activity and/or productivity,
- ability to control molecular weight distribution irrespective of density and molecular weight,
- high efficiency in copolymerization, i.e. low reactivity ratios,
- a good ex-reactor powder morphology enabling high space-time

yields in the manufacturing process,

- and - last but not least - the catalyst should be simple to manufacture, but above all be consistent in its behaviour.

9.3.1 First generation TiCl$_3$

These catalysts are similar to the ones described in the first chapter for polypropylene, although it must be remembered that polyethylene was the first polymer to be made with such catalysts. The first catalysts comprised the reaction product between TiCl$_4$ and an aluminium alkyl, usually a trialkyl. The catalyst could be formed in-situ or be separately made. The latter procedure allows control over the morphology as well as the activity. In contrast to propylene polymerization where tacticity control dictates the use of diethylaluminium halide co-catalysts, in ethylene polymerization trialkyl aluminiums are often preferred due to the higher activity observed. There is some preference for the longer chain trialkyls, such as the tri-octyl. A dimethyl siloxane modified tri-alkyl cocatalyst is described by Machon et al[17].

A much more active catalyst was discovered by Hoechst[18] in 1966, this catalyst is based on TiCl$_2$(OiPr)$_2$/AlEt$_2$Cl + Al(iHex)$_3$. The use of these catalysts was restricted to high density polyethylene (HDPE) as these titanium catalysts gave a very linear polymer. Similar catalysts were later tested for their high-temperature performance[19].

9.3.2 Supported catalysts

In this section the two most important commercial catalysts, namely chromium supported on silica, and titanium on magnesium chloride or a magnesium chloride containing support are discussed. Furthermore, supported organometallic compounds will be mentioned in addition to the ones included with the chromium catalysts. Especially for the supported type of catalyst there is a proliferating patent literature, an - undoubtedly incomplete - patent review can be found in references 13 and 16. References 20 and 21 give the most recent overviews of chromium-supported-catalysts.

(i) chromium-containing catalysts

The chromium-containing Phillips catalyst, as discovered by Hogan and Banks[22], is made by impregnating a high-surface-area particulate silica with an aqueous solution of CrO$_3$, drying in air and activating in a dry atmosphere at 500 to 800 °C. In polymerizations this catalyst generates polymers with a density up to 0.96 g/cc. This catalyst has been widely studied with significant contributions from Phillips investigators, as would be expected. The main emphasis in these studies is the surface chemistry of the transition metal compounds on the supports. This leads to very interesting phenomena and theories as the following paragraphs will illustrate; the link with the true polymerization is however difficult to make. Especially so, as the common opinion is that only a few percent - at best - of the surface metal compounds are really active in the polymerization. Thus the study of the surface chemistry would have to be very precise in order that the catalytically important compounds present in very low fractions should also be distinguished. This is never the case, so most of the hypotheses regarding the polymerization mechanism are based on intelligent guesses and on analogies with similar systems.

The surface chemistry is strongly dependent on the composition of the supports used. Surface hydroxyl groups are the most important species, as they will react with most of the metal compounds added. On silica it is known that these hydroxyl groups (silanol groups) are partly present as pairs (geminal or vicinal, i.e. on the same or adjacent silicon atoms) and partly single. The concentration of silanol groups decreases upon heating (calcination) and the fraction of pairs decreases at the same time. Silanol groups on a surface are also mobile. Apart from the chemistry, the physical characteristics of a support are of the utmost importance. Firstly its surface area has to be very large as it is usually necessary to load the support with a large amount of metal compound (supports with 200 to 600 m^2/g are commonly used), secondly the pore volume and pore diameter are of importance as the catalyst ingredients, since the monomer and the polymer must be able to migrate to or from the metal-containing sites. For instance in one commercial catalyst, particles in the range of 30 to 150 μm with a pore volume of about 2.2 ml/g are applied (ref. 23). This physical structure of the support is stable only to a certain temperature limit at which sintering starts, lowering the surface area, the pore volume and diameter and consequently the accessability of the active centres. It is clear that the catalyst preparation conditions should exclude sintering conditions.

Recently studies[24,25] on widely different supports - AlPO$_4$ with varying Al/P ratio - were reported. This type of investigation will certainly enhance the understanding of this class of heterogeneous catalysts. Some observations from these studies will be mentioned later.

In 1970 Hogan[26] gave the first detailed report on the chemistry involved in the preparation of the Phillips catalyst. Upon mixing and drying, a reaction occurs between the surface silanol groups of the support and the chromium species, and this esterification is made complete by removing the water formed (see reaction (1) and (2) in Fig. 9.2). Two silanol groups are necessary for the formation of a new surface compound. On activation at high temperature the chromium survives as a hexavalent species, which is surprising as bulk CrO_3 loses half of its oxygen already at temperatures as low as 480 °C with the formation of the reduced species

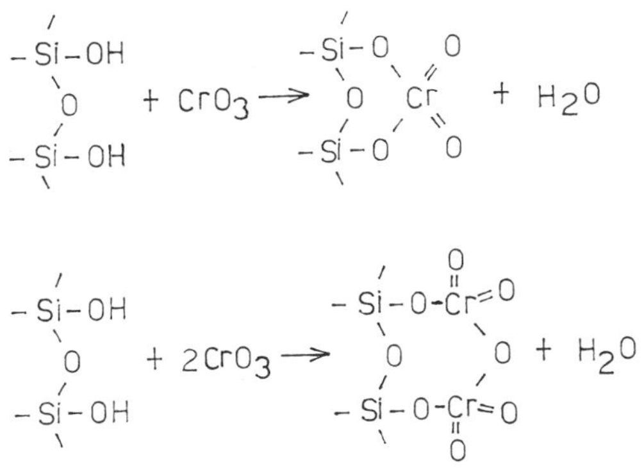

$$
\begin{array}{c}
-Si-OH \\
\quad\backslash O \\
-Si-OH
\end{array}
+ CrO_3 \longrightarrow
\begin{array}{c}
-Si-O \quad O \\
\quad\backslash O \diagup Cr \diagdown \\
-Si-O \quad O
\end{array}
+ H_2O
$$

$$
\begin{array}{c}
-Si-OH \\
\quad\backslash O \\
-Si-OH
\end{array}
+ 2CrO_3 \longrightarrow
\begin{array}{c}
\quad\quad\quad O \\
-Si-O-Cr \diagup\!\!=O \\
\quad\backslash O \quad\quad O \diagup \\
-Si-O-Cr\!\!=O \\
\quad\quad\quad O
\end{array}
+ H_2O
$$

Fig. 9.2: *Reactions of CrO₃ with surface-silanols on silica*

Cr_2O_3. Apparently the new surface compound is much more stable. Only at very high chromium loadings is some reduction (i.e. bulk behaviour) observed. The hexavalent chromium can however be reduced to a bivalent state by reaction with, for instance, hydrogen or carbon monoxide. The latter reactant is preferred as it is not only faster but the reaction product CO_2 is inert, whilst the reaction product of hydrogen reduction, water, is able to hydrolyse the Si-O-Cr bond. Colour changes accompany these reduction steps: starting with an orange hexavalent material, the CO reduction gives a blue or purple reduced material depending on the conditions chosen. When CO is removed by nitrogen stripping at higher tempera-

tures a blue colour results while at room temperature a strong
CO-complex, which can not be decomposed by simple evacuation,
endows the reduced chromium species with a purple colour. Such a
purple catalyst is able to polymerize ethylene (turning indigo blue
in the process) but the kinetics show an induction period. This is
due to the presence of oxygenated compounds formed by insertion of
the CO in the growing chain, which slow down the polymerization
rate, but are physically removed from the active site on further
polymerization.

Finch[27] studied the <u>reduction behaviour</u> of chromium on different supports and after various calcination treatments. His findings included the order of reaction in the reduction with hydrogen
which he reported to be between two and three, an observation for
which no explanation can be given. At calcination temperatures of
up to 500 °C about 4.2 g.equivalent H/g.atom Cr is consumed in the
reduction when a silica support is used and about 3.6 and 3.0 on
mixed alumina/silica and pure alumina support respectively; whilst
above 500 °C these ratios decline, the decrease being least for the
silica support. In a temperature-programmed reduction of the fresh
catalyst the rate maximum is highest for silica as a support
material. Similar reaction behaviour is shown when using CO as
reductant. At lower chromium loadings reduction is reported to be
less facile.

A few general conclusions can be drawn from the above. Firstly
the silica support stabilizes chromium(VI) better against thermal
reduction than either mixed $Al_2O_3.SiO_2$ or pure alumina. This
follows the same order as the catalytic activity, suggesting a
relationship with the stability of the surface chromium compound.
Of importance also is the last finding, which indicates that there
are various surface species differing in stability. This could lead
to catalysts with more than one type of active site, for example
showing wide molecular weight distributions, which indeed appears
to be the case (see below).

In a similar study[28] Schuyt and co-workers also characterized
chromium oxide on silica. In addition to a fair number of similar
observations to those mentioned above, they observed a re-oxida-
tion of Cr(II) to Cr(III) by the remaining silanol groups at higher
temperatures, a process in which the conversion is proportional to
the square root of the time elapsed. Secondly they also stress the
importance of the procedure of water removal from the catalyst and
the ability to keep the chromium in its highest valence state. For

instance carrying out the reduction at low hydrogen flow, implying a high partial pressure of water in the system, leads to Cr(III)-oxide clusters, which are very difficult to reduce. Despite the small size of the clusters, less than 5 nm, they are kinetically very stable.

A recurring question in this type of catalyst is related to the surface chromium-containing compound that is initially formed on the catalyst: is it chromate or bichromate? Many research groups have studied this question using a variety of techniques. From IR and UV-reflectance spectra Zecchina et al[29] conclude that dichromate is the main product. Phillips researchers come to the opposite conclusion and used many different approaches to prove this point. McDaniel[30] started this work by following the silanol concentration as a function of the chromium loading and the calcination temperature. The difference in surface silanol concentration (before and after chromium addition) at a given calcination temperature is directly proportional to the chromium loading diminishing strongly with the activation temperature employed. The difference is about 2 OH/Cr at 200 °C, falling to about 1 OH/Cr at 500 °C, being as low as 0.4 OH/Cr at 800 °C. Taking into account the fact that remaining free silanol groups are still dehydroxylated by the calcination treatment, he arrived at the conclusion that chromate is formed to a very large extent. This was substantiated by experiments using tert.butyl chromate under anhydrous conditions.

In another elegant study[31], McDaniel and Welch tried to answer the above question using a different approach. The route followed was to use the reaction of chromyl chloride with the support at 200 °C under anhydrous conditions. As shown in Fig. 9.3 the ratio of Cr and Cl on the support gives a clue as to which surface species is formed. The variables studied by them were the calcination temperature and the chromium loading. The results of their study can be summarised as follows:

- chromate is one of the products, its formation decreasing with increasing calcination temperatures, indeed no chromate is observed at all above 850 °C; this behaviour is similar to the reaction of $SiCl_4$, $TiCl_4$, BCl_3 with supports. Chromate formation increases at decreasing chromium loading, and more chromate is formed by a subsequent heating in nitrogen at 300 °C by further reaction with silanol groups (see Fig. 9.3). A maximum of 80% chromate is observed.

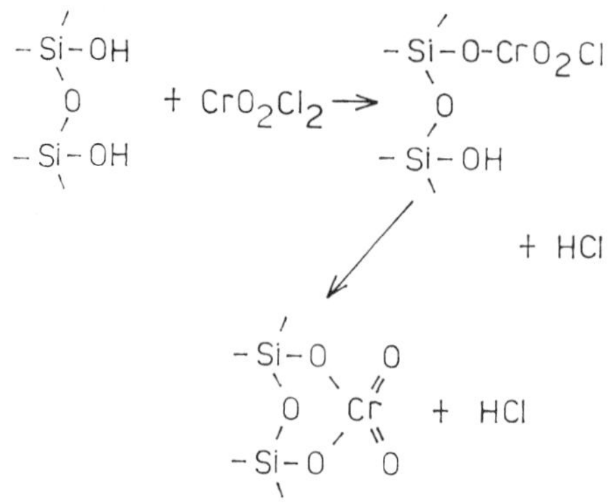

Fig. 9.3: *Reaction of chromyl chloride with surface-silanols on silica*

- by stripping the reacted support with HCl it appeared that more silanol is present than at the start (for compounds made at calcination temperatures exceeding 400 °C). This suggests that siloxane groups are also active in forming surface chromium compounds.

- a "chromate" catalyst made at 400 °C is identical in its polymerization behaviour (both activity _and_ molecular weight) to a normal catalyst made at the same calcination temperature, suggesting that in the latter case chromate species are also present and are necessary for active site formation.

- a "non-chromate" catalyst made at 800 °C is totally polymerization inactive, which means that the chloride-containing chromium compound (see figure 9.3) cannot be converted to an active site. Treating this catalyst with oxygen at 400 °C to burn away the chloride gives an active catalyst with properties again similar to a normal catalyst made at 800 °C.

The above gives ample evidence for the presence of surface chromate compounds and the polymerization activity of catalysts containing these groups. As stated above, this is not conclusive evidence that the active site is derived from chromate.

Other investigations[32] from the same group also relate to the same problem. Using normally prepared catalysts and treating them with HCl followed by measurement of the silanol concentration also leads to a diagnostic method for distinguishing chromate and bichromate surface species. The evidence points again to chromate as the main species on the catalyst, not only formed from the silanol but also from siloxanes. A further test method involved measuring the silanol concentration by reaction with methyl magnesium iodide after reduction of Cr(VI) to Cr(II) - this latter step being necessary since Cr(VI) also reacts. A consumption of silanol is found, although it decreases with increasing calcination temperature. This fits in with the above: chromate is formed both by reaction with silanols and siloxanes, the latter process being more important at the higher temperatures. Additional evidence was obtained by treating a HCl stripped catalyst with chromyl chloride. The resulting loading with chromium was almost identical to the parent catalyst, showing as well identical polymerization behaviour. This is to be expected when the parent catalyst indeed has a high fraction of chromate, as the stripping with HCl would leave the silanols in the right configuration for reaction with chromyl chloride to chromate, at the same loading.

Calcination always leads to loss of surface hydroxyls, it has been found however that this process is enhanced at temperatures above 600 °C when CO is present. Explanation of this fact is difficult, but the water gas shift reaction is thought to play a role. This reaction might well be an additional reason why reduction with CO gives more active catalysts than hydrogen reduction, as the remaining concentration of hydroxyls plays a contributory role in determining catalyst activity, as is discussed below.

The saturation coverage for hexavalent chromium appears to decrease with increasing calcination temperature[33]. Exceeding this concentration leads to the formation of chromium(III)-oxide as shown by XRD and UV reflectance spectroscopy. The saturation concentration is independent of the starting chromium compound used in the catalyst preparation over a wide variety of compounds, including ammonium chromate, chromium(III) nitrate or chromium acetate. Additionally removal of silanol with $(NH_4)_2SiF_6$ for example leads to lower saturation coverage values. The amount of chromium which can be attached to the support is evidently only dependent on the calcination temperature of the support, i.e. on its silanol concentration and reactive siloxane groups (provided the catalyst preparation is carried out under anhydrous conditions) and is independent of the activation temperature applied later. The activity of the catalyst is inversely related to the saturation coverage possible.

To explain all the above data it is assumed that the formation of an ester between silanol and CrO_3 stabilizes the SiOH for condensation but only to a certain extent. Upon heating, these esters will decompose to siloxane and CrO_3, the latter when it is mobile can either react with a silanol group or agglomerate with other CrO_3 molecules to form clusters.

A number of studies have investigated the <u>reduced</u> catalysts with physico-chemical methods, especially UV-reflectance spectroscopy. In all cases more than one reduced species was observed[29,34,35]. Reduction is mainly to the Cr(II) valence state although a few percent of Cr(III) is frequently observed. Pairs of reduced chromium species are sometimes also found. Testing the complexation behaviour of reduced catalysts Krauss[36] found 1:1 and 1:2 complexes with water, alcohols, ethers, trimethylsilanol and trimethylsilylether. It is noteworthy that a complex with silanol is indeed possible, in the case of surface silanols this is sometimes put forward as a prerequisite for the formation of active sites. Reduced catalysts also react with tungsten-carbene complexes, the resulting product shows[37] a reasonable metathesis activity but is not active in polymerization.

Weist et al[38] studied the changes in the void structure (by mercury porosimetry) in relation to the activation and reduction stages applied to the catalyst (three commercial Phillips catalysts were investigated). A shift to larger pore sizes (e.g. from 9 to 13.5 nm) and a decrease in surface area was observed upon calcination. Strangely enough the surface area increased again in the reduction with CO, most significantly in the catalyst with the lowest initial pore volume of about 1 ml/g. Such studies are important, as the nature of the pores must have a large bearing on the catalyst behaviour, but our knowledge of this area remains rather meagre.

Only low melt index polymers can be generated easily on the original Phillips catalyst. A greater range of molecular weights, and thus melt indices, is obtainable in modified catalysts, which are treated with a $Ti(OR)_4$ compound[39] before the activation step. A 10-fold increase in melt index can be reached at comparable reaction conditions. Additionally, changes occur in the molecular weight distribution. The optimum in the MW regulation is observed at an applied Ti/Cr ratio in the catalyst preparation of about 3. This is probably linked to the reaction stoichiometry of $Ti(OR)_4$ with CrO_3 (which is 3) and leads possibly to $[(RO)_3TiO]_3Cr$, i.e. a reduced trivalent chromium species. This is subsequently anchored to the support in the following steps leading to insertion of a Ti-O linkage between the Si and the Cr. When applying a two step activation procedure (first reduction with CO of an air-dried catalyst followed by partial oxidation with air) leads to even higher melt indices. Again an optimum at a Ti/Cr ratio of about 3 is observed. The valence of the chromium compound after CO-reduction is three. Another interesting observation is the effect of chromium loading on the carrier on the melt index: by halving the loading the melt index is significantly lowered, suggesting an involvement of two chromium species in the transfer.

In a related study[40] by the Phillips group an even greater enhancement of the melt index range is obtained by a sequential treatment of the support with the ingredients and applying the chromium oxide in the dry state. However they disclaimed the optimum in the effects at a Ti/Cr ratio of three, although this could be due to the very different method of catalyst preparation.

The following patent[41] example describes a catalyst made using both chromium and titanium compounds: A particulate silica material, showing a particle size of around 70 μm and a specific surface area of 300 m^2/g, is mixed with a solution of CrO_3 in distilled water. After filtration the solid is dried in nitrogen at 200 °C for 4 hours. This dry material is suspended in dry isopentane and the desired amount of tetraisopropyl titanate is added, mixed well, and subsequently the isopentane is distilled off. The last ingredient, ammonium hexafluorosilicate is added in the dry state, well mixed, whereafter the solid is heated at 50 °C under nitrogen for 1 hour, followed by 1 hour at 150 °C (to remove the last traces of the isopentane and decompose the titanate before air is allowed to react with the catalyst). The final activation is carried out in air, for 2 hours at 300 °C and next 8 hours at 750 °C. The compositions applied are as %m/m on support: 0.8 CrO_3, 35 $Ti(OiPr)_4$, 1.5 $(NH_4)_2SiF_6$; this translates to the following elemental composition of the catalyst: 0.4 %m/m Cr, 5.6 %m/m Ti and 0.7 %m/m F. It is stated in the patent that the fluorine content of the catalyst is very important; at increasing F-levels

-the fraction of low molecular weight material in copolymers
is lowered (n-hexane extractables)
-the copolymerization parameters are lowered
-the molecular weight is increased.

Two, more recent, chromium containing catalysts were developed
using organo-chromium compounds as the starting material. The first
one is bis(triphenylsilyl)chromate, which as such has a low activity in ethylene polymerization, requiring high pressures to form
polymer[42]. This despite the fact that its structure is very similar
to the ones formed on a silica surface (see above).

Adding this compound to a support increases the activity
considerably, and this is further enhanced by adding co-catalysts
of the type $AlR_2(OR)$. In this case activities of the order of 6 kg
PE/g Cr.h.bar are observed at 90 °C. The pore size of the support
must be over 6.7 nm to allow a good penetration and reaction of the
ingredients. The valence state of the reduced chromium species is
not known. It is suggested that the reaction of the silanol group
with the organo-chromium compound plays a decisive role in the
enhancement of the activity. The molecular weight is, as in all
cases, controlled by the polymerization temperature and the application of hydrogen. However in this case a full range of melt
indices up to 100 can be readily made. The molecular weight distribution shows a "hump" on the high molecular weight side, making it
a broad distribution, indicating more kinds of active sites than
just one. The behaviour of this catalyst is different from the
"inorganic" one mentioned above, which indicates that at least one
ligand on the chromium is differing. This ligand is presumably the
silyl group.

The second organo-chromium based catalyst[43] is derived from
chromocene (i.e. di-cyclopentadienyl-chromium). Adding chromocene
to a high surface area silica, leads to evolution of cyclopentadi-
ene(Cp) even at ambient temperature. The reaction is not complete
however and some Cp-groups remain bound to the catalyst. The
cyclopentadiene-evolving reaction is truly due to the silanol
groups, as their removal by pretreating the support with an aluminium alkyl gives an unreactive support and catalyst. The valence
state of the chromium on the catalyst is two. This catalyst is
highly active, and readily forms high molecular weight polymers,
the melt index of which is easily controlled by hydrogen (even

better than with the silyl chromium catalyst). Many properties of this catalyst are different from other supported chromium catalysts, leading to the conclusion that one cyclopentadienyl group is still present on this catalyst. Evidence for this hypothesis is to be found in studies[44] in which the chromocene catalyst was heat-treated at different temperatures. The activity decreased by an order of magnitude when heating up to 350 °C, some activity being regained when heating above this temperature. The molecular weight increased considerably in the same temperature range. Using chromatographic methods it was shown that cyclopentadiene was one of the main products of pyrolysis, lending support to the hypothesis that this group was originally present on the active catalyst. This was further substantiated in a later study[45] in which various other chromium sandwich compounds were tested for their catalytic activity. In addition to bis-cyclopentadienyl also indenyl, fluorenyl and 9-methylfluorenyl chromium compounds were used to make catalysts and compared with respect to activity and chain transfer behaviour with hydrogen and their copolymerization parameters. A decreasing reactivity towards hydrogen was noted with decreasing electron density on the chromium ion, i.e. of the ligands tested cyclopentadienyl is the best for melt index control and 9-methylfluorenyl the worst.

More data on the relation of catalyst composition and polymerization performance are given in the section dealing with kinetics and molecular weight (see 9.4).

Patent examples of the above two catalysts are now given to illustrate in some detail how they are prepared:
- for the silylchromate catalyst[46]: A silica, showing a specific surface area of 370 to 400 m^2/g and average pore diameter of 25 to 27 nm, is mixed with an aqueous solution of aluminium nitrate, filtered and dried at 200 °C. This product is suspended in pentane and mixed with tetraisopropyl titanate, subsequently the pentane is evaporated and the residue heated in oxygen at 810 °C for 17 hours. This modified support contains 0.5 % m/m aluminium and 7.5 % m/m titanium (both calculated as oxides). The catalyst is simply made by suspending the modified support in hexane and adding 3 %m/m of bis-triphenylsilyl chromate, stirring for one hour under nitrogen. This suspension is directly used in the polymerization.
- for the chromocene catalyst: the simplest method is described in

reference 47, in which the catalyst is made in-situ by mixing a suspension of silica in a hydrocarbon solvent with a solution of chromocene yielding a slurry which is used as such in polymerization. A more sophisticated version[48] was reported more recently: A microspherical silica of intermediate density, a specific surface of about 300 m^2/g and an average pore diameter of about 20 nm is dried for 8 hours at 200 °C. This material is mixed in dry isopentane with bis-(cyclopentadienyl)-chromium (1% on support) and tetraisopropyl titanate (35% on support). After evaporation of the isopentane, ammonium hexafluorosilicate (0.3 % on support) is added and mixed. The activation is done by first heating under nitrogen for two hours at 150 °C followed by two hours at 300 °C. Subsequently the nitrogen is replaced by air and after intial heating for 2 hours at 300°C the final step requires 8 hours at 800 °C. Using higher specific surface area supports with larger pore volumes leads to higher melt indexes at identical polymerization conditions.

(ii) titanium supported on magnesium chloride

We turn now to the second important class of commercial supported catalysts, the titanium on magnesium chloride containing supports; recent reviews are to be found in references 13, 15 and 16. The simple ball-milling of $TiCl_4$ and $MgCl_2$ already leads to a very active catalyst for ethylene polymerization. Examples are given by Hsieh[49] and Kashiwa[50]. Usually catalyst recipes are much more elaborate, the major objective being not activity but other, end-use related properties. For the magnesium compound a whole range comprising MgR_2 and RMgX (Grignards), MgX_2, MgO, $Mg(OH)_2$, Mg(OH)X, $Mg(OR)_2$ can be used in conjunction with $TiCl_4$ and possibly other ingredients. In all cases surface chlorination will lead to a layer of $MgCl_2$ on which the titanium species are bound (see especially the discussion in chapter 1). For instance in the reaction of $Mg(OR)_2$ with $TiCl_4$ the alkoxide is completely transformed into a new crystalline species[18]. The same holds for the catalyst made from MgO and $TiCl_4$ for example, for which surface chlorination has also been proved[51]. One can also simply react $MgCl_2$ with a donor, such as an alcohol, and subsequently treat the resulting compound with $TiCl_4$, activities of the order of 36 kg PE/g Ti.bar.h are reported[49,50].

The higher activity is due to more active sites[11,12] and a larger propagation rate constant, although on the latter no unanimity exists (see below). The molecular weight distribution is generally narrower on this type of catalyst compared to simple $TiCl_3$-type catalysts.

Sometimes the combination of ingredients leads to extra effects besides activity; for instance using $Mg(OEt)_2$ + $Ti(OiPr)_4$ + $TiCl_4$ as in "Solvay" and "Hoechst" catalysts gives a compound which, without adding a co-monomer, generates a polymer of reduced density as a function of the Mg/Ti ratio[52]. Furthermore, Ti on Mg catalysts copolymerize much more effectively with 1-hexene than the CrO_3/SiO_2 catalyst, giving densities as low as 0.92 at lower hexene levels in the feed[53] (see also section 9.5).

In a more fundamental study[54] into the relationships of catalytic activity and composition of $TiCl_x$-complexes, well defined crystalline mixed complexes were made of $TiCl_4$ combined with group II chlorides and assisted by $POCl_3$ as an extra ligand. Structures such as $(TiCl_6)MgL_6$ and $(Ti_2Cl_{10})MgL_6$ were obtained and found - in combination with co-catalysts - to be active in ethylene polymerization although at one order of magnitude below commercial catalysts. However in changing the group II element in the complexes, magnesium very clearly gave the most active catalyst.

The newly developed polypropylene supported catalysts have also been applied in ethylene polymerization[55]. A major advantage in this case is the preparation of saleable powders directly from the polymerization, i.e. not using a melt-extrusion step for homogeneization and nibbing. It must be noted however that the first spherical morphology $MgCl_2$ supported catalysts were developed by Montedison for ethylene polymerization and involved spray-quenching of $MgCl_2 \cdot 3EtOH$ and treating with $TiCl_4$.

Patent examples of a few catalysts of this class are given below. An early Hoechst patent[52] describes the catalyst preparation from $Mg(OR)_2$, $TiCl_4$ and $Ti(OR)_4$. The procedure is fairly simple, starting with adding a $TiCl_4$ solution to a suspension of $Mg(OEt)_2$ and $Ti(OiPr)_4$ in a high boiling hydrocarbon diluent at 50 °C. Some time is allowed to complete the reaction, then the solid is isolated by decantation and purified by washing at 60 °C. The titanium/magnesium ratio in the final catalyst is controlled by the amounts of the ingredients applied in the first stage. The patent gives examples of various Ti/Mg ratios, ranging from 6 to 0.2.

Within this range the density of the polymer decreases from 0.965 to 0.944, whilst the activity increases 2.5-fold. The activity is not very high however, being in the order of 0.6 to 1.5 kg PE/gTi.bar.h. In a more recent patent of the Nippon Oil Company[56], catalysts useful for gas-phase polymerization are claimed. The patented improvement lies in the pre-contacting of the catalyst with α-olefins, such as propylene or 1-butene in the presence of a diluent. The catalyst is made by ball-milling at room temperature for 16 hours a mixture of anhydrous $MgCl_2$, dichloroethane and $TiCl_3.0.33AlCl_3$. Using tri-isobutyl aluminium as co-catalyst and the pre-reaction described above, an activity of 10 kg PE/g Ti.h.bar is claimed. A different version[57] is the ball-milling of $MgCl_2$ with electron-donor complexes of $TiCl_4$, as electron donor tetrahydrofuran and the well-known ethylanisate are described. Catalyst productivity is very high (2000 kg PE/g Ti at about 1 hour polymerization at 20 bar), and its copolymerization behaviour is also attractive. The catalyst can also be mixed with a normal support, such as silica.

(iii) supported organometallic compounds

The third class of supported catalysts reviewed here are the supported organometallic compounds (excepting those of chromium as they were treated above). Reviews can be found in references 12,13 and 58. It is useful to summarize here the behaviour of the relevant organometallics in a homogeneous system as well. The π-compounds of transition metals such as titanocene and zirconocene are inactive in ethylene polymerization, the allyl-compounds however are active without the application of a co-catalyst, albeit at a low level. The activity can be influenced by the ligand attached to the transition metal centre, the highest activity being obtained with the most electron donating groups. Thus 2-methylallyl is more active than the allyl, and 2-phenyl-allyl is far less active, see also Table 9.1. The σ-compounds such as tetra-benzyl zirconium also show a low polymerization activity. These activities increase by some orders of magnitude when the σ-compounds or the allyl-compounds are added to a high surface area support, as the data in the table illustrate. The reaction with the support is analogous to the ones described above for the chromium catalysts, two silanols replace two hydrocarbyl groups on the metal. In halide substituted metal compounds, the hydrocarbyl groups are removed selectively in the reaction with the support. The catalysts are used as such, i.e.

TABLE 9.1

Activities of organometallic catalysts in ethylene polymerization (ref 58)

Compound	Support	Activity, g/g.h.bar
$Zr(allyl)_4$	-	2.0
$Zr(2\text{-methallyl})_4$	-	4.0
$Cr(allyl)_3$	-	0.3
$Cr(2\text{-methallyl})_3$	-	2.1
$Cr(2\text{-phenylallyl})_3$	-	0.04
$Zr(benzyl)_4$	-	0.8
ditto	alumina	>390
$Cr(2\text{-methallyl})_3$	silica	45
$Zr(allyl)_4$	silica	38
$Zr(allyl)_3Br$	silica	210
$Zr(allyl)_4$	alumina	>600

they require no co-catalyst; the activities are generally of the same order as first generation Ziegler catalysts. One of the main reasons that the supported version is much more active is the prevention of the dimerization of zirconium hydrides, which - when mobile - form a strong, polymerization inactive complex. The hydride when single can add a monomer under formation of an active alkylated species[59]. Except for the chromium compounds none of these catalysts reached a commercial application.

The pentadienyl derivatives of Ti, V, and Cr supported on alumina or other carriers show also a high activity towards ethylene polymerization[60]. They behave however very differently compared with their cyclopentadienyl cousins. Vanadium is the least reactive of the three, the molecular weight is lowest with chromium as chromium-alkyls give more readily β-hydrogen elimination.

The catalyst preparation as described in patents is relatively simple once the transition-metal-organic compound has been synthesized. In reference 61 the use of tetra neophyl zirconium (neophyl is $-CH_2-C(CH_3)_2-C_6H_5$, a substituent carrying no β-hydrogens). The catalyst is simply made by stirring the zirconium compound in a diluent with the support at room temperature. After some 20 hours the zirconium is already reduced to a large extent to a lower valence. The activity is very sensitive to the type of support used, here also alumina supports gives the most active catalysts. A further step in the catalyst preparation is sometimes the pre-reduction of the reaction product of organometallic and support with hydrogen as for instance described in reference 62 for tetrabenzyl or tetra neophyl zirconium. Activities are not too different from the directly obtained catalysts however. The range of supports studied was extended with titania, which gives an even lower activity than silica. More details about this catalyst are given in reference 63.

9.3.3 homogeneous catalysts

Ever since the discovery of the polymerization activity of the homogeneous combination Cp_2TiCl_2/Al-alkyl (see e.g. references 64 and 65), a lot of effort has been put into this type of catalyst systems. A large part of the interest stems from the hope that these homogeneous systems could form a model system for the important heterogeneous catalysts. Therefore mechanistic and especially kinetic studies are abundant, but attempts have also been made to devise commercially attractive combinations. Reviews on the titanium and zirconium catalysts are to be found in references 66 to 68. A few homogeneous catalysts have been mentioned earlier in this book, such as the vanadium catalysts for syndiotactic polypropylene polymerization (chapter 2) or for the copolymerization of ethylene and propylene to rubbers (chapter 3). Furthermore some soluble catalysts were mentioned in the previous sections; for instance detailed studies of tetraallyl zirconium led Yermakov[12] to the conclusion that this compound is not a good model for the active site.

Even with the seemingly simple system consisting of Cp_2TiCl_2 and an aluminium alkyl the kinetics and the chemistry are very complex, and different views on the active complex still abound[10,68,69]. The polymerization activity is rather low, and in non-oxidizing solvents the decay of the activity is rapid. Using methylene chloride as a solvent gives better activities, due to the reoxidation of Ti(III) to Ti(IV). This latter valence is regarded as the valence of titanium in the active complex. Surprisingly the activity is increased by the addition of water (!) to this catalyst system[70,71], and the optimum activity is observed at $Al:H_2O$ of about 2:1. This effect of water appears to be independent of the specific catalyst combination chosen as it is observed with $Cp_2TiEtCl/AlEtCl_2$ (ref 70), in $Cp_2TiCl_2/AlMe_2Cl$ (ref 71) and in $Cp_2TiMe_2/AlMe_3$ (ref 67 and 72). The effect of water is very large especially in the halide-free catalyst combinations (see below). A whole range of other polar and reactive compounds have been studied for their effects[71], but no rate increase was noticed at all, indicating that water or the reaction product of water and the catalyst/co-catalyst is a very specific ingredient. The reaction product of trimethyl aluminium and water is an aluminoxane $[Al(CH_3)O]_n$ with a degree of oligomerization (n) depending on the reaction conditions. Pre-reacting the alkyl with water gives the best results[73], especially when a "slow release agent" for water such as $CuSO_4 \cdot 5H_2O$ or $Al_2(SO_4)_3 \cdot 6H_2O$ (ref 74) is used. Using this pre-reacted co-catalyst leads to activities of 2 kg PE/g Ti.h.bar at 25 °C when using $Cp_2Ti(CH_3)_2$ as the transition metal compound[67]. Note that this catalyst is halide free and still very active, contrary to all previous beliefs that for an active catalyst at least one halide ligand should be present. The highest activities were observed when applying very low concentrations of the titanium compound (10^{-5} molar) and high Al/Ti ratios of 100 for example.

Using chloride-free <u>zirconium</u> derivatives and methyl aluminoxanes extremely high activities were obtained[72,74-80], which are stable over long time periods. Here also the transition metal concentration is very low (10^{-7} to 10^{-9} molar) and high aluminium concentrations of 10^{-2} molar (as aluminium). Then activities (at 70 °C) of up to 1000 kg PE/g Zr.h.bar were found, the activity increasing with the oligomerization grade of the aluminoxane. A stable complex between the two ingredients was suggested. The zirconocene dichloride was subsequently reported to be even more active than the halide-free system. Chien and Wang[81] have recently shown that in the catalyst system Cp_2ZrCl_2/aluminoxane the major part of the methylaluminoxane can be replaced by simple trimethylaluminium without losing much of the activity.

Polymerization active cationic complexes were described by Jordan[82], they are of the type $[Cp_2Zr(R)(THF)]^+[B(C_6H_5)_4]^-$. The activity is low in apolar solvents and no activity is found in polar solvents.

By changing the ligand on the transition metal these homogeneous catalyst systems allow modification of the catalytic properties. An example has already been given in the adjustment of the copolymerization parameters of ethylene and propylene in chapter 3. But also for instance the molecular weight can be affected in the same manner. In a patent of Ewen and Welborn[83] a series of examples is given in which at identical polymerization conditions and in the absence of hydrogen, the molecular weight is varied by a factor of 6 just by changing the ligand on the zirconium. When more electron donating groups are present on the cyclopentadienyl group the molecular weight is reduced. The molecular weight distribution is not different to any great extent, and the activity decreases only at high substitution (steric hindrance?) on the cyclopentadiene ring.

Gianetti et al[74] studied both changes in the σ- and the π-ligands of the zirconium species. Varying the σ-ligand to benzyl, phenyl or trimethylsilyl gave in all three cases a more active catalyst system compared to the methyl compound. Changing cyclopentadienyl to indenyl and fluorenyl gave a more active system in the former case, whilst fluorenyl gave lower activity. Surprisingly the pure sigma compounds such as tetrabenzyl zirconium are also active in the combination with aluminoxane. Comparing the metals from the group IVB in the periodic system zirconium is by far the most active, by a factor five over titanium and as much as a factor of ten over hafnium.

Another option[84] with this type of catalysis is the use of mixtures of different transition metal complexes, each with their own values of propagation, transfer and copolymerization constants, thus making directly intimate blends of (co)-polymers in a one-reactor set-up. In polyethylene, blends are frequently made for reasons of tailoring the end-use properties, in our opinion the use of highly active homogeneous catalysts will allow a greater control. This has indeed been carried out in studies of Ahlers and Kaminsky[85], for instance when using the combination of Et(Ind)$_2$ZrCl$_2$ with Cp$_2$HfCl$_2$ and aluminoxane, broad MWD's, with Q values up to 10 were obtained. The ratio of the two transition metal compounds is an easy regulator on the width of the distribution.

The group of Soga attempted a very simple preparation of homogeneous catalysts by reacting hydrates of inorganic salts of titanium and zirconium with trimethylaluminium, and checking for polymerization activity in the filtrates of the reaction products[86]. Indeed some activity is found, and the polydispersity is high, implying multiple active species; for propylene, as expected, only atactic polymer is formed.

The studies of Soga, Doi et al gave rise to a number of different homogeneous <u>chromium</u> catalysts: $Cr(stearate)_3/AlEt_2Cl$ (ref 87), $Cr(acetate)_3 \cdot$acetic anhydride/$AlEt_2Cl$ (ref 88), tetra-t-butoxy chromium/$AlEt_2Cl$ (ref 89). Activities were not very high and in some cases over-reduction to Cr(I) valence state led to fast decay. Reoxidation to the active valence state of Cr(II) was easily accomplished with air[87]. In a number of cases these catalysts were combined with metal chlorides, of which $MgCl_2$ again led to the largest increases in activity (values of 17 kg PE/g Cr.h.bar are reported). The structure of the modified Cr(III)-acetate was studied by infrared, most probably a polymeric system is formed containing both unidentate acetate groups and bidentate ones[90].

High temperature homogeneous catalysis was studied by Yano et al[91,92] using catalyst systems generated by in-situ mixing of ingredients. The true homogeneous system Mg(2-ethyl hexanoate)$_2$ pre-reacted with 2-ethyl hexanol reacted in-situ with $Ti(OR)_4$ and DEAC, shows at 200 °C a productivity of about 10 kg/g Ti in 5 minutes, provided all MgR_2 is converted to the alkoxide by reaction with the alcohol. Even higher activities of up to 40 kg/g Ti per minute are observed in a system containing Mg-octanoate/H_2O/$Ti(OBu)_4$/DEAC. This activity is reached at 70 °C in an "aged" system (i.e. pre-reacted) or at 150 °C in a freshly mixed system.

A totally different but very interesting set of new homogeneous catalysts were described by Klabunde et al[93]. In this study the well-known ethylene oligomerization catalysts based on nickel (such as structure a in the scheme below), were modified by adding strong phosphine acceptors such as di-cyclo-octadiene nickel or the

a   b   c

rhodium complex Rh(acac)(ethylene)$_2$. Then active ethylene polymerization catalysts were formed. Even in polar solvents such as alcohols or acetone there is some activity. By changing the substituents in structure a, such as shown by the examples b or c higher active systems are formed, in which c is active by itself as the pyridine ligand is only very weakly bound and is easily replaced by olefin. The interest in these systems stems from the possibility of (co)polymerizing polar monomers, which would broaden the scope of modified polyolefins greatly. A similar approach has been taken by Ostoja Starzewski et al by applying bis(ylid)nickel complexes[94]. The products are very dependent on the ligand chosen and, for example, chain length can be controlled between oligomers and high MW polymer.

9.3.4 overview of activities

In Table 9.2 a compilation is given of some of the reported activities of various catalysts in ethylene polymerization.

The observed activities span a range of over five orders of magnitude. The homogeneous zirconium systems are the most active, with the titanium on magnesium chlorides as a good second. The use of the homogeneous catalysts in industry will depend not so much on their activity but on their performance in other respects, such as:
- bulk density obtained in slurry polymerization (i.e. catalyst morphology control - see chapter 1) as this determines the space-time yield of the process,
- the freedom in molecular weight control, which should preferably be done at temperatures above say 60 °C, since at lower temperatures the removal of the heat of polymerization is relatively expensive,
- the behaviour in copolymerization (see below),
- the temperature stability of the catalyst system, as this determines the process choice as well (see below, solution polymerizations are operated at above 100 °C, this of course necessitates catalyst stability at these temperatures).

TABLE 9.2

Comparison of the activities of different catalysts

Catalyst	Activity, kg PE per g transition metal per hour per bar	Polymerization temperature used in activity determination °C	reference
$TiCl_3/AlEt_2Cl$	0.09 - 0.6	70	own data
CrO_3/SiO_2	40	80	67
$((C_6H_5)_3SiO)_2CrO_2/SiO_2$	6	90	42
Cp_2Cr/SiO_2	3	90	43
$Cr(CH_3CO_2)_3 \cdot AA \cdot MgCl_2/AlEt_2Cl^*$	17	30	87
$TiCl_4/MgCl_2+BuOH//AlEt_3$	36	70	50
$TiCl_4/Mg(OH)Cl/cocat.$	200	80	67
$TiCl_4/MgCl_2//AlEt_3$	75 - 150	80	12
$Zr(allyl)_4$	0.04	80	12
ditto $/SiO_2$	1.6	80	,,
ditto $/Al_2O_3$	4	80	,,
$Cp_2TiCl_2/Al(CH_3)_2Cl$	1 - 4	30	67
,, / ,, $/H_2O$	40	30	,,
$Cp_2Ti(CH_3)_2/[Al(CH_3)O]_n$	4	20	,,
$Cp_2Zr(CH_3)_2/[Al(CH_3)O]_n$	1000 - 3000	70	75,79
$Cp_2ZrCl_2/[Al(CH_3)O]_n$	5000	90	80

* : AA = acetic anhydride

9.4 POLYMERIZATION BEHAVIOUR, KINETICS AND MOLECULAR WEIGHT

We will deal with these aspects using the same catalyst sequence as used above.

9.4.1 First generation TiCl$_3$

In terms of the polymerization reactions and the kinetics, all the rules that apply for propylene also apply for ethylene, only the individual rate constants differ. A good kinetic study on this type of catalyst with regard to both activity and molecular weight relationships has been reported by Berger and Grieveson[95,96]. They simplified the catalyst system in the sense that no side reactions affecting the rate to any extent were present in the system chosen, of γ-TiCl$_3$/AlEt$_2$Cl. Also mass transfer aspects were carefully controlled, and found to be absent up to polymerization temperatures of about 60 °C. As expected it was observed that the rate of polymerization was directly proportional to the catalyst and monomer concentration, whereas the rate was found to be independent of time and co-catalyst level. However in the case of the co-catalyst concentration a certain minimum concentration was required just as in many other systems. The activation energy measured was 73 kJ/mol. Hydrogen was found to decrease the rate considerably even at low concentrations of about 10 % volume. For the molecular weight of the PE no catalyst or co-catalyst dependence was found; it increased with time up to an asymptotic value, showed a linear relationship with monomer concentration and decreased strongly with increasing temperature. Of course hydrogen also lowered the MW, the reaction order observed being one. All these data are similar to those described for PP.

In chapter 5 the inverse relationship of activity and macroscopic catalyst particle size was mentioned for this type of catalyst. This is probably caused by the high reactivity of ethylene leading to low ethylene concentrations in the inside of the particle, and this transport problem obviously becomes worse with increasing catalyst particle size.

The TiCl$_3$-type of catalyst has also been applied in high temperature/high pressure ethylene polymerizations[17]. The kinetic parameters of TiCl$_3$.0.33AlCl$_3$ were measured up to 260 °C. The activity was highest with a new type of aluminium alkyl: siloxalane (a reaction product of triethyl aluminium and a cyclic dimethylsiloxane, its chemical formula being $Me_2(Et)SiOAlEt_2$). Deactivation of the catalyst is rapid under these drastic conditions, i.e. the lifetime is short and decreases with increasing temperature. This is not dependent on the type of co-catalyst however, the lifetime (time required to halve the rate) is about 10 seconds at 250 °C. The activation energy for the polymerization is very low (24 kJ/mol), indicating diffusion control. Machon[97] also reported a study of the effect of polymerization conditions on the properties of the polymer. The most linear polymer is made with the siloxalane catalyst, even at the highest temperature studied (280°C) where the standard catalyst system already gives a distinctly lower density. The MWD is broadened by an increase in polymerization temperature (becoming distinctly bimodal) and an increase in the residence time. Narrowing occurs when hydrogen is applied to control molecular weight. Pressure is not a true controlling parameter as changing it does not bring about any alterations in the properties of the polymer such as MW, its distribution and density. This means that in the mechanism determining these values the monomer concentration is not present in the equations due to the fact that most of them will be governed by a ratio of two or more processes, each of which is itself monomer-concentration related.

At these high temperatures radical polymerization of ethylene can also occur, probably dependent on the reaction path of reduction of Ti(III) to lower valent titanium species since some catalyst combinations appear to be more susceptible to this mode of reaction[19]. The density range covered in these conditions is wide, ranging from 0.92 to 0.95 g/cc. This is not only due to the radical polymerization but also through copolymerization with 1-butene generated in-situ in a parallel dimerization reaction.

As regards the morphology of the ex-reactor polymer powder (from suspension processes) one generally observes the so-called cobweb structure[98] at higher yields. An example of this has been given in chapter 5, see Fig. 5.3. The strands are thought to be formed in a process which starts with polymerizing ethylene first on the surface of the catalyst particle, followed by reaction inside the particle, this leads to volume increase and thus drawing of the initially formed polymer. Marchessault[99] in his review of powder morphology compiles data on the observed higher melting point of ex-reactor polyethylene, which at about 140 °C is appreciably higher than after remelting (132 °C, just as the normal

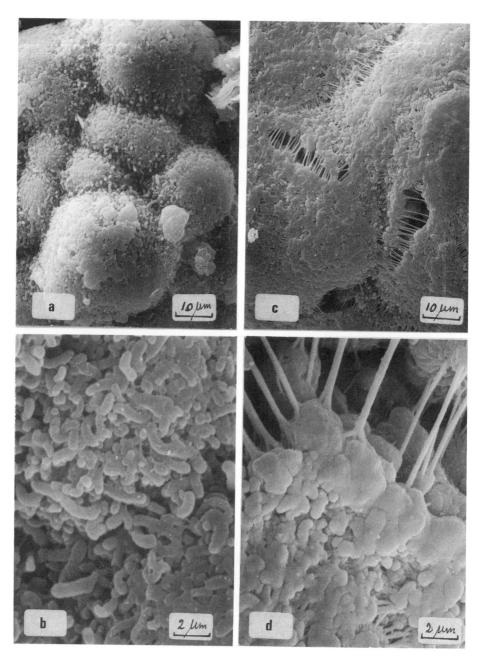

Fig. 9.4: *Scanning electron micrographs of polyethylene powders, varying in yield (in gPE/gTiCl₃) and catalyst: 190 (a,b); 330 (c,d; the β-form of the catalyst used in (a,b) was applied)*

polyethylene). The extent of this high melting variety increases with polymerization time, and is therefore not directly related to the polymer quality arising from the active sites. This high melting material is believed to be an extended crystal form of the polymer, evidence of this being obtained by X-ray diffraction (see below). At lower yields of polymer on catalyst the cobweb structure is not yet present, as is shown in the photographs of Fig. 9.4.

9.4.2 supported catalysts

(i) chromium containing catalysts

The general behaviour[20,21,100,101] of the chromium-on-silica catalyst in polymerization with ethylene is given in Fig. 9.5. The

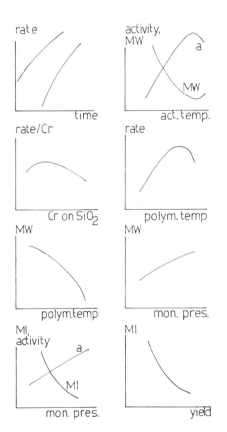

Fig. 9.5: Generalized kinetic aspects of CrO_3/SiO_2-catalysts

main features are

i)   the induction period shown by catalysts which are activated under oxygen,

ii)   the fact that with all catalysts the rate increases with time up to a final constant rate and

iii) the most important fact that the rate increase is faster the higher the calcination temperature employed, peaking at 925 °C, and thus the overall activity is higher as well.

The induction period is dependent on the temperature of calcination, but also on the polymerization temperature. At low polymerization temperatures (50 °C) almost no activity is exhibited, at slightly above 100 °C the above described behaviour is observed, whilst at 150 °C neither an induction period nor the gradual rate increase is noted. The induction period phenomenon is also found in continuous polymerization; it has been reported that after adding a spike of catalyst to a large scale reactor it took some 8 minutes before an increased heat load was observed[23]. At constant chromium loading a change in the calcination temperature leads only to differences in the hydroxyl population, a lower concentration being conducive for higher activity. This could be explained by complexation of surface silanol with the reduced chromium species. Removing the hydroxyls through fluorination leads to higher activity but also to higher molecular weights. No induction period is shown[102] when a catalyst is first activated via CO-reduction, although the polymerization rate still increases with time. This latter point could be due to slow initiation (rarely observed in other systems) or to accessibility problems at the start of the polymerization. The pre-reduction temperature should be below 500 °C since higher temperatures afford low activity catalysts, which show reduced coordinative unsaturation (down from 2 CO molecules per chromium atom at 350 °C reduction to 0.3 at 800 °C). This loss of vacancies is probably caused by agglomeration of the chromium oxide after detaching itself from the surface. With CO-reduced catalysts the polymerization activity is maintained to much lower temperatures, even as low as 0 °C. Propylene is found to be a much stronger reducing agent than ethylene[103] (complete reduction at 1 minute at 90 °C versus > 15 minutes at 145 °C for ethylene). Activation can be conveniently carried out using a wide variety of agents. Spitz et al[104] used hydrogen, ethylene, propylene, propane in a comparative study. All show a maximum in the polymerization activity as function of activation time and temperature, although sometimes this maximum is very sensitive to slight changes in these two variables. Propane is a less efficient activator than is propylene.

The activity as expressed per equivalent of chromium increases strongly with decreasing chromium loading[23], indicating a non-uniform type of active site. For instance at 0.01 %m/m Cr catalyst loading, the activity per Cr is 5 times the activity at the 1.0 % Cr level. The valence of the chromium in the active catalyst has been found to be two[105]. XPS (X-ray Photoelectron Spectroscopy) studies show only Cr(II) signals both in a CO-reduced and an ethylene-reduced catalyst. Moreover catalysts made with a Cr(III) species on the carrier are totally inactive in the polymerization[101]. A fully reduced catalyst in its Cr(II) form is able to polymerize α-olefins to high molecular weight amorphous polymers at low temperature[106]. No such behaviour is observed with the catalyst in the Cr(VI) form, which is almost totally inactive. Moreover[107], this reduced catalyst stays active in higher α-olefin polymerizations for weeks (at 25 °C, solution polymerizations), which is very different from the behaviour with ethylene where the activity does decay.

Despite the above observations recent studies have reopened the discussion on the valence of the active site[108-111]. These studies tended first to indicate Cr(III) as the true active species but subsequently became less decisive. Beck and Lunsford[108] measured the electron spin resonance (ESR) of catalyst samples after adding NO, the signal observed being unique to an isolated chromium-III complex. A good correlation of the intensity of the ESR-signal and the activity of the catalyst is found. Further evidence is obtained by starting with a Cr(III)-containing catalyst and either oxidizing or reducing it; in both cases the activity decreases in good correlation with the ESR-intensity. A finding which argues against the previously accepted Cr(II) as active species is the fact that exposing a Cr(VI) starting material to long reduction times leads to a completely dead catalyst. Moreover, following the chemiluminescence intensity when reacting the catalyst with oxygen - a reaction which is thought to be specific for chromium in its (II)-valent state - does not correlate at all with the observed activity of the same catalysts in ethylene polymerization; the luminescence increases to a maximum at increasing reduction times of Cr(VI) precursor catalysts whilst the activity decreases to zero. Rebenstorf and Larson[109] pose a more complicated picture of the catalyst, only dinuclear complexes are active and the valency can be either (II) or (III). The two different active species have a very different temperature behaviour. In their 1985 paper Myers and Lunsford[110] measured the number of active sites by a CO-poisoning technique. Very low fractions of the available chromium were indicated to be polymerization active. Even on a catalyst containing extremely low chromium levels (0.11 %m/m) only 4 to 7 % of the metal present is thought to be active, depending on the activation conditions. These findings must make the investigator cautious in making firm statements as to the valency of the active site. To do this with confidence one should be able to make a complete mass balance of all possible chromium species (isolated and di- and polynuclear) and correlate this with observed kinetics as a function of catalyst preparation conditions. This stage has not yet been reached. Moreover one has to realize that the activity tests in these studies (subatmospheric pressure, low temperature) are very far removed from commercially relevant conditions. McDaniel reports[21] that Cr(III)-based catalysts have zero or only microscopic activity at high pressure in the temperature range of 80 to 110 °C, thus making any conclusions on active site valency under commercial conditions even more shaky. An interesting observation from the above study is that in catalysts activated at 400 and 900 °C respectively, the fraction of active sites is 4 and 7% respectively, whilst the polymerization activity differs by a factor 10. This can only be due to a higher propagation rate constant; the authors reason that in the high temperature activation step, complexes between the active site and silanol groups are destroyed, leading to the higher intrinsic activity.

Using $AlPO_4$ as a support instead of SiO_2 changes the kinetic behaviour drastically[24]:
- the polymerization starts instantaneously after introducing the ethylene, i.e. no induction period is observed, reduction to the active valence is apparently very rapid,
- the rate shows a maximum rather early in the polymerization (at about 30 minutes),
- the activity is very susceptible to addition of metal alkyls,

with BR_3 giving the greatest increase in activity (factor of 5), on silica-supported catalysts the addition of a metal alkyl removes the induction period but does not increases the ultimate rate,
- pre-reduction with CO does not alter the kinetic profile but does increases the maximum rate (by a factor of 2 approximately).
A kinetic scheme in which three steps are present was put forward, which allows one to describe the observed kinetic behaviour: reduction of Cr(VI), the alkylation of the reduced species to give an active site, and decay of these sites to dead ones. Reasonable dependencies of the calculated rate constants on for example monomer pressure and temperature were reported.

Returning to the polymerization behaviour, the relationship of activity and temperature is strange, as a maximum is normally observed at about 70 to 80 °C, and at the low temperature end the apparent activation energy is 28 to 56 kJ/mol. The order of reaction in monomer changes from 1 at 145 °C to very small at -50 °C, this being explained by assuming a Langmuir-Hinshelwood mechanism[112]. The molecular weight behaviour is also given in Fig. 9.5, the temperature of polymerization has a profound effect on the MW, pressure has the expected effect, and the MW is lowered by increasing the calcination temperature of the support (which goes hand in hand with lower chromium loadings as discussed above). The MWD's are very broad with Q values of the order of 15, again indicating the presence of more than one active site. (This can also be concluded from polymerization of propylene with these catalysts, where some highly isotactic polymer is formed together with large fractions of non-isotactic material.) Another illustration of the "spectrum of sites" is given in Fig. 9.5 where the melt index is plotted against polymerization time (or yield). This shows a decreasing trend. Thus the first sites made active in the catalyst reduction step show a lower k_p/k_t ratio than later ones.

Zakharov and Ermakov[113] reported an elaborate kinetic study of supported chromium catalysts using, among other techniques, the quenching of the polymerization reaction with tritiated methanol. Due to the fact that this type of catalyst is co-catalyst free, no ambiguities exist regarding the number of metal polymer bonds and clear conclusions as regards the fraction of active sites and their propagation rate constants can be drawn. In the increasing rate period they observed an increase in the number of active sites, strongly suggesting a slow initiation reaction (which might be connected with the removal of the oxidation products). As usual the rate is directly proportional to the monomer pressure; the fraction of active sites is independent of the ethylene pressure indicating that support break up is not a limiting factor in developing activity. Only a low fraction of the available chromium is active in the polymerization, the order of magnitude is about 1 %. This is much lower than the value indicated by Hogan on the basis of chemical poisoning experiments[21,26]. The propagation rate constant is 280 l/mol.s at 75 °C on an alumina-silica support, on pure silica it is about 2.5 times higher. The activation energy for the propagation step is low (about 21 kJ/mol). However the temperature dependence of the polymerization rate is very complex, for instance a large difference is observed in the rates when changing the temperature during the course of a single run compared to separate single temperature runs.

The polymerization mechanism can be formulated in an analogous manner to other Ziegler-Natta catalysts, i.e. monomer coordination followed by insertion in a Cr-alkyl or Cr-hydride bond. Only the origin of the initial H or R group remains elusive. For the reduction reaction of Cr(VI) with ethylene, formaldehyde has been identified as a product[114]. This polar compound is probably slowly removed from the active site. One possible mode of initiation is shown below (after Krauss) in which a hydride and an allyl-derivative are formed. They can both lead to propagation. A different explanation is put forward by Ghiotti et al [115], who in polymeriza-

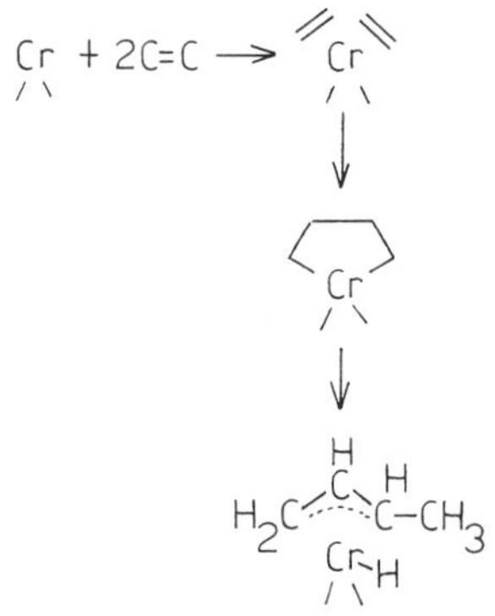

$$Cr + 2C=C \longrightarrow \overset{/\!/}{\underset{/\ \ \backslash}{Cr}}\overset{\backslash\!\backslash}{}$$

$$\downarrow$$

$$\underset{/\ \backslash}{Cr}$$

$$\downarrow$$

$$\underset{\underset{/\ \backslash}{Cr\!-\!H}}{H_2C\overset{\overset{H}{\overset{|}{C}}\overset{H}{}}{\cdots\cdots}C\!-\!CH_3}$$

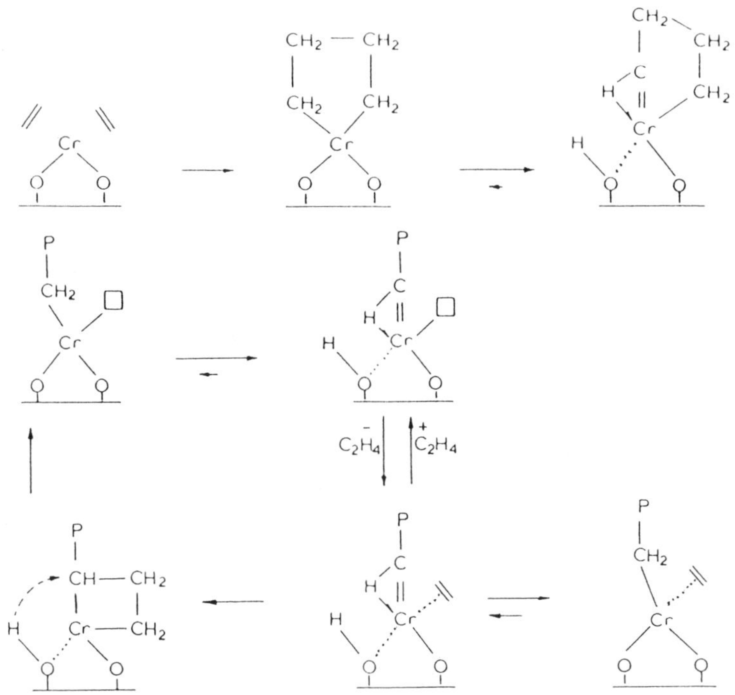

Fig. 9.6: Initiation and propagation on a CrO_3-silica catalyst according to Ghiotti et al [from reference 115, copied with permission fom Elsevier]

tions under very mild conditions (25 °C, low yield) could not detect any end-groups in the polyethylene formed. This led them to the suggestion that the polymer was cyclic by terminating on the chromium, as shown in Fig. 9.6. The propagation is thought to involve a transformation to a carbene species via hydrogen transfer to an oxygen atom. Such a mechanism is in accordance with McDaniel's observation that no scrambling of hydrogens occurred during propagation (see below).

By using deuterated ethylenes as monomer, it has been shown recently that a 4-ring metallocarbene is not an intermediate in the mechanism of the polymerization[116] since no scrambling of hydrogen or deuterium was detected (see also chapter 1.7). The formation of chains with odd-numbers of carbon[35] needs explanation, albeit that they are only found at high polymerization temperatures (over 150 °C).

Many studies of the Phillips catalyst have the aim of lowering the molecular weight in polymerizations without hydrogen. For instance, better performance with respect to molecular weight is found[100] when making a catalyst in two steps by first calcining the support at high temperature (say 870 °C) and then bringing on the chromium under anhydrous conditions followed by activation in air. This gives a melt index 2 to 2.5 times higher at the optimum compared to the standard catalyst. Even better results were obtained when such a catalyst was made on a support which was dehydroxylated in a CO atmosphere, giving up to 6 times the melt index; or when using carbonyl sulfide, COS, the improvement is a factor of 150! The dehydration in CO is thought to occur via the watergas shift reaction:

$$2 \text{ SiOH} + \text{CO} \longleftrightarrow \text{SiOSi} + \text{CO}_2 + \text{H}_2$$

At temperatures over 600 °C this leads to lower silanol populations than mere heating under nitrogen or air. The above catalysts had their silanol concentrations fixed before the addition of the chromium compound, and the later activation does not change this. The observed lower molecular weights have to arise from a different species, probably one formed from siloxanes (highly strained ones?). The mode of action of the silanol group on the active site remains unknown; unravelling this would require extensive kinetic measurements, including active site counts, and propagation and transfer rate constants.

A better MW performance can also be obtained by reducing a catalyst at high temperature (e.g. 900 °C in CO or CO+CS$_2$) and reoxidizing it at a lower temperature (e.g. 600 °C). Under these conditions a 10 to 15 times higher melt index is obtained compared to a normal catalyst. It is thought that this treatment increases the fraction of new, low molecular weight generating sites as the reoxidized chromium species can to a large extent react only with siloxane groups.

A further method of lowering molecular weight is to incorporate titania in the catalyst[39,40] (see also section 9.2.2). Making catalysts and incorporating TiO$_2$ by co-precipitation of titanium hydroxides, leads to a more active catalyst as the induction period is decreased and the rate is increased. Moreover the melt index increases. This points to direct effects of titanium on the active site. The extent of this desired effect on MW depends on subtle factors in the catalyst preparation. Optimum recipes include first fixing the titanium to the support (e.g. via $Ti(OR)_4$), then calcining followed by dry impregnation with CrO_3 (e.g. from acetonitrile), and finally activating in oxygen. This can result in a 500 times increase in melt index at equal or higher activity. Reversing the order of introducing the metal centres (first fixing chromium then titanium) gives no improvement at all. That a O-Ti-O-Cr bond is necessary for the above behaviour was proved by dehydroxylating a silica support with $TiCl_4$, hydrolyzing all Ti-Cl, and reacting the Ti-OH groups with chromyl chloride (see scheme in Fig. 9.7 for one possible mode of reaction). This gave a catalyst with similar properties to the best ones described above.

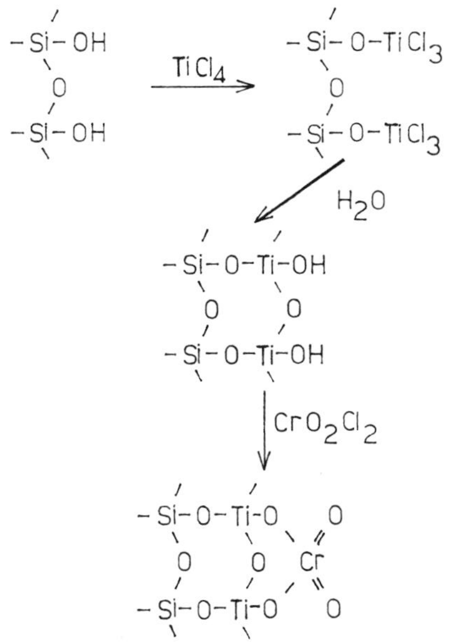

$$
\begin{array}{c}
-Si-OH \\
\quad \backslash \\
\quad O \\
\quad / \\
-Si-OH
\end{array}
\xrightarrow{\ TiCl_4\ }
\begin{array}{c}
-Si-O-TiCl_3 \\
\quad \backslash \\
\quad O \\
\quad / \\
-Si-O-TiCl_3
\end{array}
$$

$$\searrow H_2O$$

$$
\begin{array}{c}
-Si-O-Ti-OH \\
\quad O \qquad O \\
-Si-O-Ti-OH
\end{array}
$$

$$\downarrow CrO_2Cl_2$$

$$
\begin{array}{c}
-Si-O-Ti-O \qquad O \\
\quad O \qquad O \quad Cr \\
-Si-O-Ti-O \qquad O
\end{array}
$$

Fig. 9.7: *Consecutive reactions of* $TiCl_4$, H_2O *and* CrO_2Cl_2 *with surface-silanols on silica*

Hydrogen can be used in some cases to lower the MW, as is normally practiced on other catalysts. On CrO_3/SiO_2 this is not really effective, on a $CrO_3/AlPO_4$ catalyst however it is. Strangely enough the concentration of vinyl end groups in the polymer - from β-hydrogen transfer - does not however decline and thus a different, still unknown, mechanism must be operating[21].

For the chromocene catalyst some data on its polymerization behaviour have been disclosed by Karol et al[43,117]. The activity is proportional to the monomer pressure, but only after correcting this by some 7 bar. This effect could be related to the catalyst break-up, see the next paragraph. As a function of temperature, the activity peaks at a rather low temperature of 60 °C, an observation which needs many more kinetic measurements to elucidate its causes. The activation energy appeared to be 42 kJ/mol in a rather narrow temperature range of 30 to 65 °C. The molecular weight distribution is, with a Q-value of 7 to 10, of intermediate width. The best feature of this catalyst is the fast transfer rates with hydrogen, at 20% hydrogen in the gas-phase a melt index of 145 dg/min is already reached. The ratio of the hydrogen transfer over the transfer with monomer is one order of magnitude larger for this catalyst[117] than for $TiCl_3$. There are indications that all Cr(II) in the chromocene catalyst is active in the polymerization.

The support used in the catalyst preparation is of paramount importance in determining the ultimate polymerization properties. The following observations illustrate this statement:
- the chemical nature of the support affects the activity, for instance alumina versus silica supports,
- the activity increases with increasing pore volume and/or pore size, whereas low pore volume/narrow pore size supports gives nearly inactive catalysts. Activity appears to increase with pore volume over two orders of magnitude up to a plateau at pore volumes above about 2 ml/g (see ref. 38 and 118),
- the pore size also influences the molecular weight, the wider the pore the lower the resulting MW[118,119], and this could be due to the higher probability of copolymerization with vinyl-terminated chains in small pores due to restricted transport of polymer and proximity of active sites. This copolymerization, of course, leads to higher molecular weight (lower MI),
- in Phillips catalysts the silanol population has a very large effect on the activity.

Studies related to the morphology of the powder directly from the polymerization were published. Davidson[120] describes the visible aspects of fractured polyethylene particles stemming from a "particle form" polymerization process. In the scanning electron microscope he observes the following features: threads, platelets, helices and nodules/worms/polyps. With the exception of the platelets this is similar to the findings of Graf et al[98] (mentioned above); the threads are less clear in the micrographs given, but should be expected to be more on the outside of the particle in view of the way they are formed. In a more elaborate study McDaniel investigated the fracturing of the Phillips catalyst[118] by taking intermediate samples, burning away the polymer and measuring the particle size distribution of the remaining solids. Both the standard chromium catalyst as well as experimental ones (reacting $TiCl_4$ with dibutyl magnesium absorbed on silica) were studied. In both cases, for catalysts having near optimum activity, the median size of the catalyst remnants was 8 μm, independent of polymerization time (20 minutes or longer), starting particle size and temperature. This held even at 150 °C, i.e. a condition of solution polymerization. The catalyst particle is thus fragmented rapidly, being constant in size whilst the rate of polymerization still increases. It would therefore appear that this rate increase is not related to accessibility of sites for monomer or for easy diffusional transport of the formed polymer to the outside of the particle, but could be caused by a slow initiation reaction. All this holds only for fragile catalysts, for which a pore size of over 15 nm is a prerequisite (it is assumed that breaking will happen more easily through the widest pores, which constitute the weakest part of the particle). Supports with pore sizes of 2.3 nm give dead catalysts, activity gradually improving at larger pore sizes.

These findings are very surprising in view of other observations in which supports fragment down to much smaller sizes. With the sizes observed by McDaniel direct proof of the size of the catalyst fragments could have been obtained by viewing fractured polymer particles or by melting as polymerized powder on a hot-stage microscope (preferably using a low molecular weight polymer). Such experiments could prove unambiguously that the experimental technique followed was sound. Later it was stated[21] that possibly smaller sizes of 0.1 to 1.0 μm are formed upon fracturing the catalyst.

(ii) <u>titanium supported on magnesium chloride</u>

A number of Russian groups have studied this type of catalyst in some depth. Baulin et al[121] determined the kinetic constants including the fraction of active sites on a $MgO/TiCl_4//AlEt_3$ catalysts by quenching with tritiated methanol. For a catalyst containing 0.6 % m/m titanium about 20 % of the supported titanium is active, showing a propagation rate constant of about 2200 l/mol.s at 70 °C. Also all the rate constants controlling the molecular weight were determined showing chain transfer with monomer to be the main mechanism for MW control under practical conditions: the ratio of the half lives of a growing polymer molecule is 1:85:150 with respect to transfer with monomer, aluminium alkyl and β-hydrogen tranfer respectively, assuming the polymerization conditions are 1 molar in ethylene and 1 mmol/l $AlEt_3$. The MWD on this catalyst was rather narrow (Q about 3), at very short reaction times some broadening was observed.

In the case of $TiCl_4$ on $MgCl_2$ catalysts Zakharov et al[122] studied the effect of co-catalyst type and the Al/Ti ratio. The greatest activities were obtained with $AliBu_3$ and with $AlEt_3$ provided the Al/Ti ratio was high (>150). The activity was very low with $AlEt_2Cl$ as co-catalyst. The time dependence of the rate was also a function of type of co-catalyst and its concentration. Constant rates were observed with $AliBu_3$. Up to 40% of the titanium present was found to be active in the polymerization.

Ivanchev et al[123] tested different supports for $TiCl_4$ and came to a similar conclusion as observed elsewhere: $MgCl_2$ is a special case, bestowing high activity on the $TiCl_4$ attached to the support. The changes in activity were acccompanied by changes in the molecular weight distributions; for example $TiCl_4$/MgO showed a Q value of about 3, while $TiCl_4/Al_2O_3.SiO_2$ gave 10.

Böhm published an extensive study[124,125] on the kinetic properties of a catalyst derived from $Mg(OEt)_2$ and $TiCl_4$. This catalyst had a large pore volume (1.2 ml/g) and a surface area of 60 m^2/g, the pore volume in particular is much greater than that found in conventional $TiCl_3$ catalysts. The co-catalyst of choice is triethyl aluminium with a high Al/Ti ratio of about 20 being required for optimal rates. This catalyst only develops high activity when the applied ethylene concentration exceeds a certain minimum value, consequently the rate is not directly linear with monomer concentration. This is not found for the first generation $TiCl_3$ catalysts. This phenomenon is explained by the need for a minimum polymerization <u>rate</u> for total break up of the catalyst particles[11]. This observation also leads to the requirement of a highly divided support of a very open structure, to assist in the fast break up of the particle (see also chapter 1). Compare this with the polymerizations mentioned earlier (chapter 5) with TiCl$_3$ catalysts of low porosity, which also prove this point. No catalyst remnants can be observed by electron microscopy on the polymer powders generated, thus they have to be under 5 nm. Macroscopically also the cobweb structure is found.

From a polymerization kinetics analysis[124] Böhm concluded that up to 75 % of the titanium atoms are active in the polymerization; using this value leads to propagation rate values of around 60 to 80 l/mol.s, which are of the same order as on first generation catalysts. In a study of the molecular weight distribution[125] after only 15 seconds polymerization time, the MW was found to be high and the high end tail of the distribution did not shift to higher values with time. This observation leads to a minimum propagation rate constant of 2900 l/mol.s, which is much higher than the kinetically derived one and more in line with the findings of Yermakov and Zakharov[12]. Chain transfer with aluminium alkyl and via β-hydrogen transfer occur approximately to the same extent. The Q-value of the distributions are around 7.5. This catalyst is more reactive in chain transfer with hydrogen than the first generation Ziegler catalysts, moreover the rate-lowering effect of hydrogen is less pronounced. Böhm[126] also determined the fraction of active sites as a function of the polymerization temperature and found an optimum of 75% of all titanium active at temperatures of between 65 and 85 °C, falling off drastically both at lower and at higher temperature.

Yermakov et al[12,127] also studied the polymerization kinetics and used their radioactive labelling techniques for estimating the number of active sites and thus also the propagation rate constants. The rate of polymerization is directly proportional to the titanium and the monomer concentration, the activation energy is variable - depending on the temperature range, catalyst type and titanium loading. With respect to chain transfer with hydrogen the order of reaction is 1 and not 0.5 as in the corresponding TiCl$_3$-catalysts. These measurements lead to propagation rate constants of about 12,000 l/mol.s , equal in first and second generation catalysts. Thus the activity increase is solely due to the increased number of active sites. Also the rate constants of chain transfer with aluminium alkyl and β-hydrogen removal are similar, suggesting that the active sites in both titanium catalysts are nearly identical. This outcome obviously contradicts the Böhm studies, and further work will be necessary before any definitive conclusion can be drawn.

(iii) supported organometallic compounds

Ballard found[59] that the kinetics of ethylene polymerization with alumina supported tetrabenzyl zirconium were well described by

$$\text{rate} = k_p . K . [\text{monomer}] . [\text{active chains}]$$

in which k_p is the propagation rate constant and K the complexation constant of the active site with monomer. The kinetic analysis led to the conclusion that only 1 % of the zirconium is truly active. The MW is directly proportional to the monomer concentration and with hydrogen the inverse of the number average polymerization grade is proportional to the hydrogen concentration. A polymerization mechanism similar to the ones for Ziegler catalysts was proposed, this is given in Fig. 9.8.

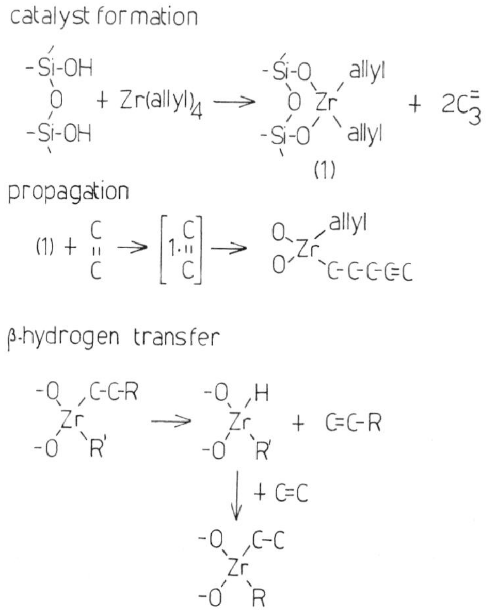

catalyst formation

propagation

β-hydrogen transfer

Fig. 9.8: Polymerization mechanism on Zr(allyl)$_4$ on SiO$_2$, after Ballard

Yermakov[12] attempted to systematize the data on this class of catalysts as follows:

- Ti, Zr and Cr-derived catalysts show approximately the same activity, only the thermal stability of the Ti and Zr systems is much lower. Hafnium-containing catalysts are definitely of lower activity. Similar rules apply in the case of propylene as monomer.

- regarding the effect of the ligand: β-methallyl complexes are the most active in the zirconium system and give also the highest loading on a support. After hydrogen activation the difference between catalysts, originally differing in organic ligand, becomes very small. This suggests that the ligand is hydrogenated off, leaving a metal hydride as active centre in each case.

- alumina as a support gives considerably higher activities than silica.

- the transfer rate constants (in their ratio with the propagation rate constant) vary in a wide range, both with transition metal and ligand. This ratio spans 2 to 4 orders of magnitude. Commercially important is the rate of transfer with hydrogen for melt index control, and cyclopentadienyl chromium complexes are the best in this regard. Generally the chain transfer increases across the series Ti < Zr < Hf < Cr. Also, catalysts containing lower valent species transfer faster.

- Q-values of the molecular weight distribution are well above 2, thus more than one type of site is to be found in these catalysts.

- the number of active sites is relatively low (about 1 % of the metal present). The propagation rate constants vary with the ligand

benzyl > allyl(alumina) > allyl(silica)

in a zirconium series. Chromium complexes show one order of magnitude higher propagation rate constant than Zr- or Ti-containing catalysts (the latter two show comparable values for k_p). The propagation rate constant for tetrabenzyl titanium on alumina catalyst is 2 orders of magnitude lower than that of a $TiCl_4/MgCl_2$ catalyst. This huge difference must be due to support and ligand effects.

- using monohalide substituted complexes a decreasing molecular weight with decreasing electronegativity of the halide ligand is observed, with iodine giving the lowest molecular weight at roughly equal activity. All aspects which govern the electron density at the metal ion apparently affect the chain transfer rate, similar relationships were already mentioned above for chromium species.

Turning now again to the morphology of the powder resulting from slurry polymerizations, Ballard[59] observed helices and "worms" in the powder ex-polymerization, the catalyst particles breaking up early in the polymerization into fragments smaller than 100 nm. The "worms" contain orientated polyethylene, with the chain parallel to the axis. There is some evidence that the catalyst fragment is pushed outward with the growing polymer, in that way the barrier to monomer transport stays constant throughout the whole polymerization.

In a study on the change in morphology of a catalyst to a polymer particle Billingham and coworkers[128] used tetrabenzyl zirconium on an alumina support of well-known properties (macroscopic size 20 to 70 μm, subparticles 0.1 to 0.6 μm and the primary particles were 0.01 μm - after ultrasonic disruption). In the polymerization of ethylene the particle initially forms hollow spheres by an unequal expansion, the central void still containing relatively large aggregates derived from the catalyst particle. This unequal growth is due to a rapid initial filling of the pores after which the polymerization is more rapid on the more accessible parts of the particle, i.e. the outside layers. Once the yield in these outer layers increases, the monomer transfer to the active sites on these subparticles also becomes impaired leading to increasing monomer concentration in the inner regions of the particle, allowing a more even growth thereafter. The subparticles grow further apart, leading to drawing of the polymer formed. This drawing probably accounts for the apparent crystallization of the polyethylene in a form with an extended chain, as suggested by the observed high melting points of around 140 to 142 °C. The ultimate morphology is the spaghetti-like structure common to all polyolefins made in suspension polymerization, the individual strands being about 0.6 micron in diameter. The subparticles of the catalyst are broken down to about 0.04 micron which is still appreciably larger than the dimensions of the primary particles in the alumina support which was of the order of 0.01 micron, but clearly very much smaller than in McDaniels' study with chromium on silica.

9.4.3 Homogeneous catalysts

The kinetics and molecular weight relationships in the earlier homogeneous catalysts (Cp_2TiCl_2/Al-alkyl) are not dealt with in this book, the reader being referred to a number of good review articles[65,68]. Only the aluminoxane systems will be highlighted.

The effect of water on the activity of the homogeneous catalysts has been mentioned above, but it is not only the activity that is affected by water. The effects on the molecular weight of the polymer parallels its effect on the activity, see Fig. 9.9, this suggests that it is especially the propagation rate constant which is influenced[68,129]. Rough calculations on the basis of the molecular weights observed show a 15 to 35 fold increase of the

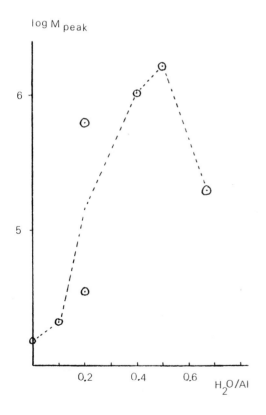

Fig. 9.9: Molecular weight of polyethylene in the system $Cp_2TiEtCl/AlEtCl_2+H_2O$ as function of the H_2O/Al ratio (after Cihlar, see reference 129)

propagation rate constant[130]. The increase of activity by adding water is common to all co-catalysts, the extent is largest however at higher halide contents of the co-catalysts[129].

Some of the most important observations with the highly active zirconium/aluminoxane systems, especially with $Cp_2Zr(CH_3)_2$, based mainly on the work of Kaminsky and co-workers, are mentioned below:
for the <u>activity</u>:

- the life time of the catalyst is very long at temperatures of <40 °C, i.e. these systems show little or no decay under these conditions,
- the activity increases with the degree of oligomerization of the aluminoxane[74,75],
- Lewis bases are strong poisons[74],
- a very large excess of aluminium compound is necessary, the order of the polymerization reaction w.r.t the aluminium is initially 0.5 and decreases to 0.3 at higher concentrations[76],
- the rate is proportional to the transition metal concentration[77], although later reports show the specific activity to increase at decreasing zirconium concentration, see below,
- the rate decreases with increasing size of the alkyl group on the aluminoxane[79,131],
- with ethylene as monomer the polymerization starts directly at a high rate, with propylene it takes hours to reach the rate maximum[77],
- a considerable influence of the diluent in the polymerization has been observed in Cp_2ZrCl_2 systems, with toluene giving much higher propagation constant and every zirconium species being active, whilst in decane the propagation rate constant was only one-sixth with 60% of the zirconium species active[132],
- zirconium compounds are more active than titanium or hafnium systems[74,79,131],
- the activity is lowered by methyl substitution in the cyclopentadienyl ring, but the molecular weight increases. In the pentamethyl substituted compound it is found that the activity increases upon pre-reacting the catalyst ingredients[79].

for the <u>molecular weight</u>:

- temperature is the main controlling variable[75], increasing this from 10 °C to 90 °C lowers the molecular weight from 1.5

million to 90,000, (see also Fig. 1.22),

- the molecular weight distributions are normally narrow, with Q-values of around 2 (as expected from a homogeneous, one-active-site-type catalyst), when methyl substituents are present in the Cp-group much broader distributions are found,

- the molecular weight is also lowered at higher zirconium concentrations, indicating a bimolecular form of transfer[79,131],

- with methyl-substituted complexes one observes lower molecular weight at lower Al/Zr ratios[79],

- at -20°C the catalyst system is a living one, in that the molecular weight increases with polymerization time[78].

Checking on the number of active centers in Cp_2ZrCl_2/aluminoxane systems at 60 °C and 10 minutes polymerization time, Tait et al[133] find 100 % activity; this number is only obtained at short reaction times of the CO used to measure the number of active sites, at longer times values far over 100 % were found. Chien and Wang[134] find [14]CO not suitable for number of active sites determination, they prefer tritiated methanol. Using this they also arrive at 100% participation of zirconium when polymerizing ethylene with Cp_2ZrCl_2/methylaluminoxane at 70 °C.

Using bis(neomenthylcyclopentadienyl)$ZrCl_2$ with methylaluminoxane kinetic studies were carried out[135]. The rate of polymerization decays less with this catalyst system than with Cp_2ZrCl_2 or $MgCl_2$-supported titanium catalysts. The bulky substituent on the Cp-ring might slow the reaction between two Zr-species, a reaction thought to be responsible for the termination reaction. The dependence on the Al and Zr concentrations has also been determined:

$$R_p \approx k.[Al]^{1.75}.[Zr]^{-1.2}.[monomer]$$

This is based, for instance, on a 100-fold increase in the specific rate with a 10-fold lower zirconium concentration. The activation energy for propagation was found to be 31 kJ/mol. About 50 % of the zirconium was active under a specific condition. The melting points of the polyethylene produced were on the low side, and molecular weight measurements should be done, eventually coupled to NMR or IR to study the composition of the products made.

With respect to the need for the high aluminium concentration no definite interpretation has been put forward up till now, nor

for the structure of the active complex. A few observations illustrate the difficulties in formulating a hypothesis: using ethyl
aluminoxane gives a lower rate than the methyl analogue[79] - pre-reacting the ethyl aluminoxane with $(CpMe_5)_2ZrCl_2$ gives lower rates
whereas in the corresponding case with the methyl form an increased
rate is observed on pre-reaction[79]; another difficult feature is
the effect of the degree of oligomerization on the activity.

Possible models of the active complex in these systems have
been given by the group of Kaminsky and by Gianetti et al[74]; they
are illustrated in Fig. 9.10. Kaminsky proposes a relatively simple
complex of the two components of the catalyst system whilst Gianetti et al describe a removal of a σ-ligand of the zirconium species,
this being replaced by the oligomeric aluminoxane which leads to a
highly polarized compound making the Zr center rather cationic.

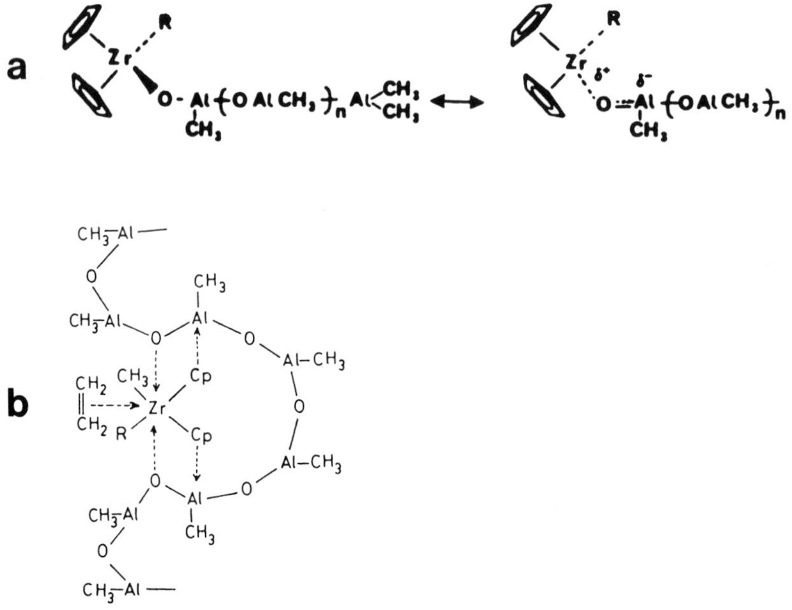

Fig. 9.10: Proposed structure of the active species in the Cp_2ZrR_2/aluminoxane
system: after Gianetti (a); after Kaminsky (b) [(a) from reference 74, copied
with permission form John Wiley and Sons; (b) from reference 76, copied with
permission form Harwood Academic Publishers]

9.4.4 Molecular weight distribution control

The control of the molecular weight distribution is highly desirable and quite a number of practical routes to effect this have been found. However explanations for the observations are lacking. A very good recent review of aspects thought to govern the distribution is the one by Zucchini and Cecchin[136]. They show that the reason for the mostly very broad distributions is chemical by virtue of the presence of a number of active site types each with their own rate constants. The distribution can certainly be influenced by the transition metal in the catalyst; co-catalyst and polar additive effects are much less clear. The polymerization conditions also affect the distribution, which becomes broader at lower temperatures and shorter reaction times; the effect of hydrogen is small or non-existent. However the transition metal compound and the choice of support have the greatest effects on the MWD. For instance narrowing is observed in the series:

$$silylchromate/SiO_2 > chromium/SiO_2 > first\ generation\ TiCl_3 >$$
$$titanium\ on\ MgCl_2$$

Even within the titanium on magnesium chloride series a range of MWD's is attainable. From the compilation made by Yermakov et al[12] a Q range of 1.5 to 10 is observed. The narrowest distribution is in $TiCl_4/MgR_2$, the broadest with $TiCl_4/Mg(OR)_2$.

Ligand effects were noted in very early studies by Wesslau[137], in a series of catalysts made from $TiCl_3/TiX_n(OR)_{4-n}/AlRX_2$ the Q-value decreases with decreasing n. On $MgO/TiCl_4//AlEt_3$ catalysts the polymerization conditions affected the distribution in the same way as described above[121].

Broader distributions are found in copolymers compared to homopolymers (see section 9.5).

It is obvious that broader distributions can always be generated from mixtures of polymers with (widely) differing molecular weights, the only difficulty might be in the attainment of a truly homogeneous mixture (see chapter 7).

9.5 COPOLYMERIZATION

Modifying the properties of polyethylene through copolymerization is of utmost importance to the total polyethylene business. The fraction of polyethylenes sold as a true homopolymer is relatively small, being about 35 %. The properties of the polyethylene are largely determined by the molecular weight (or rather Melt Index) and density. It is the latter which is changed by copolymerization. Decreasing the regularity in the chain leads to lower crystallinity, lower melting points, and - as said - lower density (this will be further discussed in the characterization chapter). In addition to the crystalline properties, the mechanical properties such as stiffness and impact strength are also affected. Above all the environmental stress cracking is highly dependent on the polymer structure, being much greater in copolymers.

Considering their commercial importance, it is surprising that the open literature contains limited data on copolymerization, especially on chromium catalysts. The number of good kinetic parameter determinations is very low, as are the number of good characterization papers. These facts combined with the many different catalyst systems which are used commercially, each with its own copolymerization behaviour, inevitably means that this review will be rather sketchy.

In the following review we will use the same catalyst sequence as in the earlier sections.

9.5.1 First generation TiCl$_3$

A large amount of data has already been discussed in chapter 3, which dealt with copolymerizations of propylene with other monomers where ethylene as comonomer featured rather heavily. In table 9.4 at the end of this section some data on kinetic parameters of ethylene/1-butene copolymerization are mentioned[138-140], the values being very high for the r_1, in the order of 60 to 80. On vanadium catalysts these numbers are lower, about 30. The most interesting extra data on this type of catalyst stems from the high temperature copolymerization[141] work carried out in France by Charbonages de France (CdF) who apply a solution polymerization process at temperatures around 250 °C (see sections 9.2.5 and 9.6). The copolymerization parameters are relatively low, with values for r_1 of 4 for propylene as co-monomer and 14 for 1-hexene as co-monomer (see Table 9.4). Low values are welcome in this process as this means that relativily low amounts of co-monomer suffice to bring about the desired level of copolymerization, no large excess of the higher olefin being required in this case compared to nearly all other processes (see below). The low numbers are in line with the earlier mentioned fact that (at least in the temperature range of 40 to 80 °C) the r value decreased with increasing temperature (see chapter 3). The copolymerization parameters are even better in a newer catalyst which uses a modified alkyl as co-catalyst. The product r_1*r_2 is much larger than 1, indicating a tendency to blocky copolymer formation. The MW is strongly decreased by copolymerization, and this feature can - especially at these high polymerization temperatures - be problematic, in that one is not able to make high molecular weight copolymers. By a judicious choice of the catalyst components, CdF is able to make fractional Melt Index polymers in densities as low as 0.92. The MWD's, with values for Q of 4 to 7, are relatively narrow for $TiCl_3$-derived polymers.

On the catalyst system which is claimed to show the highest stereospecificity, Cp_2TiMe_2 with Solvay-$TiCl_3$, copolymerization studies were carried out[142]. The kinetic parameters are given in Table 9.4, the values for r_1 are high but less so than intuitively expected for a catalyst which has a very high demand on stereoregularity, and thus, for example, on the complexation mode of the α-olefin. For this catalyst the ethylene polymerization rate is increased upon addition of other α-olefins (propylene and 1-hexene were used)[143]. This behaviour appears to be fairly general as the following pages will show. Soga et al explain this by the higher rate of monomer transfer to the active site due to the lower crystallinity of the polymer produced.

Using $TiCl_3.0.33AlCl_3$ and $AlEt_2Cl$ and $AliBu_3$ co-catalysts, Tait et al[144] showed in copolymerization with both 1-octene or 4-methyl-1-pentene that rate enhancement for ethylene in the presence of another α-olefin is a common feature in these systems. The rate increase levels off at higher co-monomer concentrations whilst the co-monomer conversion is roughly proportional to its own concentration. Measuring the number of active sites under conditions of maximum rate gives a considerable increase in this number compared to homopolymerization - no explanation can as yet be given for this phenomenon (see also below).

An interesting observation has been made by Seppälä[145-148]. In copolymerizations of ethylene and longer chain α-olefins, the addition of 1-butene to these systems led to a considerable increase in the incorporation of the longer olefin. The effect is larger the longer the olefin as Table 9.3 illustrates. The increase in incorporation is proportional to the 1-butene concentration. Reactivity ratios have been calculated in these terpolymerizations, leading to figures of 40 for 1-butene, 230 for 1-decene, 360 for 1-dodecene and 420 for 1-hexadecene. In pyrolysis GC analysis one observes some evidence for coupled incorporation of the butene and the longer olefin[149], but whether this "explains" the above effect is dubious as it would be expected that a long olefin would add even faster to a chain ending in ethylene than in butene. Similar behaviour is found by Ojala and Fink[150] using a very different catalyst (see section 9.5.3).

Ethylene is the only monomer in combination with which copolymers can be made with internal olefins, such as cyclo-olefins and 2-butene[151,152]. The latter are unable to homopolymerize, but can add to a chain ending in ethylene. Thus, when applying a very low concentration of ethylene, alternating polymers can indeed be made, some of them in highly crystalline form.

TABLE 9.3

Comparison of "butene effect" in higher α-olefin copolymerization with ethylene.

long α-olefin	C_6	C_8	C_{10}	C_{12}	C_{16}
increment in incorporation at equimolar concentrations of all three monomers, %m/m	0	≈ 1	≈ 4	≈ 6	≈ 11

TABLE 9.4

Reactivity ratios in ethylene copolymerizations with Ziegler catalysts.

Catalyst	reactivity ratios						tempe-rature °C	reference
	C_2	C_3	C_2	C_4	C_2	C_6		
TiCl$_3$.0.33AlCl$_3$/Et$_2$AlCl			85	0.021			23	138
TiCl$_3$/Et$_2$AlCl			60	0.025				139
ditto	16.7	0.036						chapter 5
TiCl$_3$.0.33AlCl$_3$/AlEt$_3$	4.9	0.25					70	123
Solvay-TiCl$_3$/Cp$_2$TiMe$_2$	10	0.22	69	0.058	69	0.033	40	142
VCl$_4$/AlEt$_2$Cl			32.5	0.016				140
TiCl$_3$.0.33AlCl$_3$/Al(octyl)$_3$	4	0.3	6	0.4	14	(0.07)	220-	141
ditto/Me$_2$(Et)SiOAlEt$_2$	3	0.45	6.2	1.4	9	1.2	250	141

An overview of the copolymerization parameters for propylene, 1-butene and 1-hexene with Ziegler catalysts is given in Table 9.4.

9.5.2 Supported catalysts

(i) chromium-containing catalysts

It is very surprising to discover that almost no copolymerization kinetic parameters at all have been described for this very important catalyst although some general data have been published. Hogan reported[153] that in solution polymerization with the chromium catalyst the co-monomer polymerization rate is greatly increased in the presence of ethylene. Factors of ten and over were observed. This is fairly easily explained by the more rapid addition of the higher olefin to an ethylene-ending chain than on the chain ending with the higher olefin, the reason being the steric hindrance. Zakharov et al studied the copolymerization kinetics of various chromium catalyst which were given different activation treatments[154]. The copolymerizations were carried out in the gas-phase, and the kinetic data were evaluated by taking into account the monomer concentration (or rather the ratio) in the copolymer particle. The r-values are given in table 9.5.

TABLE 9.5

Reactivity ratios for ethylene on chromium catalysts, from ref. 154

catalyst + treatment	propylene comonomer	1-butene comonomer
CrO_3/SiO_2, act. in O_2, 800 °C	4.4	13
$CrO_3/TiO_2/SiO_2$, ditto	4.4	13
CrO_3/SiO_2, red.w.CO at 400 °C	5.6	-
CrO_3/SiO_2, red.w. Et_2AlOEt at 25 °C	8.0	43
$((C_6H_5O)_3SiO)_2CrO_2$, red.w. Et_2AlOEt	8.0	43
Cp_2Cr/SiO_2, act. at 800 °C	>24	-
for comparison: $TiCl_4/MgCl_2//TEA$	12.8	24

The Phillips catalyst shows very low values for the reactivity ratio, which makes it very suitable in copolymerization, as low ratios of the co-monomer are required. Apparently the different catalyst treaments give rise to different behaviour, but an explanation is difficult without extensive further study. It might be that there is always a spread in catalytic species (as in heterogeneous titanium catalysts), but that different treaments give different ratios between these species.

Just as in the case with other catalyst types, the MW's are lower in copolymers. Fig. 9.11 gives the reactions involved in the chain transfer and clearly k_2 is larger than k_1. The lowering of MW by co-monomer holds generally and will not be mentioned again with the other catalysts. The increased transfer is also evidenced by

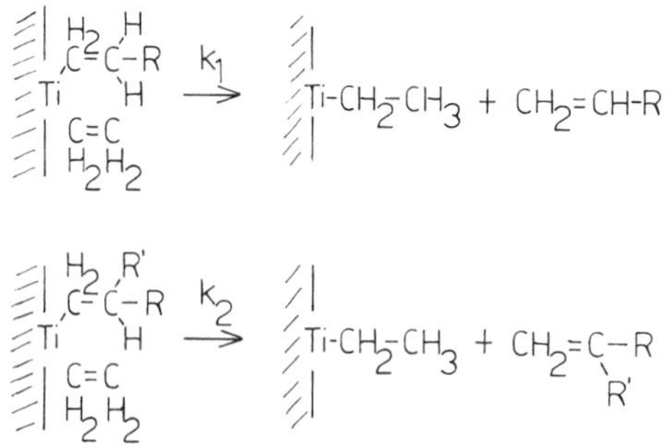

$$\overset{\text{H}_2\text{ H}}{\underset{\text{Ti}}{\overset{|}{\text{C}=\text{C}-\text{R}}}} \quad \xrightarrow{k_1} \quad \text{Ti-CH}_2\text{CH}_3 + \text{CH}_2=\text{CH-R}$$

$$\overset{\text{H}_2\text{ R}'}{\underset{\text{Ti}}{\overset{|}{\text{C}=\text{C}-\text{R}}}} \quad \xrightarrow{k_2} \quad \text{Ti-CH}_2\text{CH}_3 + \text{CH}_2=\overset{\text{R}'}{\underset{}{\text{C-R}}}$$

Fig. 9.11: Chain transfer in copolymerization

the higher level of vinylidene endgroups in the polymer[26]. Spitz[155] describes a limited study in which a co-catalyst had to be used to make the system active in copolymerization of ethylene and 1-hexene. The activity showed a maximum in the applied co-catalyst concentration. The polymer contained very high amounts of room temperature soluble material of high co-monomer content. The non-soluble polymer shows very low hexene contents, and no relationships with the monomer ratios were established. The MWD's were extremely broad, with Q-values of 18 for ethylene homopolymer and 30 for the copolymers. In a later study by Mabilon and Spitz[156] the copolymerization of ethylene and propylene was studied on various CrO_3/SiO_2 catalysts. No reactivity ratios were determined but at standard copolymerization conditions the composition of the copolymer was compared. It appeared that neither monomer concentration, nor chromium loading on the support were leading to alterations in the relative rates, but when using either a fluorinated support or a catalyst containing also titanium species, an increased propylene incorporation was noted. Hogan and Witt report[23] acceleration of ethylene polymerization by the presence of small amounts of 1-hexene in slurry polymerizations, the effect being dependent on the hexene concentration applied; at levels over 5 % mol in the feed a slight decrease in rate is noted. The Phillips investigators call it an activation effect.

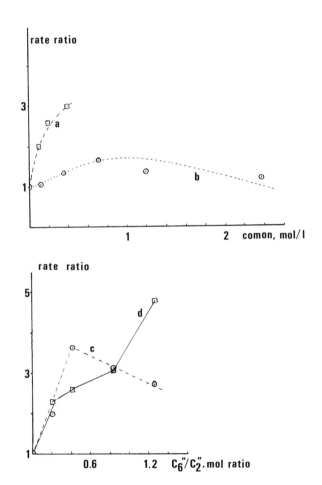

Fig. 9.12: Acceleration of ethylene polymerization in the presence of higher α-olefins: $TiCl_3$, 1-octene, ethylene pressure ≈1bar, reference 144 (a); "Zr-system", 1-hexene, ethylene pressure ≈ 4bar, reference 80 (b); $CrO_3/SiO_2 \cdot Al_2O_3$, reference 53, (c); $TiCl_4/MgCl_2$, reference 53 (d)

A more elaborate study is reported by Finogenova et al[53] who used chromium on a silica/alumina carrier of the required high surface area and wide pore type. They also report a rate increase for ethylene polymerization (up to three times) at low 1-hexene concentrations, see Fig. 9.12. At the same time the induction period, which is common with this type of catalyst, is decreased. These effects are only observed if both monomers are present right from the start. Neither the addition of 1-hexene to a running ethylene polymerization leads to higher rates, nor does the reverse situation - adding ethylene to a 1-hexene polymerization - affect this. Thus the effect must be due to the specific features of the reduction of Cr(VI) to the polymerization-active chromium species, perhaps hexene is more rapid in this, for instance due to the easier removal of the C_6-oxidation products compared to the ones derived from ethylene? The rate increase is general for linear α-olefins; branched ones, especially those on the 3-position, show decreased rates. The molecular weight distribution - as measured by fractionation - increases with increasing co-monomer incorporation, from Q's of 9 up to 22.

Gas phase copolymerizations are more efficient than slurry ones as lower co-monomer concentrations are required, moreover this is amplified by adsorption of the higher boiling co-monomers on the growing catalyst/polymer particles[157]. This phenomenon explains the fact that identical densities can be obtained by using either 4.5 % 1-butene or 2.4 % 1-hexene in the gas phase at about 85 °C (the co-monomer contents of the copolymers are of course nearly equal: 4.0 %mol for the butene and 3.7 % mol for the hexene copolymer).

Modified chromium catalysts which allow controlled copolymerization using only ethylene as feed have been developed[158]. In a recent version of this a standard CrO_3/SiO_2 catalyst is reacted with an organochromium compound, such as octakis(trimethylsilylmethyl)tetrachromium(II). This serves two purposes, firstly this compound reduces the Cr(VI) and thus activates this catalyst and removes the induction period, secondly it attaches itself to the support and is active in oligomerization. Choosing a support with isolated silanol groups is very helpful in this respect. This mixed catalyst, with a judicious choice of the organochromium compound, when used with ethylene as the sole feed gives rise to 1-butene and 1-hexene only, which of course are copolymerized with the ethylene. The extent of copolymerization, and thus the density, can be controlled by the ratio of the two catalyst components.

(ii) titanium supported on magnesium chloride

Somewhat more detailed knowledge is available for this catalyst type, mainly due to studies of the Hoechst group.

From patent examples Karol[20] calculated that on a
$TiCl_4/Mg(OH)Cl$ catalyst the reactivity ratio for ethylene was about
14 in copolymerization with propylene. This value is close to those
reported earlier for first generation $TiCl_3$ catalysts, suggesting a
similar structure of the active site(s) in the two types of catalyst.

A comparison of reactivity ratios for ethylene/propylene
copolymerizations was made by Ivanchev et al[123] showing a considerable variation with type of support. The data are given in Table 9.7
at the end of this paragraph. One should also note the difference
of copolymerization parameters when changing the co-catalyst from
$AlEt_2Cl$ to $AlEt_3$ in a first generation $TiCl_3$ catalyst, r_1 decreases
from far above 10 to about 5.

In a series of studies[11,126,159] Böhm reported an in-depth
kinetic investigation of ethylene copolymerization with 1-butene,
using the same catalyst as in his homopolymerization kinetics
study, viz. $Mg(OEt)_2/TiCl_4$. The results can be summarized as
follows:

- the copolymer composition is independent of the hydrogen
concentration,
- the polymerization rate does not significantly increase in
the presence of butene,
- the reactivity ratios are $r_1 = 67 \pm 8$ and $r_2 = 0.1 \pm 0.05$,
which gives r_1*r_2 equals 6.7 ($\pm$ 60 % relative), i.e. certainly
showing a blocky tendency,
- the fraction of reactor solubles at 85 °C increases strongly
with increasing co-monomer concentration in the feed (it
reaches 50% at 3.5 %mol butene incorporation in the copolymer),
- all rate constants k_{ij} were determined, k_{12} and k_{22} are
nearly equal and only the k_{1j} show a decrease in the presence
of hydrogen, k_{22} is also nearly equal to the homopolymerization rate of 1-butene,
- the molecular weight is lowered by both the co-monomer and
hydrogen, i.e. less hydrogen is required in copolymerization

to reach identical melt indices compared to homopolymer; the effects of these two are however not additive but synergistic, - the copolymer showed the same molecular weight distribution as the homopolymer (Q = 7.5).

In a later study these kinetic data were determined over a fairly wide temperature range, including a solution polymerization condition. The reactivity ratios observed are given in Table 9.7, no break in the values is found when passing from suspension to solution polymerization. The activation energies calculated from the temperature dependencies of the individual rate constants are rather low, in the order of 12 to 19 kJ/mol. From an analysis of these data, Böhm concludes that the complexation of the monomer with the active site is the rate determining process, showing a large entropy decrease.

Finogenova et al[53] also used a highly dispersed $TiCl_4/MgCl_2$ catalyst in copolymerization with 1-hexene, and again found a considerable rate increase for ethylene with increasing hexene in the feed. Furthermore the rate reached its steady-state condition sooner. In this case the effect on molecular weight is only very slight, with no effect at all on the distribution. A comparison of VCl_4 and $TiCl_4$ supported on $MgCl_2$ was made[160], with iBu_3Al as co-catalyst. In all studied monomer pairs the vanadium catalyst gave substantially lower values for the reactivity ratios (for the data see table 9.7), in line with other observations for vanadium. In a paper on a similar topic the authors suggested[161] that the generally observed increased ethylene polymerization rate in the presence of a co-monomer is due to higher ethylene concentrations in the copolymer surrounding the catalyst. This in its turn is due to the fact that monomer solubility in polymers is proportional to the fraction of amorphous polymer present, a fraction which is of course larger in copolymerizations.

For the supported polypropylene catalysts Luciani et al[55] studied the copolymerization of a number of ethylene/α-olefin pairs. They mentioned no rate enhancement for ethylene in the presence of higher olefins. A number of reactivity ratios have been determined, which show very large numbers from 1-butene upwards (see Table 9.7) and moreover r_1*r_2 is always very much larger than one, indicating blocky copolymerization behaviour. These large reactivity ratios necessitate fairly large fractions of 1-butene in the monomer feed in order to generate commercial products. The reactivity ratios are strongly dependent on the catalyst, the spherical type having lower values. The molecular weight distributions are fairly narrow, Q-values of about 4 are reported. Using the catalysts in the "granular" and "spherical" form, copolymer powders of sizes of 0.2 to 1 mm can be made directly, obviating the pelletization step.

Soga et al[162] studied the effect of added ethyl benzoate (EB) as electron donor on the reactivity ratios in ethylene/propylene copolymerization on the $MgCl_2/TiCl_4//AlEt_3$ catalyst system. The propylene incorporation decreases strongly, and r_1 (for ethylene) increases correspondingly, with increasing EB/Aluminium alkyl ratios. This is in line with the assumption that EB blocks multivacancy sites and thereby makes propylene incorporation more difficult (this being the monomer with the greater steric requirement). Similar trends were observed in ethylene/1-butene copolymerizations.

Using the same catalyst system in solution copolymerization at 170 °C, Kashiwa et al[163] found the following order in rate of co-monomer incorporation:

propylene > 1-butene > 4-methyl-1-pentene,

in all cases the yield easily exceeded 2.10^5 g/g Ti (in 40 minutes, at 25 bar monomer pressure). The MWD's broaden in all cases, but with propylene as co-monomer the broadening was greatest (Q from 4 to 6.5 at 6 % mol propylene incorporation). Similar broadening trends have been reported by Schlund[164] on commercial LLDPE's.

Using supported catalysts of the $MgCl_2$/ethyl benzoate/$TiCl_4$ and $MgCl_2$/di isobutyl phthalate/$TiCl_4$ type with $AliBu_3$ as co-catalyst, Tait et al[144] found the same general behaviour of acceleration of ethylene polymerization by the presence of α-olefin co-monomer. Also in this case the fraction of active sites in copolymerization is larger than in homopolymerization, being proportional to the applied co-monomer concentration.

Catalysts made by reacting $TiCl_4$ with magnesium alkyls were studied by Munoz-Escalona et al[165] in copolymerization of ethylene with 1-hexene. Decay kinetics were observed in copolymerization but not in homopolymerization, triethylaluminium giving the best kinetic results as compared to the isobutyl derivative. Again a strong enhancement in the rate is observed in the presence of the co-monomer. This group highlights differences in the morphology of the nascent polymer particles. In the copolymer case they are more porous, which could in their opinion explain the larger copolymerization rate.

A supported catalyst made by a different recipe was used by Ojala and Fink[150]: magnesium hydride is reacted with $TiCl_4$ and TEA is used as co-catalyst. A large number of co-monomer pairs were studied, and the reactivity ratios are given in table 9.6. The general trend is a decreasing copolymerization tendency with increasing chain length but not too regularly however. Remarkably, in NMR a fraction of pairs of co-monomers are always observed even at very low co-monomer contents of 2 to 3 %mol, suggesting that the first insertion makes the following insertion of the larger monomer easier. In line with this they also observe, just as Seppälä[145-148], that using 1-butene as third co-monomer tremendously increases the incorporation of longer olefins such as 1-hexadecene. The temperature dependence of the copolymerization kinetics for ethylene/1-octene was also reported[166]. In line with observa-

Table 9.6
Reactivity ratios at 20 °C on $MgH_2/TiCl_4$//TEA catalyst, from ref. 150

olefin carbon number	4	5	6	7	8	10	12	14	16
r_1		55	50	47	75	90	550	1500	1400	1050
r_2		0.02	0.02	0.02	0.013	0.01	0.004	0.004	0.0004	0.008
$r_1 * r_2$		1.1	1.0	0.94	0.98	0.90				

tions the ethylene reactivity ratio decreased and the other increased with increasing temperature, for instance at 20 °C r_1=115 and at 60 °C r_1=64. Using the assumption that 1% of the titanium is active in the polymerization (a value found in homopolymerizations) a propagation rate constant for ethylene (k_{11}) is calculated which is two times greater than in homopolymerization. This again illustrates the rate increase of ethylene in the presence of higher olefins.

Using $MgCl_2/TiCl_4$ supported on SiO_2, either in the presence or absence of ethylbenzoate, Spitz et al[167] observed a considerable increase in the 1-butene reactivity ratio with increasing 1-butene content in the feed (and thus the copolymer). The system behaves as if incorporated butene enhances the copolymerization rate, just as observed in other studies.

High pressure/high temperature copolymerizations have been carried out with a Ti/Mg catalyst with tetraisobutyl-di-aluminoxane as co-catalyst[168]. The co-monomer incorporation increases with increasing co-monomer chain length and with increasing pressure and temperature. The r_1 of ethylene decreases from 22.7 to 9.9 in going from 150 to 230 °C, the 1-butene reactivity ratio decreases with the same factor (0.10 to 0.04) leading to very different r_1*r_2's over this temperature range. This would lead to a change of blocky to alternating copolymerization in contrast to the work of CdF reported above on first generation $TiCl_3$ catalysts.

Recently studies appear which address the question of controlling the reactivity ratio by changing the catalyst, aiming at low reactivity ratios so as to simplify the copolymerization process and lower its cost. For instance a comparison was made[169] of catalysts prepared by reacting EtMgCl with $TiCl_4$, applying different Ti/Mg ratios and using in some cases a SiO_2 carrier, in their copolymerization with 1-hexene. Widely different kinetics are indeed observed, the highest reactivity towards 1-hexene being observed with a catalyst on a carrier and a Ti/Mg ration of 2.6. This catalyst shows an apparent reactivity ratio of 26, whilst the other catalysts show values of 50 up to 125 - this highest value belonging to a catalyst in which no silica carrier was used. All copolymers were shown to have a composition distribution, the catalyst with the highest copolymerization tendency giving a copolymer with the widest distribution. Another approach was followed by Muhlhaupt et al[170], who use in a supported type of catalyst sterically hindered phenols or esters both as internal as external donor. In the solid catalyst preparation 4-methyl-2,6-t-butylphenol is used. The homopolymerization activities depend strongly on the co-catalyst composition, which is varied by using the same phenol with in addition varying amounts of ethyltoluate. Conditions are for example found for which the activity in the homopolymerization of 4-methyl-1-pentene(4MP) is larger than either that for ethylene or propylene; or in other cases where the activity for ethylene and propylene is about equal. These conditions can nicely be used for copolymerization, for the couple ethylene and 4MP a wide density range is then accessible. The molecular weight distributions are relatively narrow (Q ≈ 4). Also EPR's have been made which according to NMR are rather random; no fractionation data are however given. For the copolymerization of ethylene or propylene with 4MP a considerable acceleration of the incorporation of 4MP is observed, this is similar to the acceleration reported in chapter 3. The above studies are interesting and show that in principle the tailoring of the reactivity ratio is a possibility, but more study is required to achieve good control at the required high activities.

TABLE 9.7

Reactivity ratios in ethylene copolymerization with supported catalysts.

Catalyst	reactivity ratios						tempe-rature °C	reference
	C_2	C_3	C_2	C_4	C_2	C_6		
TiCl$_4$/Mg(OH)Cl//AliBu$_3$	14	0.07						20
TiCl$_4$/MgO//AlEt$_3$	7.8	0.13					70	123
TiCl$_4$/Al$_2$O$_3$.SiO$_2$//AlEt$_3$	18.5	0.24					70	123
TiCl$_4$/MgCl$_2$//AliBu$_3$	16.9		28.6		55.5		70	160
VCl$_4$/MgCl$_2$//AliBu$_3$	3.7		12.7		38.5		70	160
Mg(OR)$_2$/TiCl$_4$//AlEt$_3$			51	0.04			120	126
			67	0.08			85	,,
			110	0.1			65	,,
			130	0.1			50	,,
"TiCl$_4$/MgCl$_2$/donor/AlR$_3$"[*]	10.7	0.17	115	0.116	655	0.17[**]	70	55

[*] : granular form of Montedison catalyst

[**] : for ethylene/1-octene 1500 and 0.18 resp.

In Fig. 9.13 we have collected a number of the above data, plotting the density of the copolymer against the feed ratio. Clearly the difference between various catalysts is enormous, especially when including a relationship found in a patent[57] which shows that a catalyst based on TiCl$_4$-electron donor complexes milled with MgCl$_2$ has very attractive copolymerization behaviour. Relatively low fractions of butene are required to reach the desired densities.

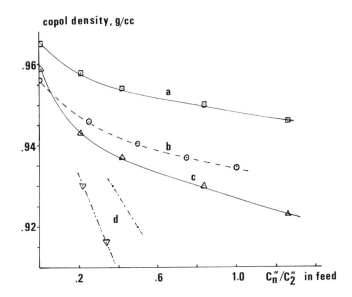

Fig. 9.13: Polymer density versus feed composition for different catalysts and co-monomers: $CrO_3/SiO_2.Al_2O_3$, 1-hexene, reference 53 (a); $Mg(OR)_2/TiCl_4$, 1-butene, reference 11 (b); $TiCl_4/MgCl_2$, 1-hexene, reference 53 (c); $MgCl_2/TiCl_4$.donor, 1-butene, reference 57 (d)

(iii) <u>supported organometallic compounds</u>

The few data known on the copolymerization behaviour show exceptionally high values for the reactivity ratio. Using chromocene on silica, Karol[43] reported a reactivity ratio of 72 in copolymerization with propylene. For ethylene-propylene this is one of the highest values recorded, only in the case of a homogeneous catalyst is a higher value described (see below). Using a patent example[61] of the neophyl-zirconium catalyst, and assuming ideal copolymerization, leads to a reactivity ratio for ethylene of 54, again in a propylene copolymerization. Such high values require extremely high co-monomer levels in order to achieve any significant degree of α-olefin incorporation.

9.5.3 <u>Homogeneous catalysts</u>

Surprisingly, in this class of (as yet non-commercial) catalysts more data are available than in many others. Wiman and Rubin[171] studied copolymerization of ethylene and 1-butene with Cp_2TiCl_2/Et_2AlCl in benzene as solvent from a mechanistic viewpoint, trying to answer in particular the question whether one or more active species are involved in this catalyst system. Their experiments are not suited to the determination of the copolymerization kinetics as the conversions reached were too high, but they observe the lower polymerization rate of 1-butene versus ethylene, the common molecular weight effects, and more importantly a very pronounced effect of aging of the catalyst system. An aged catalyst gives lower overall rates, lower butene incorporation and much higher molecular weights. From these observations they conclude the existence of at least two different active species. A similar conclusion was reached by Garcia Marti and Reichert[172] in their study of copolymerization of ethylene and styrene. A complicated kinetics was observed in this case, and the behaviour could be described by assuming that the reactivity ratios were a function of the monomer concentrations.

Other data in this catalyst class deal with the systems utilizing aluminoxanes as co-catalyst. Kaminsky's group in particular is active in this area. In 1981 it was reported[76] that with $Cp_2Ti(CH_3)_2$ and ethylene propylene mixtures a wide range of copolymers can be made (E_c range of 10 to 90 %), although the molecular weights at the propylene end of the range are very low (oils). Butene copolymers were mentioned as well as copolymers with alpha-omega-dienes which gave crosslinked copolymers, more detailed kinetic data were published in 1986. For ethylene-1-hexene copolymerization[78] the rate of ethylene polymerization at 60 °C in toluene shows a maximum as a function of the 1-hexene concentration in the polymerization feed, see Fig. 9.12. This behaviour is similar to the observations of Finogenova with other catalysts, whilst Tait observed only increasing rates, but in a smaller range of comonomer concentrations. The maximum was found to be independent of polymerization temperature and zirconium concentration. The activation energy measured was rather high (47.5 kJ/mol). The reactivity ratios were found to be r_1 = 55 and r_2 = 0.005 and nearly independent of temperature in the range 20 to 60 °C. These values are similar to those found in the case of $TiCl_3$ catalysts. The product of the reactivity ratios is clearly less than one, indicating an alternating tendency in the insertion of the co-monomer. Data on the homopolymerization of 1-hexene were also given but the conditions not specified, making it impossible to calculate the difference of homo- and copolymerization rate of 1-hexene, thus it is not clear whether the rate of hexene polymerization increases in the presence of ethylene. Copolymerization in this system is a complicated process since the zirconium concentration also has an effect on the incorporation of the hexene, which decreases at lower [Zr]. The order of reaction in aluminium compound is found to be 0.8, considerably higher than in homopolymerization. The narrow molecular weight distribution (Q≈2) indicates that only one active site is present.

Similar studies[80] were reported for 1-butene as co-monomer, here the rate decreases with increasing co-monomer concentration in the feed. In this case the reactivity ratios were determined as a function of temperature, the values are in the same range as for 1-hexene (r_1 = 65 at 60 °C). The individual rate constants were also calculated, these are high and thus in line with the high activities observed. The activation energies for these propagation rate constants are only 20 to 40 kJ/mol, highest for k_{11} and lowest for k_{22}. The r_1*r_2 value is about 1. Narrow molecular weight distributions were again found, with Q-values of 2 to 3. Propylene was also investigated, this followed roughly the same behaviour as the butene described above. The activities found are indeed very high with rates of 80 to 200 kg copolymer per g Zr and per hour (at 8 bars of ethylene pressure). Terpolymerization with ethylidene-norbornene was carried out successfully, thus the usual EPDM structure can be made on these homogeneous catalysts with a tremendously high yield. A peculiar feature of these copolymerizations is the increase in rate for the first 10 (!) hours, after which it stays constant for days on end.

The copolymerization with cyclo-olefins has been reported by Kaminsky and Spiehl[173], using the ethylene(bisindenyl)zirconium dichloride. Note that also homopolymerization of cyclopentene ($CyC_5^=$) is well possible[174] leading to a high melting, highly crystalline polymer. The rate of polymerization is somewhat lower in copolymerization than in homopolymerization of ethylene; the rate is a function of the monomer ratio however, a maximum in the rate is observed at a high cyclo-olefin/ethylene ratio. The cyclopentene ring stays intact in the polymer, i.e. there is no metathesis reaction. The $CyC_5^=$ incorporation is linear with the mol ratio in the feed, higher at lower polymerization temperature and increases at higher zirconium concentrations, i.e. this is a kinetically complex system. The reactivity ratios have been calculated using the Fineman-Ross approach, and are given in Table 9.8.

TABLE 9.8

Reactivity ratios in copolymerization with cycloolefins, from ref 173.

temperature °C	cyclopentene r_1	r_2	cycloheptene r_1
-10	80	<0.05	
0	120	<0.02	
10	250	<0.02	380
20	300	<0.007	
50			600

The temperature dependence is indeed strong. In this case, the co-monomer has almost no effects on the MW, although the MWD is broader than in homopolymerization, viz. a Q of 3 to 4.5 instead of the value of 2 which otherwise is observed. For other cyclo-olefins the following data apply: cyclohexene is inactive in this system, but both the heptene and the octene show considerable activity. The effects on the rate are larger in these cases. $CyC_7^=$ also shows only a marginal effect on the MW, whilst $CyC_8^=$ lowers this considerably, just as other linear α-olefins do.

Using the same isospecific catalyst the copolymerization kinetics of ethylene with propylene were determined[175]. Evaluation via Fineman-Ross plots lead to the values of

$$r_1=6.61 \qquad r_2=0.06 \qquad r_1r_2=0.40 \text{ at 50 °C, and}$$
$$r_1=6.26 \qquad r_2=0.11 \qquad r_1r_2=0.69 \text{ at 25 °C.}$$

These values are relatively low compared to most other systems. The MW increases with the ethylene content in the copolymer, but only beyond the 50/50 composition.

With Cp_2ZrCl_2/aluminoxane in toluene[132] the ethylene polymerization rate increases twofold upon addition of 1-hexene, and 1.5-fold when using propylene. These effects are reversible, i.e. when the propylene is depleted the rate returns to its previous homopolymerization value. The remarkable fact is that in the described system already all zirconium species are active in the polymerization, and thus a rate increase can then only(?) be rationalized by assuming an increase in the propagation rate constant.

Ewen[84,176] studied the effect of changes in the organic ligand of homogeneous catalyst systems on the reactivity ratio in ethylene/propylene copolymerizations. This has been discussed in chapter 3.

9.5.4 Block copolymerization

Cp_2TiCl_2/AlEt$_3$ in methylene chloride was used in an attempt to prepare block copolymers[69] by first polymerizing ethylene, degassing, and after a short time (not specified) followed by polymerizing propylene or butadiene. It is claimed that a mixture of blockcopolymer (insoluble) and homopolymer (atactic, thus soluble) is formed; the blockcopolymer could be separated from pure polyethylene by extraction at temperatures determined from turbidity measurements. The maximum yield of blockcopolymer was reported to be 50%. It was even possible to re-initiate ethylene polymerization and thus in principle prepare three-block copolymers PE-aPP-PE. However, neither a detailed description of the preparation of the diblockcopolymers nor the proof of their composition has been given. Considering the instability of this catalyst system these results must be viewed with some scepticism.

A completely different technique was applied by Fontanille and Siove[177,178] which involved transforming a poly-butadienyl lithium into an active Ziegler catalyst followed by polymerization of ethylene. This transformation was done by reacting the lithium compound with TiCl$_4$, the product being soluble in hexane provided the molecular weight of the butadiene chain is over 5000. The same holds for polystyrenyl lithium in toluene at molecular weights of over 10,000. Upon polymerization of ethylene on this complex, initially a homogeneous system was still present; at higher yields of polyethylene precipitation took place with a decrease in activity. Maximum activity was observed at a Li/Ti ratio of two. The efficiency of block copolymer formation could easily be determined after conversions which yielded a heterogeneous system, since the free polybutadiene stays in solution. Furthermore ozonization can isolate the polyethylene chains for a check on their molecular weight. It was found that in the PB-PE system crosslinks were formed, probably via addition of some 1,2-polymerized butadiene units to the titanium species, leading to erroneous values of the efficiency (see Fig. 9.14 for the reactions involved). Replacing PB by polystyrene led to an active system, but making no blockcopolymer at all, probably due to β-hydrogen elimination after substituting Ti for Li. This problem was overcome by capping the polystyrene with a few butadiene units before the titanium tetrachloride addition. In this way a well defined system was obtained showing a good agreement with the determined molecular weight of the polyethylene after ozonization and the molecular weight calculated from the observed conversions and the measured efficiency of block copolymer formation. The latter was found to be 0.22 ± 0.03. The structure of the active complex is thought to be $RTiCl_2 \cdot TiCl_3 \cdot 3LiCl$, and assuming this is a binuclear chloride and alkyl bridged titanium complex it contains just one vacancy suitable for monomer coordination. This structure fits in with the determined efficiency of the reaction, see figure 9.14, since a maximum of 25 % is attainable. The polymerization with the $PS-B_3-Ti$ system in toluene behaves as a living system, with no decay being

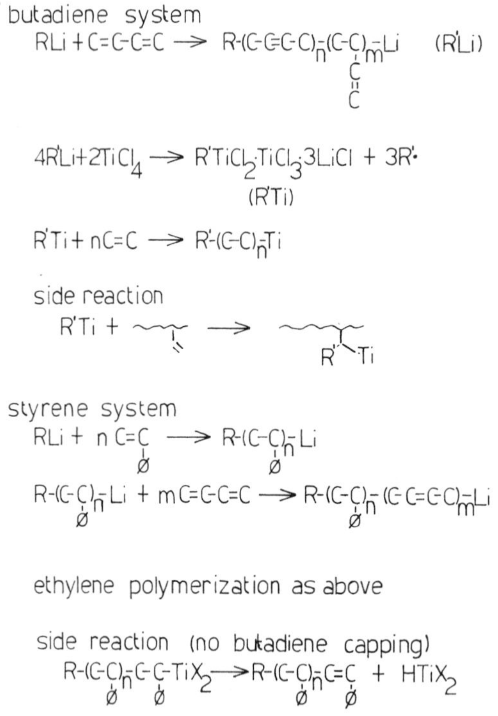

Fig. 9.14: Block copolymerization of butadiene (styrene) with ethylene

observed and the rates being proportional to the monomer and the active site concentration. This allows the calculation of the propagation rate constant. A value of 15 l/mol.s was found at 22 °C, which is rather similar to the one found in Cp_2TiCl_2/Al-alkyl combinations (about 50 l/mol.s, ref. 68) but much lower than in methyl aluminoxane systems (about 1000 l/mol.s, ref. 80). Note that no other (co-)catalyst components are present. This catalyst belongs in principle to the same class as the metal organic ones, mentioned in section 9.2.2 (e.g. the tetra-allyl complexes). This mode of polymerization has been used for preparing block copolymers of ethylene and butadiene as described in chapters 5 and 6.

9.5.5 Peculiarities in copolymerization

A number of observations are made in copolymerizations which are unexplained at the moment. The first one is the increase in the ethylene polymerization rate in the presence of a co-monomer. The observation is fairly general with the higher α-olefins (1-hexene and longer), and it is found with $TiCl_3$, supported $TiCl_4$ and homogeneous catalysts. Tait[179] put forward a number of possible explanations, which are:
- further fragmentation of catalyst in copolymerization,
- more active sites initiated in copolymerizations,
- displacement of site-blocking molecules by the co-monomer, and
- higher monomer concentration at the site due to the higher solubility of the monomer in the more amorphous phase. The fact that polyethylene appears mostly in a very highly crystalline form after a polymerization might be a case in point.

Some of these, particularly the first two are not, or are less, applicable to homogeneous catalysts (no fragmentation; sometimes already every transition metal atom present is active), whilst the latter should only hold in gas-phase or slurry polymerizations and not in true solution polymerizations. Maybe this constitutes a test for (part of) the rate-enhancing mechanism.

The second peculiarity is the increased incorporation of long-chain α-olefins when carrying out a terpolymerization with ethylene and 1-butene. No explanations at all have been put forward.

9.6 POLYETHYLENE PROCESSES

Limiting ourselves to the polyethylene processes using Ziegler-Natta or related catalysts, the general process consists of a feed purification section, the polymerization proper, the removal of monomer and possibly wax and/or diluent, and a polymer treatment including the dosing of additives (see for instance references 180 to 183). See also Fig. 9.15 and Fig. 9.16 for schemes of an old and a modern process. The feed purification is a very important item, just as in the polypropylene case, because of the high sensitivity of the catalysts employed towards certain impurities in the system. This section has to remove compounds such as water, CO, oxygen, and volatile sulfur, nitrogen and oxygen compounds to a very low level. The largest differences between the various polyethylene processes are found in the polymerization section. In principle the reactor systems chosen all have the same functions, namely to allow a controlled heat removal, and to maintain a well mixed system in order to have consistently identical conditions in the reactor as regards monomer, chain transfer agent and co-monomer. Broadly speaking four different kinds of processes can be distinguished:

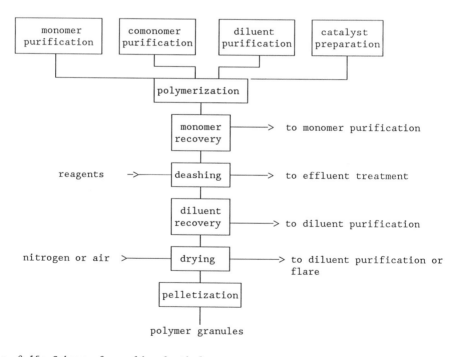

Fig. 9.15: Scheme of an old polyethylene process

- the slurry process
- the solution process
- the gas phase process
- the high pressure process.

The general features of these processes are similar to the ones described for polypropylene (see chapter 1.8), therefore only a short survey will suffice.

In the <u>suspension</u> process the conditions are chosen such that the polymer generated stays insoluble, a slurry is then formed on the heterogeneous catalysts employed. With the normal range of hydrocarbon diluents used the solubility of the polymer limits the temperature of polymerization to a maximum of 110 °C for homopolymers in low boiling alkanes such as isobutane; the allowed temperature is lowered in cycloparaffins as diluent and of course in the manufacture of copolymers as well (as in that case the dissolution temperatures are even lower). Initially the Phillips particle form process operated with n-pentane as diluent. This limited the

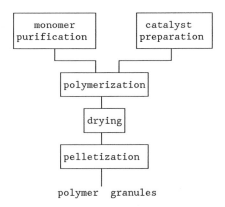

Fig. 9.16: Scheme of a modern polyethylene process (gas-phase)

operational temperature for homopolymer to 105 °C and allowed only the manufacture of polymers with a melt index below 0.1. This limit was increased by a number of measures[119], mostly catalyst modifications (as mentioned above factors such as pore size and activation temperature affect the molecular weight performance of the catalyst). However one change was independent of the catalyst: the use of a lower boiling, branched hydrocarbon allowed higher operating temperatures; isobutane gives as maximum temperature 111 °C, and is the diluent of choice (good performance and commercial availability). Note that an increase in operating temperature from 104 to 110 °C more than doubles the melt index. The heat of polymerization is always removed via the reactor walls. In the Phillips particle form process loop reactors are used with a low boiling hydrocarbon as diluent, in which all the reaction conditions are kept constant, which necessitates monomer recovery after flashing the slurry (see for instance reference 184). Other processes use multiple reactors, feeding ethylene only to the first reactor, this procedure ensuring a high ethylene conversion, thus obviating its recovery. General conditions are 80 to 100 °C and up to 35 bar in pressure, employing residence times in the order of 1 to 3 hours. The work-up of the polymer is usually executed in one of two significantly different ways, namely by filtration of the slurry or by evaporation (via flashing or by adding steam). In the filtration case, usually carried out in centrifuges, one has the ability to remove the very low molecular weight product (mostly called wax). Since this is more expensive than straight evaporation, catalysts with an acceptable wax formation are usually selected. In the Hoechst slurry process, diluent is recycled directly after the centrifugation of the slurry without any purification[185]. The yields of polymer on catalyst are so high (e.g. >3000 g/g cat for Phillips[119], i.e. >300,000 g PE/g Cr and thus less than 3 ppm Cr in polymer is present) that deashing is unnecessary.

In _solution_ processes higher temperatures are used, of up to 250 °C, mostly using a cycloparaffinic solvent, in which the polymer readily dissolves under those conditions. Due to the high temperatures employed, coupled to pressures in the order of 80 bars, the reactions are very fast and the residence times are in the order of minutes. In the Dutch State Mines version of this process the reaction is run adiabatically, which simplifies operation. The price to be paid for this process simplification is that the feed must be pre-cooled and added cold (about -40 °C) to control the final temperature.

Finally, the _gas-phase_ process uses gas as the medium for the polymerization, functioning both as heat removal medium and as feed. Different technological approaches have been used, the best known and commercially very succesful process is the Unipol process (see Fig. 1.69) of Union Carbide (see e.g. ref 186). This uses a fluidized-bed reactor, operating at around 90 °C and 20 bars total pressure. Economically this process is claimed to have an edge over the other ones. Another version is the high-pressure process which in effect uses the equipment of the radical process for making polyethylene with Ziegler-Natta catalysts[187]. Pressures range from a minimum of 300 up to about 1500 bars and temperatures are in the range of 220 to 260 °C, very short residence times in the order of tens of seconds are used. The true advantage of this process is the manufacture of a different type of polyethylene out of a LDPE plant. One could even make in-situ blends of (nearly) linear and very branched material in this process.

The type of co-monomer used differs for the different processes. The low boiling 1-butene is applied in the gas phase process and in high-pressure polymerizations, 1-butene, 1-pentene and 1-hexene in suspension processes, and the higher boiling ones (1-hexene and 1-octene) in the solution processes.

All the different technological forms of the processes have their particular stronger and weaker points. All are able to manufacture a range of polymers, but not all to the same extent or in the same range of melt index and density. As the polyethylene market is huge in some geographical areas, one could build a plant suitable for some specific grade, without necessitating that one plant has the process flexibility to make all grades. This would truly give economy of scale. Some of the weaker points of the various processes are:
- in the suspension process, the rather low polymerization temperatures when making copolymers, caused by copolymer solubility,
- in the solution process, the lower throughput for high molecular weight polymers, due to their higher viscosity; secondly higher molecular weight polymers are difficult to make in a high temperature process due to the fast transfer under those conditions; thirdly higher olefins are always formed in the normal Aufbau reaction of ethylene and the aluminium alkyl applied as co-catalyst - their incorporation into the polymer chain limits the density of the product,
- in the gas phase process it is difficult to incorporate large amounts of the higher co-monomers, such as 1-hexene or 1-octene, since this requires high feed concentrations which

438

in turn introduce more of a liquid phase, endangering the mixing in the fluid bed.

REFERENCES

1. H.R.Sailors and J.P.Hogan, History of polyolefins, *J.Macromol.Sci.-Chem.*, *A15 (1981) 1377-1402*
2. F.M.McMillan, The chain straighteners, Fruitful innovation, the discovery of linear and stereoregular synthetic polymers, The MacMillan Press Ltd, 1979
3. K.E.Bett, B.Crossland, H.Ford and A.K.Gardner, Review of the engineering developments in the high pressure polyethylene process 1933 - 1983, in Golden Jubilee Conference, Polyethylenes 1933 - 1983, The Plastics and Rubber Institute, London
4. T.O.J.Kresser, Polyethylene, Reinhold, New York, 1957
5. R.A.V.Raff and J.B.Allison, Polyethylene, Interscience, High Polymer Series XI, 1956
6. A.Hill and K.W.Doak, Mechanism of free radical polymerization of ethylene, in Crystalline olefin polymers, Part I, R.A.V.Raff and K.W.Doak editors, Interscience, High Polymer Series XX, 1964
7. P.Ehrlich, Fundamentals of the free-radical polymerization of ethylene, *Advances in Polymer Science, 7 (1970) 386-448*
8. R.L.Boysen, High pressure (low and intermediate density) polyethylene, in Kirk-Othmer Encyclopedia of Science and Technology, 1981 edition, volume 16, p 402-420
9. J.N.Hay, D.S.Harris and M.Wiles, Polymerization of ethylene by n-butyl lithium, *Polymer, 17 (1976) 613-617*
10. P.J.T.Tait, Ziegler-Natta and related catalysts, chapter 3 in "Developments in polymerisation-2. Free-radical, condensation, transition metal and template polymerisations", R.N.Haward editor, Applied Science Publishers, London, 1979
11. L.L.Böhm, Fortschritte bei der Polymerisation von Ethylen mit Ziegler-Katalysatoren, *Angew.Makrom.Chemie, 89 (1980) 1-32*
12. Yu.I.Yermakov, B.N.Kuznetsov and V.A.Zakharov, Catalysis by supported complexes, (especially chapters 3 and 5), Elsevier, Amsterdam, 1981
13. P.D.Gavens, M.Bottrill, J.W.Kelland and J.McMeeking, Ziegler-Natta catalysis, chapter 22.5 in Comprehensive Organometallic Chemistry, Volume 3, G.Wilkinson, F.G.A.Stone and E.W.Abel editors, Pergamon Press, Oxford, 1982
14. F.J.Karol, Studies with high activity catalysts for olefin polymerization, *Catal. Rev.- Sci.Eng., 26 (1984) 557-595*
15. T.E.Nowlin, Low pressure manufacture of polyethylenes, Progr. Polym. Sci., Volume 1, 1985, p 29-55
16. K-Y.Choi and W.H.Ray, Recent developments in transition metal catalyzed olefin polymerizations. A survey. I. Ethylene Polymerization, *J.Macromol.Sci.- Rev. Macromol.Chem.Phys., C25 (1985) 1-55*
17. J.P.Machon, R.Hermant and J.P.Houzeaux, Etude des activités catalytiques du trichlorure de titane violet dans la polymerisation haute température de l'ethylene, *J.Pol.Sc., Pol.Symp., 52 (1975) 107-117*
18. B.Diedrich, Second generation Ziegler polyethylene processes, p316-321 in Polymer Preprints, Volume 16, number 1, 1975, ACS, Washington
19. D.Constantin, M.Hert and J-P.Machon, Structure of polyethylene produced at high pressure by simultaneous ionic and radical mechanisms, *Makrom.Chemie, 179 (1987) 1581-1591*
20. F.J.Karol, Catalysts supported, p 120 - 146 in Encyclopedia of polymer science and technology, supplement Volume 1, J.Wiley, 1976
21. M.P.McDaniel, Supported chromium catalysts for ethylene polymerization, *Adv. in Catal., 33 (1985) 47-98*
22. US Patent 2 825 721, published March 1958, to Phillips Petroleum Company (J.P.Hogan and R.L.Banks as inventors)

23. J.P.Hogan and D.R.Witt, Supported chromium catalysts for ethylene polymerization, in Preprints, Division of Petroleum Chemistry, ACS, Volume 24, part 2, 1979

24. M.P.McDaniel and M.M.Johnson, Comparison of Cr/SiO2 and Cr/AlPO4 polymerization catalysts. 1. Kinetics, *J.Catal.*, *101* (1986) 446-457

25. M.P.McDaniel and M.M.Johnson, Comparison of Cr/SiO2 and Cr/AlPO4 polymerization catalysts. 2. Chain transfer, *Macromolecules, 20* (1987) 773-778

26. J.P.Hogan, Ethylene polymerization catalysis over chromium oxide, *J.Pol.Sc.*, *A1, 8* (1970) 2637-2652

27. J.N.Finch, Reaction studies on supported chromic anhydride catalysts, *J.Catal,, 43* (1976) 111-121

28. C.Groeneveld, G.C.Schuyt et al, Hydrogenation of olefins and polymerization of ethene over chromium oxide/silica catalysts. I. Preparation and structure of the catalyst, *J.Catal.*, *59* (1979) 153-167

29. A.Zecchina, E.Garrone, G.Ghiotti, C.Morterra and E.Borello, On the chemistry of silica supported chromium ions. I. Characterization of the samples, *J.Phys.Chem.*, *79* (1975) 966-972

30. M.P.McDaniel, The state of Cr(VI) on the Cr/silica polymerization catalysts, *J.Catal.*, *67* (1981) 71-76

31. M.P.McDaniel, The state of Cr(VI) on the Phillips polymerization catalyst. II. The reaction between silica and CrO2Cl2, *J.Catal.*, *76* (1982) 17-28

32. ditto, III. The reaction between CrO3/silica and HCl, *ibid.* 29-36

33. ditto, IV. Saturation coverage, *ibid.* 37-47

34. B.Fubini, G.Ghiotti, L.Stradella, E.Garrone and C.Morterra, The chemistry of silica-supported Cr ions: A characterization of the reduced and oxidized forms of chromia/silica catalyst by calorimetry and ultraviolet-visible spectroscopy, *J.Catal.*, *66* (1980) 200-213

35. H.L.Krauss, Key intermediates in heterogeneous olefin polymerization, IUPAC, 28th Macromolecular Symposium, 12-16 july, 1982

36. H.L.Krauss and D.Xaumann, Die Complexbildung von Oberflächen-Cr(II)/silicagel mit Wasser, Alkohol, Silanol, Siloxan und mit benachbarten Oberflächengruppen vom Typ XOY, *Z.Anorg.Allg.Chem.*, *446* (1978) 23-38

37. K.Weiss et al, Investigation of polymerization and metathesis reactions. part XIII: heterogeneous bimetallic metathesis catalysts by reaction of differently substituted Fischer-type tungsten carbene complexes with reduced Phillips catalyst, *J.Molec.Catal.*, *46* (1988) 341-349

38. E.L.Weist, A.H.Ali and W.C.Conner, Morphological study of supported chromium polymerization catalysts. 1. Activation, *Macromolecules, 20* (1987) 689-693

39. T.J.Pullukat, M.Shida and R.E.Hoff, Titanium modified chromium catalysts for ethylene polymerization, p 697-712 in Transition metal catalyzed polymerizations, Alkenes and Dienes, Part B, R.P.Quirck editor, Harwood Academic Publishers, London, 1981

40. M.P.McDaniel, M.B.Welch and M.J.Dreiling, The activation of the Phillips polymerization catalyst, III, Promotion by titania, *J.Catal.*, *82* (1983) 118-126

41. US Patent, 4 011 382, published 8 march 1977, to Union Carbide

42. W.L.Carrick, R.J.Turbett, F.J.Karol, G.L.Karapinka, A.S.Fox and R.N.Johnson, Ethylene polymerization with supported bis(triphenylsilyl) chromate catalysts, *J.Pol.Sc.*, *A1, 10* (1972) 2609-2620

43. F.J.Karol, G.L.Karapinka, C.Wu, A.W.Dow, R.N.Johnson and W.L.Carrick, Chromocene catalyst for ethylene polymerization: Scope of the polymerization, *J.Pol.Sc.*, *A1, 10* (1972) 2621-2637

44. F.J.Karol and C.Wu, Chromocene-based catalysts for ethylene polymerization: Thermal removal of the cyclopentadienyl ligand, *J.Pol.Sc.*, *Pol.Chem.*, *12* (1974) 1549-1558

45. F.J.Karol, et al, Supported bis(indenyl)- and bis(fluorenyl)- chromium catalysts for ethylene polymerization, *J.Pol.Sc.,Pol.Chem.*, *16* (1978) 771-778

440

46. German Patent, OLS 2 802 517, published 27 july 1978, to Union Carbide
47. US Patent, US 3 709 853, published 9 jan. 1973, to Union Carbide
48. German Patent, OLS 2 742 543, published 23 march 1978, to Union Carbide
49. H.L.Hsieh, Recent developments on transition metal catalysts for olefin polymerization, *Polymer Journal*, <u>12</u> *(1980) 597-602*
50. N.Kashiwa, Super active catalyst for olefin polymerization, *Polymer Journal*, <u>12</u> *(1980) 603-608*
51. N.Kashiwa and T.Tsutsui, Highly active MgO-supported TiCl4 catalysts for the ethylene polymerization, *Polymer Bulletin*, <u>11</u> *(1984) 313-317*
52. British Patent, Br 1 306 001, published 7 Feb 1973, to Hoechst
53. L.T.Finogenova, V.A.Zakharov, A.A.Buniyat-Zade, G,D.Bukatov and T.K.Plaksunov, Study of copolymerization of ethylene with hex-1-ene on applied catalysts, *Pol.Sc.USSR*, <u>22</u> *(1980) 448-454*
54. A.Greco, G.Bertolini and S.Cesca, New crystalline complex chlorides containing transition metal chlorides or oxychlorides as components of coordination catalysts, *J.Appl.Pol.Sc.*, <u>25</u> *(1980) 2045-2061*
55. L.Luciani, G.Foschini, C.Cipriani and C.A.Trischman, Linear low density polyethylene by a new economical slurry process. Part I. Catalysts and catalysis, kinetics and correlation between structure and polymer properties, presented at Antec conference (the 40th annual technical conference of the SPE), may 10-13 1982, San Fransisco
56. Br. Patent, 2 024 832, published 16 jan 1978, to Nippon Oil Company
57. Eur.Patent, 0 011 308, published 28 may 1980, to Union Carbide Corporation
58. D.G.H.Ballard, Pi and Sigma transition metal carbon compounds as catalysts for the polymerization of vinyl monomers and olefins, *Adv. in Catalysis*, <u>23</u> *(1973) 263-325*
59. D.G.H.Ballard, E.Jones, R.J.WYatt, R.T.Murray and P.A Robinson, Highly active polymerization catalysts of long life derived from σ- and π-bonded transition metal alkyl compounds, *Polymer*, <u>15</u> *(1974) 169-174*
60. P.D.Smith and M.P.McDaniel, Ethylene polymerization catalysts from supported organotransition metal complexes: I. Pentadienyl derivatives of Ti, V and Cr, *J.Pol.Sc., Pol.Chem.*, <u>27</u> *(1989) 2695-2710*
61. US Patent, 3 932 307, published 13 Jan 1976, to DuPont (R.A.Setterquist)
62. US Patent, 3 950 269, published 13 April 1976, to DuPont (R.A.Setterquist)
63. R.A.Setterquist, F.N.Tebbe and W.G.Peet, Tetraneophylzirconium and its use in the polymerization of olefins, p167sqq. in Coordination Polymerization, C.C.Price and E.J.Vandenberg editors, Polymer Science and Technology Series, Vol. 19, Plenum Press, New York, 1983
64. US Patent 2 827 446, published 1958, to D.S.Breslow
65. G.Natta, P.Pino, G.Mazzanti and U.Giannini, *J.Am.Chem.Soc.*, <u>79</u> *(1954) 2975*
66. G.Henrici-Olivé and S.Olivé, Koordinative Polymerisation an löslichen Übergangsmetall Katalysatoren, *Advances in Pol.Sc.*, <u>6</u> *(1969) 421-472*
67. H.Sinn and W.Kaminsky, Ziegler-Natta catalysis, p99-149 in Adv.in Organometallic Chemistry, Volume 18, F.G.A.Stone and R.West editors, Academic Press, New York, 1980
68. K.H.Reichert, Polymerization of α-olefins with soluble Ziegler catalysts, p 465-495 in Transition metal catalyzed polymerizations, Alkenes and dienes, Part B, R.P.Quirk editor, Harwood Academic Publishers, London, 1981
69. J.N.Hay and R.M.S.Obaid, The polymerization of ethylene with a soluble titanium-aluminium complex, *Europ.Pol.J.*, <u>14</u> *(1978) 965-969*
70. K.H.Reichert and K.R.Meyer, Zur Kinetik der Niederdruckpolymerisation von Äthylen mit löslichen Ziegler-Katalysatoren, *Makrom.Chemie*, <u>169</u> *(1973) 163-176*
71. W.P.Long and D.S.Breslow, Der Einfluss von Wasser auf die katalytische Aktivität von Bis(π-cyclopentadienyl)titandichlorid - Dimethylaluminiumchlorid zur Polymerisation von Äthylen, *Liebigs Ann.Chem.*, *1975, 463-469*
72. A.Andresen et al, Halogen-free soluble Ziegler catalysts for the polymerization of ethylene. Control of molecular weight by choice of temperature, *Angew, Chemie, Int. Ed.*, <u>15</u> *(1976) 630-631*
73. OLS 2 608 933 and OLS 2 608 863 to Sinn, Kaminsky and others

74. E.Gianetti, G.M.Nicoletti and R.Mazzocchi, Homogeneous Ziegler-Natta catalysis. II. Ethylene polymerization by IVB transition metal complexes/methyl aluminoxane catalyst systems, *J.Pol.Sc., Pol.Chem., 23* (1985) 2117-2133

75. H.Sinn, W.Kaminsky, H.J.Vollmer and R.Woldt, "Living polymers" on polymerization with extremely productive Ziegler catalysts, *Angew.Chemie, Int.Ed.Engl., 19* (1980) 390-392

76. W.Kaminsky, Polymerization and copolymerization with a highly active, soluble Ziegler-Natta catalyst, p 225-244 in Transition metal catalyzed polymerizations, Alkenes and dienes, Part A, R.P.Quirk editor, Harwood Academic Publishers, London, 1981

77. J.Herwig and W.Kaminsky, Halogen-free soluble Ziegler catalysts with methylaluminoxane as catalyst, *Polymer Bulletin, 9* (1983) 464-469

78. W.Kaminsky, Preparation of special polyolefins from soluble zirconium compounds with aluminoxane as co-catalyst, p 293-304 in Catalytic polymerization of olefins, T.Keii and K.Soga editors, Kodansha/Elsevier, Tokyo/Amsterdam, 1986

79. W.Kaminsky, K.Külper and S.Niedoba, Olefin polymerization with highly active soluble zirconium compounds using aluminoxane as co-catalyst, *Makrom.Chemie, Macromol.Symp., 3* (1986) 377-387

80. W.Kaminsky and M.Schlobohm, Elastomers by atactic linkage of α-olefins using soluble Ziegler catalysts, *Makrom.Chemie, Macromol.Symp., 4* (1986) 108-118

81. J.W.C.Chien and B-P.Wang, Metallocene-methylaluminoxane catalysts for olefin polymerization. I. Trimethylaluminium as co-activator, *J.Polym.Sc.,Pol.Chem., 26* (1988) 3089-3102

82. R.F.Jordan et al, Ethylene polymerization by a cationic dicyclopentadienyl zirconium(IV) alkyl complex, *J.Am.Chem.Soc., 108* (1986) 7410-7411 and ibid *109* (1987) 4111sqq

83. European Patent, Eur 0 129 368, published 27 dec. 1984, to Exxon (J.A.Ewen and H.C.Welborn)

84. European Patent, Eur 0 128 046, published 12 dec. 1984, to Exxon (J.A.Ewen and H.C.Welborn)

85. A.Ahlers and W.Kaminsky, Variation of molecular weight distribution of polyethylenes obtained with homomgeneous Ziegler-Natta catalysts, *Makrom.Chemie, Rapid Commun., 9* (1988) 457-461

86. K.Soga, C-H.Yu and T.Shiono, Polymerization of α-olefins with the catalyst system prepared from a hydrated transition metal compound and trimethylaluminium, *Makrom.Chemie, Rapid Commun., 9* (1988) 141-144

87. K.Soga, S.Chen, T.Shiono and Y.Doi, Homo and copolymerization of ethylene and propylene with a soluble chromium catalyst, *Polymer, 26* (1985) 1888-1890

88. K.Soga, S.Chen, T.Shiono and Y.Doi, Preparation of highly active Cr-catalysts for ethylene polymerization, *Polymer, 26* (1985) 1891-1894

89. K.Soga, S.Chen, T.Shiono and Y.Doi, Polymerization of ethylene with Cr[OC(CH3)3]4 / AlEt2Cl as catalyst modified with some metal chlorides, *Makrom.Chemie, 187* (1986) 351-356

90. S-N.Gan et al, Modified chromium(III) acetate used in the polymerization of α-olefins, *J.Catal., 105* (1987) 249-253

91. T.Yano, S.Ikai, M.Shimuzu and K.Washio, Soluble magnesium-titanium catalysts for ethylene polymerization, I., *J.Pol.Sc., Pol.Chem., 23* (1985) 1455-1467

92. T.Yano, S.Ikai and K.Washio, Soluble magnesium and titanium catalyst components for ethylene polymerization, *J.Pol.Sc., Pol.Chem., 23* (1985) 3069-3080

93. U.Klabunde, R.Mulhaupt, et al, Ethylene homopolymerization with P,O-chelated nickel catalysts, *J.Pol.Sc., Pol.Chem., 25* (1987) 1989-2003

94. K.A.Ostoja Starzewski et al, Linear and branched polyethylenes by new coordination catalysts, p349-360 in Transition metals and organometallics as catalysts for olefin polymerization, W.Kaminsky and H.Sinn editors, Springer, Berlin, 1988

95. M.N.Berger and B.M.Grieveson, Kinetics of the polymerization of ethylene

with a Ziegler-Natta catalyst. I. Principal kinetic features, *Makrom.Chemie, 83 (1965) 80-99*

96. B.M.Grieveson, ditto, II. Factors controlling molecular weight, *Makrom.Chemie, 84 (1965) 93-107*

97. J.P.Machon, Polyéthylènes préparés par catalyse Ziegler a haute température, *Europ.Pol.J., 12 (1976) 805-811*

98. R.L.J.Graf, G.Kortleve and C.G.Vonk, On the size of the primary particles in Ziegler catalysts, *Polymer Letters, 8 (1970) 735-739*

99. R.H.Marchessault, B.Fisa and H.D.Chanzy, Nascent morphology of polyolefins, *Critical reviews in Macromol. Sc., october 1972, 315-349*

100. M.P.McDaniel and M.B.Welch, The activation of the Phillips polymerization catalyst. I. Influence of the hydroxyl population., *J.Catal., 82 (1983) 98-109*

101. R.Merryfield, M.P.McDaniel and G.Parks, An XPS study of the Phillips Cr/silica polymerization catalyst, *J.Catal., 77 (1982) 348-359*

102. M.B.Welch and M.P.McDaniel, ditto. II. Activation by reduction/reoxidation, *J.Catal., 82 (1983) 110-117*

103. R.Spitz, A.Revillon and A.Guyot, Supported chromium oxide catalyst for olefin polymerization. XI. Comparison between ethylene and propylene polymerization, *J.Macromol.Sci.,Chem., A8 (1974) 1129-1136*

104. R.Spitz, G.Vuillaume, A.Revillon and A.Guyot, Supported chromium oxide catalysts for olefin polymerization. VI. Activation process and catalytic activity versus ethylene polymerization, *J.Macromol.Sci., Chem., A6 (1972) 153-168*

105. M.P.McDaniel, The state of Cr on the Phillips polymerization catalyst, p 713-736 in Transition metal catalyzed polymerizations, Alkenes and Dienes, Part B, R.P.Quirk editor, Harwood Academic Publishers, 1981

106. K.Weiss and H-L.Krauss, Surface compounds of transition metals. XXVIII. Polymerization of n-1-alkenes with surface chromium(II) on silica gel support at low temperatures and normal pressure, *J.Catal., 88 (1984) 424-430*

107. H.L.Krauss, New results with Phillips systems: surface compounds of transition metals, part XXXII, *J.Molec.Catal., 46 (1988) 97-108*

108. D.D.Beck and J.H.Lunsford, The active site for ethylene polymerization over chromium supported on silica, *J.Catal., 68 (1981) 121-131*

109. B.Rebenstorf and R.Larson, Polymerization of ethylene by chromium surface compounds on silica gel, *J.Catal., 84 (1983) 240-243*

110. D.L.Myers and J.H.Lunsford, Silica-supported chromium catalysts for ethylene polymerization, *J.Catal., 92 (1985) 260-271*

111. D.L.Myers and J.H.Lunsford, Silica-supported chromium catalysts for ethylene polymerization: the active oxidation states of chromium, *J.Catal., 99 (1986) 140-148*

112. R.Spitz, Supported chromium oxide catalysts for olefin polymerization. VII. Kinetics of ethylene polymerization, *J.Macromol.Sci., Chem., A6 (1972) 169-176*

113. V.A.Zakharov and Yu.I.Ermakov, Kinetic study of ethylene polymerization with chromium oxide catalysts by a radiotracer technique, *J.Pol.Sc., A1 (1971) 3129-3146*

114. L.M.Baker and W.L.Carrick, *J.Org.Chem., 33 (1968) 616*

115. G.Ghiotti, E.Garrone and A.Zecchina, IR investigation of polymerization centres of the Phillips catalyst, *J.Molec.Catal., 46 (1988) 61-77*

116. M.P.McDaniel and D.M.Cantor, Hydrogen transfer during propagation of the Phillips catalyst, *J.Pol.Sc.,Pol.Chem., 21 (1983) 1217-1221*

117. F.J.Karol, G.L.Brown and J.M.Davison, Chromocene-based catalysts for ethylene polymerization: Kinetic parameters, *J.Pol.Sc., Pol.Chem., 11 (1973) 413-424*

118. M.P.McDaniel, Fracturing silica-based catalysts during ethylene polymerization, *J.Pol.Sc., Pol.Chem., 19 (1981) 1967-1976*

119. J.P.Hogan, D.D.Norwood and C.A.Ayres, Phillips petroleum company loop reactor polyethylene technology, *J.Appl.Pol.Sc., Applied Polymer Symposia, 36 (1981) 49-60*

120. T.Davidson, Microstructure of particle-form polyethylene, *Pol.Letters,* _8_ (1970) 855-859

121. A.A.Baulin, V.N.Sokolov, A.S.Semenova, N.M.Chirkov and L.F.Shalayeva, Determination of the concentration of active centers and the rate constants of the separate stages in polymerization of ethylene on "deposited" catalysts based on TiCl4 and AlEt3, *Pol.Sc.USSR.,* _17_ (1975) 51-60

122. V.A.Zakharov, S.I.Makhtarulin and Yu.I.Yermakov, Kinetic behaviour of ethylene polymerization in the presence of highly active titanium-magnesium catalysts, *React.Kinet.Catal.Letters,* _9_ (1978) 137-142

123. S.S.Ivanchev, A.A.Baulin and A.G.Rodionov, Promotion by supports of the reactivity of propagating species of Ziegler-supported catalytic systems for the polymerization and copolymerization of olefins, *J.Pol.Sc., Pol.Chem.,* _18_ (1980) 2045-2050

124. L.L.Böhm, Ethylene polymerization process with a highly-active Ziegler-Natta catalyst: 1. Kinetics, *Polymer,* _19_ (1978) 553-561

125. ditto, 2. Molecular weight regulation, *ibid,* 562-566

126. L.L.Böhm, Homo- and copolymerization with a highly active Ziegler-Natta catalyst, *J.Appl.Pol.Sc.,* _29_ (1984) 279-289

127. V.A.Zakharov, G.D.Bukatov and Yu.I.Yermakov, On the mechanism of olefin polymerization by Ziegler-Natta catalysts, Adv. in Polymer Sc., Volume 51, Springer Verlag, 1983

128. A.Akar, N.C.Billingham and P.D.Calvert, The morphology of polyethylenes produced by alumina-supported transition metal alkyls, p 779-798 in Transition metal catalyzed polymerizations, alkenes and dienes, Part B, R.P.Quirk editor, Harwood Academic Publishers, London, 1981

129. J.Cihlar, J.Mezjlik and O.Hamrik, Influence of water on ethylene polymerization catalyzed by titanocene systems, *Makrom.Chemie,* _179_ (1978) 2553-2558

130. J.Chilar, J.Mejzlik, O.Hamlik, P.Hudec and J.Majer, Polymerization of ethylene catalyzed by titanocene systems. I. Catalytic systems Cp2TiEtCl/AlEtCl2 and Cp2TiEtCl/(AlEtCl2 + H2O), *Makrom.Chemie,* _181_ (1980) 2549-2561

131. W.Kaminsky, M.Miri, H.Sinn and R.Woldt, Bis(cyclopentadienyl)zirkon-Verbindungen und Aluminoxan als Ziegler-Katalysatoren für die Polymerisation und Copolymerisation von Olefinen, *Makrom.Chem., Rapid Commun.,* _4_ (1983) 417-421

132. T.Tsusui and N.Kashiwa, Kinetic study on ethylene polymerization with Cp2ZrCl2/methylaluminoxane catalyst system, *Polymer Commun.,* _29_ (1988) 180-183

133. P.J.Tait et al, The effect of 14CO contact times on active centre determinations for the polymerization of ethylene catalysed by Cp2ZrCl2/methylaluminoxane, *Makrom.Chemie, Rapid Commun.,* _9_ (1988) 393-398

134. J.C.W.Chien and B-P.Wang, Metallocene-methylaluminoxane catalysts for olefin polymerizations. IV. Active site determinations and limitation of the 14CO radiolabelling technique, *J.Pol.Sc., Pol.Chem.,* _27_ (1989) 1539-1557

135. J.C.W.Chien and A.Razavi, Metallocene-methylaluminoxane catalysts for olefin polymerization. II. bis(neomenthyl cyclopentadienyl)zirconium dichloride, *J.Pol.Sc., Pol.Chem.,* _26_ (1988) 2369-2380

136. U.Zucchini and G.Cecchin, Control of molecular-weight-distribution in polyolefins synthesized with Ziegler-Natta catalytic systems, *Advances in Polymer Science, volume* _51_, 1983, 101-153

137. H.Wesslau, Über Molekulargewichtsverteilungen von Niederdruckpolyäthylene, 3. Mitteilung : Niederdruckpolyäthylene mit enger molekulargewichtsverteilungen, *Makrom.Chemie,* _26_ (1958) 102-118

138. R.D.A.Lippman, Copolymerization kinetics of α-olefins using Natta catalysts, *Polymer preprints,* 8 (1967) 396-399, ACS, Washington

139. S.Davison and G.L.Taylor, Sequence length and crystallinity in α-olefin terpolymers, *Br.Polym.J.,* _4_ (1972) 65-82

140. W.Cooper, Kinetics of polymerization initiated by Ziegler-Natta and related catalysts, chapter 3 in "Chemical Kinetics", C.H.Bamford and C.F.H.Tipper

444

editors, Volume 15, Non-radical polymerization.

141. J.P.Machon, Some aspects of the ethylene copolymerization in the high
     pressure Ziegler process, p 639-649 in Transition metal catalyzed
     polymerizations, alkenes and dienes, Part B, R.P.Quirk editor, Harwood
     Academic Publishers, London, 1981

142. K.Soga, H.Yanagihara and D-H.Lee, Evaluation of olefin reactivity ratios
     over highly isospecific Ziegler-Natta catalyst, *Makrom.Chemie, 190 (1989)*
     37-44

143. K.Soga, H.Yanagihara and D-h.Lee, Effect of monomer diffusion in the
     polymerization of olefins over Ziegler-Natta catalysts, *Makrom.Chem., 190*
     (1989) 995-1006

144. P.J.T.Tait, G.W.Downs and A.A.Akinbami, Copolymerization of ethylene and
     α-olefins - a kinetic consideration, presented at the Akron conference,
     1986

145. J.V.Seppälä, Copolymers of ethylene with butene-1 and long chain α-olefins.
     I. Decene-1 as long chain olefin, *J.Appl.Pol.Sc., 30 (1985) 3545-3556*

146. J.V.Seppälä, Copolymers of ethylene with butene-1 and long chain
     α-olefins, II. Dodecene-1 as long chain α-olefin, *J.Appl.Pol.Sc., 31*
     (1986) 657-665

147. J.V.Seppälä, ditto, III. Hexadecene-1 as long chain α-olefin,
     J.Appl.Pol.Sc., 31 (1986) 699-707

148. J.V.Seppälä, Ethylene terpolymerization with 1-butene and long chain
     α-olefins: Reactivity ratios, *Macromolecules, 18 (1985) 2409-2412*

149. J.Tulisalo, J.Seppälä and K.Hästbacka, Determination of branches in
     terpolymers of ethylene, 1-butene, and long α-olefin by
     pyrolysis-hydrogenation gas chromatography, *Macromolecules, 18 (1985)*
     1144-1147

150. T.Ojala and G.Fink, Structure study of ethene/α-olefin copolymers,
     Makrom.Chemie, Rapid Commun., 9 (1988) 85-89

151. G.Natta, G.Dall'Asta, G.Mazzanti, I.Pasquon, A.Valvassori and A.Zambelli,
     Crystalline alternating ethylene-cyclopentene copolymers and other
     ethylene-cycloolefin copolymers, *Makrom.Chemie, 54 (1962) 95-101*

152. F.Danusso, Recent results of stereospecific polymerization by heterogeneous
     catalysts, *J.Pol.Sc., C, 4 (1963) 1497-1509*

153. J.P.Hogan, Olefin copolymerization with supported metal oxide catalysts,
     chapter III in Copolymerization, G.E.Ham editor, Interscience, New York,
     1964 (Volume XVIII in the High Polymer series)

154. V.A.Zakharov et al, Gas-phase copolymerization of ethylene with propene and
     with 1-butene using supported chromium and titanium containing catalysts,
     Makrom.Chemie, 190 (1989) 559-566

155. R.Spitz, B.Florin and A.Guyot, Ethylene-hexene-1 copolymers through
     modified Phillips catalysis, *Europ. Pol.J., 15 (1979) 441-444*

156. G.Mabilon and R.Spitz, Copolymérisation éthylène-propylène par des
     catalysateurs à l'oxyde de chrome, mesure de rapport de réactivité à l'aide
     d'un reacteur simulant un lit fluidisé, *Eur Pol.J., 21 (1985) 245-249*

157. J.P.Hogan, Ethylene α-olefin copolymers made in gas phase, IUPAC, 28th
     Macromolecular Symposium, 12-16 july, 1982

158. E.A.Benham, P.D.Smith and M.P.McDaniel, A process for the simultaneous
     oligomerization and copolymerization of ethylene, *Polym.Eng.Sc., 28 (1988)*
     1469-1472

159. L.L.Böhm, Zur Copolymerisation von Ethylen und α-Olefinen mit
     Ziegler-Katalysatoren, *Makrom.Chemie, 182 (1981) 3291-3310*

160. L.G.Echevskaya, V.A.Zakharov and G.D.Bukatov, Composition effect of highly
     active supported Ziegler catalysts for ethylene copolymerization with
     α-olefins, *React.Kinet.Catal.Lett., 34 (1987) 99-104*

161. G.D.Bukatov, L.G.Echevskaya and V.A.Zakharov, Copolymerization of ethylene
     with α-olefins by highly active supported catalysts of various composition,
     p 101-108 in Transition metals and organometallics as catalysts for olefin
     polymerization, W.Kaminsky and H.Sinn editors, Springer, Berlin, 1988

162. K.Soga, T.Shiono and Y.Doi, Effect of ethyl benzoate on the
     copolymerization of ethylene with higher α-olefins over TiCl4/MgCl2

catalytic systems, *Polymer Bulletin, 10 (1983) 168-174*

163. N.Kashiwa, T.Tsutsui and A.Toyota, Solution copolymerization of ethylene with α-olefins by a MgCl2 supported TiCl4 catalyst, *Polymer Bulletin, 12 (1984) 111-117*

164. B.Schlund and L.A.Utracki, Linear low density polyethylenes and their blends: part 1. Molecular characterization, *Pol.Engin. and Sc., 27 (1987) 359-365*

165. A.Munoz-Escalona, H.Garcia and A.Albornoz, Homo- and copolymerization of ethylene with highly active catalysts based on TiCl4 and Grignard compounds, *J.Appl.Pol.Sc., 34 (1987) 977-988*

166. G.Fink and T.A.Ojala, Copolymerization of ethylene and higher α-olefins with MgH2 supported Ziegler catalysts, p169-181 in Transition metals and organometallics as catalysts for olefin polymerization, W.Kaminsky and H.Sinn editors, Springer, Berlin, 1988

167. R.Spitz, V.Pasquet and A.Guyot, Linear low density PE prepared in gasphase with bisupported SiO2-MgCl2 Ziegler-Natta catalysts, p 405-416 in Transition metals and organometallics as catalysts for olefin polymerization, W.Kaminsky and H.Sinn editors, Springer, Berlin, 1988

168. G.Luft, H.Grünig and R.Mehner, Copolymerization of ethylene and 1-olefins with Ziegler catalysts under high pressure, p183-188 in Transition metals and organometallics as catalysts for olefin polymerization, W.Kaminsky and H.Sinn editors, Springer, Heidelberg, 1988

169. T.E.Nowlin, Y.V.Kissin and K.P.Wagner, High activity Ziegler-Natta catalysts for the preparation of ethylene copolymers, *J.Pol.Sc.,Pol.Chem., 26 (1988) 755-764*

170. R.Muhlhaupt, D.W.Ovenall and S.D.Ittel, Control of composition in ethylene copolymerizations using magnesium chloride supported Ziegler-Natta catalysts, *J.Pol.Sc., Pol.Chem., 26 (1988) 2487-2500*

171. R.E.Wiman and I.D.Rubin, Mechanism of coordination polymerization. Copolymerization of ethylene and butene-1 with bis-(cyclopentadienyl)-titanium dichloride and diethylaluminium chloride, *Makrom.Chemie, 94 (1966) 160-171*

172. M.Garcia Marti and K.H.Reichert, Zur koordinativen Copolymerisation von Äthylen and Styrol, *Makrom.Chemie, 144 (1971) 17-27*

173. W.Kaminsky and R.Spiehl, Copolymerization of cycloalkenes with ethylene in presence of chiral zirconocene catalysts, *Makrom.Chemie, 190 (1989) 515-526*

174. W.Kaminsky et al, Isotactic polymerization of olefins with homogeneous zirconium catlaysts, p 291-301 in Transition metals and organometallics as catalysts for olefin polymerization, W.Kaminsky and H.Sinn editors, Springer, Berlin, 1988

175. H.Drögemüller, K.Heiland and W.Kaminsky, Copolymerization of ethene and α-olefins with a chiral zirconium/aluminoxane catalyst, p 303-308 in Transition metals and organometallics as catalysts for olefin polymerization, W.Kaminsky and H.Sinn editors, Springer, Berlin, 1988

176. J.A.Ewen, Ligand effects on metallocene catalyzed Ziegler-Natta polymerizations, p271-292 in Catalytic polymerization of olefins, T.Keii and K.Soga editors, Kodansha and Elsevier, 1986

177. M.Fontanille and A.Siove, Activity of titanium/lithium based macromolecular complexes, as initiator for Ziegler-Natta polymerization of ethylene, p313-321 in Transition metal catalyzed polymerizations, Alkenes and Dienes, Part A, R.P.Quirk editor, Harwood Academic Publishers, 1981

178. A.Siove and M.Fontanille, Intrinsic reactivity of active species in Ziegler-Natta polymerization of ethylene initiated with Li/Ti-based macromolecular complexes, *J.Pol.Sc., Pol.Chem., 22 (1984) 3877-3884*

179. P.J.T.Tait, Newer aspects of active centre determination in Ziegler-Natta polymerization using 14CO radio-tagging, p309-327 in Transition metals and organometallics as catalysts for olefin polymerization, W.Kaminsky and H.Sinn editors, Springer, Berlin, 1988

180. J.N.Short, Low pressure ethylene polymerization processes, p651 - 669 in Transition metal catalyzed polymerizations, Alkenes and Dienes, Part B, R.P.Quirk editor, Harwood Academic Publishers, London, 1981

446

181. J.N.Short, Low pressure linear (low density) polyethylene, p385 - 401 in
     Kirk-Othmer Encyclopedia of science and technology, volume 16, 1981 edition
182. J.P.Hogan, Linear (high density) polyethylene, p421 - 433, ibid
183. E.Paschke, Ziegler process polyethylene, p 433 - 452, ibid
184. J.P.Hogan, HDPE : from laboratory breakthroughs to a maturing industry,
     Golden Jubilee Conference, Polyethylenes 1933-1983, The Plastics and Rubber
     Institute, London, 1983
185. B.Diedrich, Second generation Ziegler polyethylene processes,
     J.Appl.Pol.Sc., Applied Polymer Symposia, 26 (1975) 1-11
186. R.B.Staub, The Unipol process - Technology to serve the world's
     polyethylene markets, Golden Jubilee Conference, Polyethylenes 1933-1983,
     The Plastics and Rubber Institute, London, 1983
187. J-P.Machon, Actual and potential development of high pressure process for
     linear polyethylene, Golden Jubilee Conference, Polyethylenes 1933-1983,
     The Plastics and Rubber Institute, London 1983

Chapter 10

CHARACTERIZATION OF POLYETHYLENES

10.1 INTRODUCTION

Characterization of polyethylenes is the subject of this chapter, the preparative details were given in the previous chapter 9. Not only the linear polymers made on the Ziegler-Natta and other catalysts mentioned there will be reviewed, also LDPE will be included where appropriate. Certainly from a historical viewpoint LDPE belongs in this chapter; being the first commercial polyethylene, all characterization started of course with this polymer type.

The methods used in the characterization are nearly all identical to those used elsewhere in this book, for a more detailed discussion of these the reader is referred to chapter 13.

The main topics to be reviewed are the composition and crystallinity related effects, and it will be attempted to relate the data to preparative conditions. Minor attention will be given to dynamic mechanical analysis and mechanical properties. It will turn out that LDPE is the most complex product in the whole spectrum of polyethylenes. Unraveling its detailed composition in terms of its short and long chain "branching" appears to be an especially difficult problem. The recent technique of ^{13}C-NMR spectroscopy appears again a very powerful tool in studying this.

10.2 COMPOSITION, ITS MEASUREMENT AND THE RELATION WITH THE
                              PREPARATION PROCESS

Five techniques are described which are able to measure various aspects of the composition of polyethylenes. These are infrared, radiolysis, pyrolysis, NMR and fractionation. Their outcome is discussed in the following.

10.2.1   Composition via infrared.

Infrared spectroscopy is the oldest tool used in elucidating the structure of polyethylenes. The very first realization that side groups were present in LDPE stems from this type of measurement. For good reviews of the application of IR techniques see references 1 to 3. The article of Cudby[2] gives also a good, concise, historic survey. Some features of the infrared method and its applicability are described in chapter 13.

Quantitative determination of co-monomer content in linear copolymers (LLDPE's) can relatively easy be done as only one type of side chain is expected to be present and thus no ambiguity exists in the value of the extinction coefficient. For LDPE the situation is more problematic. The total unravelling of both type and amount of side-chains is not possible with this technique. The total concentration of methyl can however readily and rather accurately be measured. After compensating for the normal chain ends of the polymer (either two methyls in hydrogen-controlled polymerizations or one methyl in cases without hydrogen) a number results which is related to branching only. (This compensation can only be done via a separate technique measuring number average molecular weight, e.g. GPC.) The measurement of total branch concentration is described in detail in, for example, the ASTM method D 2238-68. As will be shown later the combination of IR and ^{13}C-NMR is able to give a good description of the branches in LDPE, which neither technique can do on its own or only with difficulty (for NMR). Qualitatively ethyl branches are positively identified, furthermore methyl directly bound to the main chain is absent. Using the methyl deformation band as a sensitive probe in FTIR measurements on various LDPE's, Usami and Takayama[4] conclude that the branching distribution is constant as peak maximum and shape stay the same. The same authors reach an identical conclusion by applying ^{13}C-NMR (see section 10.2.4).

Apart from the alkyl side chains, the evaluation of the unsaturated groups can very well be done with IR and is very valuable for distinguishing various types of polyethylenes. A number of examples are given in Table 10.1. For instance "Phillips polyethylene" contains mostly one vinyl group per molecule, as no hydrogen is employed in the polymerization to control MW on their chromium catalyst. When using hydrogen in the polymerization, saturated endgroups are of course expected (although that is not always the case, e.g. on CrO_3/silica - see chapter 9). This is indeed found, see "Ziegler polyethylene" and other examples in Table 10.1. Upon introduction of co-monomers the concentration of vinylidene groups increases at the expense of the vinyl groups. This holds fairly generally, independent of catalyst type; the cause is the additional termination possibility with the co-monomer. The effect is naturally less pronounced in polymerizations where hydrogen is present. Sometimes very small concentrations of internal double bonds in trans configuration are observed. They probably stem from isomerization of the vinyl groups, as in higher temperature polymerizations a clear trend with the polymerization temperature is observed.

In LDPE low concentrations of vinyl and vinylidene groups are normally observed stemming from the disproportionation and transfer reactions.

TABLE 10.1

Examples of end-groups in polyethylenes

Catalyst	Hydro-gen?	MI g/10min	Infrared data, per 1000 C methyl	vinyl	vinylidene	Remarks	Ref
"Ziegler"	yes		3.6	0.09	0.06	MW=50,000	5
"Phillips"	no		3.1	1.58	0.08	MW=50,000	5
do,hexene copol	no		3.7	0.83	0.08	MW=150,000	5
LDPE			20-33	0.08-0.25	0.17-0.33		
CrO_3/silica	no		3.4	1.0	0.1		6
do,hexene copol	no		39	1.4	1.0		6
Chromocene/SiO_2	yes	0.21	1.58	0.057	0.017		7
	yes	5.5	2.24	0.052	0.036		7
Chromocene/SiO_2	no	2.1	12.7	0.52	n.d.		8
solution polym.	no	16.0	16.1	1.5	n.d.		8
$TiCl_3$/siloxalane	no		5.9	0.556	<0.005	d=0.954	9
high T, high p	,,	0.9		0.27	0.005	d=0.924,E/B	10
	,,	25		0.35	0.09	d=0.918, ,,	10

n.d. = not detected

In two related studies Kashiwa et al investigated the copolymerization of ethylene with propylene, 1-butene and 4-methyl-1-pentene using either $MgCl_2$/$TiCl_4$//$AlEt_3$ (ref 11) or $VOCl_3$/Al(octyl)$_3$//$AlEt_3$/$CHCl_3$ (ref 12). Referring only to the generation of unsaturated structures, the $TiCl_4$/$MgCl_2$ catalyst combination gives essentially a constant concentration of double bonds per polymer molecule (about 0.65) independent of co-monomer type or its level of incorporation. However the distribution over the various types of unsaturation changes appreciably from only vinyl in the homopolymer to increasing fractions of vinylidene (more in the propylene than

in the 1-butene case). Surprisingly the 4-methyl-1-pentene copolymers showed only vinyl unsaturation again. In the vanadium catalyst
case the situation is different, as there also internal trans
double bonds can easily form from structures arising from secondary
insertion, i.e. with the "branched" carbon bound to the active
center:

V-R + C=C-C-C --> V-C(C-C)-C-R (secondary insertion)

V-C(C-C)-C-R --> V-H + C-C-C=C-R

as opposed to:

V-R + C=C-C-C --> V-C-C(C-C)-R (primary insertion)

V-C-C(C-C)-R --> V-H + C=C(C-C)-R

This is most notable in the 1-butene-containing copolymers; leaving
out the hydrogen for molecular weight control gives increased
levels of internal double bonds as expected.

10.2.2. Radiolysis

Numerous investigators studied the radiolysis of especially
LDPE for various reasons. In terms of the composition there are
optimistic reports[1,13] describing the use of this technique for
determination of the branch type and the branch distribution in
LDPE. These are based on the formation of volatile products containing next to hydrogen as the main product, a number of smaller
hydrocarbons either saturated or unsaturated. The evidence pointed
to the splitting off of the total branch at the tertiary carbon
atom, thus allowing for a determination of type and concentration
of the branches. These claims were put on a more realistic basis in
for instance a recent study of Bowmer and O'Donnell[14]. They investigated the radiolysis in depth, for example as function of the
temperature of the reaction, the procedure for measuring the
volatile fraction and sample quantity. Optimal measuring conditions
were specified. From previous studies it was already known that the
radiolysis yield depends on the branch length: it decreases with
the branch length. Chain ends (i.e. longer branches) also contribute to the volatile products, especially a C_2 fraction. One should
take all these factors into account in evaluating the gaseous
radiolysis product composition and relating this to a branch

distribution. In that way the radiolysis data agreed fully with the distribution found via ^{13}C-NMR measurement. The main conclusion of the above study is this: since the advent of the ^{13}C-NMR technique, radiolysis has no advantage for the determination of the branching distribution.

10.2.3. Pyrolysis studies

Identification of pyrolysis fragments can also be used in qualitative and quantitative studies of polymer structure. Two widely different pyrolysis conditions have been used, one at high temperature[15-17] (600 - 800 °C) leading to almost total volatilization, the other[18,19] at a far lower temperature (360 °C) which reportedly would mainly give fragments from reactions starting from the branch points and almost no main chain scission is observed. The yield of volatile material is of course much lower in the second case (about 2%).

The branch points are the weak spots in the chain and scission reactions have a high probability of starting there; additionally the tertiary-hydrogen is reactive in hydrogen abstraction reacti-

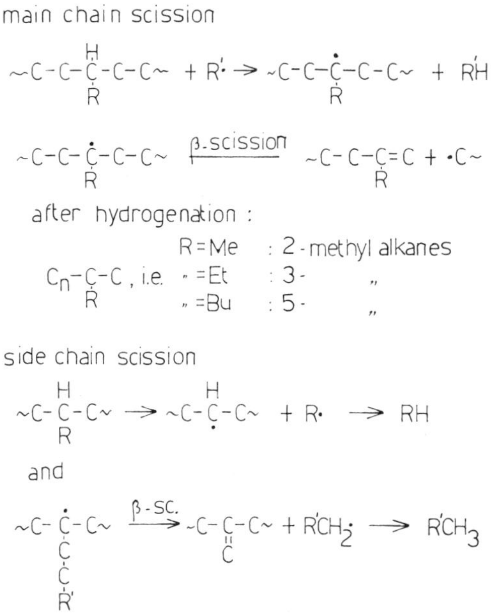

main chain scission

$$\sim C-C-\underset{\underset{R}{|}}{\overset{\overset{H}{|}}{C}}-C-C\sim \; + \; R\cdot \; \rightarrow \; \sim C-C-\underset{\underset{R}{|}}{\overset{\cdot}{C}}-C-C\sim \; + \; R\overset{'}{H}$$

$$\sim C-C-\underset{\underset{R}{|}}{\overset{\cdot}{C}}-C-C\sim \; \xrightarrow{\beta\text{-scission}} \; \sim C-C-\underset{\underset{R}{|}}{\overset{}{C}}=C \; + \; \cdot C\sim$$

after hydrogenation :

$$C_n-\underset{\underset{R}{|}}{\overset{}{C}}-C \; , \; i.e. \begin{array}{l} R=Me \quad : \; 2\text{-methyl alkanes} \\ \;\; \textquotedbl = Et \quad : \; 3\text{-} \quad\quad \textquotedbl \\ \;\; \textquotedbl = Bu \quad : \; 5\text{-} \quad\quad \textquotedbl \end{array}$$

side chain scission

$$\sim C-\underset{\underset{R}{|}}{\overset{\overset{H}{|}}{C}}-C\sim \; \rightarrow \; \sim C-\underset{\underset{}{\cdot}}{\overset{\overset{H}{|}}{C}}-C\sim \; + \; R\cdot \; \rightarrow \; RH$$

and

$$\sim C-\underset{\underset{\underset{R'}{|}}{\overset{}{C}}}{\overset{\cdot}{\underset{|}{C}}}-C\sim \; \xrightarrow{\beta\text{-sc.}} \; -C-\underset{\overset{\|}{C}}{\overset{}{C}}-C\sim \; + \; R'\overset{\cdot}{C}H_2 \; \rightarrow \; R'CH_3$$

Fig. 10.1: *Some decomposition paths and pyrolysis products in polyethylenes*

ons, the resulting tertiary radical will show a facile β-scission. In all studies the pyrolysis products are hydrogenated before analysis by gas chromatography coupled to flame ionization detection or mass spectrometer. Model ethylene copolymers give, apart from the n-alkanes as main products, a series of substituted alkanes specific for each type of co-monomer. Fig. 10.1 illustrates a number of possible decomposition routes showing that a methyl, ethyl, propyl and butyl side group would be distinguishable (the figure is far from exhaustive in all decomposition pathways, for example at higher temperature more random type of fragmentation occurs by scission further away from the tertiary carbon than the low temperature β-scission). In this way Seeger and Barrall[15] qualitatively identify ethyl and butyl groups in LDPE, Sugimura and Tsuge[16] add methyl to this list. The latter group later[17] attempted to quantify the technique by assuming random placement of the side groups in LDPE, and they concluded that butyl is the most abundant side group followed by ethyl and methyl.

Using the low temperature pyrolysis method Haney et al[18] investigated the presence of the 2-ethyl-hexyl and other more complicated branches predicted to be formed by the backbiting mechanism of 1,5 hydrogen transfer (see below, section 10.2.4). They use mass spectrometry for identification. Due to the low temperature the only products observed arise from either α- or β-scission from the branch points. From model copolymers containing branched sidegroups the structure of the side group is easily derived from the pyrogram. The direct splitting off of the side group is also generally observed.

LDPE's show qualitatively the same pattern, although peak ratios differ widely between individual polymers. A large number of branched fragments are found, in fact many more than expected from the study of the model polymers. For instance 3-methyl pentane is very prominent, as is 3-methyl heptane and 3-methyl octane. In analyzing the non-linear fragments more closely (determining their structure mainly from their mass spectrum) the investigators found a number of homologous series as products. These are given in Fig. 10.2. Of these, some can be reconciled from the established Roedel and modified Roedel mechanism of backbiting (see below), but others need additional rearrangement, which could be a 1,3 hydrogen shift. Apart from the mechanism of formation their study shows that the number of different branch types could be much higher than previously expected. However, as many different structures lead to

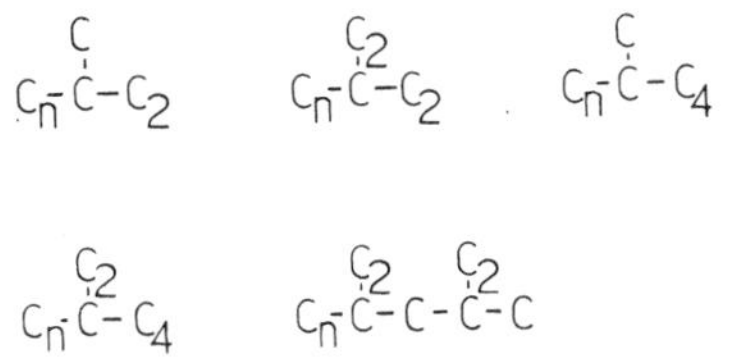

$$C_{\bar{n}}\overset{\overset{C}{|}}{C}-C_2 \qquad C_{\bar{n}}\overset{\overset{C_2}{|}}{C}-C_2 \qquad C_{\bar{n}}\overset{\overset{C}{|}}{C}-C_4$$

$$C_{\bar{n}}\overset{\overset{C_2}{|}}{C}-C_4 \qquad C_{\bar{n}}\overset{\overset{C_2}{|}}{C}-C-\overset{\overset{C_2}{|}}{C}-C$$

Fig. 10.2: Homologous decomposition product series detected by Haney et al (reference 18)

identical decomposition products a direct, unequivocal proof that a certain type of branch is present cannot be given. On the other hand, the presence of 2-ethyl hexyl branches is regarded as very likely, as the expected decomposition products are observed, whilst the formation of this type of branch is readily explained by 1,5 hydrogen transfer.

10.2.4. Composition via ^{13}C-NMR spectroscopy

As already discussed in the case of polypropylene and its copolymers, the use of ^{13}C-NMR is a powerful tool for determining detailed characteristics of the polymer. This holds also for polyethylene.

(i) HDPE

The most simple polymer is the linear homopolymer of ethylene. One expects only methylene carbons and - at high sensitivity - the resonances from the end-groups. The latter can be methyl (with, in principle, the two neighbouring carbons separately discernible) and vinyl. Up till now only one detailed study on HDPE has been reported[20]. This study addressed the question of the presence of long chain branching in HDPE. It has been found that particularly chromium-catalyzed polyethylene ("Phillips" PE) showed larger activation energies in meltflow than titanium catalyzed ones. The level of branching is of course very low, thus requiring a high sensitivity, obtainable when using the high-field generating superconductive magnets in NMR. In the NMR-spectrum of the chromium catalyzed samples Randall could indeed identify resonances attributable to the carbons located α and β w.r.t. the branchpoint and found a methine carbon absorption as well. Additionally the end-groups (both methyl and vinyl - via the allylic carbon) were of course observed. The relative intensities are such that less than one branch per polymer molecule is present. Further evidence is obtained when calculating M_n from the endgroup concentration ex NMR, and comparing this with the M_n from GPC measurements (which assumed the PE was linear). The M_n ex GPC is then definitely smaller - as expected when branched molecules are present (see also section 10.2.5).

Surprisingly, some of the polymers tested showed also short chain branches despite the fact that no co-monomer was added in the polymerization. For instance the titanium catalyzed sample showed ethyl side groups. The spectra can be quantitatively evaluated. As, regrettably, no preparative details on the samples were given, no relation can be established between the extent of branching and catalyst used and/or the polymerization conditions applied.

(ii) <u>linear copolymers</u>

Reports on NMR measurements on α-olefin copolymers are not abundant and mostly deal with qualitative measurements only. Spectra for copolymers with low co-monomer fraction, leading to isolated units only, were given by Randall[20]. They constituted the series propene to octene and as an illustration of the power of ^{13}C-NMR they are reproduced in Fig. 10.3. More recently copolymers were also described of higher co-monomer content, with as co-monomers 1-butene[21,22], 1-hexene[23], 1-octene[24] and 4-methyl-1-pentene[24].

Assignment of the resonances is mostly not a problem although some difficulties arise when tacticity effects can play a role. In a series of ethylene/1-butene copolymers made on either a vanadium or a titanium catalyst, one observes tail-to-tail butene diads in the former case[22]. This is not surprising in view of the experience with ethylene/propylene copolymerizations (see chapters 3 and 4). The study of Hsieh and Randall[21] on ethylene/1-butene copolymers observed considerable difference between those spectra and the ones from hydrogenated polybutadiene of similar "butene" content, differences in the sequence distribution are thus indicated (and probably in tacticity as well). Not all the peaks were resolved nor could all of them be assigned. The triad and tetrad distributions can be calculated from the spectra, and are very useful for characterization of the copolymer as regards its sequence distribution. In the ethylene/1-hexene case[23] the observed triad distribu-

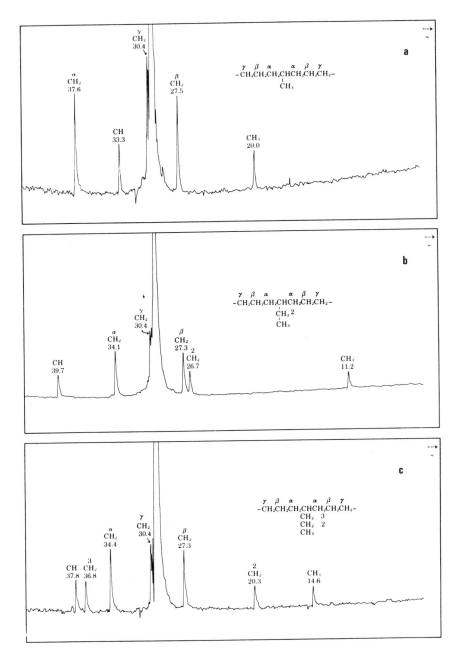

Fig. 10.3: ^{13}C-NMR spectra of ethylene-α-olefin copolymers of low co-monomer content: ethylene/propylene (a); ethylene/1-butene (b); ethylene/1-pentene (c)

456

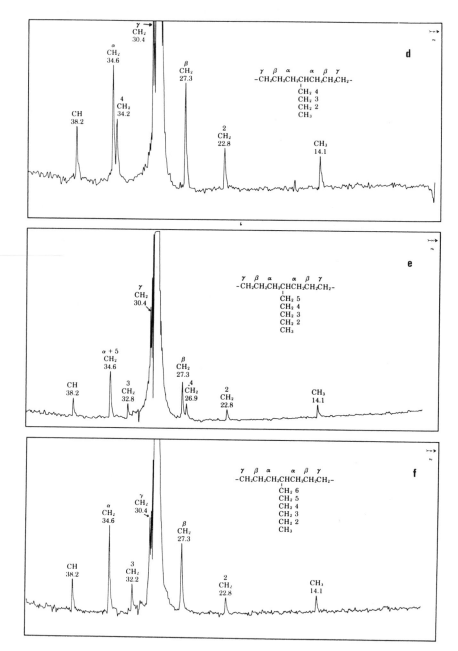

Fig. 10.3 continued: ethylene/1-hexene (d); ethylene/1-heptene (e); and ethylene/1-octene (f) [from reference 20. Reprinted with permission from ACS Symposium Series. Copyright 1980 American Chemical Society]

tion was compared to the predicted one assuming Bernouillian statistics. Triads containing isolated hexene units show a lower frequency than predicted and hexene triads are higher, i.e. the copolymers appear to be blocky. The catalyst used is not disclosed in that paper, but is expected to be chromium oxide on silica. No copolymerization parameters were calculated from the triad distributions in the papers, an own attempt to use the data failed to give consistent values for r_1*r_2. The study of Kimura et al[24] used 1-octene and 4-methyl-1-pentene copolymers made on "supported titanium" as catalyst. The copolymers were fractionated by simple extraction with different solvents. The co-monomer content was relatively low. The triad distribution evaluated for both copolymers showed - contrary to the case mentioned above - that the observed fraction of isolated co-monomer units in triads is <u>higher</u> than calculated via Bernouilian statistics. This is rather surprising, but the difference might be due to the fractionation applied.

In gas-phase prepared copolymers on the Phillips catalyst a considerable fraction of inversions is noted in ethylene/propylene copolymers[25], whether this is a common feature of chromium catalysis is not clear. The sequence distribution of the co-monomer is also more random and less alternating compared to titanium catalysts.

In a characterization study on commercial LLDPE samples Schlund and Utracki[26] only observed isolated co-monomer units in their NMR analysis, provided that the catalyst contained titanium. In the case of vanadium-containing catalysts also pairs of the co-monomer units were detected even at the low co-monomer levels normally present in LLDPE's.

A slightly different approach for describing the observed sequence distribution was taken by Randall and Hsieh[27]. They define the "monomer dispersity" (MD):

$$MD = \text{run number} / \%\text{mol co-monomer incorporated}$$
$$= 100 * 0.5[EX]/[X]$$

with the run number as defined by Harwood and Ritchey[28] (i.e. the average number of like monomer sequences - runs - per 100 monomer units) and X denoting the co-monomer used. It turns out that MD is different for different co-monomers, with 1-butene being the least random (MD ranging from 60 to 99 %) and 1-octene being totally random. So again this leads to the observation that butene gives more "clusters" of co-monomer units. If one selects copolymers at equal monomer dispersity the correlations of copolymer properties show much less spread than using just the co-monomer content in the correlation.

In a very interesting study of Usami et al[29], the technique of Temperature Rising Elution Fracionation (TREF, see chapter 13) and NMR were combined in the study of the sequence distribution in ethylene/1-butene copolymers. These copolymers were made on $TiCl_4$/$MgCl_2$ catalyst under various polymerization conditions. Fractionation showed the copolymers to have a bimodal branch distribution in the plot of fraction eluted versus elution temperature. A first peak is found around 70 °C, the second at about 90 °C. For LDPE only one peak is observed at about 70 °C. A limited number of fractions were analyzed by NMR, and the copolymerization parameters calculated from the observed composition. A change in r_1*r_2 value is observed, coinciding with the two peaks in the fractionation. At the highly branched end (i.e. at low elution temperature) r_1*r_2 is 0.5 to 0.7 whilst at the higher temperature peak r_1*r_2 is about 1. This suggests strongly two kinds (or families) of active sites on the catalyst, each with their own kinetic parameters (a similar conclusion is reached from homopolymer PP fractionation studies). The step change in the r_1*r_2 product finds its parallel in the vinyl concentration (which is very low - ranging from 0.05 to 0.15 per 1000 C). Splitting the copolymers at an intermediate fractionation temperature of 84 °C gives two fractions of distinctly differing molecular weight, the higher-branched ones are lower by a factor of 1.5 to 2. This type of study leads to very interesting facts and warrants further investigation both with catalyst type as parameter as well as polymerization conditions.

(iii) LDPE

LDPE turns out to be a very complex copolymer containing a considerable number of different types of branches in widely different concentrations. The NMR studies progressed from being initially qualitative, discussing the types of branches identified, to - much later, and only rather recently - attempts at quantification. The history of these investigations will not be given here, only their outcome will be summarized.

Qualitatively the following branches have been identified, see especially the papers of Randall[30], Axelson et al[31], Usami and Takayama[4,32], and Freche and Grenier-Loustalot[33]:

methyl, ethyl, propyl, n-butyl, n-amyl, 2-ethyl-hexyl.

Branches longer than 5 carbon atoms can not be distinguished individually, only the third carbon from the end in a long branch gives a recognizable own resonance. There is still no unanimity over the presence of the methyl, propyl and 2-ethyl-hexyl branches,

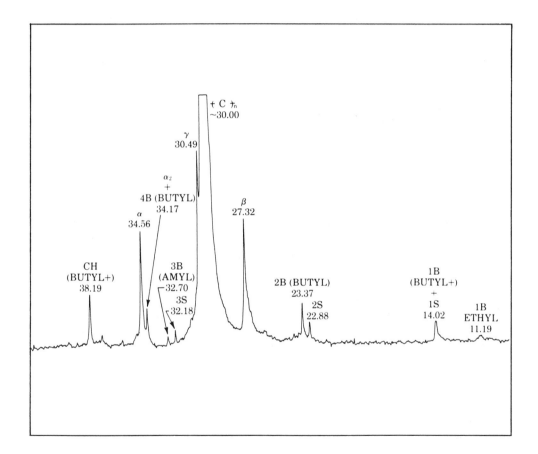

Fig. 10.4: Example of ^{13}C-NMR spectrum of LDPE (resonances identified by distance from branch point in main chain (α, β, γ); and carbon number in branch starting with the methyl end-group (1, 2, 3,..); B stands for branch of 5 or less carbon atoms, S for saturated end-groups of long chain branches) [from reference 20. Reprinted with permission form ACS Symposium Series. Copyright 1980 American Chemical Society]

although the more recent investigations give evidence for the more complex branch structures such as 2-ethylhexyl or quaternary ethyl-butyl branches[33]. Most investigators conclude from their spectra that ethyl branches do also occur in pairs, but not in longer sequences. The pairs have probably a racemic configuration. The fraction of ethyl side groups present in pairs is a question not fully resolved. The trend is to exclusive presence of paired ethyls. Usami also observed paired ethyl branches where one ethyl is connected to a quaternary carbon. When some oxygen-containing groups are present in polyethylene these give rise to their own specific resonances as observed by Usami and Takayama. Spectra from various samples are mostly clearly different, no "typical" LDPE appears to exist in terms of branching. NMR is thus a very useful tool for fingerprinting different samples. An example is given in Fig. 10.4.

In a quantitative sense the studies of Axelson et al[31] show that using all resonances in a multiple regression analysis leads to large standard deviations of about 30 % and sometimes negative branching concentrations were calculated. They subsequently used unique resonances to arrive at quantitative results. The same approach was used by Usami and Takayama[32] who surprisingly did arrive at a constant distribution of the side chains for a variety of LDPE's. Table 10.2 gives their values. The ratio of butyl over ethyl is slightly above 1. Using this criterion Axelson et al[31] observed a wider range of this value and therefore the conclusion of the existence of a typical LDPE needs further proof (see also below). A simple linear regression using the integral of 6 specific resonances was used by Hay et al[34] to assess the fractions of ethyl, butyl, amyl and longer branches (and thus neglecting the minor ones). For the LDPE's measured in that study the branch frequencies vary widely, with for example butyl/ethyl ratios ranging from 3 to 19 and amyl side-groups mostly being more abundant than ethyl. Again using all twelve observed resonances, Cheng[35], using a computerized optimalization of peak intensities, calculated ethyl, butyl and amyl branch content and two kinds of data on long chain branching, neglecting the other possible side-groups. The procedure works, no negative values are calculated, and the difference between separate samples is large.

TABLE 10.2

Branch distribution in LDPE's observed by Usami and Takayama[32]

Branch	Relative Abundance %
methyl	2
ethyl(paired)	23
ethyl(paired and one quaternary)	11
propyl	2
n-butyl	36
n-amyl	12
longer branches	14

(2-ethyl-hexyl was not observed by these authors)

Only in one case a relation between the polymerization conditions and structure was established. Cudby[2] used a combined NMR and
IR technique, with the latter giving the total number of branches
(corrected for 2 end-groups per molecule) and the former the
relative levels of the branches, limiting these to ethyl, butyl and
longer-than-butyl. The branch concentration is mainly temperature
dependent, with butyl and longer branches present from the lowest
temperature mentioned onwards (140 °C), whilst ethyl appears later
at about 200 to 220 °C. The pressure dependence for the ethyl
branch is very pronounced, for example at 320 °C as polymerization
temperature the number of ethyl/1000 C decreases from 18 at 1000
bar to zero at 2500 bar. Also the other branches show a pressure
relation in the same direction but to a much lower extent, polymers
without butyl or longer branches cannot be generated in the temperature range 140 to 320 °C and the pressure range 1000 to 2500 bars.

The observed branch distribution (type and frequency) can be
explained to a large extent by the backbiting mechanism of Roedel[36]. This, in the first instance, would only give isolated butyl,
amyl, etc branches. But - as first mentioned by Simpson and Turner
independently (see Willbourn's paper[1]) - one can assume a second
step following the first one after adding one ethylene, i.e after
converting the secondary radical into a primary, more reactive one.
See Fig. 10.5. This can give rise to both the paired ethyl branches
and to a 2-ethyl-hexyl side-group. In both cases the backbiting
process competes with the propagation and thus the extent of
branching should decreases with increasing pressure (i.e. increa-

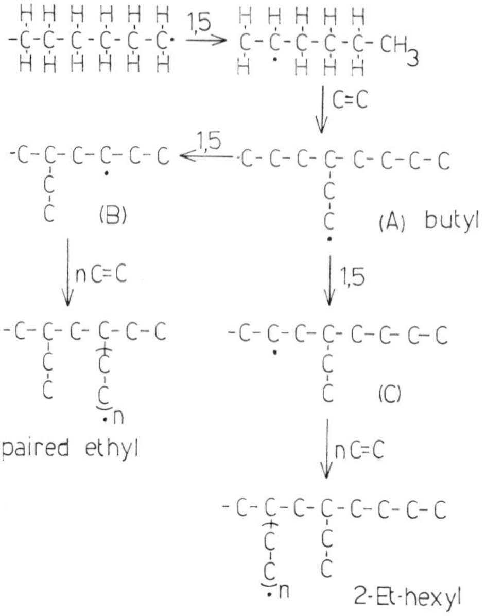

$$\begin{array}{c}
\text{H H H H H H} \quad \xrightarrow{1,5} \quad \text{H H H H H} \\
-\text{C}-\text{C}-\text{C}-\text{C}-\text{C}-\text{C}\cdot \quad\quad \text{C}-\text{C}-\text{C}-\text{C}-\text{C}-\text{CH}_3 \\
\text{H H H H H H} \quad\quad \text{H} \;\cdot\; \text{H H H}
\end{array}$$

$\downarrow C=C$

-C-C-C-C-C-C $\xleftarrow{1,5}$ -C-C-C-C- C-C-C-C
C   C
C (B)   C (A) butyl

$\downarrow n\,C=C$ $\downarrow 1,5$

-C-C-C-C-C-C   -C-C-C-C- C-C-C- C-C
C   C   C
C   C $_n$ (C)

paired ethyl $\downarrow n\,C=C$

- C-C-C-C-C-C-C-C
C   C
C   C $_n$ 2-Et-hexyl

Fig. 10.5: Roedels extended backbiting mechanism

sing propagation rate). It is not clear why the formation of the ethyl branch should be more sensitive to the ethylene pressure than the others. Using conformational chain statistics Mattice and Stehling[37] calculated the relative probabilities of the formation of side groups by backbiting, assuming that next to a close proximity of the reacting groups, a nearly rectilinear geometry is also required. For isolated side branches the prediction is that butyl is by far the most abundant, but up to a 13-carbon-branch the probabilities are all above zero. Especially amyl and hexyl groups are calculated to be of equal weight. Increasing the temperature leads to more branching particularly for branch lengths 4 to 9 carbon atoms long. The subsequent backbiting from radical A in Fig. 10.5 has a considerable larger probability than the initial one due to the configurational state the chain has at that moment. Moreover the two modes of reaction have an identical calculated chance. Thus the concentration of paired ethyl and 2-ethyl-hexyl groups are predicted to be equal. Additionally further backbiting from radicals B or C should also show this higher probability. All this would lead to clustering of the branches. There is some evidence that this is indeed the case from the higher than expected melting

point for LDPE compared to true random copolymers such as hydrogenated polybutadiene (see below). The calculations certainly can
explain qualitatively the observed structures in LDPE, but for a
quantitative evaluation much more work is needed both on the
calculation model as in the analysis of LDPE's.

As an overall conclusion for the NMR characterization of
polyethylenes it can be said that as far as the published reports
go, the status on the qualitative assignment of the resonances is
fairly good. This holds certainly for LLDPE's, in the case of LDPE
more work is needed to clarify the situation especially around the
2-ethyl-hexyl groups and the extent of clustering of the branches.
For quantitative analysis, the right machine settings have been
determined and the near future will hopefully see the appearance of
NMR analysis on LLDPE's to elucidate the copolymerization behaviour
of various catalysts and the effect of polymerization conditions on
the structure of these copolymers.

10.2.5 Fractionation
    The methods used in studying fractionation of polyethylene,
whether simple or elaborate, are similar to those described for
polypropylene in chapter 13. The main difference lies in the
temperature (range) applied.

A few highlights from fractionation studies will be given
here, but this section is by no means exhaustive on this topic.
Molecular weight distribution effects of the catalyst applied have
been mentioned in the previous chapter. The main topic here is the
distribution of co-monomer over the chains in LLDPE or the branch
distribution in LDPE, the question being, is it homogeneous or not?
There is overwhelming evidence of the intermolecular heterogeneity
in both ethylene copolymers as in LDPE. For instance Shirayama et
al[38] fractionated a whole set of ethylene/α-olefin copolymers made
on $VOCl_3/AlEt_2Cl$. All were very heterogeneous, even when separating
the polymer into only 4 to 7 fractions the co-monomer content
showed ratios of 10 and higher between first and last sample.
Although the fractionation method should only distinguish on
co-monomer fraction it appears that the molecular weight decreases
monotonously at increasing co-monomer level. This is generally
found also on commercial samples[39]. Even a simple split into two
fractions by a Soxhlet extraction has been used to analyze for

heterogeneity in commercial copolymer samples[40]. All LLDPE's tested were found to be heterogeneous.

The copolymer heterogeneity is found for a variety of catalysts and polymerization conditions, for example:
- $VOCl_3/AlEt_2Cl$ at 30 to 60 °C (ref. 38)
- $TiCl_3.0.33AlCl_3$/siloxalane at 250 °C (ref 9 and 41)
- "Solvay" catalyst, $Mg(OR)_2/TiCl_4//AlEt_3$ at 85 °C (ref 42)
- Supported $TiCl_4$ at around 75 °C (ref 24, 29, 43 and 44).

In the case of the high temperature polymerization[9] the presence of mixtures of HDPE and LDPE (in-situ generated!) is observed, sometimes even together with a third type of polymer structure which is probably best described as a graft-polymer of LDPE on HDPE.

With the "Solvay" catalyst the observed spread in the composition is less than in the one described for vanadium (only a factor of 2 in co-monomer content at the end fractions). However this ratio is higher when comparing reactor solubles to the reactor insolubles in the polymerization[42], then - especially at the low overall co-monomer contents - the ratio increases to a maximum of about 5. For the "supported" $TiCl_4$ catalyzed systems the heterogeneity is very large, for a butene copolymer Kimura et al[43] reports fractions containing 20 and 0.5 %m/m butene at the extremes.

The interesting outcome of combining temperature rising elution fractionation (TREF) and NMR for ethylene/1-butene copolymers w.r.t. the copolymerization parameters has been mentioned in the NMR section.

Two cross-fractionation studies have been reported[45,46]. In the first one of Hosoda[45] a molecular weight fractionation was carried out at 120 °C initially, followed for some fractions by a crystallinity fractionation (TREF). Laboratory and commercial LLDPE were tested as samples. From the first fractionation a relation of co-monomer content to molecular weight is obtained, this decreases monotonously with MW. The density of these fractions shows a minimum at about 100,000 MW, due to the opposing tendencies of increasing MW leading to lower crystallinity and thus density, whereas at higher MW also the co-monomer content decreases leading to higher crystallinities. Using the observed MWD's of the TREF fractions, a three-dimensional plot was generated with as axes, co-monomer content, molecular weight and the quantity. Such plots show a number of peaks: the low MW material is high in co-monomer content, which decreases towards the higher MW end, not monotonously but in peaks. This points again to a limited number of (classes of) active sites. When testing LDPE in the same manner only one peak is observed. Schouterden et al[46] use a similar approach, and again the lower MW fractions are highly branched and broader in composition distribution than the higher molecular weight ones. An analytical approach was reported by Housaki et al[47], in which FTIR is coupled to GPC. In this way one obtains in one go both the molecular weight distribution as the distribution in co-monomer content (they use 2955 cm^{-1} for the methyl of the co-monomer and 2928 cm^{-1} for methylene). In line with previous observations they observe, in a commercial ethylene/1-octene copolymer, a monotomous decrease in co-monomer content with MW.

An exception to the above mentioned heterogeneity of the copolymer composition is the study of Hunter et al[48] using $VOCl_3$/ $Al_2Et_3Cl_3$ at room temperature. All copolymers generated on this homogeneous catalyst were proven to be homogeneous in composition by fractionation, also the molecular weight distribution is narrow (Q=2). This is in line with the earlier reported homogeneity of some vanadium catalyzed ethylene/propylene copolymers. It is a pity that no fractionation studies have yet been reported on copolymers made with the new, highly active catalyst systems based on $Cp_2Zr(CH_3)_2$/aluminoxane. One expects them to be fairly uniform in composition as, for instance, also the MWD's are narrow, but proof is lacking.

LDPE has of course also been studied with these techniques. For instance Shirayama et al[49] used a true cross-fractionation technique by first carrying out a molecular weight fractionation followed by a TREF on a number of fractions from the first fractionation. Checking the fractions from the first fractionation on their branching level (by IR) the samples showed already widely different branching distributions. Some were truly narrow, others wide (up to a factor 2 to 3 difference), always the lower MW's are more branched. The subsequent TREF showed the fractions had again a branching distribution, being wider the lower the overall MW. This distribution could well be described by a statistical function. No relationships were established with the polymerization conditions however.

Usami et al[50] fractionate a LDPE (made in a tubular reactor with propylene as co-monomer) on a molecular weight basis; the fractions are then characterized by NMR and pyrolysis-GC. For short-chain branching these methods are generally in agreement and show that only ethyl and butyl branching are decreasing with increasing MW, with the ethyl-group showing the strongest dependence. There is a rather steep increase in the ethyl content at very low MW's, suggesting that the lower MW fraction of this LDPE is made at a higher temperature than the rest. The long-chain branching slightly increases with MW.

In the paper by Wild et al[51] on the analytical TREF method examples are given of branching distributions of LDPE's made under various conditions. The method has a considerable discriminating power, showing for instance clearly much wider branching distributions for polymers made in a tubular reactor as compared to autoclave polymers. This is undoubtedly due to the narrower temperature and pressure range experienced by the reactants in the latter type of reactor.

It is felt that a combination of ^{13}C-NMR and TREF is very powerful in characterizing polyethylenes. Systematic studies are however not yet reported.

One extra effect in LDPE fractionation deserves mentioning. As LDPE contains long chain branches the GPC-determined distribution is shifted to lower molecular weights. This is due to the separation on a hydrodynamic volume basis in GPC, which is considerable lower for branched versus linear molecules of identical molecular weight. Conversely, the difference in molecular weight parameters derived from GPC and, for example, those from LVN or light scattering can be used to judge the presence of long chain branches. Examples can be found in references 52 and 53.

A summarizing statement on LDPE structure has to be made after surveying the various analytical techniques related to composition. Not discussing molecular weight, nor its distribution and neither the long chain branching but only the shorter branch types: which branches have unequivocally been identified and does this form a uniform picture? IR and especially NMR techniques are the only techniques giving positive identification, unfortunately radiolysis and pyrolysis are too non-specific. Qualitatively methyl, paired ethyl, butyl, amyl, quaternary ethyl-butyl and 2-ethyl hexyl have been identified. Quantitative analysis appears to be very diffi-

cult. By comparison with theoretical calculations[37] and the pyrolysis experiments of Haney et al[18] one could expect more structures
in either slightly longer branches (hexyl, heptyl,..) or more
complicated branches. Thus the present status is probably still
only one of partial success in elucidating LDPE's structure. It is
expected that the future will see more sensitive NMR measurements
extending the number of branch types and leading to a better
quantitative description.

10.3 CRYSTALLINITY RELATED PROPERTIES, STRUCTURE OF CRYSTALS

The polyethylenes described in this and the preceding chapter,
whether homopolymer, copolymer or the low density version, are all
crystalline polymers. (The copolymers which are amorphous have
already been discussed in chapters 3 and 4.) The extent of crystallinity varies strongly, being mainly dependent on the composition
of the polymer. The general rule is that structural defects (co-monomer, branches in LDPE) make the formation of the crystalline
phase more difficult, leading to lower crystallinities and less
perfect crystals. This is reflected of course in properties such as
the melting temperature (or its range), the crystallinity and thus

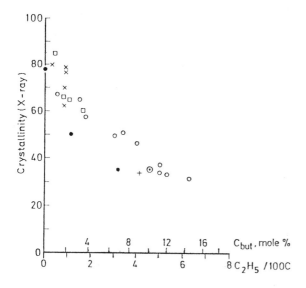

Fig. 10.6: X-ray crystallinity of various ethylene-1-butene copolymers as
function of the 1-butene content [from reference 54, copied with permission from
Springer Verlag]

the density of the polymer. An example of decreasing crystallinity is given in Fig. 10.6 for ethylene-1-butene copolymers (taken from reference 54).

The thermodynamic data for the melting of pure polyethylene have been reported most recently by Wunderlich and Czornyj[55] and are given in Table 10.3.

In this section first the observed properties such as the melting point, melting enthalpy and the density will be reviewed in relation to the composition of the polymer. In a last sub-section the general ideas on the structure of the crystals and the effect of alien units will be summarized.

TABLE 10.3

Equilibrium thermodynamic data of the melting of pure polyethylene (ex ref.55)

Property	Value
equil. melting temperature	414.6 ± 0.5 (141.5 °C)
heat of fusion at T_m	4.1 ± 0.2 kJ/mol CH_2
entropy of fusion at T_m	9.9 ± 0.4 J/K/mol CH_2
crystal volume at T_m	1.0038 m^3/Mg
amorphous volume at T_m	1.2765 ,,
crystal volume at 298 °C	0.9970 ,,
amorphous volume at 298 °C	1.1739 ,,
crystal density at 298 °C	1.003 Mg/m^3
amorphous density at 298 °C	0.8519 ,,

The methods used for measuring the properties of the crystalline polymers are not different from the ones described for polypropylene (see chapter 13). Thus the melting point (range) and the melting enthalpies are mostly determined by DSC, and the crystallinity is based on density or wide angle X-ray spectroscopy (assuming for the density of pure crystalline polyethylene 1.003 and for amorphous 0.8519 - although other values have been used in the past). For a rapid characterization mostly DSC is used, but regrettably most investigators only report the maximum as the melting point and rarely mention the width of the melting peak. The latter, especially for copolymers, is a relevant characterization feature. For instance Schouterden et al[56] carry out isothermal crystallizations in the DSC with LLDPE's. They observe in the heating cycle that a relatively large fraction of the copolymer did not crystallize at that temperature, and the amount increases with increasing co-monomer content. This is another illustration of the heterogeneity of these copolymers. Examples of some DSC-thermograms are given in Fig. 10.7. Hosoda[45] defines a chemical composition distribution index derived from the DSC-thermogram: Usually the thermogram is composed of more than one peak, and he suggests as index the ratio

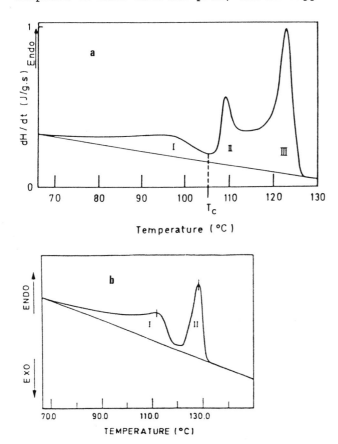

Fig. 10.7: Examples of DSC-heating traces on LLDPE's containing 1-octene as co-monomer as function of the crystallization temperature: 105 °C (a); 120 °C (b) [from reference 56, copied with permission from Springer Verlag]

470

of the area of the highest melting peak to that of the others. This index was shown to correspond nicely with the heterogeneity as determined from fractionations (see above). Moreover it was found that the index increases (i.e. the copolymer becomes more homogeneous) with increasing co-monomer size, in line with the NMR observations of Randall and Hsieh[27].

As numerous mechanical properties are directly dependent on the crystallinity and the crystalline morphology one would expect a large number of publications on this topic. This is recently indeed the case. In addition more fundamental studies of polymer crystals, for which polyethylene is considered to be a model polymer are fairly abundant.

10.3.1 <u>Melting temperature, melting enthalpies and density</u>
    A few individual literature reports will first be discussed

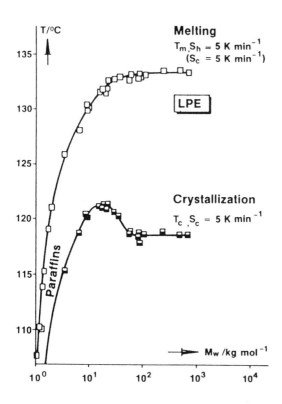

Fig. 10.8: Crystallization and melting peak temperatures ex DSC for linear PE-fractions and paraffins as function of MW, scan rate 5K/min [from reference 57, copied with permission from Springer Verlag]

before attempting a more comprehensive survey. Mathot and Pijpers[57] studied with DSC the melting and crystallization behaviour of polyethylene fractions of widely varying molecular weight (and narrow distribution). Fig. 10.8 gives a number of their data, from which the melting behaviour appears fully normal, increasing with MW up to a maximum. This relationship can be expressed as follows

$$T_m = 133.4 - 27.1/M_w$$

in °C and M_w in kg/mol. The crystallization behaviour is very different: after an initial increase in crystallization temperature up to a MW of 20,000 a decrease is noted up to a MW of 60,000 whereafter the crystallization temperature stays constant. The explanation given takes the delaying effect of entanglements into account from a MW of 20,000 onwards, whilst at the higher molecular weight of 60,000 the chains are so large that parts of them can independently crystallize. Similar behaviour is shown by fractions of LDPE, although at a lower level of temperature of course.

A number of systematic studies have been reported on the effect of co-monomer type (i.e. branch length and structure) on the crystalline properties of ethylene copolymers. Shirayama et al[38], realizing the heterogeneity of the copolymers generated on the $VOCl_3/AlEt_2Cl$ catalyst, used fractions for determining the melting behaviour and the crystallinity. The co-monomer types tested ranged from propylene to dodecene as linear olefins and additionally a few branched olefins were used such as 4-methyl-1-pentene. The composition was determined by IR. The difference in melting point at identical mol-fractions is not very large, being about 10 °C at co-monomer contents under 7.5 %mol. The larger branches are more effective in lowering the melting point, whilst there is no considerable difference between the shorter branches, i.e. CH_3 is as effective as C_3H_7. In crystallinity (by X-ray) a much more complicated relationship with branch length is reported, showing the following order of effectiveness in decreasing the crystallinity :

$$C_1 \ll nC_4 < iC_4 < C_6 < C_3 < C_2 < nC_{10}$$

The smallest branch has far less effect on crystallinity than on the melting temperature (see also paragraph 10.3.2), whilst already ethyl and propyl branches have large effects on both properties. In the case of the density a spread of 0.01 to 0.015 g/cm^3 is found

for various branch types at equal molar fraction. The larger
branches lower the density more than smaller ones, and branched
structures are less effective (see below for an opposite observation). For some co-monomers the effect on density and melting point
diminishes at higher loadings, this being especially the case for
1-butene and 1-pentene (in line with the clustering tendency of
these monomers as described elsewhere).

Phillips investigators[58] describe for copolymers with linear
α-olefins a considerable difference between propylene and 1-butene
with respect to the density. All larger co-monomers fall on the
same curve however, which is only slightly lower than the one for
1-butene. The same conclusion regarding propylene and 1-butene was
reached by Kashiwa et al[11].

Although similar conclusions were reached in later studies, on
details there are differences. For instance Hogan[59] reports that
branched co-monomers are more effective in lowering the density,
illustrated by the fact that for a density of 0.924 either 3.7 %mol
of 1-hexene or only 2.9 %mol of 4-methyl-1-pentene is required
(copolymers made on a chromium catalyst). Similar conclusions were
reached by Kashiwa et al[11] and Burfield and Kashiwa[60] on copolymers
made on a $TiCl_4/MgCl_2$-supported catalyst at 170 °C. Fig. 10.9 gives
some of the data obtained and clearly the 4-methyl-pentene-1
copolymer scores the lowest density at equal molar composition.

Alamo, Domszy and Mandelkern[61] prove in an elaborate and very
careful study that except for methyl branches, the other branches
tested (ethyl, hexyl, acetate) all follow one relation of melting
point versus co-monomer content (when determining the melting point
under near-equilibrium conditions in a dilatometer on high-temperature annealed samples). The difference in melting temperature is
about 10 °C at equal incorporation. Using the DSC method the same
conclusion is reached for truly random copolymers.

The heterogeneous nature of LLDPE's is also reflected in their
melting behaviour. Taking commercial samples for example, one
observes[39,62] multi-peaked melting behaviour over a broad range,
similar to the behaviour shown by ethylene/propylene copolymers
mentioned in chapter 4 (see also Fig. 10.7). On using fractions,
for instance ex TREF-fractionations, a single peak results as
expected.

The enthalpy of fusion decreases with increasing co-monomer

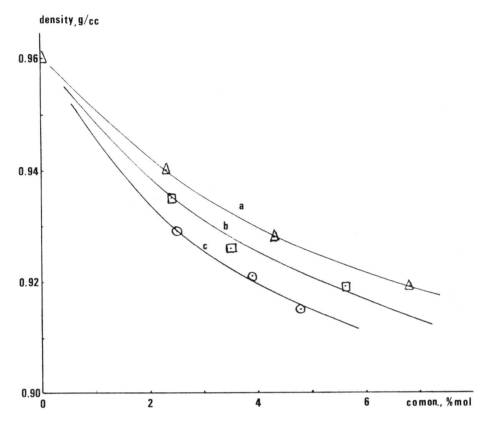

Fig. 10.9: Density of ethylene copolymers as function of co-monomer content and co-monomer structure: propylene (a); 1-butene (b); 4-methyl-1-pentene (c)

content (see e.g. ref. 60 and 61), rapidly up to 3 %mol, more slowly at higher levels of co-monomer. This phenomenon is again independent of co-monomer type when comparing ethyl, hexyl or acetate branches and when compensating for molecular weight effects (see below).

On closer study, investigating copolymers made with different catalysts or under different conditions, one can observe changes in behaviour. A case in point is the study of Kimura et al[43] in which ethylene/1-butene copolymers were compared made on a "supported titanium" catalyst either made in a slurry polymerization at 70 °C and 20 bar, or in a high pressure tubular reactor (and thus at much higher temperature). A combination of characterization techniques was applied. In TREF one observed at identical elution temperatures

up to 2 %mol difference in butene content of the fractions. Differences in melting behaviour, with the high pressure copolymer
melting at a lower temperature at identical co-monomer level, were
also observed. Finally the tetrad distribution of a few samples
were determined by ^{13}C-NMR: the high pressure copolymer is more
random, containing more isolated butene units and considerably less
paired butene units. This is in line with the observed behaviour,
the relation with copolymer parameters was however not made.

Similar effects of sequence distribution were described by
Alamo et al[61] and Brady and Thomas[63]. The latter show the large
difference in behaviour of true random copolymers (made by hydrogenation of polybutadiene) and commercial LLDPE's; for instance in
the melting the former melt at lower temperatures and have a
steeper dependency on the co-monomer content, in line with previous
studies[61]. Also the difference between the melting points of LDPE
and linear ethylene/1-butene copolymers, in which LDPE melts at a
higher temperature at identical branching densities[37,64], has been
explained by sequence distribution differences[36]. This is in line
with the mechanism of branching, where once a branching reaction
occurs the configuration is conducive for subsequent branching
reactions[37].

That homogeneous catalysts allow the preparation of copolymers
with a more random sequence distribution and a narrow compositional
distribution is shown, for example, by the data of Kaminsky and
Spiehl[65] on ethylene/cyclopentene copolymers. A _single_ peak is
observed in the DSC, and the slope of the DSC peak melting point
with co-monomer content is about 7 °C/%mol incorporated. This is
fairly effective as will be clear when comparing this with the data
in Fig. 10.12 and Fig. 10.13 (see below). Alamo and Mandelkern
recently also reported on the same type of homogeneously made
copolymers[66]; their copolymers were made with Cp_2ZrCl_2. A compli-
cating feature arose in that the melting temperature and the
crystallinity appeared to be dependent on the MW at the high end,
which was unexpected. However, when comparing copolymers within a
narrow MW range, a unique relation between melting temperature and
co-monomer content is found, the slope in this case being about
11.5 °C per %mol incorporated co-monomer. The data from Hosoda[45],
based on fractions of LLDPE's made on a titanium supported catalyst, show slightly lower slopes, but different for different
co-monomers: for 1-butene about 7.6 °C, for 1-hexene 8.9 °C and for
1-octene and 4-methyl-1-pentene even 10.7 °C per %mol incorporated

in the copolymer.

An attempt has been made to collect data on melting phenomena and density for different polymers, preparation conditions, etc..

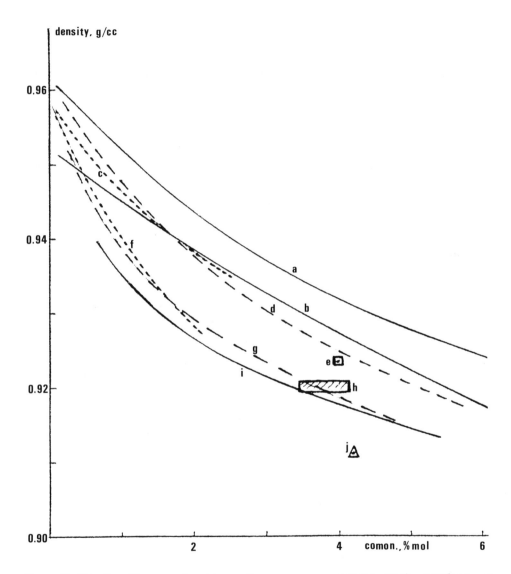

Fig. 10.10: Density as function of co-monomer content with catalyst as parameter; propylene and 1-butene as co-monomers: propylene, ref.67 (a); propylene, 1-butene, ref.38 (b); propylene, ref.68 (c); 1-butene, ref.67 (d); 1-butene, ref.59, (e); 1-butene, ref.68, (f); 1-butene, ref.10, (g); 1-butene, ref.29, (h), 1-butene, ref.69, (i); 1-butene, ref.70,71, (j)

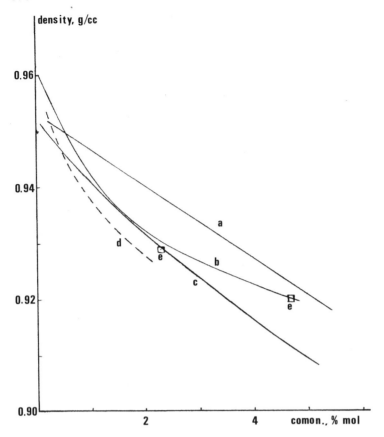

Fig. 10.11: Density as function of co-monomer content with catalyst as parameter, 1-hexene as co-monomer: ref.72, (a); ref.59 (b); ref.38 (c); ref68 (d); ref.41 (e)

The outcome is rather meagre, despite the large commercial activity in linear copolymers no substantial set of data could be extracted from the literature. In a large number of cases some of the required characterization data are lacking, for instance giving only density and melting point but no composition, and this makes a relevant correlation even more difficult. This is the case in many patent examples but also in the open literature. Assuming that all data are dependable, which particularly with the difficulty of compositional analysis (see also chapter 13) is debatable, the following facts have been found.

Fig. 10.10 and Fig. 10.11 give respectively the most relevant

data of <u>density</u> <u>versus</u> <u>composition</u> for propylene and 1-butene (Fig.
10.10) and 1-hexene (Fig. 10.11). Propylene is obviously less
effective in lowering the density and for the other co-monomers
there is a clear effect of catalyst type and possibly also of the
polymerization conditions. With 1-butene as co-monomer the lowest
densities are observed for the supported titanium catalyst[69] and
the $TiCl_3$/siloxalane catalyst used at very high temperature[10]. The
$VOCl_3$-derived polymers are, in this case considerably above the
other two. In the 1981 edition of the Kirk-Othmer in two related
chapters Phillips investigators report[67,68] on the density of
butene copolymers, the two curves are given in Fig. 10.10 and are
clearly different, for reasons unknown. It has not been disclosed
for which catalyst nor under which polymerization conditions these
copolymers were made, the only clue comes from a separate report[59]
in which a chromium catalyst gives a copolymer with a density
falling on the highest line. Assuming that this holds generally,
then for reaching a density of 0.93 in ethylene/1-butene copolymers
at the applied polymerization conditions one needs:

about 1.5 % mol incorporated 1-butene on supported titanium catalyst
about 1.8 ,,   ,,   ,,   ,, $TiCl_3$/siloxalane
about 3.0 ,,   ,,   ,,   ,,   chromium oxide on
silica
about 3.5 ,,   ,,   ,,   ,, $VOCl_3$/$AlEt_2Cl$.

Interestingly the very lowest densities are observed for hydrogenated polybutadienes[70,71], and this fits in with the observed
alternating tendency which must lead to maximum disturbance compared to more random or blocky sequence distributions.

In the case of 1-hexene copolymers (Fig. 10.11) the relationships are different again. The "modified chromium" and $VOCl_3$
catalyzed samples show the lowest density, with the results of the
homogeneous catalyst clearly above this. One would have expected
otherwise as a truly random copolymer would be thought to form on a
catalyst having only one type of active site (MWD is narrow!). To
resolve this, more detailed characterization studies are required.
The $TiCl_3$/siloxalane-made copolymers follow the chromium catalyst
line. The one line from the Kirk-Othmer is given as well, this also
follows closely the chromium catalyst line. Comparing again at a
density of 0.93 in ethylene/1-hexene copolymers one needs:

about 2.2 % mol incorporated 1-hexene on $VOCl_3$, "modified chromium"
and $TiCl_3$/siloxalane
about 3.6 ,, ,, ,, ,, $Cp_2Ti(CH_3)_2$/alumino-
xane.

For certain catalysts 1-butene appears to be the more effective
co-monomer in reaching a fixed density, but this conclusion is
admittedly based on a rather limited set of data.

Fig. 10.12 and Fig. 10.13 give a number of data on <u>melting</u>
<u>temperature</u> <u>versus</u> <u>composition</u>. The melting point is a less well
defined property than density and different approaches in its

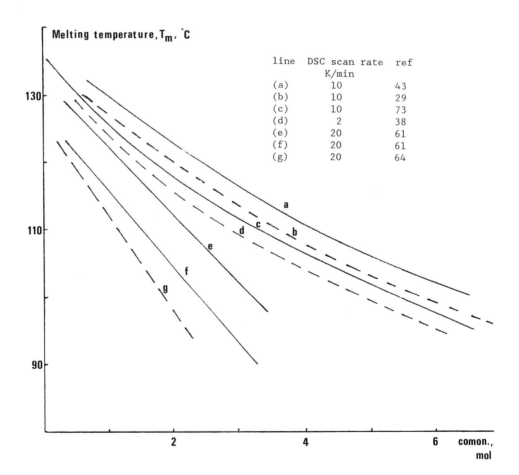

Fig. 10.12: DSC-melting temperatures of ethylene-1-butene copolymers as function
of the 1-butene content (measurements on fractions except for line (c))

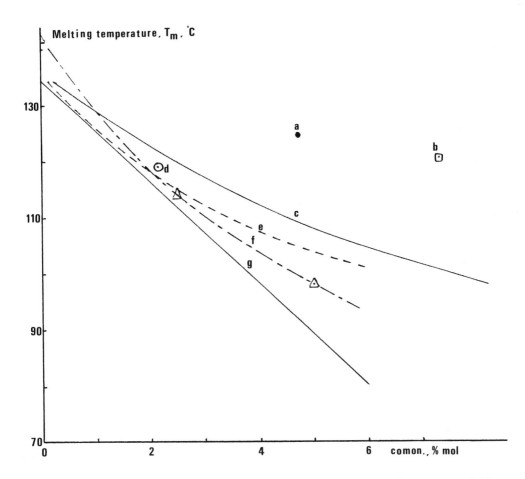

Fig. 10.13: DSC-melting temperatures of ethylene copolymers: 1-hexene, ref.41, (a); propylene, ref.41, (b); propylene, ref.38, (c); 1-hexene, ref.74, (d); 1-hexene, ref.38, (e); 1-octene, ref.48, (f); 1-hexene, ref.45, (g)

measurement lead to widely differing values. Also some groups studied fractions instead of total polymers when they realized their compositional heterogeneity[61,38]. As shown by Shirayama et al[38] the effect of the heterogeneity of the samples on the melting point determination can be large, especially at the higher co-mono-mer contents; melting temperature levels off at around 120 °C for the total polymer whilst in fractions a linear decreasing melting point is observed down to at least 20 %mol co-monomer. All this makes comparison very difficult except within one research group. For the 1-butene copolymers a number of parallel lines are given for the different catalysts. Except for one line all data in Fig. 10.12 are based on measurements on fractions. A very wide range of melting points is indicated at equal compositions. The comparison of the hydrogenated polybutadiene and ethylene/1-butene made on a titanium catalyst shows the probable effect of sequence distribution. Other comparisons are difficult, however the homogeneously catalyzed copolymers (with the melting point determined on the whole sample) are lower in melting temperature than the fractionated supported titanium polymers, indicating a good degree of randomness for the former. For ethylene/1-hexene and propylene copolymers (see Fig. 10.13) only a very limited number of data have been found, the surprising feature is the very different behaviour of $TiCl_3$/siloxalane with these monomer pairs, whilst in other cases no such deviating behaviour is noted. Heterogeneity of the samples could be a cause for this.

In trying to fit LDPE on the same type of graph Casey et al[64] found higher melting temperatures for LDPE compared with ethylene copolymers at identical branching levels (based on IR measurement of methyl groups). Casey et al used fractions of the model copolymers to overcome the problems of heterogeneity; if one uses total polymers – as for example Bastien et al[75] - then the observation is just opposite, with LDPE melting lower. Using the data of the fractions as a base, the conclusion has to be that the sequence distribution in LDPE is less random and more blocky than in ideal random copolymers. This fits in with the accepted mechanism of branch formation (see above).

Some data will now be given of <u>density versus melting point</u>. This is not a very useful relationship from a structure-property type of view, but it allows one to use more data in a comparative way. The available data are represented in Fig. 10.14. The two sets of ethylene/1-hexene copolymers made by Finogenova et al[77] show considerable lower melting temperatures for the CrO_3/silica-alumina catalyzed copolymers at equal density. Assuming that the melting point is more affected by differences in the sequence distribution than the crystallinity and density (see ref 61), this probably means that the chromium oxide catalyzed samples are more random, and more homogeneous in their composition. The remainder of the data in Fig. 10.14 are based on commercial samples, all show only a slight dependence of the melting point on the density, which very probably indicates a high degree of heterogeneity in copolymer

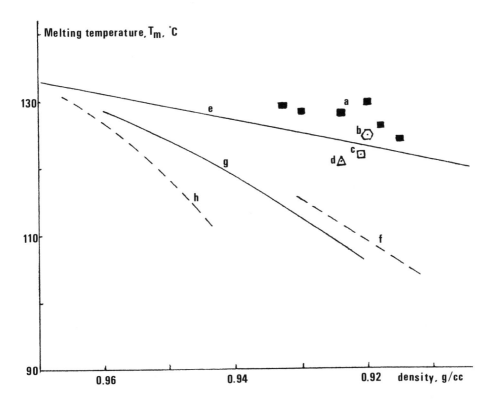

Fig. 10.14: *Density versus melting temperature for various ethylene copolymers:*
1-butene, ref.69, (a); 1-hexene, ref.41 (b); 1-butene, ref.41, (c); propylene,
ref.41, (d); 1-butene, ref.76, (e); "LDPE", ref.76, (f); 1-hexene, ref.77, Ti/Mg
catalyst, (g); 1-hexene, ref.77, Chromium catalyst, (h)

composition. LDPE takes a separate position in this graph, showing
much lower melting temperatures at equal density when compared with
LLDPE.

Molecular weight has a large influence on crystallization and
thus on crystallinity, density and melting enthalpy. Lower values
of these parameters are observed with an increase in chain length.
An illustration of this has already been given in chapter 6 in
which the melting behaviour of blends of polypropylene and ethyle-
ne/propylene copolymers of widely varying molecular weight was
given. For homopolymer the density ranges from 0.97 g/cm^3 for wax
to 0.935 g/cm^3 for ultra-high molecular weight polyethylene[5], this

behaviour is illustrated in Fig. 10.15.

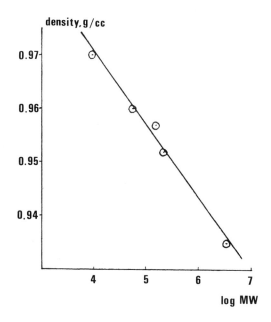

Fig. 10.15: Density of homopolymer polyethylene as function of its molecular weight

    As a last interesting observation, the melting and crystalli-
zation behaviour of small polyethylene blocks between butadiene
blocks is mentioned[78,79]. At a block length below 40 CH_2-units no
crystallization was observed, contrary to the behaviour of isolated
chains which would readily crystallize. At longer block lengths
differences in the type of polyethylene crystals were found. The
intermediate sized blocks (70 to 90 CH_2-units) crystallized with
extended chain crystals, whereas above 90 units block length the
normal chain folded crystals were observed. The melting point
decreased sharply in the transition.

10.3.2 Crystal structure, fate of alien groups

    In this sub-section a short resumé will be given on the
structure of poylethylene crystals and the way this is affected by
the presence of non-methylene groups. Only an introductory treatment of this subject will be followed, thus no in-depth exposé
should be expected. Still this topic is of interest to polymer
chemists as the strong relation between the crystalline nature of

polyethylene and its mechanical properties is well known, thus when aiming at tailoring of the properties, knowledge on the more detailed structure of the crystal is also required. As a general reference the reader is referred to Mandelkern's book[80]. Very illustrative reviews were given on the occasion of the Golden Jubilee Conference of Polyethylene by Mandelkern[81], Keller[82] and Vonk[83]. A later paper by Mandelkern[84] is also worth referring to for an excellent and concise review.

Polymers crystallize in lamellar crystals, i.e. in thin platelets in which the polymer chains are parallel to the smallest dimension. The thickness of the lamellae is normally much smaller than the polymer chain length, necessitating a folding of the chains to allow (re)entry into the crystal. Only a small fraction of the chains show an adjacent re-entry, the largest fraction re-enters the crystal irregularly. The lamellae in bulk crystallized polymer samples form a "higher" organisational structure in the well known spherulites, which have a radial symmetry. Different organisational structures can be realized by changing for instance the conditions of crystallization, see for example the references 82,85 and 86.

Using linear homopolymer polyethylene as a study material very high crystallinities can be obtained of up to 90 %, lamellar thickness can be substantially increased by, for instance annealing near the melting temperature, and sizes up to 10 μm have been observed. The crystallization behaviour is however strongly molecular weight dependent.

The question arises, what is the effect of end-groups or co-monomer/branch units? The fate of end groups was initially studied by etching of crystalline samples (crystallized from solution) with fuming nitric acid (see e.g. the studies by Keller and Priest[87,88]). The observed nearly-complete disappearance of the vinyl groups in particular after such a treatment was taken as proof that almost no endgroups were present in the crystalline regions. More recent NMR studies[89] on melt crystallized samples showed a far lower selectivity, with some 45 % of all vinyls present in the lattice. In the case of side chains stemming from co-monomers, the behaviour depends on the size of the side group. For small side groups (such as CH_3, Cl, =O) the view is that they can be accomodated in the crystal on an equilibrium basis, at least for a fraction. Their effect, for instance, on lowering crystallinity is relatively small as has been mentioned in section 10.3.1.

They still distort the lattice leading to a larger unit cell, especially in the "a" direction, and the melting point is considerably lowered. All larger side groups are excluded from the crystal in a thermodynamic sense, but they still can be occluded as defects. These facts lead to considerable effects of these groups on the crystallinity and melting points. For this latter case, the exclusion of the foreign groups from the crystal, means that a far lower fraction of the polymer sequences can take part in the crystallization, which leads for instance to much smaller lamellar thicknesses even under optimal conditions. For example at 0.6 %mol foreign groups the maximum lamellar thickness observed was only 1.25 μm. Also annealing effects are far less than in homopolymers. No dependence of the behaviour on molecular weight is found for these copolymers, only the composition determines the outcome of the crystallization. Obviously the sequence distribution is now of the utmost importance as a blocky type of copolymer will show nearly unhampered crystallization and a true random copolymer will show a hindered one. Although this behaviour is generally independent of the type of side-group, for long chain branches as in LDPE the behaviour is different. Here, as expected, co-crystallization of the branch can easily occur.

Proof for the above comes from different types of experiments. Lattice expansion has been measured for copolymers and found to be largest for methyl[90,91] and independent of crystallization conditions. For longer branches the lattice expansion is far less in slow crystallization, only by rapid quenching are lattice expansions realized similar to the methyl-branched case[91]. Also, oxidation studies have been done in this field[88,92], on ethylene/1-butene copolymers it was found that all butene groups were removed. This means they have all to be accessible for the oxidant, leading to the conclusion that the butene groups are indeed excluded from the crystalline parts of the sample. A similar result was obtained by Cutler et al[93] in samples of LDPE; from NMR measurements on oxidized samples it was concluded that ethyl and butyl were excluded from the crystal, long chain branches were not. Ethylene/propylene copolymers were shown to form thicker lamellae, almost equal to HDPE, compared to higher olefin copolymers crystallized at identical conditions[61]. This is in line with the above as ethylene/propylene polymers are much more like a homopolymer than the others and thus should form thicker crystals. Using NMR on solid polymer samples[89,94,95] as an analytical tool, which allows the analysis of the crystalline and non-crystalline part separately, the above concepts have been proved to be correct. For example, in the studies of van der Hart et al[89,94,96] it was shown by NMR on an ethylene/1-butene copolymer that in the amorphous regions of the sample 10 times more butene groups were present than in the lattice, which meant that only 4 to 8 % of the available branches were present in the crystals. Similarly, in a propylene copolymer slowly cooled form the melt, it was shown[96] that 50-75 % of the methyl endgroups, 42-63 % of the vinyl endgroups and only 21-27 % of the methyl branches were present in the crystalline regions. Thus melt crystallization is less discriminating, as expected, as more of the endgroups are built into the crystal compared to solution crystallization (see above). McFadden et al[97] suggest that the motionally restricted groups are to a large extent present in interfacial regions and not truly in the crystal itself. The bases for this are firstly the decrease in line-width observed when the samples are swollen in CCl_4, i.e. they are accessible for this solvent, and secondly the strong decrease in intensity of the relevant peaks upon heating to 50 °C, which is a temperature much below the expected melting point. Laupretre et al[95] showed that a constant fraction of the methyl groups were present in the crystalline phase independent of the crystallization conditions.

Still related to the same question, i.e. what can be tolerated as alien groups in a crystal, Alamo et al[98] studied the co-crystallization of nearly identical polymers using fractions of narrow molecular weight distribution and of equal molecular weight. DSC and selective extraction techniques were used as diagnostic tools. Polyethylene and a copolymer of ethylene with 2.2 %mol 1-butene do not co-crystallize either from solution or upon isothermal crystallization. When quenching from the melt, the behaviour changes and co-crystallization is indeed observed. Increasing the mismatch to 3.2 %mol still gives co-crystallization under those conditions, but not at the 5.5 %mol level. These conclusions are in line with previous studies.

Returning to the introduction to this sub-section: for tailoring the LLDPE properties a prerequisite is the ability to control the sequence distribution in the copolymer. Judging from the open literature this situation has not yet been reached, and catalyst developments are required to attain this goal; probably homogeneous catalysts can reach this goal more easily than heterogeneous ones.

10.4 Dynamic mechanical analysis

The glass transition of pure homopolymer polyethylene has long been an unresolved problem giving rise to very elaborate studies and long debates. The general agreement recently is that the so-called γ-transition at around -130 °C is to be called the glass transition. For a good review on this problem the reader is referred to the paper of Stehling and Mandelkern[99].

In their elaborate study into the relation of copolymer composition and properties Shirayama et al[38] also studied the dynamic mechanical properties. As a general rule, upon copolymerization the dynamic modulus decreases in value. For the loss modulus a more complex change accurs in that the α-transition (around 7 °C for HDPE) rapidly disappears, in its place the β-transition is observed which increases in value and shifts to lower temperature with increasing co-monomer content. The γ-transition at about -120 °C is nearly unaffected. As function of co-monomer type only the shortest (methyl) and the longest (octadecyl) ones tested showed clearly different behaviour. The temperature of the maximum in the β-transition is lowest and identical for all other branches, and is highest for octadecyl branched copolymers, with ethylene/propylene copolymers in between. This is probably a reflection of the co-crystallization of the C_{18} branches lowering the mobility at the branch point, whilst methyl is (partly) incorporated in the lattice and there also less effect would be expected. It takes almost 20 %mol co-monomer to reach the minimum temperature for the β-transition of -50 to -55 °C, in line with the curves given in chapter 4 for ethylene/propylene copolymers based on propylene-rich copolymers. For the dynamic modulus there is more gradation of the effect of branch length, it gradually decreases with its length (with ethyl and propyl branches being an exception). In Fig. 10.16 the glass transition is given as function of the composition on the basis of the measurement of Shirayama et al[38]. In the same figure the data are drawn for ethylene/butene copolymers made by hydrogenation of polybutadienes[70,71]. No distinct differences are observed despite the fact that in terms of sequence distribution and heterogeneity of the polymers large differences exist (the hydrogenated polybutadienes have a distinct alternating tendency and are homogeneous with respect to composition, the vanadium catalyzed copolymers are very heterogeneous).

It has recently been recognized[86] that the dynamic mechanical

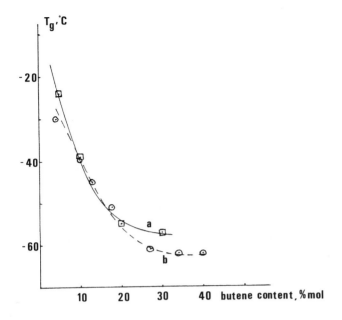

Fig. 10.16: Glass transition temperatures of ethylene/1-butene copolymers, data ex ref.38 (a) and hydrogenated polybutadiene, data ex ref.71 (b)

properties can be drastically influenced by the crystalline morphology. A number of LDPE's were either quenched or slowly crystallized, and the resulting crystallinity was not very different. The loss maxima however differed 50 to 60 °C, with the quenched samples having a loss peak ranging from -7 to +40 °C, depending on the starting polymer.

10.5 Mechanical properties

This section will be kept very short, and it will serve only to highlight the relation between a few mechanical properties and polymer characteristics. In this way the complex interrelationships playing a role in this area will be partly made visible and the difficulty of tailoring polymer properties will become apparent.

Very generally spoken the properties of polyethylenes depend mainly on density, molecular weight (always expressed as Melt Index), and molecular weight distribution. But branch length also appears to play a role, not only in the rather obvious case of melt flow, in which LDPE behaves in a drastically different manner from all other kinds of polyethylenes, but also in more subtle ways. An illustration is the difference in time-to-failure of pipes made from LLDPE's based on butene or hexene as co-monomer; the latter are 100 times better[100]. Other illustrative examples can be found in a study of Bubeck and Baker[101] on the effects of composition on environmental stress cracking resistance (ESCR). Three different polymer series were tested:

- in ethylene/propylene copolymers the ESCR showed an optimum at rather low propylene levels (density of 0.954),
- in ethylene/1-octene copolymers the ESCR increases up to the lowest density value tested (0.919),
- in a series of constant density the ESCR increases with increasing branch length from methyl to hexyl; the increase on compression moulded samples is a factor 15 at the tested density of 0.92. With increasing branch length the spherulite size also increases considerably. It is noteworthy that LDPE does not fit in this series (as one might expect because of its long chain branching), but it shows very poor stress cracking behaviour and gives the smallest spherulites.

In terms of strength properties, such as stiffness and yield stress, there exists a unique relation with density for all types of polyethylenes including LDPE (see e.g. references 38, 100, 102, 103); stiffness etc. decreases with decreasing density. Density is not a primary independent variable from the polymer synthesis point of view, but not surprisingly the crystallinity is the overriding property determining strength in this type of polymers, and density is a very good measure of this.

Payer[100] gave an extensive description of the effect of variables such as molecular weight and its distribution on mechanical properties of LLDPE's. An overview is given in Table 10.4 together with data taken from some other sources. Of special interest are the dependencies on MWD and co-monomer type. On broadening the distribution the impact strength decreases but the stress corrosion resistance (ESCR) increases. The density is also affected by this parameter. Longer sidegroups give for instance higher impact strength.

At equal density LLDPE is more stiff, has a higher elongation, a much higher ESCR and a higher melting point than LDPE. The most

TABLE 10.4

Overview of relationships of mechanical properties of LLDPE's with some polymer characteristics. Changes in properties are given when the independent parameter is increased.

Independent variable	Dependent variable					
	density	tear strength	tensile impact	elongation in tensile impact	at break	ESCR
MW	down	up	up	up	indep	up
MWD	up	down	down	down	-	up
comonomer level	down	down	up	up	down	up
branch length	-	indep.	up	up	down	-

important properties are the much improved load bearing abilities, i.e. there is less creep and longer time to failure in long-duration experiments; secondly, the ESCR improvement is of high commercial value.

REFERENCES
1. A.H.Willbourn, Polymethylene and the structure of polyethylene: Study of short chain branching, its nature and effects, *J.Pol.Sc.*, *34* (1959) 569-597
2. M.E.A.Cudby, Observations on the molecular structure of polyethylene from vibrational and magnetic resonance spectroscopy, paper D6 in Polyethylenes 1933-1983, Golden Jubilee Conference, 8-10 June 1983, London, The Plastics and Rubber Institute.
3. A.Solti, D.O.Hummel and P.Simak, Computer-supported infrared spectroscopy of polyethylene, ethene copolymers, and amorphous poly(alkyl ethylene)s, *Makrom. Chemie, Macromol. Symp.*, *5* (1986) 105-133
4. T.Usami and S.Takayama, Identification of branches in low-density polyethylenes by Fourier Transform infrared spectroscopy, *Polymer Journal*, *16* (1984) 731-738
5. E.Paschke, Ziegler Process Polyethylene in Kirk-Othmer Encyclopedia, 1981 edition
6. R.Spitz, B.Florin and A.Guyot, Ethylene-hexene-1 copolymers through modified Phillips catalysis, *Europ.Pol.J.*, *15* (1979) 441-444
7. F.J.Karol, G.L.Brown and J.M.Davison, Chromocene-based catalysts for ethylene polymerization: kinetic parameters, *J.Pol.Sc., Pol.Chem.*, *11* (1973) 413-424
8. F.J.Karol et al, Chromocene catalysts for ethylene polymerization: Scope of the polymerization, *J.Pol.Sc., A1*, *10* (1972) 2621-2637
9. D.Constantin, M.Hert and J-p.Machon, Structure of polyethylene produced at high pressure by simultaneous ionic and radical mechanisms, *Makromol.Chemie*, *179* (1978) 1581-1591
10. J.P.Machon, Actual and potential development of high pressure process for linear polyethylene, paper B2 in Polyethylenes 1933-1983, see ref.2
11. N.Kashiwa, T.Tsutsui and A.Toyota, Solution copolymerization of ethylene with α-olefins by a MgCl2 supported TiCl4 catalyst, *Polymer Bulletin*, *12* (1984) 111-117
12. N.Kashiwa, T.Tsutsui and A.Toyota, Solution copolymerization of ethylene with α-olefins by vanadium catalyst, *Polymer Bulletin*, *13* (1985) 511-517
13. P.M.Kamath and A.Barlow, Quantitative determination of short branches in

490

high pressure polyethylene by γ-radiolysis, *J.Pol.Sc.*, *A1*, _5_ (1967) 2023-2030

14. T.N.Bowmer and J.H.O'Donnell, Nature of the side chain branches in low density polyethylene: volatile products from γ-radiolysis, *Polymer*, _18_ (1977) 1032-1040

15. M.Seeger and E.M.Barrall, Pyrolysis gas chromatographic analysis of chain branching in polyethylene, *J.Pol.Sc.*, *Pol.Chem.*, _13_ (1975) 1515-1529

16. Y.Sugimura and S.Tsuge, Pyrolysis hydrogenation glass capillary gas chromatographic characterization of polyethylenes and ethylene-α-olefin copolymers, *Macromolecules*, _12_ (1979) 512-514

17. Y.Sugimura, T.Usami, T.Nagaya and S.Tsuge, Determination of short chain branching in polyethylenes by pyrolysis hydrogenation glass capillary gas chromatography, *Macromolecules*, _14_ (1981) 1787-1791

18. M.A.Haney, D.W.Johnston and H.H.Clampitt, Investigation of short-chain branches in polyethylene by pyrolysis-GCMS, *Macromolecules*, _16_ (1983) 1775-1783

19. J.Tulisalo, J.Seppälä and K.Hästbacka, Determination of branches in terpolymers of ethylene, 1-butene, and long α-olefin by pyrolysis hydrogenation gas chromatography, *Macromolecules*, _18_ (1985) 1144-1147

20. J.C.Randall, Characterization of long-chain branching in polyethylenes using high-field carbon-13 NMR, in Polymer characterization by ESR and NMR, ACS symposium series 142, ACS, Washington, 1980

21. E.T.Hsieh and J.C.Randall, Ethylene-1-butene copolymers. 1. Comonomer sequence distribution, *Macromolecules*, _15_ (1982) 353-360

22. G.J.Ray, J.Spanswick, J.R.Knox and C.Serres, Carbon-13 NMR study of ethylene-butene copolymers, *Macromolecules*, _14_ (1981) 1323-1327

23. E.T.Hsieh and J.C.Randall, Monomer sequence distributions in ethylene-1-hexene copolymers, *Macromolecules*, 15 (1982) 1402-1406

24. K.Kimura, S.Yuasa and Y.Maru, Carbon-13 NMR study of ethylene-1-octene and ethylene-4-methyl-1-pentene copolymers, *Polymer*, _25_ (1984) 441-446

25. V.A.Zakharov et al, Gas-phase copolymerization of ethylene with propene and with 1-butene using supported chromium- and titanium-containing catalysts, *Makrom.Chemie*, _190_ (1989) 559-566

26. B.Schlund and L.A.Utracki, Linear low density polyethylenes and their blends: Part I. Molecular characterization, *Polym.Eng.Sc.*, _27_ (1987) *359-366*

27. J.C.Randall and E.T.Hsieh, 13C-NMR in polymer quantitative analysis, chapter 9 in NMR and Macromolecules, ACS symposium series 247, ACS, Washington, 1984

28. H.J.Harwood and W.M.Ritchey, The characterization of sequence distribution in copolymers, *Polymer Letters*, _2_ (1964) *601-607*

29. T.Usami, Y.Gotoh and S.Takayama, Generation mechanism of short-chain branching distribution in linear low-density polyethylenes, *Macromolecules*, _19_ (1986) 2722-2726

30. J.C.Randall, The identity of the amyl branch in low-density polyethylenes, *J.Appl.Pol.Sc.*, _22_ (1978) 585-588

31. D.E.Axelson, G.C.Levy and L.Mandelkern, A quantitative analysis of low-density (branched) polyethylenes by carbon-13 Fourier transform NMR at 67.9 MHz, *Macromolecules*, _12_ (1979) 41-52

32. T.Usami and S.Takayama, Fine branching structure in high-pressure, low-density polyethylenes by 50.10 MHz 13C-NMR spectra, *Macromolecules*, _17_ (1984) 1756-1761

33. P.Freche and M-F.Grenier-Loustalot, Caractérisation des branchissements courts dans un polyèthylène basse densité par RMN 13C, *Eur.Pol.J.*, _20_ (1984) 31-35

34. J.N.Hay, P.J.Mills and R.Ognjanovic, Analysis of 13C FT-NMR spectra of low-density polyethylene by the simplex method, *Polymer*, _27_ (1986) 677-680

35. H.N.Cheng, Determination of polyethylene branching through computerized 13C-NMR analysis, *Polymer Bulletin*, _16_ (1986) 445-452

36. M.J.Roedel, The molecular structure of polyethylene. I. Chain branching in polyethylene during polymerization, *J.Am.Chem.Soc.*, _75_ (1953) 6110-6112

37. W.L.Mattice and F.C.Stehling, Branch formation in low-density polyethylenes, *Macromolecules*, *14* (1981) 1479-1484
38. K.Shirayama, S-I Kita and H.Watabe, Effects of branching on some properties of ethylene/α-olefin copolymers, *Makrom.Chemie*, *151* (1972) 97-120
39. F.Mirabella and E.Ford, Characterization of linear low density polyethylene: Cross-fractionation according to copolymer composition and molecular weight, *J.Pol.Sc., Pol.Phys.*, *25* (1987) 777-790
40. C.France, P.J.Hendra, W.F.Maddams and H.A.Willis, A study of linear low-density polyethylenes : branch content, branch distribution and crystallinity, *Polymer*, *28* (1987) 710-712
41. J.P.Machon, Some aspects of the ethylene copolymerization in the high pressure Ziegler process, in Transition metal catalyzed polymerizations, alkenes and dienes, part B, R.P.Quirk editor, Harwood Academic Publishers, London, 1981
42. L.L.Böhm, Zur Copolymerisation von Ethylen und α-Olefinen mit Ziegler-Katalysatoren, *Makrom.Chemie*, *182* (1981) 3291-3310
43. K.Kimura, T.Shigemura and S.Yuasa, Characterization of ethylene-1-butene copolymer by differential scanning calorimetry and 13C- NMR spectroscopy, *J.Appl.Pol.Sc.*, *29* (1984) 3161-3170
44. T.E.Nowlin et al, High activity Ziegler-Natta catalysts for the preparation of ethylene copolymers, *J.Pol.Sc., Pol.Chem.*, *26* (1988) 755-764
45. S.Hosoda, Structural distribution of linear low-density polyethylenes, *Polymer Journal*, *20* (1988) 383-397
46. P.Schouterden et al, Fractionation and thermal behaviour of linear low density polyethylene, *Polymer*, *28* (1987) 2099-2104
47. T.Housaki et al, Study on molecular weight dependence of short-chain branching of linear low-density polyethylene using the GPC/FTIR method, *Makrom.Chem, Rapid Commun.*, *9* (1988) 525-528
48. B.K.Hunter, K.E.Russell, M.V.Scammell and S.L.Thomson, The preparation and characterization of homogeneous copolymers of ethylene and 1-alkenes, *J.Pol.Sc., Pol.Chem.*, *22* (1984) 1383-1392
49. K.Shirayama, T.Okada and S-I Kita, Distribution of short chain branching in low-density polyethylene, *J.Pol.Sc.*, *A*, *3* (1965) 907-916
50. T.Usami et al, Branching structures in high-pressure low-density polyethylene as a function of molecular weight determined by 13C-NMR and pyrolysis-hydrogenation gas chromatography, *Macromolecules*, *20* (1987) 1557-1561
51. L.Wild, T.R.Ryle, D.C.Knobeloch and I.R.Peat, Determination of branching distributions in polyethylene and ethylene copolymers, *J.Pol.Sc., Pol.Phys.*, *20* (1982) 441-455
52. E.P.Otocka, R.J.Roe, M.Y.Hellman and P.M.Muglia, Distribution of long and short branches in low-density polyethylenes, *Macromolecules*, *4* (1971) 507-512
53. D.E.Axelson and W.C.Knapp, Size exclusion chromatography and low-angle laser light scattering. Application to the study of long chain-branched polyethylene, *J.Appl.Pol.Sc.*, *25* (1980) 119-123
54. Yu.V.Kissin, Structure of copolymers of high olefins, *Advances in Polymer Science*, *15* (1974) 91-155
55. B.Wunderlich and G.Czornyj, A study of equilibrium melting of polyethylene, *Macromolecules*, *10* (1977) 906-913
56. P.Schouterden et al, Dynamic SAXS and WAXS investigations on melt crystallized LLDPE using synchrotron radiation: crystallization and melting behaviour, *Polymer Bull.*, *13* (1985) 533-539
57. V.B.F.Mathot and M.F.J.Pijpers, Crystallization and melting behaviour of polyethylene fractions obtained by various fractionation methods, *Polymer Bulletin*, *11* (1984) 297-304
58. J.P.Hogan and D.R.Witt, Supported chromium catalysts for ethylene polymerization, in *Preprints ACS Division of Petroleum Chemistry*, *24* (2) (1979) 377-387
59. J.P.Hogan, Ethylene α-olefin copolymers made in gas phase, presented at the 28th Macromolecular Symposium, IUPAC, Amherst, Mass, 12-16 July 1982

60. D.R.Burfield and N.Kashiwa, DSC studies of linear low density polyethylene. Insights into the disrupting effect of different co-monomers and the minimum fold chain length of the polyethylene lamellae, *Makrom.Chemie, 186 (1985) 2657-2662*

61. R.Alamo, R.Domszy and L.Mandelkern, Thermodynamic and structural properties of copolymers of ethylene, *J.Phys.Chem., 88 (1984) 6587-6595*

62. G.H.Edward, Crystallinity of linear low density polyethylene and of blends with high density polyethylene, *British Polymer J., 18 (1986) 88-93*

63. J.M.Brady and E.L.Thomas, Effect of short-chain branching on the morphology of LLDPE-oriented thin films, *J.Pol.Sc.,Pol.Phys., 26 (1988) 2385-2398*

64. K.Casey, C.T.Elston and M.K.Phibbs, Methyl group clusters in free radical polyethylene, *Polymer letters, 2 (1964) 1053-1056*

65. W.Kaminsky and R.Spiehl, Copolymerization of cycloalkenes with ethylene in presence of chiral zirconocene catalysts, *Makrom.Chemie, 190 (1989) 515-526*

66. R.G.Alamo and L.Mandelkern, Thermodynamic and structural properties of ethylene copolymers, *Macromolecules, 22 (1989) 1273-1277*

67. J.P.Hogan, Linear (high density) polyethylene, In Kirk-Othmer Encyclopedia, 1981 edition

68. J.N.Short, Low pressure linear (low density) polyethylene, in Kirk-Othmer Encyclopedia, 1981 edition

69. L.Luciani et al, Linear-low density polyethylene by a new economic slurry process, part 1 and 2, presented at Antec 82, 10-13 may 1982, San Fransisco, Soc. Plastics Engineers

70. J.M.Carella, W.W.Graessley and L.J.Fetters, Effects of chain microstructure on the viscoelastic properties of linear polymer melts: Polybutadienes and hydrogenated polybutadienes, *Macromolecules, 17 (1984) 2775-2786*

71. T.M.Krigas, J.M.Carella, M.J.Struglinski, B.Crist and W.W.Graessley, Model copolymers of ethylene with butene-1 made by hydrogenation of polybutadiene: Chemical composition and selected physical properties, *J.Pol.Sc., Pol.Phys., 23 (1985) 509-520*

72. W.Kaminsky, Polymerization and copolymerization with a highly active, soluble Ziegler-Natta catalyst, in Transition metal catalyzed polymerizations, Alkenes and dienes, Part A, R. P. Quirk editor, Harwood Academic Publishers, London, 1983

73. W.Kaminsky and M.Schlobohm, Elastomers by atactic linkage of α-olefins using soluble Ziegler catalysts, *Makrom.Chemie, Macromol.Symp., 4 (1986) 103-118*

74. W.Kaminsky, Preparation of special polyolefins from soluble zirconium compounds with aluminoxane as co-catalyst, in Catalytic polymerizations of olefins, T.Keii and K.Soga editors, Kodansha, Tokyo, 1986

75. I.J.Bastien, R.W.Ford and H.D.Mak, Melting points of some polyethylene resins, *Polymer Letters, 4 (1966) 147-150*

76. R.B.Staub, The Unipol process - Technology to serve the world's polyethylene markets, in Polyethylene 1933-1983, see ref 2

77. L.T.Finogenova et al, Study of copolymerization of ethylene with hex-1-ene on applied catalysts, *Pol.Sc.USSR, 22 (1980) 448-454*

78. J.N.Hay and J.F.McCabe, Block copolymers of ethylene. I. Butadiene, *J.Pol.Sc., Pol.Chem., 16 (1978) 2893-2900*

79. J.N.Hay and M.Wiles, The crystallization characteristics of ethylene block copolymers, *J.Pol.Sc., Pol.Chem., 17 (1979) 2223-2231*

80. L.Mandelkern, Crystallization of polymers, McGrawHill, New York, 1964

81. L.Mandelkern, Influence of chain structure and crystallization conditions on polyethylene properties, Polyethylene 1933-1983, see ref 2.

82. A.Keller, Polyethylene as a paradigm of polymer crystal morphology, Polyethylenes 1933-1983, see ref 2.

83. C.G.Vonk, Crystal structure: the effect of the defects, Polyethylenes 1933-1983, see ref 2.

84. L.Mandelkern, The relation between structure and properties of crystalline polymers, *Polymer Journal, 17 (1985) 337-350*

85. L.Mandelkern and J.Maxfield, Morphology and properties of low-density (branched) polyethylene, *J.Pol.Sc., Pol.Phys., 17 (1979) 1913-1927*

86. L.Mandelkern, M.Glotin, R.Popli and R.S.Benson, The influence of crystalline morphology (supermolecular structure) on the dynamic mechanical properties of polyethylene, *Polymer Letters*, *19* (1981) 435-441

87. A.Keller and D.J.Priest, Experiments on the location of chain ends in monolayer single crystals of polyethylene, *J.Macromol.Sci.*, *B2* (1968) 479-495

88. A.Keller and D.J.Priest, On the effect of annealing on the chain ends in single crystals of polyethylene, *Polymer letters*, *8* (1970) 13-18

89. D.L.VanderHart and E.Perez, A 13C NMR method for determining the partitioning of end groups and side branches between the crystalline and noncrystalline regions in polyethylene, *Macromolecules*, *19* (1986) 1902-1909

90. P.R.Swan, Polyethylene unit cell variations with branching, *J.Pol.Sc.*, *56* (1962) 409-416

91. C.H.Baker and L.Mandelkern, The crystallization and melting of copolymers. II. Variations in unit cell dimensions in polymethylene copolymers, *Polymer 7* (1966) 71-83

92. M.Shida, H.K.Ficker and I.C.Stone, The manner of participation of ethyl branches in crystallization of ethylene-butene-1 copolymer, *Polymer Letters*, *4* (1966) 347-352

93. D.J.Cutler, P.J.Hendra, M.E.A.Cudby and H.A.Wills, Chain branching in high pressure polymerized polyethylene, *Polymer*, *18* (1977) 1005-1008

94. E.Perez, D.L.VanderHart, B.Crist Jr. and P.R.Howard, Morphological partitioning of ethyl branches in polyethylene by 13C NMR, *Macromolecules*, *20* (1987) 78-87

95. F.Laupretre et al, Crystallization conditions. Influence of crystallization conditions on the location of side-chain branches in ethylene copolymers as studied by high-resolution solid-state 13C-NMR, *Polymer Bulletin*, *15* (1986) 159-164

96. E.Perez and D.L.Vandenhart, Morphological partitioning of chain ends and methyl branches in melt-crystallized polyethylene by 13C-NMR, *J.Pol.Sc.,Pol.Phys.*, *25* (1987) 1637-1653

97. D.C.McFadden et al, Morphological location of ethyl branches in 13C-enriched ethylene/1-butene random copolymers, *Polymer Commun.*, *29* (1988) 258-260

98. R.G.Alamo, R.H.Glaser and L.Mandelkern, The co-crystallization of polymers. Polyethylene and its copolymers, *J.Pol.Sc.,Pol.Phys.*, *26* (1988) 2169-2195

99. F.C.Stehling and L.Mandelkern, The glass temperature of linear polyethylene, *Macromolecules*, *3* (1970) 242-252

100. W.Payer, Copolymerisate aus Ethylen und α-Olefinen, *Angew.Makrom.Chemie*, *94* (1981) 49-61

101. R.A.Bubeck and H.M.Baker, The influence of branch length on the deformation and microstructure of polyethylene, *Polymer*, *23* (1982) 1680-1684

102. R.W.Ford, Physical properties of some ethylene polymers, *J.Appl.Pol.Sc.*, *9* (1965) 2879-2886

103. R.Popli and L.Mandelkern, Influence of structural and morphological factors on the mechanical properties of the polyethylenes, *J.Pol.Sc.*, *Pol.Phys.*, *25* (1987) 441-483

494

Chapter 11

POLY(1-BUTENE), ITS PREPARATION AND CHARACTERIZATION

11.1 INTRODUCTION

Poly(1-butene) is the next member in the polyolefins family after polyethylene and polypropylene. Its commercial importance is still very modest, with some 30,000 tpa production capacity at present (1988). A number of companies are active in poly(1-butene) with Shell in the United States as the main producer, and interest and/or a small production from companies such as Mitsui Petrochemical, Neste Oy and Hüls. Historically speaking, poly(1-butene) and its higher homologues were studied very soon after the discovery of isotactic polypropylene, both by the Phillips group as well as by Natta's.

Mechanically it is an interesting polymer showing for instance a complete elastic recovery after modest deformations of up to 50%, this leading to the high creep resistance of the material and its high stress corrosion and impact resistance. This makes the polymer very suitable for applications as pipe, especially when also exploiting the fact that poly(1-butene) can be used up to temperatures very near to its melting temperature: service temperatures of up to 90 °C are not uncommon. Also its film properties are outstanding, film is strong and this property is coupled to high tear strength.

Considerable scientific interest is raised by the crystallization behaviour of poly(1-butene), four different crystalline forms can be distinguished of which a few are metastable.

For the purpose of this book - which attempts to link the characteristics to some basic properties of the polymer - the body of knowledge that is available turns out to be very meagre. This is of course not too surprising in view of the very small commercial activity around this polymer. This means that for the catalysts used and the polymerization behaviour experienced one has to take recourse mainly to patents. And on the characterization topic, not many coherent studies can be found. Therefore this chapter will be fairly short, highlighting in the main some of the preparative details, although most of the available studies in characterization will be summarized as well.

The literature up to 1968 has been nicely reviewed in the book on poly(1-butene) written by Rubin[1].

11.2 CATALYSTS

Being one of the α-olefins, it is not surprising that in the polymerization of butene one can use the same catalysts as described for propylene. In this case also tacticity plays a predominant role in determining the polymer properties, and the aim of the polymerization experts has been to combine high activity and good stereoselectivity in one and the same catalyst. This task is made more difficult than in the propylene case as the higher olefins all show intrinsically lower polymerization rates, for which their higher steric requirements might be mainly responsible[2].

In distinguishing the same classes of catalyst as given in chapter 1 for propylene, it turns out that many examples can be found of polymerization with the first generation $TiCl_3$-catalysts, almost none with the second generation highly active $TiCl_3$'s, whilst quite a number of patents deal with the use of supported catalysts in the polymerization of 1-butene, and quite succesfully at that! Only the first and last class will be dealt with in any detail. Additionally the results with homogeneous catalysts with 1-butene will be mentioned.

11.2.1 $\underline{TiCl_3}$ catalysts

As expected the use of $TiCl_3$ in the α, γ or δ modification leads to rather isotactic polymer just as is the case with propylene. In the butene case also the prime variable is the co-catalyst applied, both in terms of activity and selectivity. As a second variable the use of "third components" or modified co-catalysts is frequently found whose sole aim is to improve the selectivity of the chosen catalyst system.

In the simplest system of $TiCl_3$/co-catalyst the isotacticity increases in the series

$$AlEt_3 < AlEt_2Cl < AlEt_2Br < AlEt_2I.$$

The isotacticity is measured in this case by determining the fraction of the product which is soluble in boiling diethyl ether. For the co-catalysts given above, the ether-solubles range from up to 80 % with the trialkyl derivative down to 5 or 6 % with the iodide containing aluminium alkyl, or even as low as 2 %m/m in an example described by Natta's group[3]. Activity shows the opposite trend. The beneficial effect of iodide-containing co-catalysts is nicely illustrated in Fig. 11.1 in which activity and selectivity is given for polymerizations using mixtures of the chloride and the iodide as co-catalyst (data from reference 4). Pre-reacting the AlEt$_2$Cl with iodine in a separate step at an I$_2$/Al ratio of around 0.25 gives even better results in terms of the isotacticity as values of only 3 % ether solubles were reported[5].

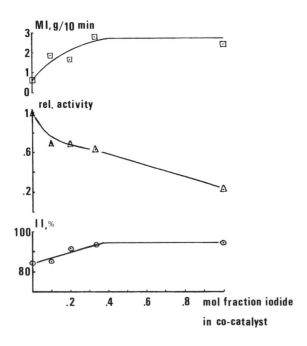

Fig. 11.1: *Effect of AlEt$_2$I in TiCl$_3$//AlEt$_2$Cl + AlEt$_2$I on polymerization of 1-butene (data from reference 4)*

Using triethylaluminium on a proprietary TiCl$_3$ catalyst Hüls reported[6] a very strong dependence of the selectivity on the Al/Ti ratio applied. This is illustrated in Fig. 11.2. The catalyst used is a "sesqui" reduced TiCl$_4$, which has as average composition TiCl$_3$.0.5AlCl$_3$. This implies that at low Al/Ti the true co-catalyst is not a tri-alkyl derivative but an AlEt$_2$Cl formed with the AlCl$_3$. Up to Al/Ti = 1 this could be the only active co-catalyst (assuming that all AlCl$_3$ present in the catalyst is accessible for reaction). At higher Al/Ti a true trialkyl co-catalyzed polymerization occurs with its concomittant high soluebles level. The decrease in solubles at the higher Al/Ti ratios is more difficult to explain, but might be due to over-reduction.

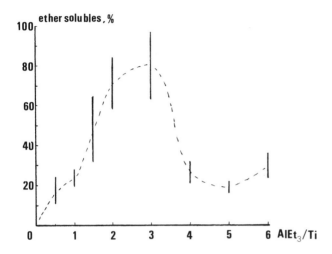

Fig. 11.2: Effect of $AlEt_3/TiCl_3$ ratio on fraction ether solubles when using a $TiCl_3.0.5AlCl_3$ catalyst (data from reference 6)

Apart from the co-catalyst variations polar additives to a standard $TiCl_3/AlEt_2Cl$ catalyst system have been claimed. Compounds as diverse as dimethyl hydrazine in synergy with hydrogen[7], lithium diphenyl phosphide[8] and triphenyl phosphite[9] have been claimed as tacticity improvers in 1-butene polymerizations. All of them will also be active in the same sense in propylene polymerizations on

TABLE 11.1
Use of lithium diphenyl phosphide as selectivity control agent (ref 8)

phosphide, %mol on $TiCl_3$	activity, g/g/h	ether solubles, %m/m	M_v*10^{-5}
0	112	11	9.3
5	86	6	9.8
11	79	5	12.6
33	53	7	12.6

General polymerization conditions : Stauffer AA as $TiCl_3$, $AlEt_2Cl$, Al/Ti = 2.6, 66 °C, no hydrogen, n-heptane diluent, 1.66 bar total pressure, 2 hours.

similar catalysts. An example is given in Table 11.1. Not only is the selectivity increasing in this case but also the molecular weight. This is in line with propylene experience: $AlEtCl_2$ is very probably made inactive by complexation or reaction with the third component.

The new highly isospecific catalyst of Soga and Yanagihara has been already mentioned in the polypropylene chapters. Also in the 1-butene case this catalyst, Solvay-$TiCl_3$ combined with Cp_2TiMe_2, gives very high isotactic material[10]. In polymerizations at 40 °C, a poly(1-butene) is generated containing 1.7 %m/m solubles (measured by recrystallization from decane) which is a very low figure. Moreover, the NMR data give very high values for the fraction of isotactic pentads, 98.8 % (see also section 11.6.1).

11.2.2 Supported catalysts

Three examples of supported catalysts from patents will be given in the following; the data will clearly show that also in the 1-butene case these catalysts are both much more active and more selective at the same time. The first example is one of Phillips[9], which is made by co-milling $TiCl_4$, $MgCl_2$, ethyl anisate and 1-hexene. The resulting solid catalyst is used in conjunction with a co-catalyst consisting of $AlEt_3$, $AlEt_2Cl$ and ethyl anisate in a mol ratio of 1.5:1.5:1. Yields of up to 4000 g polymer per gram of solid catalyst were reported at isotactic indices of 96 %. The presence of the ethyl anisate is essential - as expected - as without this compound about 60 % of ether solubles are found. The remarkable feature of this catalyst is its temperature behaviour: it displays maximum activity around 30 °C! Using the same catalyst for propylene polymerization gives a far different picture in which the activity increases at least up to a temperature of 60 °C.

Mitsui Petrochemical Industries claims in a product patent[11] a catalyst which is made via a very elaborate procedure. The solid component is derived from $TiCl_4$, $MgCl_2$ and isobutyl phthalate and as co-catalyst $AlEt_3$ and for example vinyl triethoxysilane is described. No activities are given in the patent, the selectivities are however very high, ranging from 96 to 99%. The molecular weight distribution is claimed to be fairly narrow, with Q-values around 5.

Finally, as a last example, the Toa Nenryo KK patent data will be described[12]. The aim of the patent is to describe a catalyst system for 1-butene polymerization allowing a non-deashing process delivering a high isotactic polymer in a high bulk density form. Quite a challenge, which according to the authors is completely met. The solid catalyst component is a kind of doubly supported type of catalyst as it starts by impregnating a high surface area silica with an alkyl magnesium compound. This is then reacted successively with a chlorine-substituted alcohol (such as 1,1,1 trichloro ethanol), di-n-butyl phthalate, and $TiCl_4$. It is thought that $MgCl_2$ will be formed somewhere in this preparation. A typical solid catalyst contains 56.5 %m/m silica, 4.4 %m/m Mg, 15.1 %m/m Cl and 2.4 %m/m Ti, whilst the surface area is 279 m^2/g and the pore volume 0.9 cc/g. As a co-catalyst, combinations of trialkyl aluminiums and silicium derivatives such as phenyl-tri-ethyoxy silane are described. Such catalyst systems give, in 4 hours at 37 °C, at 50 % monomer in diluent a yield of 2400 g polymer per g of solid catalyst, i.e. with the above mentioned solid only 10 ppm Ti, 19 ppm Mg, but still 64 ppm Cl are present in the ex-reactor polymer. The bulk-density of the resulting polymer powder is 0.40 g/cc. The use of an additional donor in the co-catalyst such as ethyl anisate, tetrahydrofurane or piperidine derivatives increases the yield, to as much as 75 %. The inventors compare their catalyst with other systems claimed by competitors. In all described cases (both the Phillips and Mitsui Petrochemical examples of above were used, and also a "Solvay" type of catalyst) the drawback of prior art catalysts is their inability to give a stirrable, stable slurry over the desired length of the polymerization. In their own case the silica is of paramount importance as without it difficulties arise with the stirring after a short polymerization time. Clearly, these data are quite impressive.

A high activity catalyst, but showing a poor selectivity was described by Bacskai[13]. The special trick is the use of a mixture of alkyls as co-catalyst, namely DEAC and dibutylmagnesium which, when used in conjunction with a supported catalyst, gives very high activities. Similar results were reported for propylene.

11.2.3 Homogeneous catalysts

Only two relevant examples have been found in the literature. Both deal with catalysts which have also been used in propylene polymerization. The first one[14] is the combination of a chiral zirconocene [ethylenebis(4,5,6,7-tetrahydro-1-indenyl) zirconium dichloride] and methylaluminoxane which also for 1-butene leads to high yields (about half of those achieved with propylene). The molecular weight is at similar levels as polypropylene made with these catalysts, and with 50,000 at 20 °C it is reasonably high. As with the other olefins the MW decreases on increasing the polymerization temperature. It is a pity that no isotactic index was reported, the only statement related to tacticity is "the consistency was wax-like to crystalline".

The other example is the titanium catalyst of Ewen[15] which when used at low temperature with propylene led to a highly tactic polymer made in a chain-end controlled mechanism[16]. Also with 1-butene at -60 °C a stereospecific polymer is produced, although less regular than in the propylene case. The sequence distribution derived from the NMR spectrum agrees with Bernouillian statistics, and a probability of isotactic placement of 0.6 is calculated (for propylene under comparable conditions this was 0.88). All the evidence, including that derived from the ^{13}C-labelled endgroups, points to chain-end control. In this mechanism it is logical that poly(1-butene) is less isotactic than polypropylene, as in the butene case the difference between the substituents of the chiral carbon (ethyl and polymer) is less than in the propylene case (methyl and polymer).

11.2.4. Catalyst comparison w.r.t. isotacticity

In the following table 11.2 a comparison is made in terms of isotacticity of different catalyst systems. A similar comparison of activity is much more difficult to make due to its greater sensitivity to the polymerization conditions.

The isotacticity range covered is large for both catalyst classes, clearly the supported catalyst versions are superior in reaching the highest levels. Equating ether-solubles with xylene-solubles for polypropylene, no large difference in catalyst selectivity is present between polybutene and polypropylene, with the latter being somewhat higher. A complete statement should include the quality of the isotactic fraction. As will be shown in a later section (11.6.2) there is limited evidence that the ether residue from poly(1-butene) is not as pure as the corresponding polypropylene fraction. Thus the overall conclusion is that regulating the isotacticity of polybutene is more difficult than in the case of propylene.

TABLE 11.2
Isotacticity comparison of catalyst systems

catalyst/ co-catalyst	tempera-ture, °C	hydro-gen present?	isotacticity, ether insolubles, %m/m	ref.
StaufferAA/AlEt$_2$Cl	60	no	85	7
TiCl$_3$/AlEt$_2$Cl+AlEt$_2$I(4:1)	70	no	92	4
low T red.TiCl$_3$/AlEt$_2$Cl	60	yes	95 - 97	own data
TiCl$_3$.0.5AlCl$_3$/AlEt$_2$Cl	70	yes	87	6
TiCl$_4$/MgCl$_2$/EB/1-C$_6$// AlEt$_3$/AlEt$_2$Cl/EB	27	yes	94 - 95	9
"Mitsui catalyst"	60	yes	96 - 99	10
"SiO$_2$-supported cat"	37	yes	99	11
TiCl$_4$/MgCl$_2$//AlEt$_3$/EB	50	no	<73	17

11.3 POLYMERIZATION ASPECTS

The effects of polymerization variables on the kinetics, molecular weight and the tacticity of poly(1-butene) are expected to follow to a large extent similar rules as in the case of propylene. This is indeed generally the case, for instance the rate is proportional to the catalyst and monomer concentration in kinetic studies[18] with TiCl$_3$/AliBu$_3$. The polymerizations were carried out at 40 °C in this study, which gives a semi-heterogeneous type of operation. The activities observed, with 0.3 g/g.h are very, very low. The kinetic curves show two regimes, a low rate at the start followed by a period of much faster polymerization. The change-over point was always found at the same yield of polymer (about 1.5 g/g) and the rate ratio in the two regimes was constant. This rate increase was explained by the break-up of the catalyst. The molecular weight of the polymers was independent of catalyst concentration, Al/Ti ratio and polymer yield for the conditions studied. The MW of the ethylether solubles was about 7.5 times lower than that of the residue.

There are however reports in the literature of considerably different behaviour in comparison with propylene polymerization. Examples are the effect of hydrogen on the tacticity and - in

some cases - the effect of monomer concentration. These are dealt
with in more detail below. A marked difference with both ethylene
and propylene polymerization is the wide temperature range in which
poly(1-butene) is soluble in the normally applied diluents or in
its own monomer. A lower cloud point in monomer of 43 °C was
reported by Mobil[4]. That is, above this temperature only a solution
polymerization can be done, in which the heterogeneous catalyst is
the only substance not dissolved. Mass transfer from the gas to the
liquid phase can potentially be a problem in these cases due to the
high viscosities encountered. Of course, when using better solvents
than 1-butene the cloud point will come down. Increasing the cloud
point can only be done by using alkanes with 2, 3 or 4 carbons, but
no data are known. This solubility aspect is however not enough to
explain the observed differences, and more investigations are
required to increase our knowledge.

Hydrogen has a remarkable effect: it is claimed to enhance the
rate of polymerization and at the same time to increase the tacticity[4,19,20] in systems using $TiCl_3$. The Mobil patent[4] reported a
near doubling of the yield and an increase of the isotactic index
of 87 to 93 %m/m at 54 °C using $TiCl_3$, with $AlEt_2Cl$ as co-catalyst.
Mason and Schaffhauser[19] observed up to a doubling of the rate.
Boucheron[20] studied a wider set of conditions and found a rate
enhancement at low hydrogen concentrations, but at higher levels
this turned into a rate decrease. The largest acceleration
observed was a factor of about 4! No decay of the rate was observed, not even at the highest hydrogen concentrations (polymerization was mostly done at 50 °C). The molecular weight appeared
proportional to the square root of the hydrogen pressure, similar
to observations in propylene polymerizations. An explanation for
these effects was given by assuming that the polymer formed
dissolved in the monomer and this caused a lower monomer concentration at the catalyst surface compared to true bulk conditions;
hydrogen detaches chains from the catalyst surface and this would
on average lead to a higher monomer concentration at the active
site. The ultimate decrease in rate is due to the slower re-initiation of the titanium hydride species. We think that there is
only a slight difference in the number of chains attached to the
catalyst in cases with and without hydrogen, at least far less than
needed to explain a rate enhancement up to a factor of four. In the
case of propylene also a rate enhancement is observed with hydro-

gen on $TiCl_3$ catalysts (made by reduction at low temperatures, see chapter 1 and 2), although only to a factor of roughly two at 60 °C. Decay is however nearly always present. The above effect is completely absent in polymerizations using supported catalyst.

The next peculiar effect is the increase in isotacticity with increasing monomer concentration[4,9]. This observation is mentioned both for $TiCl_3$ catalysts[4] and for the supported catalyst patented by Phillips[9]. Moreover the yield of polymer is certainly not linear with the monomer concentration as the data in Table 11.3 illustrate.

TABLE 11.3
Effects of the monomer concentration (taken from ref 9)

1-butene as %V/V in isobutane	25	50	75	100
yield, g polymer/g solid cat	135	435	1520	2075
isotactic index, %m/m	90	92	95	95

general conditions: catalyst as mentioned in section 11.2.2 (made from 18 %m/m $TiCl_4$, 60 %m/m $MgCl_2$, 14 %m/m ethyl benzoate and 8 %m/m 1-hexene), 27 °C and 1 hour.

Other interesting data are the temperature dependence of the activity and the selectivity. For $TiCl_3$/$AlEt_2Cl$ combinations, the data of the Hüls patent[21] shown in Table 11.4 give approximate apparent activation energies of 19 kJ/mol for the overall polymerization, and as difference between the atactic and isotactic polymerization about 8 kJ/mol. Just as in the propylene polymerization

TABLE 11.4
Temperature effect on $TiCl_3$-catalyzed polymerizations (data from ref 21)

temperature, °C	35	70	100
yield, g PB/g $TiCl_3$	923	2000	2727
ether solubles, %m/m	11.2	12.9	18.5

general conditions: $TiCl_3$.$0.5AlCl_3$, $AlEt_2Cl$, 6 hours, at the two lower temperatures hydrogen was applied, reduced solution viscosity roughly equal at about 2.8 dl/g.

the latter figure is positive. The strange temperature behaviour of the Phillips catalyst has already been mentioned, there is however only a small effect of temperature on the selectivity (which is maximal at around 38 °C). As a diluent cyclohexane appears to give both higher yields as well as larger selectivities compared to isobutane.

Kashiwa et al[17] describe the use of a $TiCl_4/MgCl_2$ supported catalyst in which no donor is applied in the preparation of the solid catalyst component. They investigate the polymerization behaviour (activity, selectivity and molecular weight) as function of the donor/Ti ratio when applying $AlEt_3$ and either ethylbenzoate or tetramethyl piperidine as co-catalyst. The polymer is analyzed by recrystallization from n-decane at 25 °C. It appears that the yield of the soluble fraction decreases continuously with increasing donor/Ti ratio and the yield of crystalline polymer shows a small maximum and then decreases as well. The isotacticity of the resulting polymer thus increases. This behaviour is very similar to that observed with propylene as monomer, although in that case the observed isotactic indexes are much higher. Ethyl benzoate is the better donor in terms of the level of isotacticity that can be reached (about 70 %), in terms of isotactic yield the piperidine derivative is slightly better. The molecular weight of the polymer fractions increases considerably by applying more donor. Here also the molecular weights of the solubles are lower than those of the crystalline polymer fraction, but the difference is only a factor 2.

The stereochemistry of the first insertion into a titanium-alkyl bond has been investigated for a number of different catalyst systems. The main aim was a comparison of propylene with higher α-olefins to check the effect of a greater steric requirement. A ^{13}C-enriched co-catalyst is used in the polymerization which allows the specific measurement in NMR of the stereochemistry of the endgroups formed. The results for $TiCl_3$ based catalysts are given in table 11.5. Inspection of the table shows that the first insertion into a titanium- methyl bond is not very specific in the case of 1-butene, not even when the iodide co-catalyst is applied. The latter, as has been described before, gives the highest stereoregularity for the total polymer. Using the ethyl derivative a much higher specificity is observed. Noting that the polymers generated

TABLE 11.5
Stereochemistry of first insertion step in $TiCl_3$ catalysts

catalyst system	polym. temp. °C	polymer fraction analyzed	I_e/I_t (a)		ref
			1-butene	propylene	
$TiCl_3/Al(^{13}CH_3)_3/Zn(^{13}CH_3)_2$	25	ether residue	≈ 1	1	22
$TiCl_3/Al(^{13}CH_3)_2I$	80	ether sol. ether res.	1.4 1.4	3	23
$TiCl_3/Al(^{13}CH_2CH_3)_3$	25	ether res.	2.2	3.4	24

(a) I_e/I_t is the ratio of the erythro over threo placement of the first inserted unit, as derived from NMR measurements.

on the above catalysts are very stereospecific one can conclude that the selectivity is stemming from an asymmetric catalyst complex, which needs an alkyl substituent larger than methyl to be really effective. Comparison with propylene shows that in all cases the butene insertion is less specific. Present thinking assumes that butene does not polymerize on exactly the same set of sites as propylene, but that due to the higher steric requirement it will polymerize on the more open sites having a smaller regulating power. (Compare the explanation of copolymerization behaviour in chapter 4.) This implies also that in the iodide co-catalyst, not just one chloride is replaced by iodide on the site (thus increasing its asymmetry and thus regulating ability as in the propylene case) but possibly two, giving a smaller positive effect.

Using the same technique as mentioned above also the stereospecific homogeneous ethylenebis-indenyl-titanium-dimethyl/-methylaluminoxane catalyst system was studied[25]. Both the enriched $Al(CH_3)_3$ and $Zn(CH_3)_2$ were added to the polymerization in order to introduce the endgroup labelling. For comparison the ethyl derivatives were also used. The polymers, prepared at -45 °C, were highly stereospecific. Again a large difference was noted between methyl and ethyl experiments, in the same direction as above. In the ethyl case only one type of endgroup could be detected, having the same structure as the polymer (there might be a temperature effect, note the low polymerization temperature). For the methyl case only butene showed some specificity in the first insertion, implying the effect of the size of the substituent on the incoming olefin. However the effect of the size of the alkyl substituent on the titanium is far more critical, as the ethyl experiments showed.

11.4 COPOLYMERIZATION

In a process sense there is not much to add to the general trends already mentioned in chapters 3 and 5. Almost no specific work is done in 1-butene copolymerization as regards kinetics, catalyst effects on the copolymerization parameters and related aspects. For kinetics and mainly infrared characterization the reader is referred to Kissin's excellent review[26].

The relevant data on 1-butene/propylene copolymerization have been mentioned before in chapter 3. In the accompanying table 11.6 the copolymerization parameters are given for the relevant co-monomer pairs. As expected the rate ratios follow the same pattern as in the propylene case with lower ratios at increasing bulkiness of the co-monomer.

1-Butene is the first member of the α-olefin series which can isomerize into two products: trans and cis 2-butene, the thermody-

TABLE 11.6
Copolymerization parameters of 1-butene/α-olefin copolymerizations

Comonomer	Catalyst	r_1 (1-butene)	r_2	r_1*r_2	ref
ethylene	$TiCl_3$/DEAC	0.025	60	1.5	27
propylene	,,	0.51	4.7	2.4	,,
,,	,,	"0.5"	"2"	"1.0"	28
,,	$TiCl_3$/$AlEtCl_2$/HMPTA	0.8	4.3	3.4	29
,,	$TiCl_3$/$AlEt_3$	0.5	1.6	0.8	30
,,	VCl_4/DEAC/anisol	0.7	0.7	0.5	31
1-pentene	VCl_3/$AlEt_3$	0.72*	0.3*	0.3	32
1-decene	$TiCl_3$/DEAC	1.5	0.7	1.1	33

*: in the original reference these are given as r_1=0.3 and r_2=0.72 which is regarded as a misprint.

namically more stable isomers (with trans somewhat more stable than cis). So, when a reaction pathway exists isomerization will occur. Apparently Ziegler-Natta catalysts made from $TiCl_4$ or $TiCl_3$ with AlR_3 as co-catalyst allow such a pathway, as isomerization is observed in these systems. Examples are, for example the data from Laputte and Guyot[34] who show the reversibility of the isomerization with the above catalysts. This also means of course that 2-butene can be polymerized to poly(1-butene), which is indeed observed[35]. Or otherwise, that from 2-butene and 2-pentene copolymers can be made with a structure identical to the copolymer directly made from the α-olefins[32]. In the latter, $VCl_3/AlEt_3$/nickel dimethyl glyoxime was used at 80 °C. There is only a small difference in the copolymerization parameters between using the 2- or the 1-olefins as starting material. Clearly the isomerization is not the rate determining step in this case. This is supported by the fact that the isomer distribution at the end of the experiment is very close to the thermodynamic equilibrium.

Very fundamental measurements of rates have been made by Ammendola et al[36], who were able to measure the rate ratios of the very first insertion into a Ti-methyl bond. Their technique is similar to the one used to ascertain the types of endgroups in various polymers by the same group of investigators, a ^{13}C-enriched co-catalyst is used in the polymerization and thereby the endgroups containing the labelled alkyl group are readily observed in ^{13}C-NMR. Several co-monomer pairs were studied by this group allowing them to calculate the relative rate of many α-olefins with ethylene for instance as the base. This leads to the following figures for the first insertion into a Ti-methyl bond: ethylene shows 7 times, and propylene 1.4 times the insertion rate of 1-butene. One must expect that these ratios will not apply in the succeeding insertions as there the steric requirement is greater, and thus the ratios will be larger as well. They are then expected to be more in line with the above given reactivity ratios.

11.5 POLYMERIZATION PROCESS

The only larger scale poly(1-butene) producer (Shell in the USA) uses the technology they bought from Witco, who obtained theirs from Mobil. The process is a solution polymerization process, using a first generation catalyst consisting of $TiCl_3$ and an iodide modified co-catalyst. The process is rather conventional in most aspects, and it resembles the older polypropylene processes in that a true deashing step is necessary. This is just done by a water wash of the polymer cement. A better deashing result is obtained[37] when using α-hydroxysulphonic acids which can be formed in-situ from a ketone, SO_2 and water.

The most interesting feature of the process is the monomer removal after the polymerization and catalyst wash. One makes use of the rather low upper cloud point of solutions of poly(1-butene) in its monomer. This cloud point is around 90 °C, at which temperature a two phase system forms with one phase containing a high polymer fraction, the other phase being much leaner in polymer. In the process the polymer solution is heated to about 125 °C to assist flashing of the monomer upon release of the pressure. At these higher temperatures the polymer concentration in the polymer-rich phase is as high as 50%.

An extensive study was reported by Charlet et al[38] on the measurement and theory of lower critical solution temperatures for a number of polyolefins. We mention here only a number of their results, given in table 11.7. No effect on these values is found of the tacticity of the poly(1-butene) sample. The LCST is a function of molecular weight, being proportional to $M^{-0.5}$.

TABLE 11.7
Lower critical solution temperatures of poly(1-butene) (from ref 38)

solvent	T_c °C	LCST °C
n-pentane	196.6	148
2-methyl butane	187.8	143
n-hexane	234.2	151
2,2-dimethyl butane	226.8	171
n-heptane	267.1	236

11.6 CHARACTERIZATION

There have been no systematic studies reported on poly(1-butene), and relationships between catalyst, composition and properties are at present almost non-existent. We will treat the copolymer characterization data in this same section.

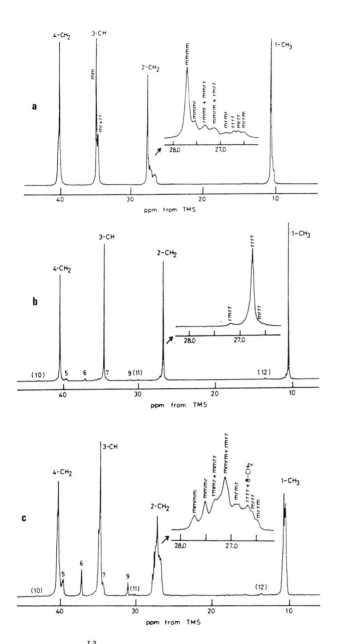

Fig. 11.3: ^{13}C-NMR spectra of isotactic (a), syndiotactic (b) and atactic poly(1-butene) (c). [from ref. 45. Reprinted with permission from Macromolecules. Copyright 1986 American Chemical Society]

11.6.1. Composition

Starting with the ^{13}C-NMR type measurements on the homopolymer, only three publications deal with the tacticity of the polymer[17,39,40]. The spectrum of Asakura and Doi[40] is used also in later publications where the total pentad distribution is evaluated in a quantitative sense[41]. An example of the spectra obtained is given in Fig. 11.3, which compares syndio, isotactic and atactic poly(1-butene). The main resonances are easily distin-

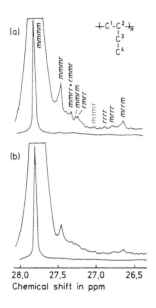

28,0   27,5   27,0   26,5
Chemical shift in ppm

Fig. 11.4: The ^{13}C-NMR spectrum of the sidechain methylene of a poly(1-butene) fraction: whole polymer (a), decane-insoluble polymer (b). [from ref. 10, with permission of Hüthig & Wepf Verlag, Basel]

guished and thus assigned. Generally speaking the sensitivity to tacticity is greatest for the sidechain methylene group, followed by the chain-methine. The best resolved spectrum of the side chain methylene group found in the literature is given in Fig. 11.4, based on the product made by the highly specific catalyst of Soga[10]. The corresponding spectrum of the solubles is also given. The assignments given in this figure are based on reference 42 with some slight changes. However, the range of the tacticity effect is distinctly less than in polypropylene. The spectrum is not totally resolved in terms of the possible pentads as the peaks are not fully separated. The quantitative pentad distribution of a total polymer was obtained[41] by an iterative procedure using the observed peak positions and the assignments on the basis of the γ-effect and the calculated conformation[42]. This latter calculation also correctly predicts the smaller range of the tacticity effects.

From the studies of Mauzac et al[39], Kashiwa et al[17] and the recent one of Icenogle and Klingensmith[43] one can conclude the following:

(a) The atactic poly(1-butene), isolated via extraction with ethylether, is not truly a random copolymer of l and d placements, as [mr] is far smaller than 2[mm] or 2[rr]. The observed range of [m] is 0.46 to 0.60. The one extract reported by Makino et al[44] has an isotacticity of about 60 %, both these observations are similar to the ones in polypropylene. When isolating atactic poly-1-butene via a dissolution/crystallization type of fractionation, the material obtained is even higher in [m], namely 72% (ref.43). A true atactic, with [mm]:[mr]:[rr] of 1:2:1, can be made by the hydrogenation of a random poly(1,2-polybutadiene)[45].

(b) After ether extraction the residual poly(1-butene) is not totally isotactic, Mauzac et al report [m] of 0.86 to 0.96 in that fraction. Also Kashiwa et al[17] report a low isotacticity for a recrystallizate - their observed pentad distribution is given in table 11.8. Clearly all the possible pentads are still present, even the rrrr pentad, thus either the fractionation method applied is not very selective or the polymer is truly very heterogeneous in sequence distribution. One would like to see a spectrum of a recrystallizate starting from a higher isotactic poly(1-butene); they themselves report that for commercial polybutenes the recrystallizates show 93-94 % mmmm pentad, the concentrations of the other pentads were not disclosed however. This is a pity as it might have allowed one to discern the most frequently occurring propagation error (as in polypropylene). The same holds for the fractionation reported by the Shell group[43], illustrated by the pentad distribution for the purest material, also given in table 11.8. The spectrum given in Fig. 11.4 suggests partly that the same error type is present in poly-1-butene as in polypropylene as next to mmmm, one observes clearly mmmr and mrrm, the presence of mmrr is however doubtful.

(c) The pentad distribution is independent of the donor/Ti ratio[17] in the supported catalyst system, probably here also the general rule is that the catalyst system used determines the amounts but not the kind of the various polymer fractions, just as in polypropylene.

TABLE 11.8
Example of sequence (pentad) distribution in soluble and recrystallized poly(1-butene)

| sequence | fraction (as %mol) in | | |
	solubles (ref 17)	recrystallizate (ref 17)	recrystallizate (ref 43)
mmmm	30.4	77.8	94.7
mmmr	13.7	5.9	1.1
rmmr + mmrr	15.7	5.9	1.8
mmrm + rmrr	13.7	3.3	1.1
mrmr	6.8	1.6	0.4
mrrm	6.0	2.5	0.4
mrrr	6.2	1.5	0.2
rrrr	6.8	1.6	0.4

In line with one of the above conclusions also Icenogle[46] reported earlier that in a commercial poly(1-butene) containing 0.6 %m/m of ethersolubles, the NMR measurement showed a [m] of 0.94; i.e. here also the "isotactic" residue is not very pure.

In the fractionation studies of Icenogle and Klingensmith[43] two catalyst systems were compared, the commercial one (based on $TiCl_3$) and the other one a supported catalyst. In the fractionation, the former gives a composition distribution like the ones found in polypropylene, i.e. two fractions of almost constant tacticity are found, one atactic the other high isotactic, and there is a fraction with a composition bridging these two (like the stereoblock material ex polypropylene). In the case of the supported catalyst an almost continuously changing tacticity is observed, suggesting that there are also more differing active centres on this catalyst than on the $TiCl_3$-one.

The calculated pentad distribution was used to check various polymerization models by Asakura et al[41,47], the bicatalytic site model could be made to fit the data best.

Relaxation times have been measured[40] on the various carbons in poly(1-butene), also checking the tacticity as a variable in this. A small, but distinct effect was observed, with the isotactic sequence being more mobile than the syndiotactic one.

For data on the syndiotactic form, see section 11.8.

We turn now to the NMR-analysis of copolymers. Early reports looked into the sequence distribution in copolymers of 1-butene and propylene[28,48,49]. In the paper[48] by Bunn and Cudby the general difference in the spectra was shown in relation to mode of polymerization ("random" versus "block") and level of co-monomer. Randall[49] studied a number of copolymers in the mid range of the composition, assigned the tetrads and evaluated the sequence statistics. Surprisingly a good fit was obtained with Bernouillian statistics which, considering the type of catalyst used, is very strange. Quantitative analysis of low levels of propylene in commercial copolymers was described by Fisch and Dannenberg[28]. The greatest sensitivity is found in the methylene of the backbone. No tacticity effects are observable in their spectra. They find a good correlation of the calculated co-monomer levels and the co-polymerization data, which lend support to their analysis. The tetrad assignments given by Bunn and Cudby and by Randall were later slightly revised by the same authors[50].

The only two reports[46,51] on the molecular weight distribution give figures of Q around 10 to 11 for commercial polymers. These data are similar to those for polypropylene, and certainly not much narrower, indicating that a solution polymerization on a heterogeneous catalyst has no effect on the width of the distribution.

11.6.2. Crystallinity related characterization.

Poly(1-butene) exists in four different crystalline modifications, and most of them are an aspect of every day life with this polymer. Therefore a short summary of this aspect will be given. Note that this is contrary to polypropylene, for which also some modifications exist, but only one is normally encountered, with as minor exception the γ-modification in copolymers.

Crystallizing from the melt at around room temperature leads to modification II which upon standing transforms (solid -->

514

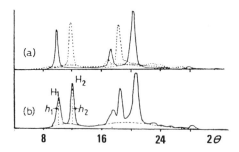

Fig. 11.5a: X-ray diffractograms of poly(1-butene) crystalline modifications I and II. [from ref.60, by permission of publishers, Butterworth & Co, Ltd]

solid) into modification I. Precipitating or crystallizing poly(1-butene) from a good solvent leads to modification III, which upon heating melts relatively low and then transforms into modification II. These three forms all show a different WAXS diffraction pattern, as shown in Fig. 11.5a and b. The last modification is I', which in WAXS is very similar to modification I but shows a much lower melting point. This modification is formed in those cases where the poly(1-butene) is formed as a solid, i.e. in low temperature polymerizations where the polymer is insoluble or in gas-phase polymerizations[52,53]. A few data on the crystal modifications are given in Table 11.9.

Many transformations of one modification into another are possible. In practical terms, the most important one is for the melt- crystallized samples from II to I. This is a solid-solid

TABLE 11.9
Melting point and thermodynamic data for poly(1-butene) crystal modifications (mainly from ref. 54 and 55)

	modification				
	I	I'	II	III	amor-
crystal form	hexagonal		tetra-	ortho-	phous
	(twinned)		gonal	rhombic	
melting temperature, °C	133	≈93	120	≈102	
melting enthalpy, kJ/mol	7.57		8.16		
melting entropy, J/degr.mol	18.6		20.8		
cryst. density, g/cc	0.9058		0.8632		(0.855)
cryst. specific volume, cc/g	1.104		1.1585		

transformation in which the more loosely wound helix of form II
(11_3) (ref 56) is "tightened" to the helix of form I (3_1), at the
same time the density increases considerably (see also table
11.9). The transformation is irreversible, which suggests that
form I is the thermodynamically most stable one, which is however

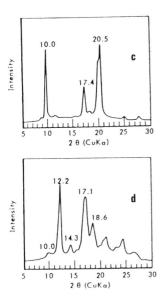

Fig. 11.5b: X-ray diffractograms of poly(1-butene) crystalline modifications I
and III: PBI (c), PBIII, crystallized from decalin (d) [from ref. 52, copied
with permission from John Wiley & Sons]

not in line with the thermodynamic data given above. The rate of
transformation is maximum around room temperature, the half life
being in the order of 24 hours for a homopolymer. The rate is
increased tremendously by applying pressure or by orienting the
sample. The above data also expain why the melting of form II can
easily be observed: at higher temperature the transformation rate
into form I is negligible. The form III transforms rapidly to II
upon heating, the conversion is fast enough to be observable in
DSC type experiments. Thus in DSC one sees first an endothermic
effect of melting of the type III modification, followed by an
exotherm of the recrystallization into form II, and subsequently
melting of this latter modification. By slow heating of form III a
solid-solid transformation into form I can be observed[57], which
shows then however a low melting point. Form I' shows a similar

transformation as III, i.e. in DSC melting, crystallization into form II and again melting is observed. It has been shown[58] that III can be converted into I′ by heating at around 93 °C, and the I′ form can be changed into form II by heating at a slightly higher temperature of 104 °C; both are crystallizations from a melt.

The effect of incorporation of various co-monomer units in the poly(1-butene) chain on the above crystal forms and their transformations has been studied fairly extensively, especially by Turner-Jones from ICI (see e.g references 59-61). Before embarking on this topic, first a description of the effect of copolymeriza- tion and copolymer composition on the overall crystallinity and melting point is given. For this the reader is referred to Fig. 11.6 and Fig. 11.7, which give the X-ray crystallinity and the melting points respectively of various 1-butene/comonomer pairs. Clearly there is a wide range of effects: a monomer such as 3-me- thyl-1-butene shows high crystallinity over the total composition range, whilst with either ethylene or the long chain 1-decene the crystallinities show a large drop. Propylene and 1-pentene show an intermediate behaviour. The 1-decene case comes closest to the behaviour expected for random copolymerization in which the "alien" unit is not accommodated in the crystal. Contrariwise 3-methyl-1- butene and some others form highly co-crystallizing systems. This does not necessarily mean that only one crystalline phase is present, for instance as shown in Fig. 11.7 in many cases two or more different phases can be distinguished in X-ray, and can be observed to melt at different temperatures under the microscope. A notable exception is the combination 1-butene/1-pentene in which only one crystal-form is observable up to over 80 %mol co-monomer content. This is the only case of true co-crystallization where, independently of the composition, the crystal structure can accom- modate the other unit easily. An expansion of the unit cell is found, linear in the pentene content, coupled to a decrease in the melting point. Note that both poly(1-butene) and poly(1-pentene) show a 3_1 helix in their form I crystals, this makes incorporation of the co-monomer units easier. In the case of 3-methyl-1-butene, with a 4_1 helix, one observes with increasing co-monomer content a gradual "unwinding" of the poly(1-butene) helix. In the other extreme the copolymers with long α-olefins show no lattice expan- sion, i.e. these large units are indeed excluded from the crystal.

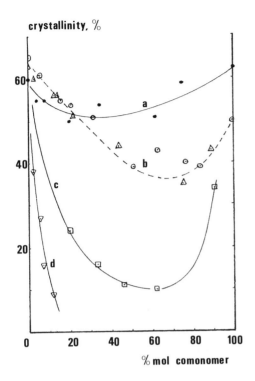

Fig. 11.6: *Crystallinity by X-ray of 1-butene copolymers as function of the composition (based on data ex references 59 and 60, samples slowly cooled at 6K/h): comonomer 3-methyl-1-butene (a); 1-pentene (b, o); propylene (b, Δ); ethylene (c); 1-decene (d)*

This then directly explains the sharp drop in crystallinity. The incorporation of propylene into the crystalline lattice also takes place in solution crystallized copolymers[62].

What are the effects of copolymerization on the crystal modification of the polybutenes? Two different actions have been observed: either destabilization of form II or, contrariwise, stabilization of form II. The former action is observed for copolymers of butene with ethylene, propylene and 1-pentene, which copolymers thus show an accelerated transformation of the type II to type I crystal form. At high co-monomer levels no form II can be obtained, even upon cooling from the melt one finds directly form I. Propylene appears to be the most effective co-monomer for this accelerated transformation.

Stabilization of form II is found for copolymers with longer chain α-olefins and with the branched-chain olefins. Completely stabilized form II is for instance observed at co-monomer levels

518

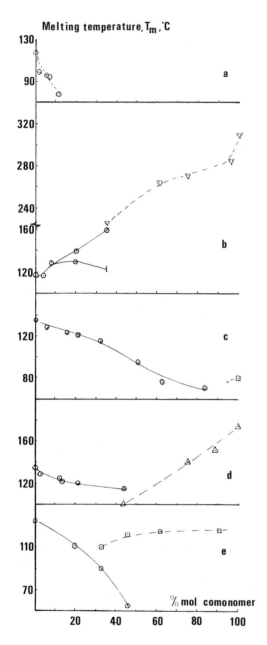

Fig. 11.7: Melting temperature of 1-butene copolymers as function of their composition (based on data ex references 59 and 60); comonomer: 1-decene (a); 3-methyl-1-butene (b); 1-pentene (c); propylene (d); ethylene (e)

of 13 %mol for 1-hexene, 7 %mol for 1-octene, 4 %mol for 1-nonene, 10 %mol 3-methyl-1-butene and 7 %mol for 4-methyl-1-pentene. Even co-monomers which do not enter the crystalline lattice will stabilize form II, as in the case of 4,4-dimethyl-1-pentene where only 3 %mol is required for stabilization as well as in the case of the longer α-olefins. It is suggested that the larger, bulkier side groups in the amorphous fold-region of the material act as barriers for the conversion of one crystal form into the other. For those co-monomers which enter the lattice AND show stabilization (e.g. 3-methyl-1-butene and 1-hexene) one assumes that these bulkier groups in the crystal will hamper the transformation.

11.7 DYNAMIC MECHANICAL ANALYSIS

The glass transition of poly(1-butene) lies around -17 °C, in dynamic mechanical analysis one finds an extra loss peak at about -160 °C which is attributed to rotation of sidegroups[1]. The transformation of form II into form I can be nicely followed by dynamic mechanical analysis as the different forms show different behaviour[63].

The rather high glass transition of poly(1-butene) makes this polymer, or its copolymers with other α-olefins, less suitable for toughening of brittle polymers.

11.8 SYNDIOTACTIC POLY(1-BUTENE)

Direct synthesis of syndiotactic poly(1-butene) has not met with success, no reports are found in the literature and the use of catalysts which generate crystalline, syndiotactic polypropylene, such as VCl_4/$AlEt_2Cl$/anisol, generate only amorphous polymers with 1-butene[31]. There is however some infrared evidence to suggest that a syndiotactic polymer was indeed formed (a repetition of these experiments, using ^{13}C-NMR as the measurement technique would be interesting). In copolymerization studies of 1-butene and propylene Zambelli et al[31] observed that propylene-rich copolymers only showed syndiotactic crystallinity, up to the high co-monomer content of 30 %mol. This led them to suggest that also in this case the butene unit is accommodated in the lattice of syndiotactic polypropylene.

An indirect synthesis was then attempted by a polymer-analogue conversion of syndiotactic 1,2-polybutadiene by hydrogenation. In 1960 this was not successful in the sense that only an amorphous polymer resulted[64]. Doi et al[45] were however successful in this synthesis by using a highly syndiotactic polybutadiene as a base material and using a very efficient hydrogenation catalyst. The NMR-spectrum is shown in Fig. 11.3 and clearly a highly stereo-regular polymer resulted. In this case the polymer is crystalline, with a low melting point of 45-48 °C and an observed enthalpy of melting of 12-16 J/g. No crystallographic studies have yet been reported. DSC-measurement of the glass transition temperatures showed it to decrease from syndio (-24°C) via atactic (-29°C) to isotactic (-34°C). That syndiotactic poly(1-butene) is higher in its glass transition fits with the longer spin-lattice relaxation times of isotactic sequences compared to syndiotactic ones, i.e. the isotactic sequences have a larger segmental mobility[40].

11.9 BLENDS CONTAINING POLY(1-BUTENE)

The interest in blends of poly(1-butene) and other polyolefins resides in the expectation that compatibility will exist, certainly in the melt, but possibly also at lower temperatures. This could lead to interesting properties. The melt compatibility is assumed by all investigators, although the experimental proof is limited to statements such as "a clear homogeneous melt is obtained" from a 50/50 blend of poly(1-butene) and polypropylene[65]. Or otherwise[66] the calculated solubility parameters are used as an argument, for instance for ethylene/propylene rubber blends the difference in δ has been calculated as <0.01. The compatibility at lower tempera-tures has been studied using DSC and dynamic mechanical analysis. In DSC of polypropylene blends Siegmann observes[65,67] separate crystallization of the components, the crystallization temperature of polypropylene is only slightly lowered, whilst that of poly(1-butene) is increased, by nucleation of its crystallization by the presence of crystallized polypropylene. The melting points are also affected, in this case polypropylene is most influenced (a 8°C lowering) and in line with this also the specific melting enthalpy is lowered. A co-crystallization is not observed. Similar conclu-sions were reached by Lee and Chen[68], who in addition to increased crystallization temperatures for poly(1-butene) also observed that the crystallinity of this polymer in the blend was considerably higher, especially in the poly(1-butene) rich blends.

In the blends studied by Siegmann the glass temperature moves gradually from 6 °C for pure polypropylene to -7 °C at the other end of the composition range. The system does not behave truly in an "additive" sense as minor amounts of polypropylene have a disproportionately large effect. Nevertheless this also cannot be regarded as a good proof of the compatability of the amorphous phases as the resolving power of the measurement is not good enough to show two individual transitions if there were no compatibility. Still Siegmann concludes tentatively, also on the basis of mechanical properties shown by the blends, that a fairly high compatibility in the amorphous state exists.

The difficulty of obtaining co-crystallization is shown in the experiments of Gohil and Petermann[69] who used a method of fast crystallization under high strain and thus orientation. In this way blends in the 30/70 to 70/30 range all showed still two crystalline phases, one of polypropylene the other of poly(1-butene). Only in the corners of the composition diagram was one phase found. By quenching 0.25 mm thin films from 230 °C into cold isopentane at -160 °C, Hsu and Geil[70] found similar results.

In blends with ethylene/propylene rubber Kallitsis and Kalfoglou[66] observed in DSC an increasing crystallinity of the poly(1-butene) through nucleation by the rubber, and a slight depression of the melting point (of 4 °C for form I or 2 °C for form II). In dynamic mechanical analysis both transitions are clearly visible, i.e. there is no mixing at the segmental level.

REFERENCES
1. I.D.Rubin, Poly(1-butene) - its preparation and properties, Gordon and Breach, New York, NY, 1968
2. Y.V.Kissin, Monomer reactivity in stereospecific polymerization with heterogeneous Ziegler-Natta catalysts, in Transition metal catalyzed polymerizations, alkenes and dienes, part B, R.P.Quirk editor, Harwood Academic Publishers, New York, 1983
3. G.Natta, I.Pasquon, A.Zambelli and G.Gatti, Highly stereospecific catalytic systems for the polymerization of α-olefins to isotactic polymers, *J.Pol.Sc.*, *51* (1961) 387-398
4. US Patent, 3 362 940, published 9 Jan 1968, to Mobil Oil Corporation
5. US Patent, 3 635 839, published 18 Jan 1972, to Mobil Oil Corporation
6. Br.Patent, 1 432 595, published 22 April 1976, to Chemische Werke Hüls AG
7. US Patent, 3 907 761, published 23 Sept. 1975, to Ethylene Plastique
8. Eur Patent, 0 006 968, published 23 Jan 1980, to Continental Oil Company
9. Eur Patent, 0 002 522, published 27 June 1979, to Phillips Petroleum Company
10. K.Soga and H.Yanagihara, Extremely highly isospecific polymerization of olefins using Solvay-type TiCl3 and Cp2TiMe2 as catalyst, *Makrom.Chemie*, *189* (1988) 2839-2846
11. Eur Patent, 0 172 961, published 5 March 1986, to Mitsui Petrochemical Industries
12. Eur Patent, 0 187 034, published 9 July 1986, to Toa Nenryo Kogyo KK
13. R.Bacskai, High activity mixed metal alkyl cocatalysts for α-olefin polymerization, *J.Appl.Pol.Sc.*, *35* (1988) 321-326
14. W.Kaminsky, K.Külper, H.H.Brintzinger and F.R.W.P.Wild, Polymerization of

522

propene and butene with a chiral zirconocene and methylaluminoxane as co-catalyst, *Angew.Chemie, Int.Ed. Engl.*, <u>24</u> (1985) 507-508

15. A.Zambelli, P.Locatelli, A.Grassi, P.Longo and A.Proto, 13C-Enriched end groups of polypropylene and poly(1-butene) prepared in the presence of bis(cyclopentadienyl)titanium diphenyl and methyl aluminoxane, *Macromolecules*, <u>19</u> (1986), 2703-2706

16. J.A.Ewen, Mechanisms of stereochemical control in propylene polymerizations with soluble group 4B metallocene/methylaluminoxane catalysts, *J.Am.Chem.Soc.*, <u>106</u> (1984) 6355

17. N.Kashiwa, J.Yoshitake, A.Mizuno and T.Tsutsui, Polymerization of butene-1 with highly active MgCl2-supported TiCl4 catalyst system, *Polymer*, <u>28</u> (1987) 1227-1231

18. M.H.Jones and M.P.Thorne, The polymerization of 1-butene by TiCl3-AliBu3 catalysts, *Can.J.Chem.*, <u>40</u> (1962) 1510-1520

19. C.D.Mason and R.J.Schafhauser, Effect of hydrogen on the rate of Ziegler-Natta polymerizations, *Pol.Letters*, <u>9</u> (1971) 661-664

20. B.Boucheron, Phénomène de transfert de matière au voisinage des centres actifs en polymérisation anionique coordinée en solution de butène-1, en relation avec l'effet de l'hydrogène, *Eur.Pol.J.*, <u>11</u> (1975) 131-138

21. German Patent, OLS 2 247 786, published 11 April 1974, to Chemisch Werke Hüls

22. A.Zambelli, M.C.Sacchi, P.Locatelli and G.Zannoni, Isotactic polymerization of α-olefins: Stereoregulation for different reactive chain ends, *Macromolecules*, <u>15</u> (1982) 211-212

23. P.Locatelli, M.C.Sacchi and I.Tritto, Ziegler-Natta polymerization of linear α-olefins. Stereochemical evidence of the existence of various isotactic sites on the TiCl3 surface, *Macromolecules*, <u>19</u> (1986) 305-307

24. P.Locatelli, I.Tritto and M.C.Sacchi, Steric control in the insertion of the first monomeric unit in the stereospecific polymerization of linear 1-alkenes, *Makrom.Chem., Rapid Commun.*, <u>5</u> (1984) 495-499

25. P.Longo, A.Grassi, C.Pellecchia and A.Zambelli, 13C-Enriched end groups of isotactic polypropylene and poly(1-butene) prepared in the presence of ethylene- di-indenyl dimethyl titanium and methylaluminoxane, *Macromolecules*, <u>20</u> (1987) 1015-1018

26. Yu.V.Kissin, Structures of copolymers of high olefins, *Advances in Polymer Science*, <u>15</u> (1974) 91-155

27. S.Davison and G.L.Taylor, Sequence length and crystallinity in α-olefin terpolymers, *Br.Pol.J.*, <u>4</u> (1972) 65-82

28. M.H.Fisch and J.J.Dannenberg, Propylene incorporation in 1-butene copolymers by carbon-13 NMR spectroscopy, *Anal.Chem.*, <u>49</u> (1977) 1405-1408

29. H.W.Coover et al, Costereosymmetric α-olefin copolymers, *J.Pol.Sc.*, A1, <u>4</u> (1966) 2563-2582

30. I.Hayashi and K.Ohno, *Chem.High Pol.(Japan)*, <u>22</u> (1965) 446

31. A.Zambelli, A.Lety, C.Tosi and I.Pasquon, Polymerization of propylene to syndiotactic polymer. V. Steric control and copolymerization, *Makrom.Chem.*, <u>115</u> (1968) 73-88

32. T.Otsu, A.Shimizu, K.Itakura and M.Imoto, Monomer isomerization polymerization, IV. Monomer-isomerization copolymerizations of β-olefins by coordinated anionic mechanism, *Makrom.Chem.*, <u>123</u> (1969) 289-292

33. R.D.A.Lipman, Copolymerization kinetics of α-olefins using Natta catalysts, *Polymer Preprints, ACS*, <u>8</u> (1967) 396-399

34. R.Laputte and A.Guyot, Sur les systèmes cataliques TiCl4-AlR3. III. Isomérisation du butène-1, *Makrom.Chem.*, <u>129</u> (1969) 250-266

35. T.Otsu, A.Shimizu and M.Imoto, *J.Pol.Sc.*, <u>4</u>, A1, (1966) 1579

36. P.Ammendola and A.Zambelli, Ziegler-Natta polymerization of 1-alkenes : relative reactivities of some monomers towards insertion into metal-methyl bonds, *Makrom.Chem.*, <u>185</u> (1984) 2451-2457

37. US Patent 4 395 356, published 26 July 1983, to Shell Oil Company

38. G.Charlet, R.Ducasse and G.Delmas, Thermodynamic properties of polyolefin solutions at high temperatures: 2. Lower critical solution temperatures for polybutene-1, polypentene-1 and poly(4-methylpentene-1) in hydrocarbon

solvents and determination of the polymer-solvent interaction parameter
for PB1 and one ethylene-propylene copolymer, *Polymer*, *22* (1981) 1190-1198

39. M.Mauzac, J.P.Vairon and P.Sigwalt, 13C-{1H} NMR determination of the
microstructure of poly(butene-1), *Polymer*, *18* (1977) 1193-1194

40. T.Asakura and Y.Doi, 13C NMR study of the chain dynamics of polypropylene
and poly(1-butene) and the stereochemical dependence of the segmental
mobility, *Macromolecules*, *16* (1983) 786-790

41. T.Asakura, M.Demura, K.Yamamoto and R.Chujo, Polymerization mechanism and
conformation of poly(1-butene), *Polymer*, *28* (1987) 1037-1040

42. T.Asakura, K.Omaki, S-N.Zhu and R.Chujo, The carbon-13 NMR chemical shift
of poly(1-butene) referring to that of 2,4,6,8,10,12,14,16,18-nonaethyl
nonadecane and a comparison of the chemical shifts between poly(1-butene)
and polypropylene, *Polymer J.*, *16* (1984) 717-726

43. R.D.Icenogle and G.B.Klingensmith, Characterization of the stereostructure
of three poly-1-butenes: discrimination between intramolecular and
intermolecular distributions of defects in stereoregularity,
Macromolecules, *20* (1987) 2788-2797

44. K.Makino, M.Ikeyama, Y.Takeuchi and Y.Tanaka, Structural characterization
of 1,2-polybutadiene by 13C-NMR spectroscopy: 1. Signal assignment in
hydrogenated polybutadienes, *Polymer*, *23* (1982) 287-290

45. Y.Doi, A.Yano, K.Soga and D.R.Burfield, Hydrogenation of polybutadienes.
Microstructural and thermal properties of hydrogenated polybutadienes,
Macromolecules, *19* (1986) 2409-2412

46. R.D.Icenogle, Temperature-dependent melt crystallization kinetics of
poly(butene-1): a new approach to the characterization of the
crystallization kinetics of semicrystalline polymers, *J.Pol.Sc.*,
Pol.Phys., *23* (1985) 1369-1391

47. T.Asakura et al, 13C-NMR spectrum of poly-1-butene and its polymerization
mechanism, *Rep.Progress Pol.Phys.Jap.*, XXIX (1986) 577-580

48. A.Bunn and M.E.A.Cudby, Monomer sequence distribution in butene/propylene
copolymers by 13C-NMR, *Polymer*, *17* (1976) 548-550

49. J.C.Randall, A 13C NMR determination of the comonomer sequence
distributions in propylene-butene-1 copolymers, *Macromolecules*, *11* (1978)
592-597

50. A.Bunn, M.E.A.Cudby and J.C.Randall, 13C-NMR of propylene/butene
copolymers. Reassignment of -PB-centred tetrads, *Polymer*, *21* (1980)
117-119

51. "Butene polymers" by A.M.Chatterjee in Encyclopedia of polymer science and
engineering, Volume 2, J.Wiley, 1985

52. J.Boor and E.A.Youngman, Polymorphism im poly-1-butene: Apparent direct
formation of modification I, *Pol.Letters*, *2* (1964) 903-907

53. J.P.Rakus and C.D.Mason, The direct formation of modification I'
poly(butene-1), *Pol.Letters*, *4* (1966) 467-468

54. U.Leute and W.Dollhopf, High pressure dilatometry on polybutene-1, *Colloid
& Polymer Science*, *261* (1983) 299-305

55. H.W.Starkweather and G.A.Jones, The heat of fusion of polybutene-1,
J.Pol.Sc., *Pol.Phys.*, *24* (1986) 1509-1514

56. A.Turner Jones, Polybutene-1 type II crystalline form, *Pol.Letters*, *1*
(1963) 455-456

57. F.Danusso, G.Gianotti and G.Polizzotti, Isotactic polybutene-1:
Modification 3 and its transformations, *Makrom.Chem.*, *80* (1964) 1-12

58. G.Goldbach and G.Peitscher, Infrared investigations of the polymorphic
modifications of polybutene-1, *Pol.Letters*, *6* (1968) 783-788

59. A.Turner Jones, Crystalline phases in copolymers of butene and
3-methylbutene, *Pol.Letters*, *3* (1965) 591-600

60. A.Turner Jones, Cocrystallization in copolymers of α-olefins. II. Butene-1
copolymers and polybutene type II/I crystal phase transition, *Polymer*, *7*
(1966) 23-59

61. A.Turner Jones, Copolymers of butene with α-olefins. Cocrystallizing
behavior and polybutene type II --> I crystal phase transition, *J.Pol.Sc.*,
C, *16* (1967) 393-404

62. P.Cavallo, E.Martuscelli and M.Pracella, Properties of solution grown crystals of isotactic propylene/butene-1 copolymers, *Polymer*, *18* (1977 42-48

63. J-Y.Decroix, J-F.May and G.Vallet, Etude comparée des propriétés mécaniques dynamiques des formes I et II du polybutène-1. Cinetique de transformation II --> I, *Makrom.Chem.*, *163* (1973) 295-307

64. G.Natta, Progress in the stereospecific polymerization, *Makrom.Chem.*, *35* (1960) 35-131

65. A.Siegmann, Crystalline/crystalline polymer blends: Some structure-property relationships, *J.Appl.Pol.Sc.*, *24* (1979) 2333-2345

66. J.K.Kallitsis and N.K.Kalifoglou, Physical characterization of blends of isotactic poly(butene-1) with ethylene-propylene copolymer, *Eur.Pol.J.*, *23* (1987) 117-124

67. A.Siegmann, Crystallization of crystalline/crystalline blends: polypropylene/polybutene-1, *J.Appl.Pol.Sc.*, *27* (1982) 1053-1065

68. M-S.Lee and S-A.Chen, The enhancement of poly(1-butene) crystallinity in poly(1-butene)/polypropylene blends, *J.Pol.Sc.*, *Pol.Letters*, *C*, *25* (1987) 37-43

69. R.M.Gohil and J.Petermann, Binare Mischkristalle in Polymeren: Das System Polypropylen-Polybuten-1, *Colloid & Polymer Science*, *259* (1981) 265-266

70. C.C.Hsu and P.H.Geil, Structure-property-processing relationships of polypropylene-polybutene blends, *Polym.Eng.Sc.*, *27* (1987) 1542-1556

Chapter 12

THE HIGHER POLY(α-OLEFINS)

12.1 INTRODUCTION

The polymerization and polymer characterization of the higher α-olefins is the subject of this chapter, i.e. dealing with all α-olefins except ethylene, propylene_ and 1-butene. Due especially to the low commercial activity with these olefins, this chapter will be very short. The only present day commercial production deals with poly(4-methyl-pentene-1), made in a 2000 tonnes per annum by Mitsui Petrochemical Industries in Japan. In the past ICI also had a production facility for this type of polymer, which combines a very high melting point (about 250 °C) with high crystallinity (and thus strength) and high transparency. This makes it for instance well suited for high temperature uses, such as in medicine where sterilisation is required.

Products derived from the longer, straight chain, α-olefins can - according to patent literature - be used in principle as base materials for lubricating oil formulations (low molecular weight) or as pour point depressants and drag-reducers (higher to very high molecular weight). The commercial activity in these areas is unknown to us.

The topics to be discussed below are all related to more scientific issues such as "what are the differences in the behaviour of various catalyst systems with different olefins; are the number of active sites constant or not; etc." and "how is the crystallographic structure dependent on the length of the side chain". Almost no instances can be found of studies which deal with the effect of catalyst type or polymerization conditions on the properties of the resulting polymer, i.e. those topics which form our main interest in this book.

12.2 POLYMERIZATION ASPECTS

Generally speaking all α-olefins can be polymerized on the various classes of Ziegler-Natta catalysts to mainly isotactic polymer. The rate decreases with increasing steric requirements, i.e. both mere chain length in straight-chain olefins leads to a decrease, as does branching near to the double bond. Especially

3-alkyl-substituted α-olefins are very slow in polymerization.

Due to the fact that higher α-olefins are convenient liquids at temperatures of interest to the polymer scientist, these monomers are often used in scientific studies in kinetics. A case in point is the study of Burfield, Tait and McKenzie[1-4], using 4-methyl-pentene-1 (4MP) as monomer. They derived the kinetic constants for propagation (including the fraction of active sites) and chain transfer for a VCl_3/AlR_3 catalyst system, and fitted this into an elaborate mechanistic scheme. This includes the complexation of both monomer and aluminium alkyl on the active sites and chain transfer with monomer and alkyl (the temperatures were too low to have β-hydride elimination). The catalyst shows an initiation period, which becomes smaller the higher the polymerization temperature, a behaviour also observed by others[5]. This period might be related to catalyst break up. The fraction of active sites on their catalyst is very low, only 0.02 to 0.06 % is active. The type of aluminium alkyl applied as co-catalyst exerts quite some influence on this fraction, as with $AlEt_3$ it is three times as large as with $Al(hexyl)_3$. Chain transfer appeared to be fastest with the triethyl and tri-isobutyl derivatives, with the n-butyl and hexyl compounds it is much slower.

The effect of hydrogen was not studied by the previous investigators. A large promoting effect of hydrogen on the activity is observed by Pijpers and Roest[6] in 4MP polymerization, i.e. an observation similar to the one in butene polymerization. This holds again up to an optimum value for the hydrogen/monomer ratio, above this the activity of the catalyst system decreases. The observed activity increase is about a factor of 2 at the optimum hydrogen level. Not only the activity changes but also the mode of decay is altered. In polymerizations without hydrogen the decay is first order, whilst in its presence this becomes second order. The suggested explanation is the presence of hindered monomer diffusion to the active sites when no hydrogen is present, this being less problematic in the case with hydrogen. Moreover a bimolecular reduction of two neighbouring titanium hydride species is proposed to explain the second order deactivation.

Similar elaborate kinetic studies have been done, using supported catalyst systems, by Tait et al[7] and Chien and Ang[8,9]. Both studies had as explicit objective the comparison of the kinetics of the higher olefin with propylene. The first study again used 4MP as monomer and employed the well known $MgCl_2/TiCl_4/donor$ solid catalyst with ethylbenzoate(EB) or diisobutyl phthalate(DIP) as the donors. Trialkyl aluminiums were used as co-catalyst. The findings can be summarised as follows:

- the dependence of the rate on time shows a maximum for the EB case and steady behaviour for DIP, whilst for propylene a decreasing rate is always observed,
- the rate increases with increasing alkyl concentration to a steady value (DIP) or to a maximum followed by a slow decrease (EB),
- in aromatic solvents the rates are higher than in aliphatic ones,
- the fraction of active centers is dependent on both the type of solid catalyst (with EB showing up to 83 % and DIP only up to 9.6 % active) and on the alkyl applied (with ethyl > isobutyl > n-hexyl > n-decyl); the phthalate-modified catalyst is obviously more difficult to alkylate, and also the higher alkyls are less reactive in the alkylation of sites. The fraction of active sites in the EB case is far higher than in the comparable propylene polymerization, but no explanation could be given for this observation.
- the rate is not directly proportional to the bulk monomer concentration but the surface concentration of the monomer has to be used in the rate equation.

In the related study by Chien and Ang[8,9] the polymerization of 1-decene was investigated with their "high mileage catalyst", which is a special form of a $MgCl_2/TiCl_4$ supported catalyst. In comparison with the data of propylene polymerization they concluded that
- the same fraction of sites was active, ranging from 1.4 % at 0 °C over 12 % at 50 °C to 9.4 % at 70 °C,
- the decay rates were similar,
- the molecular weight distribution was also similar (being rather narrow with Q≈4).

Differences were noted in the kinetic behaviour with monomer and activator concentration, similar to the case above. With 1-decene the rate is not true first order in monomer concentration and it decreases with increasing aluminium alkyl concentration, whilst for propylene these are true first order and independent of alkyl concentration at higher Al/Ti ratios. The explanation is sought in the larger coordination of the monomer with the active sites making the factor $k_M^*[M]$ not negligible in the rate equation:

$$\text{rate} = k_p \cdot [\text{Ti}^*] \cdot k_M[M]/(1 + k_M[M] + k_A[A])$$

This causes the rate to become fractional order in monomer.

Some more practical aspects are mentioned to conclude this section. In the polymerization of 4MP it is reported to be essential that a very thorough deashing of the catalyst remnants is done as otherwise the optical properties are impaired[10]. To this end a good slurry polymerization was mentioned to be beneficial, as apparently the higher solution viscosities obtained with poor stereoselective catalysts or at higher temperature impaired the removal of the catalyst ingredients. Of course, with more modern, highly active catalysts this will be less of a problem.

Isomerization can in principle occur with all of the monomers described in this chapter. Some data on this were reported by Krentsel et al[11] on both 4MP and 3-methyl-butene-1 (3MB). In the former case the highest isomerization rate was observed with TiCl_x combined with trialkyl aluminiums applied as co-catalyst, VCl_x catalyst systems were less active. The isomerization products gave only a slight rate reduction, which suggests different sites active in polymerization and isomerization. For the other branched, low reactive monomer 3MB the maximum isomerization occurred with TiCl_4/lithium alkyl. Trialkyl aluminium as co-catalyst also shows considerable isomerization, even in combination with TiCl_3. Only the $\text{TiCl}_3/\text{AlEt}_2\text{Cl}$ system is free of isomerization. Contrary to the previous case a drastic lowering of the polymerization rate is observed upon introduction of the isomerization products to the polymerization of 3MB.

An interesting development has been reported by Chung[12] who designed a synthesis method for functional polyolefin polymers. The monomers applied contain, next to an α-olefinic group as endgroup, the borabicyclo[3,3,1]nonane grouping, prepared by reacting the corresponding borane with dienes. They can be polymerized using the normal TiCl_3 catalysts. The reactivity of the monomer is dependent on the spacing between the double bond and the borane group, when there are only three methylenes as spacer the reactivity is very low, but at 4 or 5 the activity becomes nearly constant. Possibly also copolymerization can be carried out. The resulting products can be converted to polyols by treatment with $\text{NaOH}/\text{H}_2\text{O}_2$. These polymers have remarkable properties.

12.3 CHARACTERIZATION

Compositional studies in terms of tacticity, sequence distributions, etc. are almost completely absent in the literature. One obvious reason is the great difficulty one will have with the study of tacticity in the higher poly(α-olefins) by ^{13}C-NMR for example, as the sensitivity to tacticity (expressed as the range in ppm's for a specific carbon) is strongly decreasing. A case in point is poly(1-butene) as described in the previous chapter. Another example will be given below.

Most of the characterization studies reported deal with the crystalline nature of the polyolefins generated using isospecific catalysts.

The tacticity of poly(4-methyl-1-hexene) has been studied[13] on a few fractions of a polymer made with $TiCl_4$/AliBu$_3$ and the (S)-form of the monomer. The spectrum shows possible tacticity effects in the resonance of the 4-methyl group, but the range is small (0.45 ppm) and the assignment had to be done on the basis of theory as no polymers with specific tacticities are known. The estimated range of isotacticity was from about 50% in the low molecular weight solubles to 95 % in the most crystalline residue. The objective of the investigators was to establish a link between the measured isotacticity and the observed optical rotation of these polymer fractions; this failed.

We turn now to the crystallinity related effects. For ease of reference Table 12.1 gives a survey of the reported crystallographic data including the observed melting points of a large range of straight chain poly(α-olefins). Almost all of the polyolefins are crystalline, only for poly(1-hexene) no crystallinity has been reported. All have a helical conformation in the crystalline lattice, expanding from 3 monomer units per turn for propylene to four such turns for the longer α-olefins. Butene and pentene polymers show modifications with either helix. The polymers of straight chain longer α-olefins give crystals having both main chain and side chain crystallinity (modification I, formed by slow cooling) or only side chain crystallinity (modification III, formed by quenching from the melt). In DSC one can discern mostly two endothermic peaks in slowly cooled samples (i.e. those containing mainly form I), a small one at the lower temperature and the main one at a higher temperature. This first, minor transition has in the past been assigned to atactic polymer, of which one could expect only form III to form as the main chain will be of irregular sequence distribution. If true this would give an indirect measure of tacticity. Subsequently it has been shown[23,25] that this transition also contains a transformation of the side chain (an all-trans conformation is changed into a partly gauche conformation) and moreover imperfect isotactic polymer must also be expected to give the same transition. DSC-measurements on quenched samples

TABLE 12.1
Survey of crystallographic and melting data of poly(α-olefins)

α-olefin, carbon number	crystallographic data				melting temperature, °C	reference
	modification	type	helix, units per turn	crystal density, g/cc		
2		0	-	1.0	139	14
3	α	M	3	0.94	180	14
	β	Tr	3	0.92	150	
4	I	H	3	0.95	134	15
	II	T	3.67	0.90	123	
	III	O			102	
5	I	M	3	0.92	111	14,16,17
	II	M	4	0.89-0.90	80	
6		M	3.5	0.73		15,16
		O	3.5	0.91		
7					37	18
8					10	19
9					32	18
10*			4		17,31	20**,18,19,21,22
11*			4		27,38	20,18,21
12*			4		32,46	20,18,19,21,23
13*			4		37,50	20,18,21
14*			4		21,55	24,19,21,23
15*			4		25,60	24,19
16*			4		29,66	24,18,19,21,23
18*		O	4	0.95	47,72	24,16,18,19,21,23
20*			4		64,90	24
22*			4		91	19

*: all these polymers show at least two modifications, see text, the melting points are given in the order mod.III, mod. I.
**:the reference giving the quoted value is given first
M:Monoclinic, O:Orthorombic, Tr:Trigonal, H:Hexagonal, T:Tetragonal

(i.e. those containing modification III) show three first-order transitions[26]. The first one is an endotherm of the melting of modification III, this is followed by an exotherm due to the formation of crystalline modification I (sometimes also a small fraction of modification II is present), and finally of course melting of modification I is observed. The transition temperatures increase with increasing chain length. A modification II can be obtained by crystallizing from toluene solution[27], it is slightly less regular than modification I. For poly(1-docosene) table 12.2 gives the data obtained for the melting enthalpies and melting points. Clearly modification I and II are rather close.

TABLE 12.2
Melting points and enthalpies of the modifications of poly(1-dococene) (taken from reference 27)

modification	melting enthalpy J/g	melting point K
I	145	363
II	120	356
III	40	347

We turn finally to the characterization of <u>copolymers</u>. The recurring theme of heterogeneity of copolymers is also strongly present with these monomers. For instance Turner Jones[16] fractionated a 4MP/1-hexene copolymer, a copolymer for which it was shown that a fair amount of copolymerization took place, and also an easy substitution of the hexene unit in the 4MP lattice (see below). Still a simple recrystallization from xylene gives a crystalline residue having only 1/3 of the hexene content of the xylene soluble material. As expected the crystallinities of these fractions differ greatly.

A similar conclusion can be reached from DSC and X-ray studies of copolymers, which frequently show the melting temperatures and the diffraction patterns of the parent homopolymers in a slightly changed form[5,16]. A different, but very elegant approach for judging the homogeneity of a copolymer was reported by Benaboura et al[28]. They copolymerized styrene with 1-hexene on a stereoregular catalyst system. The resulting copolymer was analyzed by size exclusion chromatography using a double detection system, one based on the refractive index the other on ultraviolet absorption. In their case, a bimodal distribution was found and the high molecular weight fraction contained much less styrene than the other.

Turner Jones[16] studied the copolymerization of 4MP with α-olefins in the same manner as the studies reported in the previous chapter on 1-butene copolymers. From the known crystallographic data of the homopolymers she could predict that incorporation of 1-pentene and 1-hexene units in the 4MP crystalline lattice was expected to be fairly easy and would occur with only minor disturbance of the lattice or the crystallinity. This was predicted to be a case of isodimorphism. Longer units were expected to give much stronger disturbance of the crystallinity. The prediction is mainly on the basis of the size of the units: pentene has the same main chain number of carbons, and 1-hexene is just one carbon longer in the main chain. Moreover poly(4MP) has a helix with 3.5 units per turn, identical to that expected for 1-hexene and intermediate between the two observed forms of poly(1-pentene). The experimental results were as follows:

for 1-pentene: indeed highly crystalline over the total composition range, however both crystal forms present at intermediate compositions, the 4MP lattice shrinks and the poly(1-pentene) lattice expands, the melting point of the P4MP fraction decreases somewhat and the corresponding poly(1-pentene) fraction melts distinctly above the homopolymer melting point. In the middle range of composition the copolymer can crystallize in either the P4MP or the poly(1-pentene) form, only the thermal history and/or the orientation applied determines the outcome. Many elements of these observations are in line with expectation.

for 1-hexene: crystallinity decreases linearly with increasing hexene content, there is only one crystal phase, no change in lattice parameters, the melting point being only very slightly lowered. Thus very easy incorporation of the hexene unit in the P4MP lattice is indeed observed, however one would have expected a high crystallinity over a wide composition range upon ideal substitution in the lattice. That this is not found is thought to be due to sequence distribution effects, i.e. one hexene unit does not impair the crystallization of the helix, but a short run of these units will, and thus the presence of these are invoked. Some evidence for this is found on the other side of the composition range where higher crystallinity than expected is found, this might

be due to 4MP runs "forcing" the hexene-rich helix to crystallize.
for longer olefins: as expected a drastic decrease in crystallinity
and melting point is observed in copolymers with octene, decene and
especially octadecene. In the latter case no change in lattice
parameters is observed, just as in the butene copolymer case.

REFERENCES

1. D.R.Burfield, I.D.McKenzie and P.J.T.Tait, Ziegler-Natta catalysis: 1. A
    general kinetic scheme, *Polymer*, *13* (1972) 302-306
2. I.D.McKenzie, P.J.T.Tait and D.R.Burfield, ditto 2. A kinetic investiga-
    tion, *ibid*, *307-314*
3. D.R.Burfield and P.J.T.Tait, ditto 3. Active centre determination, *ibid*
    315-320
4. D.R.Burfield, P.J.T.Tait and I.D.McKenzie, ditto 4. Quantitative
    verification of kinetic scheme, *ibid 321-326*
5. V.Sh.Shteinbak et al, Polymerization of 4-methyl-pentene-1 and its
    copolymerization with propylene on the TiCl3 + AlEt2Cl complex catalyst,
    Eur.Pol.J., *11* (1975) 457-465
6. E.M.J.Pijpers and B.C.Roest, The effect of hydrogen on the Ziegler-Natta
    polymerization of 4-methyl-1-pentene, *Eur.Pol.J.*, *8* (1972) 1151-1158
7. P.J.T.Tait, M.Abu Eid and A.E.Enenmo, Polymerization of 4-methyl-pentene-1
    using magnesium chloride-supported catalysts, p309-321 in Advances in
    Polyolefins, R.B.Seymour and T.Cheng editors, Plenum Press, New York, 1987
8. J.C.W.Chien and T.Ang, Magnesium chloride supported high mileage catalysts
    for olefin (decene-1) polymerization. XIV. Propagation and chain transfer,
    J.Pol.Sc., *Pol.Chem.*, *25* (1987) 919-934
9. ibid, XV, Termination and energetics, *J.Pol.Sc.*, *Pol.Chem.*, *25* (1987)
    1011-1026
10. K.J.Clark and R.P.Palmer, Transparent polymers from 4-methyl-pentene-1, in
    the "Chemistry of polymerization processes", SCI monograph no. 20, Society
    of Chemical Industry, London, 1966
11. B.A.Krentsel et al, Some aspects of coordination homo- and copolymerization
    of high α-olefins, cyclic and allene hydrocarbons, in "Coordination
    polymerization", C.C.Price and E.J.Vandenberg editors, Plenum Press, New
    York, 1983
12. T.C.Chung, Synthesis of polyalcohols via Ziegler-Natta polymerization,
    Macromolecules, *21* (1988) 865-869
13. A.L.Segre and R.Solaro, Tacticity analysis of poly[(S)-4-methyl-1-hexene]
    by 13C-NMR, *Polymer Commun.*, *27* (1986) 216-217
14. A.Turner Jones, Cocrystallization in copolymers of α-olefins. II. Butene-1
    copolymers and polybutene type II/I crystal phase transition, *Polymer*, *7*
    (1966) 23-58
15. from "Crystalline olefin polymers", Part I, R.A.V.Raff and K.W.Doak
    editors, Interscience Publishers, New York, 1964
16. A.Turner Jones, Cocrystallization in copolymers of α-olefins. I. Copolymers
    of 4-methyl-pentene-1 with linear α-olefins, *Polymer*, *6* (1965) 249-268
17. A.Turner Jones and J.M.Aizlewood, Crystalline forms of polypentene-1,
    Pol.Letters, *1* (1963) 471-476
18. J.Wang, R.S.Porter and J.R.Knox, Physical properties of the
    poly(1-olefins). Thermal behaviour and dilute solution properties, *Polymer
    J.*, *10* (1978) 619-628
19. I.Modric, K.Holland-Moritz and D.O.Hummel, Raman- und Infrarotspektren
    isotaktischer Polyalkyläthylene. V. Der Wellenzahlbereich unterhalb 700
    cm^{-1}, *Colloid & Polymer Sc.*, *254* (1976) 342-347
20. G.Trafara, ditto. 5. Röntgenographische und thermoanalytische Untersuchun-
    gen an den isotaktischen Homologen Poly(1-heptylethylen), Poly(1-oktyl-

534

ethylen), Poly(1-nonylethylen), Poly(1-decylethylen) and Poly(1-undecyl-ethylen), *Makrom.Chemie, Rapid Commun.*, *1* (1980) 319-326

21. K.Holland-Moritz, I.Modric, K-U.Heinen and D.O.Hummel, Raman- und Infrarotspektren der höheren, isotaktischen Poly-alfa-olefine. I. Der Einfluss der Temperatur auf die Schwingungsspektren, *Colloid & Polymer Sc.*, *251* (1973) 913-918

22. J.C.W.Chien and T.Ang, Magnesium chloride supported high mileage catalysts for olefin (1-decene) polymerization, XI. Determination of poly(1-decene) molecular weight, chain dimensions and thermal transitions, *J.Pol.Sc.*, *Pol.Chem.*, *24* (1986) 2217-2230

23. K.Holland-Moritz, ditto III, Polydodecen-1, Polytetradecen-1, Polyhexadecen-1, Polyoktadecen-1, *Colloid & Polymer Sc.*, *253* (1975) 922-928

24. K.Blum and G.Trafara, Haupt- and Seitenkettenkristallinität bei isotaktischen Poly(alkylethylen)en. 4. Die Abhängigkeit des Kristallinitätsgrades bei den höheren Homologen von der thermischen Vorbehandlung, *Makrom.Chem.*, *181* (1980) 1097-1106

25. G.Trafara, Haupt- und Seitenkettenkristallinität bei isotaktischen Poly(1-alkylethylen)en, 3. Änderung der Ordnungszustände bei der höheren Homologen, *Makrom.Chem.*, *181* (1980) 969-977

26. ditto, 1. Thermoanalytische and röntgenografische Untersuchungen an der höheren Homologen, *Makrom.Chem.*, *177* (1976) 1089-1095

27. ditto, 2. Röntgenografische, kalorimetrische and ramanspektroskopische Untersuchungen am it-Poly(1-eicosylethylene), *Makrom.Chem.*, *179* (1978) 1837-1846

28. A.Benaboura, A.Deffieux and P.Sigwalt, Experimental evidence for the regiospecific primary mode of styrene insertion into metal-carbon bonds of an isospecific Ziegler-Natta catalyst, *Makrom.Chem.*, *188* (1987) 21-33

Chapter 13

CHARACTERIZATION METHODS AND PROCEDURES

13.1 INTRODUCTION

In the previous chapters data on the characterization of polyolefins have been discussed, based on a fair number of different methods. The present chapter will describe these methods, although in varying detail. It needs to be stressed that these methods are of utmost importance as without them characterization is not really possible, so that they form an integral part of characterization studies. It is not always easy to develop relevant procedures and methods to test performance products such as the polyolefins. A good understanding of the polymer and the analytical method is a necessary prerequisite. A case in point is the solubles determination in polypropylene (see below). Moreover aspects of speed and the desire for simple, dependable methods _and_ equipment also play a role.

Our approach here is to consider very concisely the well-established methods and to refer the reader to standard text books for further information. Only in those cases where an own method was independently developed or an existing method was significantly modified will a more elaborate discussion be given. This will be the case for:
- simple fractionation: xylene solubles(XS) and hot xylene solubles (HOXS), comparison with other solubles methods (mainly in use for propylene-based polymers),
- infrared techniques for the measurement of co-monomer contents in random and toughened propylene polymers,
- a combined technique of fractionation and NMR analysis for the compositional analysis of toughened polypropylene.

Table 13.1 gives an overview of the characterization methods and procedures used in this book and described in this chapter.

TABLE 13.1
Characterization methods and procedures

STRUCTURE/PROPERTY	CHARACTERIZATION METHOD
composition	- infrared spectroscopy(IR) (section 13.2) - nuclear magnetic resonance spectroscopy(NMR) (section 13.3) - ^{14}C- scintillation counting (section 13.4) - pyrolysis-hydrogenation gas chromatography (section 13.5) - solubility methods (section 13.6)
molecular weight, molecular weight distribution	- gel permeation chromatography(GPC), or size exclusion chromatography(SEC) (section 13.7) - viscometry, limiting viscosity number(LVN) (section 13.8) - gas liquid chromatography(GLC) (section 13.9)
crystallinity	- wide angle X-ray scattering(WAXS) (section 13.10.1) - differential scanning calorimetry(DSC) (13.10.2) - optical microscopy, final melting point(TMF) (section 13.10.3)
dispersion, morphology	- optical microscopy(OM) (section 13.11.1) - scanning electron microscopy(SEM) (section 13.11.2) - transmission electron microscopy(TEM) (ditto)
melt flow properties	- melt index(MI) (section 13.12.1) - rheology (section 13.12.2)
mechanical properties	- dynamic mechanical analysis(DMA)(section 13.13.1) - mechanical properties (section 13.13.2 and 3)

In the subsequent sections of this chapter we will follow the above table and discuss the methods/procedures in that sequence. If necessary a division will be made for applications to the various polymers in the polyolefin family. Finally we will list the methods we prefer for characterizing unknown polymers, with emphasis on the compositional aspects.

13.2 INFRARED SPECTROSCOPY

The application of infrared spectroscopy(IR) to polypropylene, its copolymers and polyethylene will be separately discussed.

13.2.1 Polypropylene

For pure homopolymers of polypropylene the only characteristic measured by IR has been the IR-stereoregularity. The name of Luongo[1] is connected with this as he was the first to describe the method. A recent version of this type of measurement is given by Burfield and Loi[2], the main improvement being in the use of ^{13}C-NMR in the calibration. A number of absorbance ratios have been tested for their relation with NMR isotacticity. A few of them show an easy linear correlation, for example the ratio A998/A973, though it is dependent on the annealing conditions; upon standardizing these a rapid method is obtained for polymers of <u>widely</u> different tacticity. In view of the scatter in the calibration, however, one wonders whether the method is applicable to polymers at the high, commercial end of the tacticity scale, for instance between say 96 and 98%.

13.2.2 <u>Ethylene-propylene copolymers</u>

From the first commercial production of ethylene-propylene copolymers (abbreviated as EPC when "random" and as TPP when "block"-copolymers are meant), infrared spectroscopy was used for their analysis. This was very much dictated by the usual presence of the required infrared instrumentation at many production locations. Since then, both infrared instrumentation and analytical demands have increased considerably. This may explain the large variety of methods which exists today for the analysis of ethylene contents in ethylene-propylene copolymers.

This section relates to the quantitative analysis of both ethylene and ethylene-containing structures which are present in polymers made up of ethylene and propylene, with emphasis on TPP. Most methods reported in the open literature deal only with total ethylene content, see section 13.2.2 (ii) below. Far less information can be found on the quantitative analysis of the individual ethylene-containing structures in TPP's, so a substantial part of this section will be devoted to this aspect (see section 13.2.2 (iii) and (iv)). IR spectroscopy has an established position in the determination of E_t or E_c in ethylene-propylene copolymers. Moreover a detection limit of 0.05 %m/m ethylene and below is difficult to match by other procedures.

(i) <u>ethylene structures observed by Infrared Spectroscopy</u>

<u>qualitative aspects</u>

When one ethylene molecule is incorporated in polypropylene, the new structure equals three consecutive methylene units bounded by two methyl substituted tertiary carbon atoms. This structure is

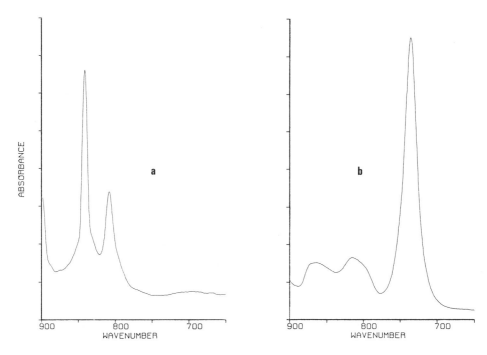

Fig. 13.1: IR-spectra of polypropylene (a) and hydrogenated polyisoprene (b), showing distinct differences in the 700-800 cm^{-1} region

nicely represented by hydrogenated polyisoprene, being the ideal alternating ethylene-propylene copolymer. Its IR spectrum is clearly different from polypropylene and the $(CH_2)_3$ unit is rather strongly absorbing in its methylene rocking vibration at 736 cm^{-1}, see Fig. 13.1. Knowing this, it is not surprising that EPC's with low ethylene contents show a similar pattern in their IR spectra, see Fig. 13.2. A distinct shift of the band-maximum towards 732 cm^{-1} reflects the small structural differences in the vicinity of the absorbing $(CH_2)_3$ unit. Increasing the content of built-in ethylene causes a gradual change of the IR spectroscopic features. The 720 cm^{-1} absorption becomes more prominent due to the increase in concentration of methylene sequences with a persistence length beyond four, see Fig. 13.3. These observations concern just the chemical structure; in addition the morphology affects the appearance of the IR spectral trace. Of course, the propylene crystallinity will change on ethylene incorporation, and also the dispersion when incorporating a separate copolymer phase. Even ethylene

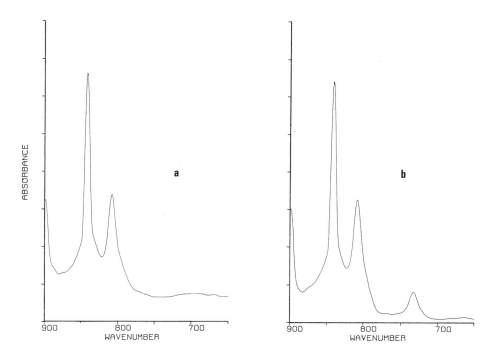

Fig. 13.2: IR-spectra of polypropylene (a) and random ethylene-propylene copolymer (b) in the methylene rocking vibration region

crystallinity may arise when the ethylene content of a polymer is sufficiently high. The presence of ethylene crystallinity in current high impact polypropylenes has already been mentioned in chapter 6, and it shows up in the IR-spectrum as characteristic doublet absorption with bands at 730 and 720 cm^{-1}, (see Fig. 13.3).

Of course, other ethylene structures can also be present in a commercial polypropylene. Additives very often contain methylene structures, sometimes also with a length of three or more consecutive methylene groups. For instance the presence of stearates has a pronounced influence on the spectrum in the 700-800 cm^{-1} range (see Fig. 13.4). Stearates can for example be present in the form of glycerol monostearate as an antistatic agent for polypropylene or as calcium stearate which may be added to neutralize any acid present or formed on processing, in order to prevent corrosion of metal processing equipment.

In this contribution on the IR analysis of ethylene-propylene copolymers, attention is strongly directed towards the methylene

540

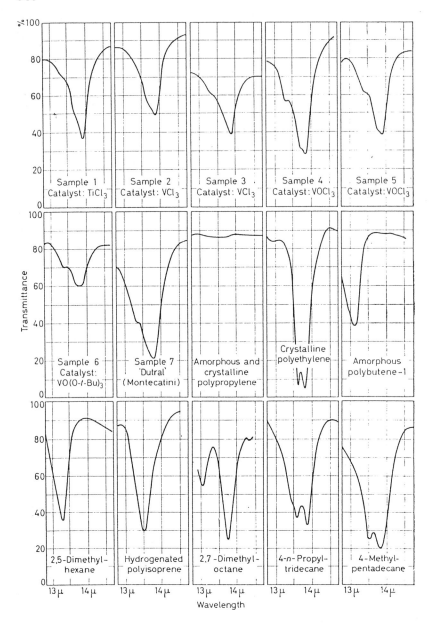

Fig. 13.3: The methylene rocking vibration in a number of different chemical components. Samples 1 to 7 are EPC's made with various catalysts and contain approximately 50 %mol ethylene. Note a more prominent 13.9 μm (720 cm^{-1}) absorption on increasing concentration of methylene sequences with a persistence length over four, and a doublet when polyethylene crystallinity is present.
[from J. van Schooten et al, Polymer 2 (1961) 357-363, by permission of publishers, Butterworth & Co, Ltd]

rocking vibration. It has the advantage of being isolated from other vibrations and of having fairly strong absorbtivity, enabling a simple analysis even at low concentration of ethylene units. However, it will also be shown that this particular vibration can be very complex due to the presence of a wide variety of ethylene containing structures. This intrinsic complexity has given us the opportunity not only to determine the total ethylene content of a sample (see (ii)), but also the amount of the ethylene-containing copolymer phase of TPP by means of E_c, see (iii), and even the amount of random ethylene-propylene copolymer therein by means of E_{cr}, see (iv).

Of course, many other spectral features arise from incorporated ethylene. In the present contribution on IR spectroscopy only a few other absorption bands will be mentioned for the analysis of ethylene-containing species.

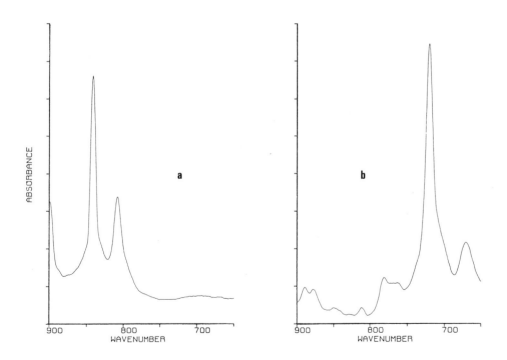

Fig. 13.4: IR-spectra of polypropylene (a) and calcium stearate (b), note the pronounced absorption band at 720 cm^{-1} in the latter

quantitative aspects

In order to enable a good quantitative analysis much care must be taken to ensure a reproducible way of preparing the sample specimen. The preparation of a sample specimen from polymers containing polypropylene crystallinity (such as TPP) is commonly done by a compression moulding technique, thus making thin self-supporting films. Usually a thin polymer film with a thickness between 0.1 and 0.3 mm is prepared at a compression moulding temperature of about 230 °C.

Another sampling option is the use of a suitable solvent. The use of polymer solutions has many advantages but only a limited range of ethylene-propylene copolymers is soluble at room temperature. For example amorphous EPC's are soluble in CCl_4 at room temperature. One has to be careful to prevent interference from solvent absorption. On the other hand, the resultant spectrum is free from all kinds of morphological effects (such as crystallinity, orientation, etc.).

To obtain an infrared spectrum of a sample specimen one can often apply a room temperature recording which is by far the easiest to perform. However, thermostatting may be required to circumvent small spectral distortions. The unpredictability of the shape of the ethylene rocking absorption band makes it necessary to take much care for an accurate wavenumber scaling. This is especially so because the absorption band maximum can vary from 732 cm^{-1} (for amorphous EPC) down to less than 720 cm^{-1} (for highly crystalline ethylene-containing samples). Most current FTIR spectrometers have a high wavelength accuracy, but for the older instruments additional measures have to be taken to ascertain accurate wavelength positioning. To this end a naphthalene solution can be inserted in the beam, thus recording its 781.3 cm^{-1} band, see Fig. 13.5.

The evaluation of a given (digital) spectrum is much simplified by the use of dedicated software. The most simple evaluation of a recorded spectrum is the spectrum matching procedure. In order to determine whether an unknown sample is matched by a particular reference sample, it is only necessary to (digitally) record both spectra and ratio them with an appropriate factor. With this procedure it is not possible to interpret any resulting mis-match easily in a quantitative manner.

All quantitative infrared analyses use the Lambert-Beer law or a modified version. The IR absorbance is linearly related to the concentration of the absorbing species. This means that a calibration must precede any quantitative analysis and its quality thus depends strongly on the calibration.

In most analysis a measure of sample content is required. For polymer films the determination of thickness is a very convenient way of obtaining information related to the amount of the sample. This film-thickness can best be obtained by measuring it at a number of predefined positions of the irradiated part of the sample and calculating the average. Another method for polymer films is the weighing of the actual part that is irradiated by the IR beam, for instance after punching out the relevant portion of the film.

Instead of relying on sample thickness or on sample weight, the amount of sample can also be taken into account by using an appropriate reference absorption band. For undiluted polymers, the

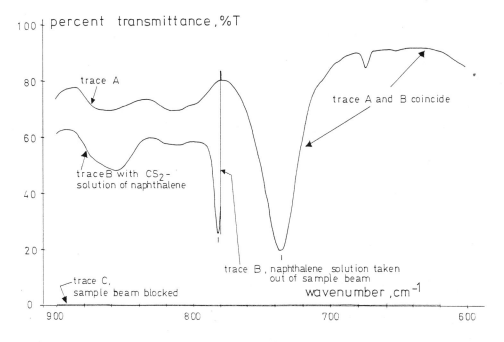

Fig. 13.5: The use of the naphthalene absorption band at 781.3 cm^{-1} for accurate wavenumber positioning

following absorption bands are often used:

    $A(4325$ cm^{-1}) due to a combination vibration of CH stretching
                   and bending fundamentals

    $A(1379$ cm^{-1}) CH_3 bending

    $A(1160$ cm^{-1}) a complex band involving both C-CH_3 stretch, C-C
                   stretch and CH bending[3]

    $A(968$ cm^{-1}) a complex band involving a CH_3 rocking and C-C
                   stretch[3].

Reference absorption bands are often advocated as they render unnecessary a separate step, thus enabling higher speed and further automation. This holds especially for FT-IR instruments. For solutions, a good measure of the amount of sample is the cell thickness.

With the often available extensive computing facilities a multiple linear regression method is preferred for measuring the total ethylene content. This procedure is based on a least-squares fitting of an unknown spectrum with a number of knowns. A similar approach can be applied for the determination of the sample content as the spectrum contains the information about the amount of sample involved. When using the amounts of sample belonging to the calibration spectra as additional calibration input, the amount of unknown sample can easily be calculated by a suitable multiple linear regression.

Another promising procedure which has already proved to be successful is based on a curve resolution procedure by which an unknown spectrum is unravelled into a number of individual absorption bands.

The accuracy of the ethylene contents derived from infrared data will never be higher than the accuracy of the data which are given to the calibration or reference samples. Moreover it is important that the total ethylene contents of the calibration samples are obtained by an absolute method, such as by NMR spectroscopy or by physical blending of well-characterized ethylene-propylene copolymers with polypropylene.

Although the methods mentioned in this section are reliable and give important information on the composition of EPC's or TPP's, more information can be obtained by combining this compositional method with a (simple) fractionation. Analysing the fractions, combined with the knowledge of the solubility behaviour gives a valuable increase in the information.

(ii) <u>determination of total ethylene content</u>

In the following, a fair number of different methods are mentioned. They all purport to determine quantitatively the total ethylene content, E_t, of an ethylene containing polypropylene by its IR spectrum. The methods are listed in an order of increasing complexity and increasing number of ethylene absorbances involved.

For earlier reviews the reader is referred to the references 4 and 5.

polymer film measured at room temperature

Only the wavenumber range between 800 and 660 cm^{-1} is required. The evaluation of the spectrum is performed by the baseline method to calculate the absorbance at 720 cm^{-1}. This method assumes the maximum absorbance to be at 720 cm^{-1}. The actual peak position can, however, vary with the actual distribution of the ethylene units over the polymer and the degree of ethylene crystallinity involved. The baseline method may become difficult to apply if any additional substance (e.g. an additive) give rise to extra absorption bands in the region where the baseline has its tangent points with the spectral trace.

Bearing in mind the calibration aspects, this simple method is restricted to cases where the ethylene distribution is invariable over all (calibration) samples. Moreover, the particle size distribution of the dispersed particles (if present) must be the same as well.

The next method uses two ethylene-related absorbances. Instead of using only the absorbance at 720 cm^{-1}, the absorbance at 736 cm^{-1} is now taken into account as well. A calibration is obtained by relating Et with A(720) + f. A(736). A typical value for f is 1.53.

This approach accounts for a varying distribution of ethylene units in the TPP whereas the wavenumber positions which are involved do reflect the positions of the maximum absorbances of the rocking vibrations of isolated (at 736 cm^{-1}) and long-sequence ethylene units (at 720 cm^{-1}). A difficulty arises over the correct fixation of the two positions. Firstly the 720 cm^{-1} position is not always identical with the band maximum, and secondly the

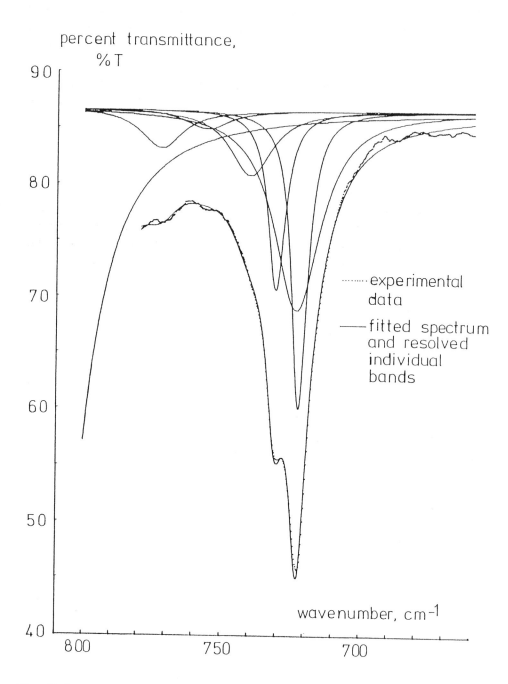

Fig. 13.6: Curve-resolved methylene rocking vibration band

magnitude of the 736 cm^{-1} absorbance is highly sensitive to minor variations in the very positioning of the 736 cm^{-1}. This is caused by the considerable slope of the broad rocking vibration band at this position. A solution to this problem is the use of an additional wavenumber calibration as mentioned above.

Of course, a strict linearity of the proposed calibration is not a necessity. Another relationship which is suggested can be formulated as:

$$100.E_t \; / \; \{ \; 1 + a.E_t + b.E_t^2 \; \} \; = \; c.A(720) \; + \; d.A(736)$$

The third method is based on a mathematical curve resolution of the complex IR absorption of the CH_2-rocking vibration, see Fig. 13.6. Total ethylene content, E_t, is determined from a calibration graph which presents a linear correlation between E_t and the summed integrated areas of the resolved bands I - V, see Fig. 13.7.

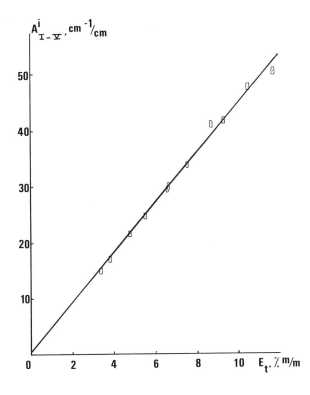

Fig. 13.7: Calibration graph relating total ethylene content (E_t) with the summation of five curve-resolved bands

The curve resolution procedure aims at the break-down of a complex IR absorption curve into its individual bands in order to abstract as much information as possible. In order to apply a proper curve resolution enabling the resolution of the individual absorption bands, it is necessary to know the number (or at least the minimum number) of bands, their profiles, their widths and their positions. The assignment of the individual absorption bands which are thought to be present under the overall absorption envelope of the broad methylene rocking vibration band is performed by the application of suitable references. This directly exposes the limitation of this method: only those structures which are incorporated in the fitting model will be discerned.

Usually infrared absorption bands are mathematically described by Lorentzian or Gaussian curves or by an expression in which both the Lorentzian and Gaussian character is incorporated, viz. by a sum- or product-function[6,7]. It seems   realistic that an initial start is based on pure Lorentzian curves, for example Fig. 13.8

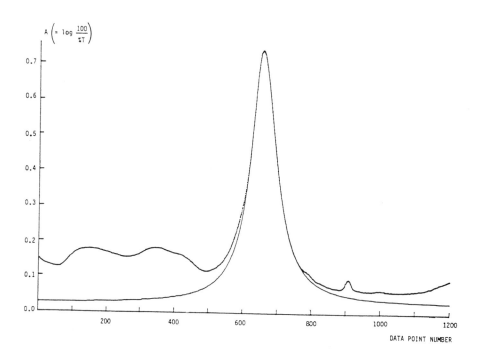

Fig. 13.8: A fully Lorentzian profile fitting the experimental methylene rocking absorption of hydrogenated polyisoprene

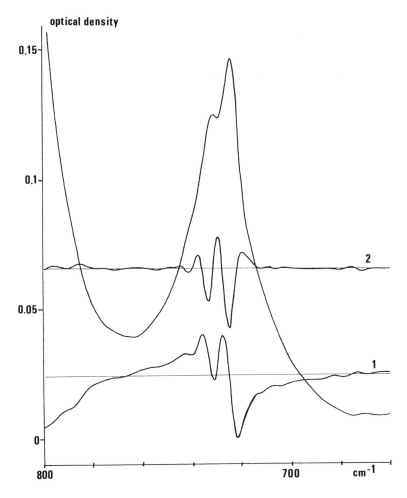

Fig. 13.9: First and second derivative of the methylene rocking profile of a TPP-sample

demonstrates that the CH_2 rocking vibration band of hydrogenated polyisoprene is perfectly fitted by a Lorentzian profile without any Gaussian character.

A usual way to establish the number and positions of the individual bands for a completely unknown spectrum is to calculate the first and subsequent derivatives of the spectrum (see Fig. 13.9). From the results thus obtained it can be concluded that at least three individual absorption bands are present under the overall envelope: namely a band at 736 cm^{-1} due to $(CH_2)_3$ and bands at 729 cm^{-1} and 720 cm^{-1}. The band at 720 cm^{-1} is usually assigned to long CH_2 sequences whereas the band at 729 cm^{-1} is clearly resulting from ethylene crystallinity in the sample. But it is known that ethylene crystallinity gives rise not to a single band at about 730 cm^{-1}, always an accompanying absorption band at about 720 cm^{-1} is observed. Thus an important step in the fitting procedure is to introduce two different absorption bands at about 720 cm^{-1}, one for long CH_2 sequences and another for ethylene crystallinity. For the particular set of reference TPP samples two additional bands also turned out to be present with absorption maximums at 750 cm^{-1} and at 768 cm^{-1} respectively. A completely different way to establish the number of different absorbing species is to intercorrelate the spectra of the whole set with each other. By this method only three different absorbing species are required in addition to the polypropylene matrix (see Fig. 13.10). Combining this finding with the seven required individual bands

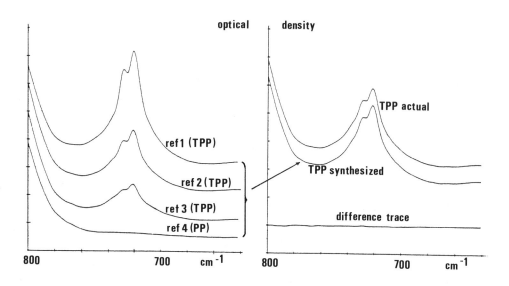

Fig. 13.10: Three different TPP-spectra are sufficient to describe an unknown TPP spectrum

leads to the conclusion that some of the individuals are strongly correlated. It is important to note that a description of the complex band with more than three individual bands is allowed but it is useless to try to extract more than 3 unknowns from the results!

At least two different bandwidths have to be incorporated in the model, namely one for the 'amorphous' CH_2 contributions and a much smaller one for the crystallinity bands. A first estimate for both is obtained from the spectra of, respectively, hydrogenated polyisoprene with a halfheight-bandwidth of 10.9 cm^{-1} and highly crystalline polyethylene with a halfheight-bandwidth of 3.0 cm^{-1}.

When a good fit is obtained (viz. only small deviations between actual and fitted spectrum, irrespective of wavenumber position), one can argue that the unknown sample contains methylene structures very similar to those which are present in the reference set. Also a good impression is obtained about their individual contributions. A completely different situation is met when the calculated spectrum does not fit to a large extent with the actual one. A misfit must have its origin in the presence of one or more additional absorbing species, compared with the samples in the reference set. In the first instance, an 'unusual' additive must be thought of.

The main advantage of this method is that the results do not rely on a very limited number of data-points. So this treatment very much resembles a noise averaging. Moreover, the fitting is based on a least squares procedure for the residuals expressed in transmittance units, which is a more realistic approach than the often applied averaging on absorbance units. Possible imperfections in the curve resolution model are not all that important for the final E_t data thus obtained.

A fourth method is especially developed for FT-IR instruments. Advantage was taken of the high wavenumber repeatability, the long and short range stability of the FT instrumentation and the inherent computing facilities. No precise thickness determination of the unknown sample is required. The thickness of a sample must be approximately 0.3 millimeter. The final calculation is based on a least square fitting of the spectral data (from 2200 cm^{-1}-1900 cm^{-1}) of the unknown with the data belonging to the reference set. This method very strongly relies on the repeatability of the recording of the spectra and the availability of a fairly powerful computing facility. Two striking advantages are the absence of a thickness determination and the full range of ethylene contents which is covered. Moreover this method has the additional advantage of using only a limited set of reference samples. Hence this method can be considered as a "spectroscopically intelligent" variation of the frequently advocated Factor Analysis approach[8].

A fifth approach is also a multiple linear regression procedure but instead of taking a selected series of calibration samples, a large set of calibration samples is generally required.
From a spectroscopic point of view, this technique is more or less
a black box method. This is particularly so while the existing
routines using this procedures have no real built-in warning
signals. If, for instance, any of the incorporated calibration
samples has been assigned a less correct ethylene content (which is
not all that unlikely within a large set!) then the final answer
for the unknown can be even more erroneous due to a disproportio-
nate counting of the spectrum of that reference sample.

polymer film measured at a temperature between 100 and 140 °C
A thin polymer film is cast on rock salt from a hot solution.
Solvents which may be used are, for example toluene, decalin or
cyclohexane. The sample is then kept at 120 °C by means of a
suitable heating cell. The absorbances at 1464 cm^{-1} and 1378 cm^{-1}
are measured by applying a single baseline from 1500-1300 cm^{-1}. A
calibration curve is obtained when E_t is plotted against
A(1378)/A(1464).

This method is an old one and is particularly suitable for
copolymers without polypropylene. Dependent on the instrumentation
and required accuracy a correction may be necessary for the slight
emission of the sample at 120 °C. It must be taken into account
that any propylene crystallinity has a pronounced influence on this
method.

A further method again uses a heated cell, the IR spectrum is
scanned from 800 cm^{-1}-660 cm^{-1} and either the integrated absorption
is calculated or a summation of three individual absorbances is
used for further evaluation. A particular equation is given[4]:

$$E_t = \{A(752) + A(734) + A(722)\}/1.375$$

Instead of 120 °C a temperature of 130 °C is also sometimes used.
The most important starting point is a sample without any remaining
ethylene crystallinity.

polymer film measured at a temperature beyond 180 °C

In order to simplify the method as much as possible, particular attention is paid to the sample preparation step. A compression moulded film of about 0.4 mm thickness is prepared in the usual way. A subsequent compression moulding step is used to fill a specially designed stainless steel sample spacer with a thickness

Fig. 13.11: Design of stainless steel micromould (thickness 0.2 mm)

of 0.2 mm. (see Fig. 13.11). To this end a piece of the 0.4 mm sample is cut, having the appropriate weight to fill this spacer.

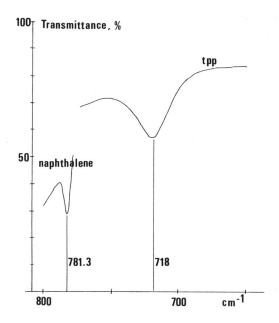

Fig. 13.12: TPP-spectral trace at 200 °C

The sample thickness is carefully measured at room temperature and the sample is subsequently placed between two KBr windows in a variable temperature cell. The temperature of the sample is raised to 200 °C, and Fig. 13.12 shows a spectral trace at this temperature. The absorption band now has its maximum at 718 cm^{-1}. The absorbances at three wavenumbers are calculated, namely at 718 cm^{-1}, 732 cm^{-1} and at 704 cm^{-1}. A quantification of the band dissymmetry (see Fig. 13.13) can now be expressed as A(732/704) = log {%T(704)/%T(732)}. The incorporation of this band dissymmetry into a final expression enables one to account for differences in ethylene sequence length distribution in the samples to be analysed. A suitable formula proved to be:

$$E_t = K_1 \cdot A(732/704) + K_2 \cdot A(718) + K_3$$

The temperature of 200 °C is chosen so that any polypropylene

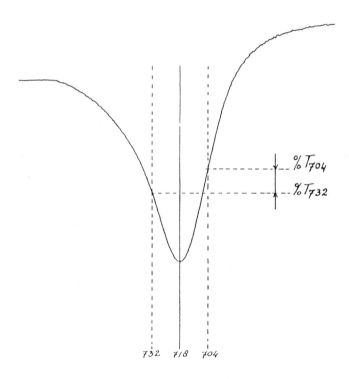

Fig. 13.13: Quantification of the band-dissymmetry

crystallinity is thus removed. Of course, only samples which have a sufficiently high melt viscosity can be measured this way since the samples are positioned in the usual (vertical) way. The method presented may appear rather cumbersome but the results are rewarding. Moreover, compared with current NMR procedures there is the advantage of requiring less sample (viz. 25-50 mg for IR compared to about 100 mg for NMR).

The levels of accuracy are comparable: about .5 % wt for E_t below 10 % and about 1.5 % wt for E_t beyond 10 %. Instead of applying the above given corrections based on the absorbances at 718 ± 14 cm^{-1}, it is usually equally possible to use only A(718) for samples which are of a similar nature, see for instance reference 9. The method then becomes comparable with the one dealt with above except for the crystallinity problems.

polymer solution

The solution method is evidently only possible if the samples are known to dissolve in an appropriate solvent. For CCl_4 solutions of polymers, there is a well-known method which is based on the absorbance at 1378 cm^{-1}, which arises completely from the (poly-)-propylene part.

Especially for a combined Gel Permeation Chromatography - IR spectroscopic analysis, the possible use of trichlorobenzene solutions of ethylene-containing propylene copolymers was investigated. A method is proposed which is based on C-H stretch vibrations. The range to be recorded is from 2700-3050 cm^{-1}. A tentative method was elaborated where the CH_2 content of normal alkanes could be determined using solutions having down to 2 ppm of the alkane in trichlorobenzene. The spectrum in this wavenumber range is built up of a number of highly overlapping absorptions. Using three n-alkanes (viz. n-C_5, n-C_7 and n-C_{15}) the hypothetical spectra for "CH_2" and "CH_3" were calculated, see Fig. 13.14. Thus a procedure may be developed using a linear regression and calculation of the "CH_2" and "CH_3" contributions for a linear n-alkane. A very similar approach may well be applicable for ethylene-propylene copolymers.

(iii) determination of copolymer content

To determine the composition of a toughened polypropylene a quantification of this polymer is required beyond its total ethylene content. A first step in this direction would be the quantita-

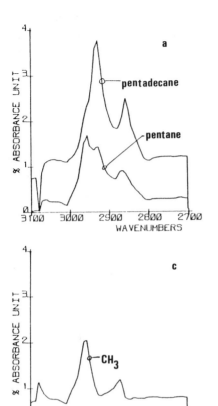

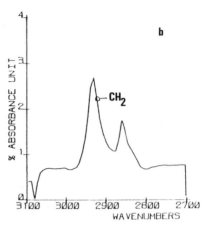

Fig. 13.14: Hypothetical spectra of the CH_2- and the CH_3-groups in n-alkanes as calculated from the spectra of n-pentane, n-heptane and n-pentadecane

tive determination of the _amount_ of copolymer. A direct analytical procedure is not easy to find. In some cases DSC methods can give an impression of the reduction in polypropylene crystallinity which occurs when part of the polypropylene is replaced by ethylene-propylene copolymer (see also chapter 8). Infrared spectroscopy can also give an indirect measure for the amount of copolymer. Assuming an equal distribution of ethylene units in the ethylene-propylene copolymer for both the unknown and the calibration samples (viz. similar copolymerisation statistics) then the ratio between isolated ethylene and long ethylene sequences will be related to E_c, the ethylene content of the copolymer. And using E_t and E_c, the fraction of copolymer is easily calculated:

$$F_c = 100 * E_t/E_c \text{ \%m/m}$$

As mentioned above the $(CH_2)_3$ sequence bounded by two tertiary $CH(CH_3)$ groups in TPP has a methylene rocking absorption band at 736 cm^{-1}. When the absorbance at 736 cm^{-1} is taken from a usual high impact polypropylene spectrum (see e.g. Fig. 13.10.) then it is obvious that this absorbance involves a considerable contribution from other overlapping bands. Nevertheless, the ratio A(736)/A(720) appears to be a fair measure for the E_c, see Fig. 13.15.

Whereas no suitable reference analytical method exists at present which can provide the absolute ethylene contents of the copolymer phase, the most obvious way of calibrating the infrared method is based on calibration samples made by blending homopolypropylene and ethylene-propylene copolymer in a known ratio. The calibration being always the most critical part for a quantitative IR method, this holds even more strongly for the E_c determination. A close correspondence is required between the distribution of ethylene units in the calibration samples and the unknown.

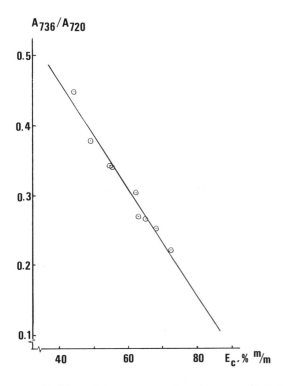

Fig. 13.15: Calibration graph relating A(736)/A(720) and E_c in TPP-samples

558

(iv) <u>determination</u> <u>of</u> <u>random</u> <u>copolymer</u> <u>content</u> <u>in</u> <u>copolymer</u>

A further distinction of the copolymer phase appears to be possible, viz. the distinction between F_{cx} and F_{cr}. F_{cx} is the percentage of the total sample which belongs to the crystallizable ethylene, for the reference samples F_{cx} data are determined by the HOXS procedure (see section 13.6.1). This HOXS procedure also determines the amount of random ethylene-propylene copolymer, F_{cr}. This random copolymer is characterised by its own ethylene content, E_{cr}. The following IR-method attempts to correlate the fractionation data from the HOXS/NMR procedure (section 13.6.1 and 13.3.1) with observations in the IR spectrum. A curve resolution approach is taken for the methylene rocking absorption, see above (iii). Of course, the method is only applicable on neat polymer films.

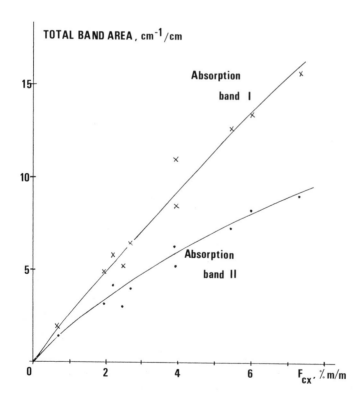

Fig. 13.16: Two individual resolved crystallinity bands in relation with F_{cx}

For the crystallizable copolymer, two absorption bands in the CH_2 rocking vibration band region are described as crystallinity bands. These bands (viz. I and III in Fig. 13.6), however, are not directly related to F_{cx} (see Fig. 13.16). This discrepancy is fairly evident, while the applied sample preparation for the IR method is completely different from the HOXS conditions. A shock cooling procedure never enables the development of an equilibrium crystallinity for all the potential crystallizable ethylene. Using only the two band-areas I and III, and assuming F_{cx} to be a parameter determining their magnitude, it is possible to get a close relationship, from which F_{cx} can be calculated, when only one additional structural parameter is influencing both I and III. This unknown additional parameter will be denoted as M. This formulation consequently supposes that both the quality of the crystalline phase and the extent of the crystallization can be expressed as a function of F_{cx} and M. The formulae used here are:

$$A_I + A_{III} = a.F_{cx} + b.M$$
$$A_{III}/A_I = c.M/F_{cx} + d$$

Elimination of M from these formulae results in:

$$(A_I + A_{III})/(A_{III}/A_I + f) = e.F_{cx}$$

with a, b, c, d, e and f as constants.

Using the available information about the structural composition of the reference samples, it is possible to make an estimate of the postulated factor M. A co-crystallization of the polyethylene fraction with some of the long ethylene sequences, present in the amorphous copolymer fraction (as determined by the HOXS procedure), is not unlikely. Fig. 13.17, relating $A_I + A_{III}$ with both F_{cx} and the calculated amount of ethylene in sequences of three (a number of three units is arbitrarily chosen) and longer, present in the amorphous copolymer fraction, shows a distinct improvement over Fig. 13.18 which relates A_I and A_{III} only with F_{cx}.

The expression used for A_{III}/A_I is based on experimental evidence and can be justified by supposing a higher crystallinity (higher A_{III}/A_I ratio) when the relative amount of the F_{cx} material within the actual crystalline structure is higher. An increase of M will lead to a higher fraction of the F_{cx}-material in the crystalline structure.

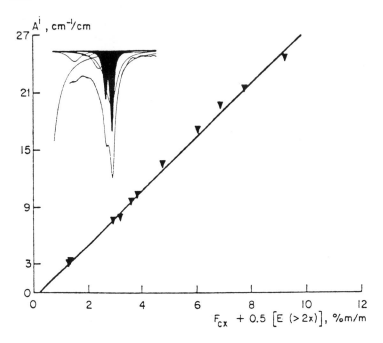

Fig. 13.17: Area of combined crystallinity bands related with F_{cx} combined with the amount of "long" ethylene sequences in the amorphous copolymer fraction

We now consider the random copolymer fraction. Assuming F_{cr} to be a single homogeneous copolymer fraction for all samples involved in which the incorporation of ethylene has occurred by one typical mechanism, it is possible to derive a statistical function which correlates the total amount of ethylene in F_{cr} (viz. E_{tr}) with the amount of ethylene present in isolated sequences. It has been found that A_{IV} relates very strongly with the amount of isolated ethylene units present in the samples (Fig. 13.19). Supposing a Bernouillian distribution of ethylene units in the random copolymer part, the ratio of isolated over non-isolated ethylene units can be expressed as:

$$E(1) \ / \ E(>1) \ = \ q/(1-q) \ ,$$

where $100 \cdot (q)^{0.5} = 100 - E_{cr,m}$ and $E_{cr,m}$ is the molar percentage of ethylene.

The present procedure is very dependent on the fixed fitting model applied. This again limits its value to the particular range

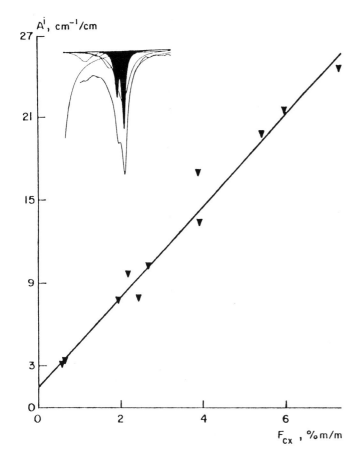

Fig. 13.18: *Area of combined crystallinity bands in relation with F_{cx}*

of compositions present in the series of calibration samples. The given F_{cx}- and E_{cr}-calibrations can only be taken as a first approximation.

It is expected that additional effort will reveal other contributing factors and/or relationships enabling a better calibration procedure. Without a correct curve resolution procedure and the use of extensively characterized calibration samples, it is impossible to make a sound judgement of the amount and character of crystalline ethylene on the basis of only an IR spectrum of a sample.

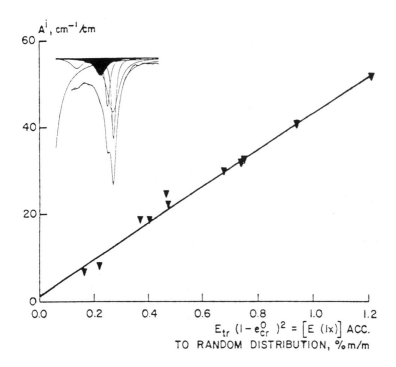

$$E_{tr} \ (1 - e_{cr}^0 \ )^2 = \left[E \ (1x) \right] \ \text{ACC.}$$
TO RANDOM DISTRIBUTION, % m/m

Fig. 13.19: Relationship between the area of resolved-band IV and the calculated amount of isolated ethylene units

13.2.3 Polyethylene products

Infrared spectroscopy is the oldest tool used in elucidating the structure of polyethylenes. For good reviews of the application of IR techniques see references 5, 10 and 11. The article of Cudby[11] also gives a concise historical survey.

The features that are discernible in the IR spectrum, apart from the absorbance related to the methylene units, are methyl and ethyl groups, unsaturation in various forms and oxygen-containing structures stemming from initiators or degradation. Table 13.2 gives the relevant wavenumbers. In Fig. 13.20 a few examples are given of IR spectra of homo- and copolymers. In recent Fourier Transform Infrared (FTIR) studies[12] the applicability of IR to ascertain the presence of butyl groups was severely questioned, as in a model ethylene/1-hexene copolymer no peak could be obtained after compensation in the 720-770 cm^{-1} range. The absorption is probably too close to the one from the ethyl groups.

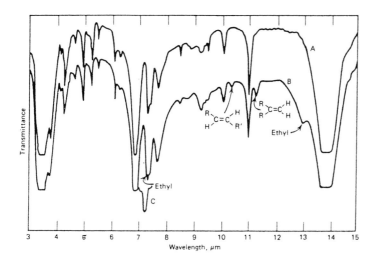

Fig. 13.20: Infrared spectra of some polyethylenes: homopolymer (A); 1-butene-copolymer (B); difference spectrum (C) [copied from the Kirk-Othmer Encyclopedia, 1981 edition, with permission from John Wiley & Sons]

Quantitative measurements using the infrared technique are far from easy, for the following reasons:
- the concentration of side groups ("branches" in the LDPE jargon) is generally low and through interference with the strong methylene absorbance gives large uncertainties,
- the extinction coefficient of the methyl group (being the main structural feature) is dependent on the chain length, examples of this are given in references 12 and 13. As in LDPE the branches are not of one type and with varying ratios between types, the quantitative analysis is difficult. The situation is however partly saved by the fact that the change in extinction coefficients is most rapid from methyl to ethyl and levels off thereafter, so by taking for LDPE an extinction coefficient for butyl might still give a reasonably close value,
- some of the absorptions are also affected by the physical state of the surroundings (crystalline or amorphous). To overcome this problem spectra can be recorded in solution or in the melt, see for example the above section 13.2.2 and ref. 14.
    Quantitative analysis has been advanced considerably by the introduction of the compensation method in double-beam instruments (which enables one to isolate the methyl absorption without any

TABLE 13.2
Relevant wavenumbers for identification and measurement
of non-methylene groups in IR-spectra of polyethylene.

Structure	Wavenumber (cm^{-1})
-CH$_3$ -CH$_2$- in ethyl	approx. 1378 (a) 767
vinyl, -CH$_2$-CH=CH$_2$	910,995
vinylidene, -CH$_2$-CR=CH$_2$	888
tr-vinylene, -CHR=CHR-	965

hydroxyl
 carbonyl
     carboxylic acid
     ketone
     aldehyde/ester
	3340-3390
	 1683-1780
	 1710
	 1715
	 1740

(a) depending on length of (side) chain

interference of the methylene absorptions). Even more progress is
made with Fourier Transform Infrared (FTIR) methods, which allow
for easy manipulation of the spectra. Sensitivities also increased
tremendously, one methyl on 10,000 carbons can easily be measured
in modern instruments.

13.3 NUCLEAR MAGNETIC RESONANCE(NMR).

13.3.1 Polypropylene

In the case of NMR it is necessary to distinguish ^{1}H (proton)
and ^{13}C (carbon) techniques. With the former the discriminating
power for sequence distributions is very low compared to the
latter, therefore only the latter is used nowadays for questions
related to the detailed structure of polyolefins. To illustrate the
resolving power of ^{13}C-NMR, recent measurements show effects on the
resonance of carbons which are 13 or even 21 carbon atoms removed
from the analyzed one[15,16]. An example of a spectrum of a very
complex kind of polypropylene is given in Fig. 13.21 (this sample
is both regio- and stereoirregular).

There has been little discussion about the identification of
the observed ^{13}C-resonances except in the initial stages of the
development of the technique. One either used empirical relationships, synthesized model compounds to check on these assignments[17],
or calculation methods based on Suter-Flory's rotational isomeric

state model coupled to the γ effect (i.e. short range interaction of carbons separated by three bonds). Examples of the latter approach are found in references 15, 18-20. An elegant proof of the identification of the peaks in polypropylene is the synthesis and subsequent NMR-analysis of hemitactic polypropylene. In this polymer every second monomer unit has a tactic relationship[16,21], which excludes a number of possible resonances as compared with normal isotactic polypropylene and enables easy identification of

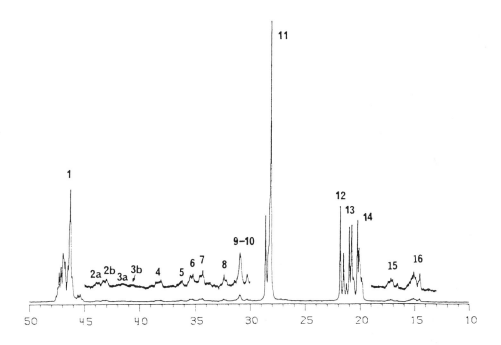

Fig. 13.21: ^{13}C-NMR spectrum of a stereo-irregular *and* regio-irregular polypropylene [from H.N.Cheng, Polymer Bulletin 14 (1985) 347-354, copied with permission form Springer Verlag, Heidelberg]

the resonances observed. In the assignment of peaks the INEPT method (insensitive nuclei enhanced by polarization transfer) can help, as in this method one can distinguish between methine, methylene and methyl carbons. This method selects the delay time (δ) prior to data acquisition and broad-band decoupling. If $\delta = 0.5$ J (J is coupling constant between ^{13}C and ^{1}H) then only the methine appears, if $\delta = 0.75J$ then methyl and methine appear upward and methylene downward. A nice example can be found in reference 22.

The information derived from the NMR spectrum can be given as pentad fraction or lower n-ads calculated from this. In a large number of cases we will compare polymers on their triad distribution ([mm],[mr] and [rr]) or even on the diads. The accuracy of these figures is about 5% relative. For a fairly recent account of the status of the NMR technique used on stereoregular polymers the reader is referred to the review of Harwood[23].

The measurement procedure involves the use of mostly 1,2,4-trichlorobenzene as a solvent, at roughly a 10 %m/m polymer concentration. The measurement itself is done at temperatures of about 110 °C for isotactic polypropylenes. There is not a large range of relaxation times for the different types of carbons in polypropylene. Randall[24] reports a maximum 32 % difference between the isotactic and syndiotactic methine carbon. This relatively small effect makes quantitative evaluation of the polypropylene spectra easy, certainly in comparison with (LD)PE (see below).

In copolymers of propylene with other olefins the polymer structure becomes already quite complex, certainly when "inverted" propylene units are also present. Numerous studies have been reported on assignment and the sequence distribution calculated on their basis. A good example is the study of Cheng[25] who identified 26 resonances in an ethylene/propylene rubber and was able to unravel this in terms of sequence composition, presence of inverted propylene units and propylene tacticity. Mostly these studies tend to extract polymerization kinetic data from the measured distribution and therefore they have already been reported in some detail in chapter 4.2.

In our own evaluation of ethylene/propylene copolymers (neat or in the presence of polypropylene homopolymer) the following approach was used. For the purpose of writing down the EP sequences that can be recognized by ^{13}C-NMR spectroscopy the following simplifying assumptions are made:
- only α, β, γ, and δ effects on the position of the carbon resonances are taken into account,
- the occurrence of inverted sequences of propylene (i.e. other than head-to-tail) is excluded,
- tacticity effects need not to be considered.
With these restrictions one can write down 8 basically different EP

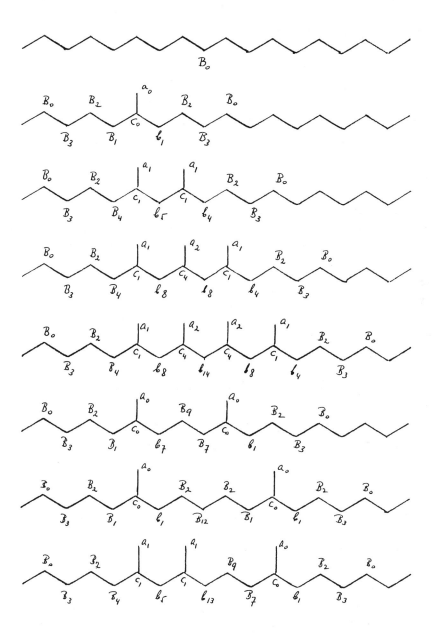

Fig. 13.22: Schematic view of the eight ethylene-propylene sequences and their distinguishable individual carbon resonances

sequences that can be distinguished by ^{13}C-NMR and no new carbon
signals can be expected from further variations in E/P sequencing.
These sequences are enumerated I through VIII in Fig. 13.22. In
each sequence the various carbons have the symbol designation as
indicated. Fig. 13.23 shows a ^{13}C-spectrum of an EP copolymer, this
figure gives also the assignments of the peaks to the various
carbon atoms.

In current NMR terms, with the exclusions mentioned above,
there are 18 different types of carbon atoms, and in principle 18
different peaks in the proton-decoupled ^{13}C-spectrum are expected
(barring impurities and accidental overlap). The integrated intensities of the peaks are, however, not mutually independent. One can
show that the following 10 relations must hold between these
quantities, leaving a total of 8 independent parameters. The 18
symbols now stand for the corresponding peak integrals; note that
capital B's refer to ethylene units and small a's, b's and c's to
propylene units and furthermore, by definition $B_1 = b_1$, $B_4 = b_4$, B_7

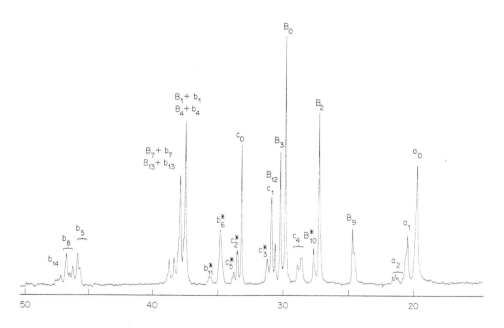

Fig. 13.23: Example of ^{13}C-NMR spectrum of an ethylene-propylene copolymer to
illustrate the various resonances from the scheme in figure 13.22 (note:
inverted propylene units designated with an asterisk)

$= b_7$ and $B_{13} = b_{13}$.

$$a_0 = c_0 \qquad\qquad (1)$$
$$a_1 = c_1 \qquad\qquad (2)$$
$$a_2 = c_4 \qquad\qquad (3)$$
$$2a_0 = B_1 + b_1 + B_7 + b_7 \qquad\qquad (4)$$
$$a_1 = B_4 + b_4 + B_{13} + b_{13} \qquad\qquad (5)$$
$$B_2 = B_1 + b_1 + B_4 + b_4 \qquad\qquad (6)$$
$$2B_9 = B_7 + b_7 + B_{13} + b_{13} \qquad\qquad (7)$$
$$B_2 = B_3 + 2B_{12} \qquad\qquad (8)$$
$$c_1 = 2b_5 + b_8 \qquad\qquad (9)$$
$$2c_4 = 2b_{14} + b_8 \qquad\qquad (10)$$

The following data are extracted from the spectra:

The method employs the following sums:

$$\Sigma B = B_0 + B_1 + B_2 + B_3 + B_4 + B_7 + B_9 + B_{12} + B_{13} \qquad (11)$$
$$\sigma B = B_1 + b_1 + B_4 + b_4 + B_7 + b_7 + B_{13} + b_{13} \qquad (12)$$
$$\Sigma a,b,c = \Sigma a + \Sigma b + \Sigma c \qquad (13)$$

The reported results are then obtained as follows:

average ethylene blocklength $\qquad N_E = \Sigma B/\sigma B \qquad\qquad (14)$

percentage of ethylene units present in single units $E_1 = (2B_9/\Sigma B)*100 \qquad (15)$

percentage of ethylene present in double units $\quad E_2 = (4B_{12}/\Sigma B)*100 \qquad (16)$

percentage of ethylene present in sequences of three or more

$$E_{\geq 3} = 100 - E_1 - E_2 \qquad\qquad (17)$$

average propylene blocklength $\qquad N_p = (\Sigma a,b,c/3)/(\sigma B/2) \qquad (18)$

percentage of propylene present in single units $P_1 = (c_0/\Sigma c)*100 \qquad (19)$

percentage of propylene present in double units $P_2 = (2b_5/\Sigma c)*100 \qquad (20)$

percentage of propylene present in sequences of three or more

$$P_{\geq 3} = 100 - P_1 - P_2 \qquad\qquad (21)$$

weight percentage of ethylene $E_c = 100*(\Sigma B)/(\Sigma B + \Sigma a,b,c) \qquad (22)$

Note that $\Sigma a,b,c$ is considered to be a better measure for the propylene content than Σa on its own, because the methyl signals of any impurity are likely to contribute in the region of Σa.

13.3.2 Polyethylene

As already discussed in the case of polypropylene and its copolymers, the use of ^{13}C-NMR is a powerful tool for determining detailed characteristics of the polymer. This holds also for polyethylene.

A few remarks are necessary about the technique. Most investigators measure in solution using for instance 1,2,4-trichloro benzene as a solvent. There are two studies[26,27] reported - both of Randall and Hsieh - checking the effect of concentration on the resulting spectrum: at values below 20 %m/m no concentration dependence is observed. They advise the use of 10 % to 15 % solutions because at lower concentrations signal-to-noise ratio is of course badly affected. The pulse spacing should exceed five times the longest relaxation time in the system - in polyethylenes these relaxation times are very dependent on the type of carbon, they increase strongly towards the end carbon, especially in longer sidechains. In one case[28] investigators measure on the molten bulk sample. Temperatures of over 135 °C should then be applied to prevent effects due to partial crystallinity, particularly line broadening. To allow for quantitative evaluation of the spectra in a Fourier Transform mode, the relaxation times for each type of carbon have to be known. These were measured in 1977 by Axelson, Mandelkern and Levy[29], and found to range from about 1.1 to 7 seconds. Especially long relaxation times are found for the methyl carbon, increasing as the side chain becomes shorter. These times are long, which means that pulse delays of 30 to 40 seconds are really required for preventing saturation and thus allowing quantitative analysis. This is certainly not practised by every investigator, and especially not in the earlier days of the ^{13}C-NMR method. The nuclear Overhauser enhancement factors are identical for all main type of carbons and equal to the maximum value of 2.

When analyzing copolymers and especially LDPE the assignments are made on the basis of the observed resonances in model polymers.

13.4 ^{14}C-SCINTILLATION COUNTING

A very accurate technique for assessing overall copolymer composition is the use of a ^{14}C-labelled co-monomer. Burning the sample, collecting the CO_2 and measuring its radioactivity in comparison with a pure homopolymer made with the same batch of labelled monomer, gives directly the desired compositional figures. This method is especially useful in areas where no other dependable methods yet exist, but can only be applied in laboratory synthesis of course. Furthermore one can generate with this technique calibration standards for other - less absolute - techniques, such as infrared.

13.5 PYROLYSIS-HYDROGENATION GAS CHROMATOGRAPHY

The general technique consists of pyrolyzing the polymer under specific controlled conditions followed by analysis of the fragments by gas chromatography. The decomposition products are usually hydrogenated in-line to simplify the chromatogram. The technique has been used to a large extent in copolymer analysis before the advent of ^{13}C-NMR or reliable infrared techniques.

Quite a number of studies on homopolymer polypropylene have been done also however, of which a few are reported in this section. In a fairly large-scale technique (employing 1 gram of polymer) Tsuchiya and Sumi[30] pyrolyzed polypropylene and observed next to the lower hydrocarbons also stereoisomers in products containing 11 carbon atoms or more. No attempt was made to correlate these with the composition of the starting material. Using pyrolysis-hydrogenation coupled to conventional gas chromatography Seeger en Cantow[31] were able to observe clear differences between isotactic, atactic and syndiotactic polypropylene. This was best observed in fragments of five and more propylene units. The isomerization by radical reactions was found to be strongly temperature dependent, for isotactic polypropylene for instance being more extensive at 500 °C pyrolysis than at 1000 °C. Later studies[32], using temperature-programmed pyrolysis, observed a series of multiplet peaks of roughly similar shapes. Identification of the peaks in the multiplet or its correlation with starting structure was not attempted. The studies of Tsuge et al[33], which used after the hydrogenation a capillary column of high resolving power, show very different chromatograms for different starting materials. For each carbon number of the fragments a triplet is mostly observed, the individual peaks of which can be equated to the stereostructure. The ratio of those peaks is very sensitive to the starting tacticity of the polymer studied. Inversions (i.e. propylene additions head-to-head or tail-to-tail instead of the normally observed head-to-tail one) can be found as well with this technique, the amount varying with both the catalyst applied and the polymer fraction studied (see chapter 2). An extension of the above work[34] (using a fused silica capillary column of even better resolving power) allowed a good correlation of calculated tacticities with the measured ones on the starting materials using NMR. Apparently the radical-induced rearrangement processes occurring during the pyrolysis are stopped far from equilibrium as otherwise no differences would be expected. Peaks of larger carbon numbers

(>12) show clearly the presence of long isotactic sequences in fragments from an original isotactic polymer, the non-isotactic structures not being discernible above the tetrad. This holds similarly for high syndiotactic polymer. So the information from these pyrograms is considerable, the main drawback being that the original order is not maintained during the analysis. In this connection it would be interesting to apply this technique to the original volatiles from polypropylene (see chapter 2).

In the case of copolymer analysis, pyrolysis-GC (coupled with hydrogenation) can be used, and it yields both the gross co-monomer content as well as information on the sequence distribution[35-37]. Of course one should know the fate of the intermediate radicals formed upon pyrolysis in order to reconstruct the copolymer chain. Identification of individual components can be done with compounds of known structure - which is easier for the shorter fragments. In the longer fragments, tacticity affects in high propylene containing copolymers have been observed[38]. Apart from obtaining data on the sequence distribution, this technique can also be used to "fingerprint" particular samples of polymers. Since the extensive development of the ^{13}C-NMR technique - which allows a non-destructive analysis, highly sensitive to the neighbouring units up to at least pentamers (a size which already gives interpretational difficulties in GC) - the method has not been used to any large extent.

13.6 SOLUBILITY METHODS, FRACTIONATION

The most frequently used technique for stereoregular polymer characterization is fractionation. Either very elaborate and cumbersome total fractionation methods are applied leading to a molecular weight distribution and a molecular weight dependent composition or more simple methods which yield only two fractions: a fraction soluble under the conditions of the test, the other insoluble, called the "solubles" and the "residue" respectively. In both cases it has been assumed that the scale of the fractionation is preparative, in the sense that on the fractions obtained the tacticity for example can be measured.

A few general remarks are in order before embarking on specific methods. The solubility of polymers is dependent on a large number of factors of which we mention a few:

- molecular weight, generally lower solubility at increasing molecular weight,
- temperature, most polymers show limited solubility both at low and high temperature, the latter phenomenon is related to the large change in solvent properties when nearing the critical temperature of the solvent. This is an important phenomenon in polymerization as a fair number of processes use low boiling hydrocarbons as diluents,
- crystallinity of the polymer, this decreases its solubility in comparison with amorphous analogues. Crystalline polymers do however dissolve below their melting point, the dissolution tempe- rature (or crystallization temperature, i.e. the temperature at which the first crystalline phase appears on cooling) is lower, the better the solvent (see below),
- structure of the solvent, expressed by Hildebrand[39] by defining his solubility parameter; it is related to the energy which is required to separate the solvent molecules. The definition is

$$\delta^2 = \text{(enthalpy of evaporation)/(molar volume)}$$

There exist also methods for calculating the solubility parameter based on group contributions, see for instance references 40 and 41. Generally speaking the solubility parameter of a solvent has a steeper, decreasing dependence with increasing temperature than that of a polymer as the density of the latter (and thus its molar volume) changes more slowly than that of a low molecular weight solvent.

By using the method of viscosity measurements, for which the viscosity of the polymer solution increases with increasing solvent quality, the solubility parameter of atactic polypropylene has been determined to be[42] equal to $16.2*10^{-3} (J/m^3)^{0.5}$ or in old units 7.9 $(cal/cc)^{0.5}$. The dissolution temperature of isotactic polypropylene has been measured for a number of solvents and leads to the value of about $16.9*10^{-3} (J/m^3)^{0.5}$ or 8.2 $(cal/cc)^{0.5}$ as the corresponding solubility parameter[43]. The same value is obtained by determining the solvent which gives the highest solvent sorption in polypropy- lene films[43,44]. For random ethylene/propylene copolymers a similar technique leads to 17.4 $(J/m^3)^{0.5}*10^{-3}$ (reference 45). Table 13.3 gives a small compilation of some solubility parameter data of polymers and solvents.

TABLE 13.3
Solubility parameters of a few polymers and solvents.

polymer or solvent	solubility parameter	
	in $(cal/cc)^{0.5}$	in $(J/m^3)^{0.5} * 10^{-3}$
polyethylene	7.9	16.2
polypropylene, isotactic	8.2	16.9
ditto	7.67±0.16[*]	15.7
polypropylene, atactic	7.9	16.2
ethylene/propylene copolymer, E_c = 40 %m/m	7.70±0.10[*]	15.8
n-pentane	7.0	14.3
n-heptane	7.4	15.1
n-decane	7.7	15.8
cyclohexane	8.2	16.8
xylene	8.8	18.0

[*]: at infinite dilution, from reference 46.

We now return to our original discussion on fractionation. Full fractionation will not receive much attention in this book, since when investigating the molecular weight distribution one will generally use GPC (gel permeation chromatography, see below) and a simpler method mostly suffices for checks on, for example the tacticity distribution. However, for a "two-dimensional" analysis of both molecular weight and crystallinity (arising from tacticity or sequence distribution) as parameter these techniques are making a comeback. The methods applied in preparative fractionation can be summarized as follows:
- employing the same principles as with non-crystalline polymers by using the coacervation at temperatures high enough to prevent crystallization. Solvent/non-solvent pairs applied are for instance xylene and 2-ethoxyethanol, decalin and phenol or propyleneglycol, o-di-chloro-benzene and diethylene glycol-mono butylether[47]. For atactic polypropylene there is no need to use higher temperatures as its crystallinity is rather low, so that solubility is mostly complete at room temperature in a wide variety of solvents. Full fractionation can in this case be effected by using low boiling solvents such as cyclohexane and acetone[48]. In these fractionations the use of column techniques makes the experimentation much simpler, the polymer to be fractionated is deposited on a carrier material such as glass beads, put into a column and eluted with a

solvent mixture of steadily increasing quality[49].
- application of techniques at temperatures at which the polymer is
allowed to crystallize, whereby the outcome of the fractionation is
dependent on both the molecular weight and the tacticity. This can
for instance be effected by using a column elution with a good
quality solvent starting at low temperatures and gradually increasing the temperature[50,51]. One observes an ever increasing melting
point of the polymer eluted, and the molecular weight increases
steadily except at the very beginning and the end of the fractionation. A similar technique is crystallization from a solution whilst
applying a shear field with a stirrer[52], the shape of the distributions obtained is similar to those mentioned above. The supercoo-
ling required for crystallization decreases with increasing molecular weight.
- another method, also dependent on both the molecular weight of
the polymer and the tacticity/crystallinity is the extraction of
the polymer in a Soxhlet with solvents of increasing boiling point.
However the effect of diffusion of the polymer out of the bulk form
may be large in this case (see below), so that one either has to
apply long extraction times or use samples of a small particle
size. Up to 14 different solvents have been used in such tests[53],
using a recrystallized polymer as a sample, thereby minimizing
diffusion effects.
- the use of intense mechanical agitation overcomes the slow rates
of the dissolution process, at least under certain conditions. When
one uses polymer particles of less than 1 mm in size and applies a
temperature of 122 °C only five minutes are required for equilibration in the case of polyethylene whilst agitating with a vibro-mi-
xer. With the usual solvent/non-solvent mixtures one can arrive at
molecular weight distributions nearly identical to those obtained
from GPC-measurements[54]. The fractions have indeed been shown to
possess a narrow MWD.

    Recently a combined GPC/tacticity instrumental fractionation
technique has been developed in the so-called cross fractionation.
A GPC apparatus is altered to allow a rising temperature fractionation of the polymer sample, which has first been put on a support
by a simple method. The eluted fractions are subsequently passed
through the GPC-part of the equipment[55]. As a tool for distinguishing polypropylenes with different structures this technique
might be very useful. Similar techniques are in use for polyethy-

lene (especially copolymers) characterization. The abbreviation TREF is often used for Temperature Rising Elution Fractionation. Instrumented techniques have been developed[56]. The temperature range normally used in polyethylene studies is from 50 °C to 100 °C. There is no effect of molecular weight on elution temperature, provided it is below 10,000.

In the case of a so-called <u>simple</u> <u>fractionation</u> the main objective is generation of a meaningful number that characterizes the polymer as regards its tacticity or composition (distribution) in the simplest manner possible. In this respect one has to distinguish the plant control type of measurement from true characterization. In the former the test is only meant to give a fast signal of deviations from the normal situation, i.e. a non-absolute method is certainly acceptable. In the latter case the requirements of the test are far more stringent, for instance the measurement should be independent of the starting form of the polymer (i.e. powder, nib, film, etc.). In the following we will focus on polypropylene, as most work done in this area has used this polymer.

TABLE 13.4
Long duration extraction of a polypropylene powder

Base powder average particle size about 30 μ. Extraction conditions: first slurried in n-hexane and filtered at room temperature, subjected to continuous Soxhlet extraction at the boiling point of n-hexane.

Time interval, h	Removed polymer, %m/m	Characterization of removed polymer						
		GPC data				DSC data		
		M_n*10^{-4}	M_w*10^{-5}	Q	R	Crystal-linity, %m/m	Melting points T_{ml}, °C	Crystal-lization temp. T_x, °C
0-1	0.95	0.14	0.29	21	27	1	38	40
1-2	1.14	0.25	0.20	8	28	15	45,83,119	60
2-5	0.35							
5-10	0.21	0.26	0.32	12	44	13	40,73,119	71 (a)
10-56	0.29	0.22	0.25	12	50	16	121,154	73,95
56-80	0.09	-	-	-	-	-	-	-
80-103	0.05	-	-	-	-	-	-	-
103-171	0.003	-	-	-	-	-	-	-

(a) analysis of the combined fractions from 2-5 and 5-10 hours

13.6.1 Polypropylene

In simple fractionation one is tempted to start with the solid polymer, however dissolution kinetics are very slow, see for example references 44 and 57. An example from our own experience is given in Table 13.4, but even with this relatively small polymer particle size the limiting value of polymer solubility is only reached in about 60 to 100 hours. This slow rate is due to the large molecules in polymers. The dissolution rate can be increased by employing a number of different measures:

- increase the temperature to lower the viscosity, increase diffusion, etc.

- minimize the dimensions of the specimen, at least in one direction,

- stir very intensively, the effect has already been mentioned above[54].

Even when using these improvements the result is still questionable, and the experimental procedure rather elaborate. A different situation arises when one starts from a solution of the polymer and allows one or more fractions to crystallize therefrom. By choosing

TABLE 13.5
Some methods used for measuring solubles in polypropylene

Used or originated by	Solvent	Temperature	Time h	Polymer form	Approx. crystallinity of solubles by X-ray, %m/m
Natta[58,*]	n-heptane	reflux(100 °C)	24	powder or ground nibs	<20
Shell	diethyl-ether	reflux(35 °C)	2	powder	3
Shell	isopentane	room temp.	0.75	powder	1-5
FDA[59]	n-hexane	reflux(68 °C)	2	any form	(<10)[**]
FDA[59]	xylene	dissolve, crystallize at 25 °C	-	any form	2-10
FDA[59]	decalin	dissolve, crystallize at 25 °C	-	any form	(2-10)[**]

*: also ISO method R922

**: predicted

(FDA: the United States Food and Drug Administration)

the right conditions one can arrive very close to equilibrium.

Many different methods, some of which are summarized in table 13.5, are used by various institutes and industries to measure the amount of solubles present in polypropylene. Broadly speaking, the methods can be divided into two classes: one involves the washing of slurries and the other involves recrystallization from a (hot) solution. The values obtained using different methods vary widely, as the examples in table 13.6 show. The reason for this is that in all slurry washes the potentially soluble polymer has to migrate from the inside of the particles in which it was formed during the polymerization. Therefore factors such as the duration and the temperature of the washes, the base morphology of the powder or the sample shape markedly influence the results obtained. Consequently, great care has to be excercised when these solubles figures are used: they should always be quoted as the amount of solubles obtained by a certain specified method. The methods using recrystallization from a hot solution, such as the

TABLE 13.6
Examples of the results obtained by different solubles measurement methods

| Method | Amount removed from | | |
| | Sample A | Sample B | Sample C |
		in %m/m	
isopentane, 25 °C, 45 min	1.0	2.2	-
isooctane, 25 °C, 45 min	0.68	2.0	0.11
n-hexane, 68 °C, 16 hours(a)	2.6	3.5	0.86
cyclohexane, 81 °C, 16 hours(a)	6.2	6.8	2.6
isooctane, 100 °C, 16 hours(a)	4.7	4.7	1.7
xylene, recrystallization	4.3	4.9	5.0

Sample A: made with low temperature reduced $TiCl_4$, particle size about 30 μ,
Sample B: made with $AlEt_2Cl$ reduced $TiCl_4$ as catalyst, particle size about 100 μ,
Sample C: made with Solvay catalyst, particle size about 200 μ.
(a): at reflux

xylene method (see table 13.5), are, however, not affected by these variables. In these methods, the separation of an isotactic fraction which is able to crystallize upon cooling and a fraction which stays in solution is obtained. More meaningful and consistent values are obtained in this way. This was already recognized early in the history of polypropylene, by scientists at Hoechst (ref 60) and others (ref 61). The Hoechst group later[62] gave additional evidence for their preference for this method.

The slurry wash methods can only be used successfully when no variations in morphology, particle size, etc. are expected. This would be the case in a manufacturing plant, although the results can only be used for plant control as they are not a true polymer characteristic.

In some of the studies reported in this book a recrystallization method will have been used. The method applied is either the standard xylene solubles method or a modified version. The xylene soluble method was originally described by Fuchs[60] and by the United States Food and Drug Administration[59]. In the xylene solubles method (abbreviated in this book by XS) one dissolves 2 grams of polymer in 200 ml of xylene (a mixture of isomers, with a boiling range of 137 to 143 °C) at around 120 °C, after dissolution this is cooled to 25 °C for 2 hours, and for analytical purposes a portion of the solution is filtered and an aliquot evaporated to dryness, the recovered weight being used in the calculation of the percentage soluble polymer. In a large number of cases one is interested not only in the soluble part of the polymer, but also in the recrystallizate. In that case one of course filters the total suspension and washes the solid with some fresh xylene.

In the modified method (called the hot xylene solubles, abbreviated to HOXS) the start of the test is the same, one cools to 25 °C for only 30 minutes after which the suspension is heated to 50 °C and kept at this temperature for another 30 minutes. Filtration is then effected through a thermostatted (50 °C) sintered glass funnel. By forcing this temperature profile on the mixture, equilibrium is very closely reached.

For a good understanding and interpretation of the data generated by this method, it is necessary to know the solubility of isotactic polypropylene under the final conditions of the test. Some data on this can be found in a publication of Natta's group[63], in which molecular weights of isotactic polypropylene soluble at the boiling points of various solvents are given. For instance in boiling ethylether a molecular weight of 2000 appears to be totally soluble (molecular weights based on osmotic pressure measurements), in pentane at 35 °C polypropylene of 3000 molecular weight is soluble, in boiling n-hexane even 10,000. Some own data using GPC are given here. The solubility of two low molecular weight polypropylene samples in various solvents and at different temperatures is given in table 13.7. In the following table, 13.8, the characteris-

TABLE 13.7
Solubility of low molecular weight isotactic polypropylene

Base material: Two thermally degraded polymers from a common feedstock. The basestock was a non-extracted, melt index 3 polymer. The isotacticities([m]) measured by NMR were 91 % for the base material and 84 % for the degraded ones. Extraction method: 2 g of polymer was stirred with 100 ml solvent for two hours.

Sample	Solvent	Temperature, °C	Solubility, %m/m
Sample 1 $M_v=1600$	isooctane	20	38
	do	70	70
	do	100	90
	isopentane	20	41
	tetrahydrofuran	20	42
	xylene	20	38
	cyclohexane	82 (reflux)	100
Sample 2 $M_v=3000$	isooctane	20	20
	do	70	49
	isopentane	20	21
	tetrahydrofuran	20	21
	xylene	20	20
	cyclohexane	82 (reflux)	100

tics of some of the fractions obtained are given. At 20 °C isotactic polymer with a molecular weight of the order of 1400 is soluble in isooctane. Similar solubilities at 20 °C are observed in the other solvents except for cyclohexane, which is the best solvent for polypropylene. The solubility of isotactic polypropylene increases sharply when the temperature reaches values of higher than 70 °C (see table 13.7), so that for a method aiming at the determination of atactic polymer only, one should apply temperatures lower than 70 °C. For the described XS and HOXS methods this condition is fullfilled, which is not the case in the very frequently applied boiling heptane solubles method, operating at 100 °C.

Another illustrative set of data on the effect of extraction temperature on both the efficiency of atactic polypropylene removal from powder and on the level of co-extracted isotactic polypropylene-

TABLE 13.8
Characterization of soluble low molecular weight isotactic polypropylene

sample	DSC data			GPC data			
	Crystal-linity, %m/m	Melting temperatures T_{ml}, °C	Crystal-lization temp. T_x °C	M_n	M_w	Q	R
SAMPLE 1							
total	31	128,161	82	1090	1420	1.30	1.42
isooctane soluble at 20 °C	27	31,46,69,81,88	-	1160	1390	1.20	1.27
isooctane insoluble at 20 °C	47	131	100	-	-	-	-
isooctane soluble at 70 °C	23	117	73	1280	1580	1.23	1.29
isooctane insoluble at 70 °C	21	135,156	101	1570	1960	1.25	1.30
SAMPLE 2							
total	35	138	91	-	-	-	-
isooctane soluble at 20 °C	22.5	10,30,43,68,81,87	-	980	1390	1.41	2.0
isooctane insoluble at 20 °C	45	141	105	-	-	-	-
isooctane soluble at 70 °C	-	-	-	1250	1780	1.43	1.62

ne is given in Table 13.9, where data are given of extractions at increasing temperature. The resulting extracts were re-extracted with isopentane at room temperature to separate them as far as possible into an isotactic and an atactic fraction. In Fig. 13.24 the molecular weight distributions of the latter are given. Clearly a fairly high temperature is required to extract a fair fraction of the atactic polymer present from the solid polymer, and at these high temperatures up to 50 % of the extract is isotactic. The parameters of the molecular weight distributions of the non-isopentane soluble part are given in table 13.10. Again increasing molecular weights with increasing temperature are noted, the absolute values are certainly lower than expected on the basis of Natta's work. There is an influence of the base polymer, which has been dealt with in chapter 2.

The efficiency of the XS method has been tested by a large number of re-extractions on residues from the first extraction. The amount in the second extraction is found to be 2 to 7 % of the

TABLE 13.9

Extractions of polypropylene at different temperatures
Base powder made with low temperature reduced $TiCl_4$, melt index 3.6, xylene solubles 4.3 %m/m (XS_0)

Solvent	Temp, °C	Amount removed, %m/m	Xylene solubles of residue, %m/m	Fraction of XS_0 removed	Solubility of extract in iC_5, %m/m	Crystallinity, %m/m	Melting point, °C	M_n	Q	R
iC8	25	0.68	3.5	0.19	100	1	51,81	2100	9.5	15
nC7	25	0.99	3.5	0.19	100	0.2	55	-	-	-
iC5	25	0.99	3.3	0.23	100	2	30,52	2100	7.4	11
CyC6	25	1.1	3.4	0.21	n.d.	2	49	-	-	-
iC8	50	1.3	3.0	0.30	n.d.	3	37,53,77	2600	5.7	8.2
CyC6	50	2.0	2.5	0.42	n.d.	5	48,70	2800	7.0	9.1
CHCl3	62*	2.8	2.2	0.49	81	-	-	4200	6.3	20
nC6	68*	2.6	2.1	0.51	91	6	50,70,105	3200	5.5	6.9
iC8	75	2.1	2.4	0.44	96	7	50,96,148	3200	5.4	6.7
nC7	75*	2.1	2.3	0.47	-	-	-	-	-	-
CCl4	77*	3.9	1.5	0.65	58	-	-	4300	5.7	8.2
CyC6	82*	6.2	1.2	0.72	51	27	132	3200	7.0	4.8
nC7	98*	5.1	1.3	0.70	56	21	130	3800	5.1	4.0
iC8	99*	4.7	1.2	0.72	60	-	-	3400	5.2	4.2

*: reflux

GPC data of xylene solubles of original sample: $\tilde{M}_n$:4600, Q=11, R=8.

TABLE 13.10

Molecular weight distribution parameters of isopentane insoluble parts of extracts removed at higher temperature

Solvent, wash conditions	M_n	Q	R	Peak mol wt
n-hexane, 68 °C	3800	20	17	2700
cyclohexane, 82 °C	3100	4	20	6140
isooctane, 99 °C	2900	9	19	4530
n-heptane, 98 °C	3000	5	18	5220

first value. The second extract is of higher crystallinity and shows a low molecular weight tail in its molecular weight distribution. A relative larger fraction of low MW isotactic polypropylene is thus present in this second extract.

A comparison of the amount of solubles obtained by both the XS

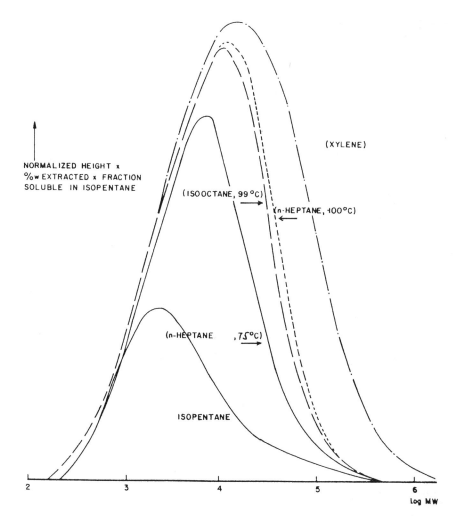

NORMALIZED HEIGHT x
%w EXTRACTED x FRACTION
SOLUBLE IN ISOPENTANE

(XYLENE)

(ISOOCTANE, 99 °C)

(n-HEPTANE, 100°C)

(n-HEPTANE , 75°C)

ISOPENTANE

2   3   4   5   6

Log MW

Fig. 13.24: Molecular weight distributions of the isopentane soluble fractions obtained from extracts made at higher temperatures

and HOXS methods is given in table 13.11. As expected the HOXS value is greater but not by much, averaging about 1 to 1.5 % absolute in a vast range range of base polymer molecular weights. This also illustrates that the isotactic polypropylene solubility under the conditions of this test is low.

Kakugo et al describe[64] another very interesting fractionation method leading to four different fractions. This probably makes it a time consuming method, but considerable information is obtained.

TABLE 13.11
Comparison of <u>XS</u> and <u>HOXS</u> on polypropylene homopolymer

Base polymer		Xylene solubles %m/m (XS)	Hot xylene solubles %m/m (HOXS)	Diffe- rence
LVN dl/g	Melt Index g/10min			
2.87	2.2	4.3	5.0	0.7
2.32	6.5	4.8	5.9	1.1
2.01	14	5.5	6.5	1.0
1.71	30	7.0	7.5	0.5
1.46	65	8.0	8.8	0.8
1.29	120	9.6	10.3	0.7
1.21	160	6.6	8.4	1.8
1.12	230	7.6	8.8	1.2

They start with a xylene recrystallization, and extend this procedure with two subsequent boiling solvent fractionations with hexane
and heptane respectively, i.e. at about 70 and 100 °C. In this way
they remove selectively the so-called stereoblock polymer, the
quantities of which differ for different catalysts and polymerization conditions (as also shown by a slightly different method in
chapter 2.4.6). It is claimed that the residue arrived at in this
method is of a constant, very high tacticity of 97% mm.

13.6.2 Polyethylene

Showing no tacticity, simple "solubles" tests for that purpose
are of course absent. However one frequently measures the low
molecular weight tail of the molecular weight distribution (the
so-called "wax") by a simple extraction. For instance extracting
PE-films with n-hexane for 2 hours at 50 °C is a much quoted
version of this test.

13.6.3 Poly(1-butene)

For isotactic poly(1-butene) also, fractionation or simple
extraction is the frequently-used tool to judge its stereoregularity. Poly(1-butene) is more readily soluble in hydrocarbon solvents compared to both polyethylene and polypropylene. For instance
the dissolution temperature in toluene is 39 °C and in n-heptane 35
°C. This means that a slurry polymerization can only be performed
at low temperatures of under about 40 °C in relatively poor solvents such as the monomer itself.

The usual method for measuring the "atactic" level is extraction of the polymer with boiling ethylether in a Soxhlet operating at the boiling point. Methylene chloride can be used to isolate very low molecular weight atactic polymer (or "oil"). An alternative measurement is proposed by Phillips investigators[65], who stir the polymer in n-hexane at room temperature for 2 hours, decant, wash once with n-hexane and once with methanol and dry. This method is quoted as giving identical results to the ethylether method. No investigations are known as to efficiency of the method, etc..

One recrystallization method has been reported in the literature[66]. The polymer is dissolved in n-decane at 110 °C, cooled to 0 °C, and left overnight at room temperature, after which the two fractions are isolated. In some earlier studies in our laboratories a similar method was used with o-dichlorobenzene as solvent, setting the cooling to 1 hour at 0 °C and 0.5 hour at 25 °C, filtering through a thermostatted filter.

13.6.4 Toughened polypropylene

Apart from the true block copolymer content for which no technique is known to us, the amount of each constituent can be calculated by using a simple fractionation technique (such as hot xylene solubles) and then measuring ^{13}C-NMR on both fractions. The fractionation splits the sample according to the crystallizing tendency in xylene at 50 °C, giving a crystalline polypropylene and ethylene/propylene copolymer next to the soluble atactic polypropylene and soluble ethylene/propylene copolymer. Sometimes when the sample is very rubbery coupled to a rather high molecular mass the crystallization is less complete and a lowering of the starting concentration is recommended to 1 or even 0.5 g/100ml. In tests at slightly different temperatures (40-60 °C) the amount of soluble material isolated slowly increases in that range, but the NMR data of the ethylene-containing species stays constant, indicating that only slightly more of the polypropylene stays in solution but that there is no change yet in the type of the ethylene-propylene copolymer.

Using some of the details in the NMR spectrum regarding the ethylene distribution the _fractions_ atactic polypropylene, soluble ethylene/propylene copolymer, the isotactic polypropylene and the non-soluble, crystalline ethylene/propylene copolymer can be derived.

The above method used as a starting point a fully unknown sample, obviously when one can get hold of the homopolymer part separately, the total analysis becomes much easier.

The calculation of the individual components out of the HOXS and the NMR data is based on the fact that from NMR one can derive the average sequence length of ethylene (N_E) as described above in section 13.2.2. This value is totally independent of the presence of propylene polymers as one only looks at ethylene-containing sequences. Assuming that the copolymers present in the respective fractions are truly random then there is a simple relation between N_E and E_C, namely

$$N_E = 100 \; / \; [100 - E_C^*]$$

in which E_C^* is the <u>molar</u> percentage ethylene in the copolymer. Knowing the E_C of the copolymer present in the HOXS fractions leads to a further, simple calculation of the other constituents. Mostly we used the data on the HOXS-extract as a basis for the method as it turned out that the E_C of the copolymer in the crystalline HOXS-residue is invariably very high (> 90%m/m). Thus knowing the ethylene content of the total product and the data on the HOXS-extract is sufficient for the total calculation.

The above concept of splitting a toughened polypropylene composition into two widely different fractions has recently also been studied by a commission of the Macromolecular Division of IUPAC[67]. Two different methods have been tested, the first being the known hot heptane extraction on 100 μm films, the second being a xylene recrystallization test in which both soluble and insoluble material were recovered. The latter test was found suitable and reproducible as the hot heptane method - as mentioned before - extracts also isotactic fractions to a far greater extent than xylene at 25 °C. Furthermore it appears from this study that even the measurement of the ethylene content is very difficult, with IR methods being worse than NMR.

A further development of such a method would be the additional fractionation of for instance the crystalline fraction as separation of polypropylene from polyethylene by fractional crystallization has been described[52]. To our knowledge this is never routinely applied however. Another, less selective way, is by measuring the melting enthalpy of both polypropylene and polyethylene and relating this to the amounts of these two polymers.

A different version of TPP-fractionation can be found in a patent of Mitsui Petrochemicals[68]. It starts with a recrystallization from kerosine (giving a soluble fraction of the copolymer and atactic polypropylene), the residue of which is further extracted with the same solvent at 110°C. The residue is mainly polypropylene or a propylene-rich copolymer. The fraction soluble at 110°C is re-extracted at a non-specified temperature with a mixture of kerosine and butyl carbitol which leads to a residue which is polyethylene or an ethylene-rich copolymer. It is a pity that no further information has ever been given on this method as the described split between the different species is a goal that everyone pursues.

13.7 GEL PERMEATION CHROMATOGRAPHY(GPC)

In a GPC analysis a very dilute solution of the polymer (to ensure that only individual polymer molecules are present in the solution) interacts with a porous solid in which the pores have different, known sizes. The porous solid is packed in a column. The separation takes place on the basis of differences in the active hydrodynamic volume (LVN*M). A small molecule will fit in nearly every available pore and therefore will take a long time to pass the column, whilst a very large molecule fits in none of the pores and elutes very rapidly.

By judicious choice of the porous solid in various sizes and using monodisperse calibration samples, one can determine the molecular weight distribution (often abbreviated as MWD) of polymers. In the GPC apparatus the detection is normally done via a very sensitive refractometer which senses small refractive index differences between eluate and pure solvent. GPC is fairly rapid and relatively easy for polymers which are soluble at room temperature. For polypropylene and other difficult-to-dissolve polymers, high temperature GPC is done (e.g. for polypropylene at 135 °C in trichlorobenzene) which is experimentally more challenging. In the case of solubles obtained by a low temperature separation one can also use room temperature GPC in for example tetrahydrofurane (THF). For a general introduction to the subject of GPC the reader is referred to ref. 69.

In presenting data on molecular weight distributions one mostly uses the ratios of the various averages of a distribution. The number average(M_n), weight average (M_w) and z-average (M_z)

molecular weights are distinguished. We denote M_w/M_n by Q and M_z/M_w by R.

In the modern literature GPC is called size exclusion chromatography (SEC).

As branched polymer molecules have a smaller radius of gyration than their linear counterparts of equal molecular weight, a GPC analysis of such a polymer leads to lower values for the molecular weight. For branched polymers different techniques such as light scattering have to be used to arrive at the right numbers. Recently the technique of GPC (SEC) has been combined with light scattering to obtain in one analysis both the molecular weight distribution as well as the branching distribution, as the first method measures the hydrodynamic volume and the other the absolute molecular weight. For an example the reader is referred to the paper by Axelson and Knapp[70].

13.8 LIMITING VISCOSITY NUMBER(LVN)

The definition of this property is

LVN = $[(V - V_0)/V]*(1/C)$ for the limiting case where C approaches zero.

In this V is the viscosity of the polymer solution, V_0 the solvent viscosity, with C being the concentration of the polymer in the solution. The resulting dimension is a reciprocal concentration, mostly expressed as dl/g. In practice one uses a single measurement after the establishment of the relation of the specific viscosity and the product of LVN and concentration. In the case of polypropylene and polyethylene the solvent often used is decalin at 135 °C, and a usual concentration for measurement is around 0.5 g/dl.

A relationship exists between the LVN and the molecular weight of the polymer, the so-called Kuhn-Mark-Houwink relation. In our own studies we used for polypropylene the following[71]:

$$LVN = 11.10^{-3} * \bar{M}_v^{0.80}$$

13.9 VOLATILES BY GAS-LIQUID CHROMATOGRAPHY

Measurement of the volatile fraction of polymers is relatively easy. The only requirement is an acceptable concentration method, as the quantitative measurement and the identification of the constituents can be readily carried out with the aid of gas-liquid chromatography (GLC). The extraction method applied in our study used a solvent blend of acetone and heptane in a 77/23 V/V ratio. In this way the amount of co-extracted higher molecular weight polymer is kept to a minimum (in separate experiments it was shown that extraction with this solvent blend removed only 20% of the normal isopentane solubles, whilst having a peak molecular weight of 1000 to 1150). Two grams of polymer powder is treated with 65 ml of the solvent blend and stirred for 45 minutes. After filtration the filtrate is concentrated on a steambath to about 10% of its original volume. Of course one loses lower boiling compounds in this way, but this does not hamper the analysis as the lower oligomers have been removed in the drying step of every process anyway. The concentrate is analyzed by GLC using for example silicone gum columns and a temperature programme up to 250 or 300 °C. Quantitative measurements are carried out using for instance n-hexadecane as an internal standard added with the solvent blend.

13.10 CRYSTALLINITY

Many ways exist for the determination of crystallinity or to measure properties of crystalline materials such as melting points, enthalpies, etc.. We will mention, very concisely, wide angle x-ray scattering (WAXS), differential scanning calorimetry (DSC) and the final melting point (TMF).

13.10.1 Wide angle X-ray scattering (WAXS)

Both for identification purposes as well as for the determination of the crystallinity WAXS is used. In this the areas arising from the crystalline and amorphous material are determined from the diffractogram and expressed as a fraction crystalline material via:

crystallinity = (area crystalline)/(area crystalline + amorphous)

For the techniques used the reader is referred to the existing literature, for instance reference 72.

13.10.2 Differential scanning calorimetry(DSC)

In this technique a sample (small, normally 10 milligram) is compared with a reference material regarding its thermal properties. The two are heated together at a constant rate and differences in heat fluxes between the two (necessary to keep the temperature of the two the same) are recorded. The deviations from the baseline can be endothermic upon melting and exothermic upon crystallization. The peak areas in DSC are directly proportional to the enthalpy involved in the transition.

The procedure most frequently applied is heating at a constant rate to well above the melting temperature of the polymer studied (e.g. 250 °C for polypropylene), cooling to below the crystalliza- tion temperature at the same constant rate, followed by reheating. The first heating run gives information on melting point (range) and enthalpy of melting of the polymer sample "as is", i.e. with its own thermal history. The second heating run has removed the external effects and allows comparison of different samples as they all start from an identical thermal history. Often one characteri- zes the DSC thermogram by the temperatures at the peak of the endo or exothermic effect. In this book they are labelled as T_{m1}, T_x, and T_{m2}, - the melting temperature observed in the first heating, the crystallization temperature in the subsequent cooling run, and the melting temperature in the last heating run respectively. Crystallinities can be calculated by measuring the heat required for melting (through integration of the appropriate peak) and dividing this value by the heat of fusion of the pure crystalline polymer. For polypropylene we took 188 J/g.

The shape of the DSC peak can also be used to extract informa- tion regarding the structure of the polymer. It has been observed for instance that the crystallization rate of polypropylene increa- ses at higher polymer tacticities[73,74]. This implies that in a normal DSC run a highly isotactic polypropylene shows the first exothermic effects upon crystallization at a lower temperature than polymers of lower tacticity. This can be used as a (relative) tacticity measurement method. The discriminating power at the high end of the isotacticity scale is however expected to be small.

13.10.3 Final melting point(TMF)

In this method, birefringence of a heated crystalline sample is measured as a function of temperature using a thin polymer sample of well-defined thermal history under a polarization micros- cope. The measured output of a photo-multiplier is extrapolated linearly, the intersection with the baseline at temperatures above the melting point being taken as the final melting point (TMF).

The sample is made into a 50 μm thin film between two glass plates, which is held at 240 °C for one minute to release any orientation present. The sample is left to cool to room temperature such that the glass at the spot of the polymer film is only in contact with air. Subsequently the sample is put on the heating table of the polarization microscope at 150 °C, and equilibrated for 5 minutes. Thereafter heating is done at a rate of 1 °C/min, at the same time recording the output of the photomultiplier. The heating is continued for a short time after the melting to generate the required baseline for the extrapolation, this baseline is of course linear as no change in light output occurs above the melting point. The TMF is defined as the point where the extrapolated baseline at high temperature crosses the extrapolated linear part of the decrasing photomultiplier output. All measurements are normally repeated three times and the average is given as the TMF, provided that the measurements do not differ more than 0.5 °C. This method proves to be very reproducible and also very sensitive to small changes in tacticity in polypropylene or small concentrations of co-monomer in for example polyethylene, as has been illustrated in the previous chapters.

The above described method is very similar to techniques described by Barrall[75] and in the British Standard Method Series[76].

13.11 DISPERSION, MORPHOLOGY

To make a dispersion visible or when studying the spherulitic crystallization one can use either optical or electron microscopy. In electron microscopy one has to distinguish between scanning electron microscopy (SEM) and (scanning) transmission electron microscopy ((S)TEM). The equipment and techniques will not be discussed (for a general reference see 72), only contrast enhancement methods will be described both for optical and electron microscopy.

13.11.1 Phase contrast/anoptral contrast microscopy

Phase contrast microscopy is a technique for enhancing contrast in specimens which are very thin (in the order of μm) and whose refractive index is very close to that of their surroundings, so that fine detail present in them produces only path differences of about a quarter of a wavelength. This small path difference has to be enlarged to approximately half the wavelength in order to bring about destructive interference and thus increase contrast. In practice this is done (for a part of the beam) by a ring in the microscope objective, which for green light represents a path length of a quarter of the wavelength; it also reduces the intensity of the incident light. The extra path difference brings the total path difference in the range for interference to occur.

Generally microscopical images of copolymer sections, nowadays at a thickness of 1 μm, are obtained via anoptral contrast(a.c.). This is almost identical to the better known negative phase contrast, the only difference being that for a.c. the phase ring has a higher absorbance. This has some advantages for random or "block" copolymers based on propylene, for instance. In a.c., structures and compounds appear darker (lighter) than their environment if they have a shorter (longer) optical path length than the surrounding matrix. Optical path length is determined by the product of thickness and refractive index.

In simple cases, for example if only random ethylene-propylene copolymer (rubber) is incorporated in polypropylene, the darker particles observed in a.c. can be ascribed to the rubber material. This is because its refractive index is lower than that of polypropylene. If, however, polyethylene type crystallinity is present, this will appear as whiter domains in the matrix. In chapter 6 we have seen a fair number of examples of this.

One can imagine that in certain situations the refractive index of the matrix equals that of the inclusion. Then no contrast is obtained. The same will occur when the particles become too small (<0.3 μm), but this is mostly due to the ultimate resolving power of the light microscope. It should be noted that the morphology of small rubber particles (< 1μm) can be seriously disturbed because of the ultimate resolving power of the microscope. Other techniques (notably transmission electron microscopy) should therefore be applied to elucidate the morphology of such small structures.

Other contrast-enhancing techniques for dispersed systems use solvents specific for the elastomeric, amorphous regions, for instance a suitable oil in which rubber swells or dissolves[77], the oil at the same time having a higher refractive index than polypropylene. Of course the optical microscope can not be used when the size of the dispersed phase is under about 1 μm, one then has to resort to either SEM (bulk specimens) or (S)TEM (ultra-thin sections).

13.11.2 <u>Electron microscopy</u>

In electron microscopy also the contrast has to be enhanced in order to make different phases easily discernible. The methods applied for SEM focus on the non-crystalline part of the sample, to etch it to some extent or remove it completely. To this end bichromate treatment at 70°C is described[78], but the more modern approach is to remove part of the amorphous regions by an air plasma or "cold burning"[78,79]. Other examples are etching with either xylene[80] or n-heptane[79,81] for fairly short times (20 seconds is mentioned) which means that only swelling can occur and possibly some rearrangement of chains. The latter is of course a risk one always runs in treating a polymer sample with a solvent, one can not be totally sure that the treatment leaves the base morphology untouched. Total dissolution of rubber phases is the objective of the treatment described by Stehling et al[82] in which microtomed samples (smooth surfaces) are treated with xylene at room temperature in an ultrasonic bath. In their experience this treatment is to be favoured over bichromate etching. Hexane extraction is also employed[83].

For TEM, the easiest, and very effective, way is the use of OsO_4[84,85] in cases where one of the constituents has unsaturated carbon-carbon bonds. This can be used for instance in blends of polypropylene and EPDM's. But unsaturation is not normally present in ethylene/propylene copolymers, thus other measures have to be taken. For polyethylene Kanig[86,87] developed a technique using chlorosulphonic acid treatment of a microtome coupe at room temperature, followed by a treatment with uranyl acetate after extensive washing of the specimen. This substance reacts with the sulphonic acid groups which are formed in the initial treatment. In this way the spacings between the crystalline lamellae become visible. Since the same holds for polypropylene one can use the difference in the spacings of the lamellae for identification purposes. A comparison of some of the above mentioned methods is to be found in the excellent paper of Bassewitz and zur Nedden[88]. RuO_4 has been found to be a suitable staining agent for both saturated and unsaturated polymer systems that contain in their unit structure ether, alcohol, aromatic or amide moieties[89,90]. Results with this agent have (also) been reported for high-density polyethylene, linear polyethylene wax, polyvinyl methyl ketone, isotactic and atactic polypropylene[89] and for blends of polystyrene, high impact polystyrene and polymethylacrylate[91]. Another method was used by Kakugo et al[92], in which first penetration of liquid 1,7-octadiene was allowed in the sample and after mild drying staining was done with a solution of OsO_4. This was followed by microtoming. In our own studies we used the technique of plasma etching[93,94] followed by shadowing with platinum under a given angle, the latter technique allows one to judge the height differences in the specimen. Recently excellent contrast was reported without staining by discriminating the elastically and inelastically scattered electrons[95].

To measure the area of dispersed phase in a microtome cut and the particle size investigators either used manual techniques by counting/measuring a fair number of particles[81] or by using a grid with a large number of points which covers a certain area and then counting those points within the dispersed phase[82] (claimed accuracy ±1%) or by using an automated image analyzer[96] on high contrast specimens (OsO_4 stained EPDM containing blends).

13.12 MELT FLOW PROPERTIES

13.12.1 Melt Index(MI)

This flow property has been measured in accordance with the usual ASTM-method D 1238. A quantity of 3 to 4 grams of polymer is used, the measurement itself is done at 230 °C, and a weight of 2.16 kg. The dimension of the Melt Index is weight extruded per unit time, mostly expressed in g/10min.

13.12.2 Rheology

The melt rheology of polymers can be measured in the usual way of capillary viscosimetry in an Instron apparatus for example. The data from these measurements are not generally used in a true characterization. The measurement of the viscosity at very low shear rates is more discriminating for dispersed systems as indications can be obtained for network formation for instance. These specific measurements can for example be done in a Weissenberg type equipment. For a general description of these methods the reader is referred, for instance, to the book of Cogswell[97].

13.13 MECHANICAL PROPERTIES

13.13.1 Dynamic mechanical analysis

In this book the technique is mainly used to measure the glass transition temperature, i.e. that temperature at which the polymer chain changes from a rather stiff behaviour to one with much more rotational freedom, this coinciding with the change from a brittle,

glassy state to a more ductile, rubbery one. For a general reference on dynamic properties of polymers the reader is referred to the
book by Read and Dean[98]. To obtain the data mentioned from our own
studies a torsion pendulum type of apparatus was used exclusively.
A description of this type of equipment and a review of the data
obtained for different polymers can be found in ref. 99.

Our automated version of the torsion pendulum apparatus
measures G' and G'' as a function of the temperature which is
continuously changed at a rate of 25 °C/hour. The measuring frequency is about 0.5 Hz. Each run consists of a cooling and a
heating run. The glass temperatures reported are the average over
both these runs. The following terms used in this book, are given
here for ease of reference:
- storage modulus, G', the component of applied shear stress which
is in phase with the shear strain, divided by the shear strain,
- loss shear modulus, G'', the component of applied shear stress
which is 90° out of phase with the shear strain divided by the
shear strain,
- loss factor (loss tangent), the damping, tan δ, is given by
G''/G', it is proportional to the ratio of energy dissipated per
cycle to the maximum potential energy stored during the cycle,
- the glass transition temperature, T_g, the temperature at which
the G'' modulus reaches a maximum value during a transition.

13.13.2 Stiffness
Both yield stress on compression moulded specimens as well as
flexural modulus on injection moulded ones have been measured.

The yield stress reported for our own measurements has been
measured on compression moulded specimens. Preferably a sheet is
used (made on a mill) to minimize the oxidation of the polymer
while in the press. The moulding conditions were: mould temperature
230 °C, heating up for 1 minute at a low pressure, followed by 5
minutes at the full pressure of 100 kg/cm^2, then cooling down to
100 °C in 20 minutes whilst under pressure. The specimen thickness
was always 1 mm.

The flexural modulus is determined on injection moulded
samples using the ASTM D790 method. The injection moulding for
toughened polypropylenes is done at temperatures which differ for
different melt flow of the polymers, but are generally in the

230-250 °C range.

13.13.3 Impact

The impact properties were mostly measured using two techniques. The Izod impact measures the impact energy absorbed in breaking a notched specimen, the samples being injection moulded. In the first instance the British Standard method BS 2782:part3:1970:method 306A was followed, in later years this was replaced by the ISO method 180-1982. The second impact method measures the falling weight impact strength in which injection moulded plates (circular in diameter) are conditioned to a certain temperature (range 23 to -50 °C), and varying weights are allowed to fall from a standard height onto the specimen which is supported by a ring. The weights applied are first increased until a specimen breaks, then decreased until it remains unbroken, increased until it breaks, and so forth, in which way the energy to fracture just half of the specimens is determined. The equipment used is the same as described in BS2782:part3:1970:method 306B.

13.14 SUGGESTED PROCEDURES FOR THE CHARACTERIZATION OF UNKNOWN SAMPLES

In the following we will mention our preferred (combinations of) techniques for quickly characterizing unknown samples of polyolefins, with the emphasis on the compositional parameters. Aspects of molecular weight and its distribution will not specifically be listed, they can be measured easily when this is regarded as relevant for the characterization desired.

13.14.1 homopolymer polypropylene

The quickest, most sensitive and discriminating way in our opinion is the measurement of both the xylene solubles (equals the level of atactic polymer) and the TMF on the total polymer (this indicates the quality of the isotactic fraction). The TMF could be related - if desired - to an absolute tacticity (e.g. expressed as % m-dyads) via calibration with NMR. However, as mentioned before, the sensitivity of the NMR-technique is not optimal at the high end of tacticities, just in the region where most of our interest necessarily lies. Thus mostly we accept the arbitrariness of the TMF value because of its high resolving power.

13.14.2 <u>copolymers</u>

We would suggest the following copolymer characterization methods:

- for overall compositional analysis, in order of importance: ^{13}C-NMR, infrared and pyrolysis gas chromatography. NMR is also able to generate data on the sequence distribution, and through this, on the kinetic polymerization parameters. IR has a limited capacity in this respect, whilst pyrolysis GC can do the same but is much more elaborate,

- for the co-monomer distribution fractionation has to be applied, coupled to compositional analysis. The fractionation can be total or very simple depending on the situation or depth of analysis required. In our own studies on propylene random copolymers the xylene solubles method is very frequently used. For polyethylenes TREF (but in a preparative sense, making compositional measurements possible on the fractions) is a very powerful tool, the application to propylene copolymers is regarded as very worthwhile,

- for effects on crystallinity and melting points DSC is a very useful tool and it is fast as well.

13.14.3 <u>toughened polypropylene</u>

This is the most difficult problem of all; as a starting point the following approach is suggested:

- the overall composition is fairly easily determined by techniques such as ^{13}C-NMR or IR (after suitable calibration),

- the amount of each constituent can be estimated by using the combination of HOXS-fractionation with NMR on the fractions,

- the tacticity of the polypropylene could be judged by measuring the final melting point of polypropylene, as said before this value is nearly independent of the presence of other phases. This has also been reported by Heggs[100]. However, in practice this proves not totally unambiguous, as a slight downward shift is noted, (a better situation will arise when the preparative fractionation of polypropylene and polyethylene is developed as an analytical method),

- the molecular weight distributions of the <u>individual</u> components cannot be measured, of course GPC analysis can be done on the fractions mentioned above, however only in cases where the molecular weights of the constituents are widely apart can one calculate the individual parameters. Approximations are not too difficult in the <u>soluble</u> fraction as a good estimate of the atactic polypropy-

lene molecular weight can be made from independent sources,
- the composition of the dispersed phase can not at present be
analytically measured on the total sample, but the future may bring
techniques such as microscopic Raman on microtome cuts, which can
at least do part of the job.

REFERENCES
1. J.P.Luongo, Infrared study of polypropylene, *J.Appl.Pol.Sc.*, *3* (1960)
    302-309
2. D.R.Burfield and P.S.T.Loi, The use of infrared spectroscopy for
    determination of polypropylene stereoregularity, *J.Appl.Pol.Sc.*, *36* (1988)
    279-293
3. R.M.Bly, P.E.Kiener and B.A.Fries, Near-infrared method for analysis of
    block and random ethylene propylene copolymers, *Anal.Chem.*, *38* (1966)
    217-220
4. C.Tosi and F.Ciampelli, Applications of infrared spectroscopy to
    ethylene-propylene copolymers, *Advances Polym.Sc.*, *12* (1973) 87-130
5. A.Solti, D.O.Hummel and P.Simak, Computer-supported infrared spectroscopy
    of polyethylene, ethene copolymers, and amorphous poly(alkyl ethylene)s,
    Makrom.Chemie, Macromol. Symp., *5* (1986) 105-133
6. B.G.M.Vandenginste and L.De Galan, Critical evaluation of curve fitting in
    infrared spectrometry, *Anal.Chem.*, *47* (1975) 2124
7. F.K.Vansant and H.O.Desseyn, Band profiles in vibration spectra,
    Ind.Chim.Belg., *39* (1974) 45-53
8. P.M.Fredericks et al, Materials characterization using factor analysis of
    FT-IR spectra, Part 1: Results; Part 2: Mathematical and statistical
    considerations, *Appl.Spectrosc.*, *39*(2) (1985) 303-310 and 311-316
9. J.M.Lomonte and G.A.Tirpak, Detection of ethylene homopolymer in
    ethylene-propylene block copolymers, *J.Pol.Sc.,A 2* (1964) 705
10. A.H.Willbourn, Polymethylene and the structure of polyethylene: Study of
    short chain branching, its nature and effects, *J.Pol.Sc.*, *34* (1959) 569-597
11. M.E.A.Cudby, Observations on the molecular structure of polyethylene from
    vibrational and magnetic resonance spectroscopy, paper D6 in Polyethylenes
    1933-1983, Golden Jubilee Conference, 8-10 june 1983, London, The Plastics
    and Rubber Institute.
12. T.Usami and S.Takayama, Identification of branches in low-density
    polyethylenes by Fourier Transform infrared spectroscopy, *Polymer Journal,*
    16 (1984) 731-738
13. P.Frêche, M-F Grenier-Loustalot and A.Cascoin, Etude par spectroscppie
    infrarouge d'alcanes ramifiés et de polymères modèles - Application à la
    determination du taux de ramifications dans les polyéthylènes basse
    densité, *Makrom.Chem.*, *183* (1982) 883-893
14. R.Alamo, R.Domszy and L.Mandelkern, Thermodynamic and structural properties
    of copolymers of ethylene, *J.Phys.Chem.*, *88* (1984) 6587-6595
15. F.C.Schilling and A.E.Tonelli, Carbon-13 NMR of atactic polypropylene,
    Macromolecules, *13* (1980) 270-275
16. M.Farina et al, Hemitactic polymers, *Macromolecules*, *18* (1985) 923-928
17. A.Zambelli et al, Model compounds and 13C NMR observation of stereose-
    quences in polypropylene, *Macromolecules, 8* (1975) 687-689
18. S-N.Zhu, X-Z.Yang and R.Chujo, 13C-NMR chemical shifts in polypropylene and
    the bi-catalytic propagation mechanism in polymerization, *Polymer J.*, *15*
    (1983) 859-868
19. S-N.Zhu, T.Asakura and R.Chujo, Carbon-13 NMR analysis of stereodefects in
    highly isotactic polypropylene by calculation of chemical shifts, *Polymer
    J.*, *16* (1984) 895-899
20. T.Asakura, Y.Nishiyama and Y.Doi, 13C NMR chemical shift of regioirregular
    polypropylene, *Macromolecules*, *20* (1987) 616-620
21. M.Farina, G.DiSilvestro and P.Sozzani, Hemitactic polypropylene: An example

of a novel kind of polymer tacticity, *Macromolecules*, *15* (1982) 1451-1452

22. K.Hikichi et al, Application of the INEPT method to 13C-NMR spectral assignments in low-density polyethylene and ethylene-propylene copolymer, in NMR and Macromolecules, Sequence, dynamic and domain structure, ACS Symposium Series, 247, ACS, Washington

23. H.J.Harwood, NMR analysis of stereoregular homopolymers, in "Preparation and properties of stereoregular polymers", R.W.Lenz and F.Ciardelli editors, D.Reidel, Dordrecht, 1979

24. J.C.Randall, Carbon-13 NMR spin-lattice relaxation times of isotactic and syndiotactic sequences in amorphous polypropylene, *J.Pol.Sc., Pol.Phys.*, *14* (1976) 1693-1700

25. H.N.Cheng, 13C-NMR analysis of ethylene-propylene rubbers, *Macromolecules*, *17* (1984) 1950-1955

26. E.T.Hsieh and J.C.Randall, Ethylene-1-butene copolymers. 1. Comonomer sequence distribution, *Macromolecules*, *15* (1982) 353-360

27. J.C.Randall and E.T.Hsieh, 13C NMR in polymer quantitative analysis, in NMR and Macromolecules , Sequence, dynamic and domain structure, ACS Symposium Series, 247, ACS, Washington

28. J.N.Hay, P.J.Mills and R.Ognjanovic, Analysis of 13C FT-NMR spectra of low-density polyethylene by the simplex method, *Polymer*, *27* (1986) 677-680

29. D.E.Axelson, L.Mandelkern and G.C.Levy, 13C Spin relaxation parameters of branched polyethylenes. Ramifications for quantitative analysis, *Macromolecules*, *10* (1977) 557-558

30. Y.Tsuchya and K.Sumi, Thermal decomposition products of polypropylene, *J.Pol.Sc., A1*, *7* (1969) 1599-1607

31. M.Seeger and H.-J. Cantow, Thermische Spaltungsmechanismen in Homo- und Copolymeren aus α-Olefinen, 2. Polypropylene mit bekannter Taktizität, Polyisobutylen und Poly(1-buten), *Makrom.Chemie*, *176* (1975) 2059-2078

32. E.Kiran and J.K.Gillham, Pyrolysis-molecular weight chromatography: A new on-line system for analysis of polymers. II. Thermal decomposition of polyolefins: Polyethylene, polypropylene, polyisobutylene, *J.Appl.Pol.Sc.*, *20* (1976) 2045-2068

33. Y.Sugimura, T.Nagaya, S.Tsuge, T.Murata and T.Takeda, Microstructural characterization of polypropylenes by high-resolution pyrolysis-hydrogenation glass capillary gas chromatography, *Macromolecules*, *13* (1980) 928-932

34. H.Ohtani, S.Tsuge, T.Ogawa and H-G.Elias, Studies on stereospecific sequence distributions in poylpropylenes by pyrolysis-hydrogenation fused-silica capillary gas chromatography, *Macromolecules*, *17* (1984) 465-473

35. A.Barlow, R.S.Lehrle and J.C.Robb, Direct examination of polymer degradation by gas chromatography, I. Applications to polymer analysis and characterization, *Polymer*, *2* (1961) 27-40

36. J.van Schooten and J.K.Evenhuis, Pyrolysis-hydrogenation GLC of poly-α-olefins, *Polymer*, *6* (1965) 343-360

37. L.Michajlow, H-J.Cantow and P.Zugenmaier, Investigation of the sequence length distribution in ethylene-propylene copolymers by pyrolysis and reaction gas chromatography, *Polymer*, *12* (1971) 70-84

38. J.C.Verdier and A.Guyot, Microstructure of propene-butene copolymers studied by flash-pyrolysis GLC, *Makrom.Chem.*, *175* (1974) 1543-1559

39. J.H.Hildebrand and R.L.Scott, Regular solutions, Prentice Hall, Englewood Cliffs, NJ, 1962

40. P.A.Small, Some factors affecting the solubility of polymers, *J.Appl.Chem* , *3* (1953) 71-80

41. S.Krause, Polymer-polymer compatibility, chapter 2 in Polymer Blends, edited by D.R.Paul and S.Newman, Academic Press, New York, 1978.

42. W.R.Moore and G.F.Boden, Heptane-soluble material from atactic polypropylene, II Interaction with liquids, *J.Appl.Pol.Sc.*, *10* (1966) 1121-1132

43. A.S.Michaels, W.R.Vieth and H.H.Alcalay, The solubility parameter of polypropylene, *J.Appl.Pol.Sc.*, *12* (1968) 1621-1624

600

44. D.A.Blackadder and G.J.LePoidevin, Dissolution of polypropylene in organic solvents: 1. Partial dissolution., *Polymer* *17* *(1976) 387-394*

45. M.C.Kirkham, Properties and microstructures of ethylene-propylene terpolymers, *J.Appl.Pol.Sc.*, *17* *(1973) 1101-1111*

46. K.Ito and J.E.Guillet, Estimation of solubility parameters for some olefin polymers and copolymers by inverse gas chromatography, *Macromolecules,* *12* *(1979) 1163-1167*

47. Polymer Handbook, 2nd edition, J.Brandrup and E.H.Immergut editors, John Wiley, 1975

48. W.R.Moore and G.F.Boden, Heptane-soluble material from atactic polypropylene. I. Fractionation and characterization of fractions, *J.Appl.Pol.Sc.*, *9* *(1965) 2019-2029*

49. R.S.Porter, M.J.R.Cantow and J.F.Johnson, The effect of tacticity on polypropylene fractionation, *Makrom.Chemie,* *94* *(1966) 143-152*

50. P.W.O.Wijga, J.van Schooten and J.Boerma, The fractionation of polypropylene, *Makrom.Chemie* *36* *(1960) 115-132*

51. D.F.Sloanaker, R.L.Combs and H.W.Coover, Melting point and composition distributions of polyolefins by fractionation, *J.Macromol.Sc.*, *A1* *(1967) 539-557*

52. A.J.Pennings, Fractionation of polymers by crystallization from solutions. II., *J.Pol.Sc.*, *C,* *16* *(1967) 1799-1812*

53. A.Nakajima and H.Fujiwara, The fractionation of polypropylene by extraction with boiling hydrocarbons with increasing boiling points, *Bulletin Chem. Soc.Japan.*, *37* *(1964) 909-915*

54. W.Holtrup, Zur Fraktionierung von Polymeren durch Direktextraktion, *Makrom. Chemie,* *178* *(1977) 2335-2349*

55. S.Nakano and Y.Goto, Development of automatic cross fractionation: Combination of crystallizability fractionation and molecular weight fractionation, *J.Appl.Pol.Sc.*, *26* *(1981) 4217-4231*

56. L.Wild, T.R.Ryle, D.C.Knobeloch and I.R.Peat, Determination of branching distributions in polyethylene and ethylene copolymers, *J.Pol.Sc.*, *Pol.Phys.*, *20* *(1982) 441-455*

57. D.A.Blackadder and G.J.LePoidevin, Dissolution of polypropylene in organic solvents: 2. The steady state dissolution process, *Polymer* *17* *(1976) 768-776*

58. G.Natta, Une nouvelle classe de polymères d'α-olefines ayant une regularité de structure exceptionelle, *J.Pol.Sc.* *16* *(1955) 143-154*

59. U.S.Federal Register §121.2501 (d) (3) and (4)

60. O.Fuchs, Zur Bestimmung des "löslichen" Anteiles in kristallinen Polymeren, *Makrom.Chemie,* *58* *(1962) 247-250*

61. P.M.Kamath and L.Wild, Fractional crystallization of polypropylene, *Pol.Engin. Sc.*, *6* *(1966) 213-216*

62. F.Kloos and H.J.Leugering, An improved method for the determination of the isotactic index of polypropylene, Preprints IUPAC international symposium on macromolecules, Florence, 1980, volume 2, p. 479-482

63. G.Natta, I.Pasquon, A.Zambelli and G.Gatti, Dependence of the melting point of isotactic polypropylene on their molecular weight and degree of stereospecificity of different catalyst systems, *Makrom. Chemie,* *70* *(1964) 191-205*

64. M.Kakugo et al, Microstructure of syndiotactic polypropylenes prepared with heterogeneous titanium-based Ziegler-Natta systems, *Makrom.Chemie,* *190* *(1989) 505-514*

65. Eur.Patent, 0 002 522, published 27 June 1979, to Phillips Petroleum company

66. N.Kashiwa, J.Yoshitaka and T.Tsutsui, Polymerization of butene-1 with highly active MgCl2-supported catalyst systems, *Polymer,* *28* *(1987) 1227-1231*

67. T.Simonazzi, Molecular characterization of ethylene-propylene block copolymers, *Pure & Applied Chem.*, *56* *(1984) 625-634*

68. Brit. Patent, 1 543 096, to Mitsui Petrochemical Industries, publication date 28 March 1979

69. W.V.Smith, Fractionation of polymers, *Rubber Chem. Techn.*, *45* (1972) 667-708
70. D.E.Axelson and W.C.Knapp, Size exclusion chromatography and low-angle laser light scattering. Application to the study of long-chain branched polyethylene, *J.Appl.Pol.Sc.*, *25* (1980) 119-123
71. J.B.Kinsinger and R.E.Hughes, *J.Phys.Chem.*, *63* (1959) 2002
72. Treatise on analytical chemistry, part I, Theory and practice, Volume 8, section H, Optical methods of analysis, Philip J.Elving editor, Interscience, John Wiley, New York, 1986
73. D.R.Burfield and Y.Doi, Polypropylene tacticity determination by DSC crystallization measurements, *Polymer Comm.*, *24* (1983) 48-50
74. D.R.Burfield and P.S.T.Loi, Approaches to the problem of tacticity determination in poylpropylene, p. 387-406 in "Catalytic polymerization of olefins", T.Keii and K.Soga editors; Kodansha, Tokyo and Elsevier, Amsterdam, 1986
75. E.M.Barrall and E.J.Gallegos, Depolarized light and differential thermal analyses of some polyolefin transitions, *J.Pol.Sc.*, A2, *5* (1967) 113-123
76. Methods of testing plastics, British Standard Institution, London, BS 2782:1970
77. S.Danesi and R.S.Porter, Blends of isotactic polypropylene and ethylene-propylene rubbers: rheology, morphology and mechanics, *Polymer*, *19* (1978) 448-457
78. M.Kojima, Stress whitening in crystalline propylene-ethylene block copolymers, *J.Macromol.Sci.,Phys.*, *B19* (1981) 523-541
79. K.Zur Nedden and K.v. Bassewitz, Elastomer-/polyolefin blends, neuere Erkenntnisse ueber den Zusammenhang zwischen Phasenaufbau and anwendungstechnische Eigenschaften, German Rubber Conference, Wiesbaden, June 1983
80. J.Karger-Kocsis, A.Kallo, A.Szafner, G.Bodor and Zs.Senyei, Morphological study on the effect of elastomeric modifiers in polypropylene systems, *Polymer*, *20* (1979) 37-43
81. J.Karger-Kocsis and V.N.Kuleznev, Scanning electron microscopic investigations of particle size and particle size distribution of EPDM impact modifier in polypropylene/EPDM blends, *Acta Polymerica*, *32* (1981) 578-581
82. F.C.Stehling, T.Huff, C.S.Speed and G.Wissler, Structure and properties of rubber-modified polypropylene impact blends, *J.Appl.Pol.Sc.*, *26* (1981) 2693-2711
83. A.Jevanoff, E.N.Kresge and L.L.Ban, Morphology of thermoplastic polyolefin blends by scanning electron microscopy, Plasticon 81, symposium Polymer Blends, The Plastic and Rubber Institute, London, 1981
84. K.Kato, Electron microscopy of ABS plastics, *J.Electron.Micr.* *14* (1965) 220-221
85. K.Kato, The osmium tetroxide procedure for light and electron microscopy of ABS plastics, *Pol.Eng.Sc.*, (1967) 38-39
86. G.Kanig, Neue elektronenmikroskopische Untersuchungen ueber die Morphologie von Polyäthylenen, *Progr.Colloid & Polymer Sci.*, *57* (1975) 176-191
87. G.Kanig, Elektronenmikroskopie an Kunststoffen am Beispiel der Polyäthylens, *Kunststoffe*, *64* (1974) 470-474
88. K.v.Bassewitz and K.zur Nedden, Elastomer-polyolefin blends - Neuere Erkenntnisse über den Zusammenhang zwischen Phasenaufbau und anwendungstechnischen Eigenschaften, *Kautschuk u Gummi*, 1985, 42-52
89. J.S.Trent, J.I.Scheinbeim and P.R.Couchman, Ruthenium tetraoxide staining of polymers for electron microscopy, *Macromolecules*, *16* (1983) 589-598
90. R.Vitali and E.Montani, Ruthenium tetroxide as a staining agent for unsaturated and saturated polymers, *Polymer*, *21* (1980) 1220-1222
91. J.S.Trent, J.I.Scheinbeim and P.R.Couchman, Electron microscopy of PS/PMMA and rubber-modified polymer blends: use of ruthenium tetroxide as a new staining agent, *J.Pol.Sc.*, *Pol.Letters*, *19* (1981) 315-319
92. M.Kakugo et al, Transmission electron microscopic observation of nascent polypropylene particles using a new staining method, *Macromolecules*, *22*

(1989) 547-551

93. R.H.Hansen et al, Effect of atomic oxygen on polymers, *J.Pol.Sc.*, *A3* *(1965)*
2205-2211

94. J.R.Hallahan and A.T.Bell, Techniques and applications of plasma chemistry,
John Wiley and Sons, New York, 1974

95. M.Kunz et al, Electron spectroscopic imaging studies on polyethylene
chain-folded and extended chain crystals, *Makrom.Chemie, Macrom.Symp.*,
20/21 *(1988) 147-158*

96. B.Z.Jang, D.R.Uhlman and J.B.Vander Sande, The rubber particle size
dependence of crazing in polypropylene, *Polym.Eng.Sc.*, *25* *(1985) 643-651*

97. F.N.Cogswell, Polymer melt rheology, Wiley, New York, 1981

98. B.E.Read and G.D.Dean, The determination of dynamic properties of polymers
and composites, Adam Hilger Ltd, Bristol, 1978

99. J.Heyboer, The torsion pendulum in the investigation of polymers, *Polymer
Eng. & Sc.*, *19* *(1979) 664-675*

100. T.G.Heggs, *Chem.Ind.*, *1969, p744*

Chapter 14

FUTURE TRENDS IN CATALYSTS, PROCESS AND PRODUCTS

14.1 INTRODUCTION

In order to put the future trends in better perspective it is helpful to consider the essence of what has been achieved in the thirty years since the beginnings of polyolefins (made by Ziegler-Natta catalysts) as a commercial product. Tracing for instance the developments in polypropylene from the first small-scale units, involving both deashing and extraction of atactic polymer, through the second generation processes employing catalysts of improved selectivity and morphology to today's plants, many of which avoid both deashing and extraction steps, it can be concluded that major advances have been made in the area of improved process economics. The same is true for the other major products polyethylene and ethylene/propylene rubber. There has at the same time been considerable activity in product developments, as exemplified by the modified homopolymers such as LLDPE or "random copolymers" and toughened products in the polypropylene field, but in essence the homopolymer polypropylene is very similar today to the one made in the very beginning of the polyolefin era. To be fair one has to state that some improvements have been made in, for instance, the control of molecular weight and its distribution to cater for different applications such as injection moulding, blow moulding, fibre processing, etc.. But, as will be elaborated below, we think there is still considerable scope in product improvement and new product development, whilst on the other side of process economics only small savings are foreseen. Indeed modern plants are already so simple in design and operation that the small savings that are still realizable could barely justify the research and development costs involved in their implementation. This statement holds only for ethylene and propylene derived polymers, in the case of higher olefins - provided they become of commercial interest - there is still a need for improvements in the process, mainly in the aspects of activity and stereoselectivity.

Probably, the only real exception to the statement on process economy involves the last remaining post-polymerization process step: granulation or extrusion of the reactor product to generate the "nibs" desired by most customers. We will deal with this in the next section.

Future advances will surely be largely concerned with improvements in the control of product quality, in product tailoring via catalyst/process technology and in the developments of new polyolefin based materials.

14.2 PROCESS IMPROVEMENTS

Present day supported catalysts possess such a high activity and selectivity that both the deashing and extraction steps of earlier processes are obviated, making further catalyst improvements in either of these key parameters meaningless as far as process economics are concerned. Therefore it is apparent that if there is room for improvement it is in the post-reaction section, i.e. in the extruder. Two different approaches to this problem can be identified:

i) ex-reactor granulate:

By producing catalysts with a dense, spherical and uniform particle shape it is possible to generate polypropylene particles which, for many applications, can be shipped and used without an extrusion step to generate the usually preferred nibs. This concept is already part of the Himont "Spheripol" process mentioned in chapter 1 (see sections 1.3 and 1.8). While attractive in terms of economy, since the energy-consuming extrusion step is obviated, various disadvantages can be pointed out such as the difficulty of homogeneously administering additives/stabilizers, the loss of the options for molecular weight modifications etc. via "extruder chemistry", and the fact that the commercially preferred large particles simply cannot be kept in suspension during polymerization, limiting the process to smaller particles than the present day industry standard.

ii) extruder polymerizations:

A second approach reducing the costs (i.e. energy consumption) of an extruder step is to combine the reactor and extruder sections, i.e. to execute the polymerization in the extruder. A major advantage of this concept is that the enormous heat of polymerization could itself be used to melt the polymer, hence eliminating the high costs involved in present day plants. In terms of catalyst technology the breakthrough required involves the development of a catalyst which is able to polymerize for example propylene to high conversions with good selectivity and activity even at temperatures in excess of the melting point of polypropylene (temperatures of around 180 °C would be envisaged in the final stages of such a process).

14.3 PRODUCT IMPROVEMENTS

It is useful to separate two classes of products in this discussion, namely the homopolymers and (random) copolymers as one class and the toughened products and blends as the other.

Starting with the homopolymers and copolymers, what is it that one would like to control when aiming at almost absolute control over the properties of the products? The following parameters come easily to mind:

- stereochemical placement of the units (for monomers propylene and higher), complete control from pure syndio to pure isotactic polymer, including control of the sequence distribution of the minor structure. The latter implies the possibility of making for example the alternating or blocky type of structures (as for instance the isotactic polypropylene fraction of the present products is alternating in the r placement as they are all isolated),

- molecular weight distribution control, i.e being able to generate products with distributions ranging from very narrow ($Q \approx 1$ or 2) to very broad ones,

- copolymerization: control of sequence distribution is the most important property in the product sense, again as with the stereochemistry one would like to be able to make alternating, random or blocky type of structures at will. In this area the kinetics of copolymerization are also important as r-values should preferably allow the use of low amounts of monomer in the feed in order to lower the cost of production. Moreover the use of polar co-monomers would extend the scope of product tailoring tremendously.

When the above would be true we could design a product at will by choosing the desired structure(s) and making the appropriate combinations of individual polymer fractions (for a smart way of doing this see below). We would have complete freedom in tailoring and not be handicapped by having to work with the "natural" product mix that a certain combination of catalyst and polymerization condition happens to generate.

In our opinion the homogeneous catalysts that are just over

the horizon, will allow most of the above. Certainly it will take a huge effort to arrive at the above ideal situation but in particular the fact that homogeneous catalysts lend themselves much more readily to study, modification and "fine-tuning" than do their heterogeneous counterparts, presents an enormous potential for steering catalyst performance and thereby tailoring product properties. Some examples of controlling reactivity ratios in copolymerization have already been given (see chapter 3). As another example, homogeneous catalysts will allow narrow molecular weight distribution to be made (Q as low as 2, this to be compared to the normally encountered 6 to 10 in polyethylene and polypropylene), and broad distribution can be generated by blending different fractions. This can be done in one reactor by blending different catalysts which show their own rate constants (this is in a way similar to the situation in the present day heterogeneous catalysts, the only, but overriding difference being that with the homogeneous catalysts mixtures one knows precisely the relevant data on rate constants, etc.). Another possibility is to make true atactic polypropylene and other atactic polyolefins, which might be attractive products in their own right. The living nature of homogeneous catalysts will lead undoubtedly to the manufacture of true block copolymers of different monomers or of different stereochemical placements. Advances in the understanding of the nature of the catalytic species could lead to catalysts capable of generating syndiotactic polymers at commercially viable temperatures i.e. at 50 °C or higher (for propylene this has already been reported). Finally by moving from the left to the right hand side in the Periodic Table (i.e. from Ti and Zr to Rh, Pd, Pt, Ni etc.) we can expect to see catalysts capable of copolymerizing polar monomers such as acrylates. This would extend the possibilities of product development to a very large extent.

As said before there is still a very long way to go in the development of the homogeneous catalyst. When assessing the situation of today one can clearly observe a number of problem areas which have to be solved before these catalysts can be commercially applied in the above sense. These are among others:

- the regiospecificity and the stereospecificity should be controllable up to a very high extent,

- the molecular weight level reached at temperatures above 50 °C is mostly too low, although using hafnium instead of zirconium already helps,

- multiple active species are often observed, one has to aim for just one active species,

- control of the morphology of the polymer product, and

- die-out i.e. side reactions of the active species should be controlled to a very large extent.

We are strongly of the opinion that the above problems may ultimately be overcome, mainly because of the extreme flexibility of a homogeneous catalytic system.

We end now with some remarks on toughened products. A number of aspects will lead to improvements for this class of products. Firstly a better fundamental understanding of the relevant parameters in this dispersed system will lead to product improvements, for example by control of dispersed phase morphology (dispersed particle size distribution and shape), and possibly also by optimizing the (in)compatibility of the ingredients. The use of well-defined building blocks, made on homogeneous catalysts, will also improve our knowledge of this complicated system, for instance into the effects of compatibility. Additionally one could possibly introduce true block copolymers into the system and check on their effects. Of course potential blockcopolymers should also be regarded in their own right as products which can be used for example at low temperatures. In principle ABA blockcopolymers with crystalline A segments could span the product range from thermoplastic elastomer to toughened plastics, such as in polypropylene. For the latter a blend approach will probably be necessary. In a process sense, homogeneous catalysts could in principle make a one-reactor toughened polypropylene production possible by using two (or more) catalysts of which one can only homopolymerize propylene, whilst the other(s) can copolymerize with ethylene and/or other co-monomers.

LIST OF ABBREVIATIONS

ABS	acrylonitrile-butadiene-styrene rubbers
acac	acetyl acetonate
ANB	acrylonitrile-butadiene rubber
Bu	butyl
Cp	cyclopentadienyl
Cp*	pentamethyl cyclopentadienyl
CSTR	continuous stirred tank reactor
DBE	di-butyl ether
DEAC	di-ethyl aluminium chloride
DEAI	di-ethyl aluminium iodide
DIP	di-isobutyl phthalate
DMA	dynamic mechanical analysis
DNBP	di-n-butyl phthalate
DSC	differential scanning calorimetry
EADC	ethyl aluminium dichloride
EB	ethyl benzoate
ENB	ethylidene norbornene
EPC	ethylene propylene copolymer
EPDM	ethylene propylene diene (monomer) rubber
EPR	ethylene propylene rubber
ESCR	environmental stress cracking resistance
ESR	electron spin resonance
FM	flexural modulus
FTIR	fourier transform infrared spectroscopy
FWIS	falling weight impact strength
G(L)C	gas (liquid) chromatography
GPC	gel permeation chromatography
HDPE	high density polyethylene
HI	(boiling) heptane insolubles
HOXS	hot xylene solubles
HOXS-E	hot xylene solubles, the extract
HOXS-R	hot xylene solubles, the residue

HMPTA   hexamethyl phosphoric acid triamide

iBu   isobutyl
II   isotactic index (=HI)
Ind   indenyl
INEPT   insensitive nuclei enhanced by polarisation transfer
iPr   isopropyl
IR   infrared spectroscopy

LCST   lower critical solution temperature
LDPE   low density polyethylene
LLDPE   linear low density polyethylene
LVN   limiting viscosity number

3MB   3-methyl-1-butene
MD   monomer dispersity
MI   melt index
4MP   4-methyl-1-pentene
MW   molecular weight
MWD   molecular weight distribution

NMR   nuclear magnetic resonance

OM   optical microscopy

PB   poly(1-butene)
PE   polyethylene
PEA   para ethyl anisate
Ph   phenyl
PIB   polyisobutylene
P4MP   poly(4-methyl-1-pentene)
PP   polypropylene
PS   polystyrene
PTES   phenyl triethoxy silane
PVC   polyvinylchloride

SEM   scanning electron microscopy
sesqui $Al_2Cl_3Et_3$
STEM   scanning transmission electron microscopy

TEA	tri ethyl aluminium
TEM	transmission electron microscopy
THF	tetra hydrofurane
TMA	tri methyl aluminium
TMF	final melting point
tpa	tonnes per annum
TPP	toughened polypropylene
TREF	temperature rising elution fractionation
WAXS	wide angle X-ray scattering
XPS	X-ray photoelectron spectroscopy
XRD	X-ray diffraction
XS	xylene solubles
YS	yield stress

I N D E X

Z